Introduction to
DISCRETE MATHEMATICS

Introduction to DISCRETE MATHEMATICS

R. Hirschfelder
University of Puget Sound

J. Hirschfelder
Honeywell, Inc.

 Brooks/Cole Publishing Company
Pacific Grove, California

Brooks/Cole Publishing Company
A Division of Wadsworth, Inc.

Printed in the United States of America

10   9   8   7   6   5   4   3   2   1

Library of Congress Cataloging-in-Publication Data

Hirschfelder, R.
    Introduction to discrete mathematics / R. Hirschfelder, J.
Hirschfelder.
        p.   cm.
    Includes index.
    Includes bibliographical references.
    ISBN 0-534-13896-9
    1. Electronic data processing—Mathematics.   I. Hirschfelder, J.
(John), [date]   II. Title.
QA76.9.M35H57   1990
510—dc20   90-36341
                                                                              CIP

Sponsoring Editor: *Jeremy Hayhurst*
Marketing Representatives: *Karen Buttles, Hester Winn*
Editorial Assistant: *Nancy Champlin*
Production Editor: *Marjorie Z. Sanders*
Manuscript Editor: *David Hoyt*
Interior Design: *E. Kelly Shoemaker*
Cover Design and Illustration: *Cloyce J. Wall*
Art Coordinator: *Lisa Torri*
Interior Illustration: *Alexander Teshin Associates*
Typesetting: *Polyglot Compositors*
Cover Printing: *Phoenix Color Corp.*
Printing and Binding: *R. R. Donnelley & Sons, Inc. Crawfordsville*

For Liberata Susie DeMaio
R. H.

In memory of Darreld Hirschfelder
J. H.

About the Cover

The cover image is a computer interpretation of an 1809 map of the old town of Königsberg, on the Baltic Coast of East Prussia. Part of the townsite occupied an island in the Pregel River, which was accessible by crossing seven bridges. In the eighteenth century, Königsberg's residents used to amuse themselves by attempting to cross each of the seven bridges once and only once. The problem remained unsolved until the Swiss mathematician Leonhard Euler (1707–1783) generalized the problem by considering which configurations of bridges and land masses would permit such a walk. His result is now known as a Euler circuit and is considered the beginning of graph theory, one of the central topics of discrete mathematics.

The image was created on a Macintosh® IIx color computer. A copy of the original Königsberg map was input with a digital scanner. The graphic devices, typography, and colors were created/added using the computer graphics program Freehand®, and the completed image was then output in the form of film separations on a Linotronic® 300 high-resolution laser printer.

Modern-day Königsberg, now called Kaliningrad, is part of the Soviet Union. Kaliningrad currently has more than the original seven bridges.

Preface

The rise of computers has affected nearly every aspect of modern life, and the undergraduate mathematics curriculum is no exception. Initially, computers were viewed as tools to do numerical computation, so their impact on mathematics teaching was limited to speeding up the process of problem solving. The problem domains themselves, and the problem-solving methods taught, were at first little affected.

As computer science has developed into a mature discipline in its own right, the relationship between computing and mathematics has become more symmetric. Now computing is not only a tool for solving problems in mathematics; mathematics has become a tool for solving problems in computing. The design of efficient computer programs and data structures has become more and more dependent on choosing the right mathematical model of the problem. Understanding of mathematical concepts has become a prerequisite to advanced computer science courses.

Thus, there have evolved mathematics courses centered on the mathematics most directly applicable to computing. The topics taught have collectively become known as **discrete mathematics.** The term is sometimes interpreted as mathematics that does not involve the real numbers. We take a broader yet more concrete view: Discrete mathematics is the mathematics of computation, regardless whether it is used to solve scientific, business, or even abstract mathematical problems. The use of continuous methods (including properties of the real and complex number systems) to solve discrete problems is an essential feature of the subject.

The topics treated in this text have been adapted from the 1984 recommendations of the MAA Panel on Discrete Mathematics in the First Two Years. The order of topics proceeds from the concrete to the most abstract. Many topics are motivated by reference to computing, and computer representations of mathematical structures are illustrated throughout.

Many students taking this course have some computer experience and will benefit from the computer exercises included at the end of each chapter. Other computer-oriented material is distributed throughout.

The book is divided into three parts:

   I   Fundamentals
  II   Basic Discrete Mathematics
 III   Advanced Topics

Part I consists of Chapters 1 through 3 and contains essential introductory material on number systems, sets, functions, and relations. Despite the fact that these topics are introduced repeatedly in mathematics courses from elementary school through calculus, it has been our experience that students' knowledge of this material varies widely. The topics in this part may be covered in detail or used for a rapid review, depending on the instructor's judgment of the students' preparation.

Part II consists of Chapters 4 through 9 and constitutes the core of the course. It includes the topics that have been found most useful in introductory computer science courses. Unique features of the treatment include the following:

Chapter 4, Logic and Proof, includes application to everyday mathematical reasoning, especially understanding and manipulating mathematical statements containing multiple quantifiers.

Chapter 5, Boolean Algebra and Logic Circuits, includes the application of systems of simultaneous equations to logic feedback circuits.

Chapter 9, Recurrence Relations and Dynamical Systems, emphasizes understanding the qualitative character of solutions of recurrence relations through numerical examples. The use of spreadsheet programs for numerical experimentation is illustrated.

Part III consists of Chapters 10 through 12 and presents several advanced topics. The selection of topics will depend on the instructor's and students' interests, the relation of this course to the particular curriculum, and the amount of time available. Chapter 11 is a prerequisite for Chapter 12; otherwise, the chapters are independent.

Suggested course structures are shown in the following table.

	Course Duration		
	One Quarter	*One Semester*	*Two Quarters*
Minimal Student Preparation	Part I: 1, 2, 3	Part I: 1, 2, 3	Part I: 1, 2, 3
	Part II: 4, 6, 7, 8	Part II: 4, 6, 7, 8, parts of 9	Part II: 4, 6, 7, 8, 9
		Part III: parts of 10; 11 or 12	Part III: 10; 11 or 12
Strong Student Preparation	Part I: Review terminology and notation; Induction	Part I: Review terminology and notation; Induction	Part I: Review terminology and notation; Induction
	Part II: 4, 5, 6, 7, 8; parts of 9	Part II: 4, 5, 6, 7, 8, 9	Part II: 4, 5, 6, 7, 8, 9
		Part III: One chapter	Part III: 10; 11 or 12

ACKNOWLEDGMENTS

We would like to thank the people who helped in the preparation of this text. Chuck Hommel of the University of Puget Sound read the entire manuscript, checked all references, and verified all answers to exercises. Ron Van Enkevort of the University of Puget Sound, Steve Cunningham of the California State University, Stanislaus, and David Pincus read early versions of several chapters and provided many helpful suggestions. Carol Moyer helped with the preparation of the manuscript, and Lori Ricigliano of the UPS library dug up any requested information. Special thanks are due to students in discrete math classes at the University of Puget Sound who read the manuscript, drew our attention to several problems and errors, and provided answers to many of the exercises. We would especially like to thank Shawnell Stevens, Della Sparks, Lisa Mancuso, and Bob Chapman.

We are particularly grateful to the following reviewers, who made many useful comments and suggestions. Donald Leake, University of Wisconsin; John Beidler, University of Scranton; Richard Mansfield, Penn State University; Richard Weida, Lycoming College; Ron Baker, University of Delaware; Paul Edelman, University of Minnesota; George Luger, University of New Mexico; Rick Sutcliffe, Trinity Western College; Mike Krozan, DeVry Institute of Technology; Tom Ralley, Ohio State University; Frederick Fuglister, John Carroll University; Catherine Murphy, Purdue University–Calumet; John Konvalina, University of Nebraska–Omaha; Ken Harris, Western Illinois University; Donald Morrison, University of New Mexico; Ron Smith, University of Tennessee, Chattanooga; John Buoni, Youngstown University; and Elwyn Davis, Pittsburg State University.

Finally, we express our appreciation to Jeremy Hayhurst, Nancy Champlin, and Marjorie Sanders of Brooks/Cole for their encouragement and support.

R. Hirschfelder
J. Hirschfelder

TO THE STUDENT

Discrete mathematics is normally studied for its application to computing. However, the objectives of this course extend beyond such direct application; indeed, the material is directly applicable to further study in mathematics. Among the goals of the course are understanding standard mathematical terminology and language, understanding the concept of proof, and improving mathematical reasoning skills—in particular, the ability to recognize valid and invalid mathematical arguments. This text will review familiar terminology and symbols and introduce you to many new ones. Each concept introduced is illustrated with many examples that should be studied carefully.

Each section is followed by both routine problems and more challenging ones. Since mathematics cannot be learned by observation alone, you should do as many routine exercises as necessary until the concepts are fully understood, then tackle the challenging ones.

The appendixes include a useful glossary and answers to all odd-numbered problems. The mathematical ideas presented in this text will provide a firm foundation for your future study of both mathematics and computer science.

Contents

Part I FUNDAMENTALS

12   Machines and Computation   413

Appendixes

Introduction to
DISCRETE MATHEMATICS

Part I

FUNDAMENTALS

1

NUMBER SYSTEMS
AND REPRESENTATIONS

In this chapter, we will review some important characteristics of number systems. The term **number system** refers to any familiar set of objects with which we can do the operations of addition and multiplication. This means that the sum of any two numbers in the system, and the product of any two numbers in the system, is some number in the same system. The systems we are most familiar with are the natural numbers, the integers, the rational numbers, the real numbers, and the complex numbers.

Computers do arithmetic with numbers. In fact, doing long arithmetic computations rapidly was the original reason for building computers. (Of course, they have since been applied to a much wider class of problems.) We will also examine carefully how numbers are represented—that is, how we write down symbols or combinations of symbols that identify numbers. Since computers manipulate only symbols, understanding how symbols represent numbers is the key to understanding how computers can be used to solve mathematical problems.

The real numbers and the complex numbers are normally considered to be the subject of mathematical analysis rather than discrete mathematics. However, there are many problems in discrete mathematics whose solutions require some use of real or complex numbers. We will therefore include some basic facts about the real and complex number systems in this chapter.

1.1 • NATURAL NUMBERS

The **natural numbers** (also called *counting numbers* or *positive integers*) are the numbers 1, 2, 3, 4, 5, . . . , where the ellipsis (. . .) indicates continuation indefinitely. (Some authors include 0 in the natural numbers.) The natural numbers taken together form a **set**. We will discuss sets in general in Chapter 2. For now, it will suffice to give a name to the set of natural numbers; the set of natural numbers is denoted by **N**. If n is a natural number, we say that n is an **element** of **N** and write $n \in \mathbf{N}$.

Representations of Natural Numbers

The most common written representation for natural numbers is the positional system with base 10. In this system, a number is represented by a sequence of the digits 0 through 9, and each digit position corresponds to a power of ten.

In principle, any natural number greater than 1 can be used as the base. That is, if b is a natural number greater than 1 and if n is any natural number, then there is a way of writing n in the form

$$n = a_k b^k + a_{k-1} b^{k-1} + \cdots + a_1 b + a_0$$

where $0 \le a_i < b$ and $a_k \ne 0$.

EXAMPLE 1.1   Write the number 19 in bases 2, 3, 4, and 5.

It is easy to check that

$$\begin{aligned}
19 &= 1 \times 2^4 + 0 \times 2^3 + 0 \times 2^2 + 1 \times 2 + 1 \\
&= 2 \times 3^2 + 0 \times 3 + 1 \\
&= 1 \times 4^2 + 0 \times 4 + 3 \\
&= 3 \times 5 + 4
\end{aligned}$$

Thus, the number 19 can be written as 10011 in base 2, as 201 in base 3, as 103 in base 4, and as 34 in base 5, as well as 19 in base 10.   ○

Clearly, in writing such representations of numbers, we need to identify the base. We will do so by writing the name of the base as a subscript on the number. The result of Example 1.1 can then be written as

$$\begin{aligned}
19_{\text{ten}} &= 10011_{\text{two}} \\
&= 201_{\text{three}} \\
&= 103_{\text{four}} \\
&= 34_{\text{five}}
\end{aligned}$$

If the subscript is omitted, base 10 is assumed.

The bases 2, 8, and 16 are particularly important in computer work. These bases are called the *binary, octal,* and *hexadecimal* number systems, respectively. The digits in the hexadecimal number system are 0, 1, 2, 3, 4, 5, 6, 7, 8, 9, A, B, C, D, E, and F, so that A represents 10, B represents 11, and so on. We will indicate hexadecimal representations of numbers with the subscript "hex."

Converting from One Base to Another

If a representation of a number in a base other than 10 is given, the base-10 representation can be derived simply by doing the indicated arithmetic.

EXAMPLE 1.2   Convert 237_{eight} to decimal.

The given representation means $2 \times 8^2 + 3 \times 8 + 7 = 159$.   ○

In Example 1.1, we wrote down several different representations of the number 19. By doing some arithmetic (in base 10, of course), we could verify that the representations were correct. But how were the representations found? We would like to have a computational method, or **algorithm,** for finding the base-b representation of a number, given its base-10 representation. More generally, we would like to have a method of converting a representation in any base to a representation in any other base.

First, consider this division problem in base 10. When 6789 is repeatedly divided by 10, the result is

$$
\begin{array}{r|l}
10 & 6789 \quad \text{remainder} \\ \hline
10 & 678 \qquad 9 \\ \hline
10 & 67 \qquad 8 \\ \hline
10 & 6 \qquad 7 \\ \hline
 & 0 \qquad 6
\end{array}
$$

Notice that the remainders are the digits of the decimal number 6789, but in the reverse order. This algorithm works for any base: When we repeatedly divide a decimal number by b, the remainders will be the digits (in the reverse order) of the base-b representation of that number.

EXAMPLE 1.3   Find the binary representation of 59.
Using the algorithm given above, we compute

$$
\begin{array}{r|l}
2 & 59 \quad \text{remainder} \\ \hline
2 & 29 \qquad 1 \\ \hline
2 & 14 \qquad 1 \\ \hline
2 & 7 \qquad 0 \\ \hline
2 & 3 \qquad 1 \\ \hline
2 & 1 \qquad 1 \\ \hline
 & 0 \qquad 1
\end{array}
$$

obtaining 111011_{two} (writing the answer from bottom to top).   o

Relations among Binary, Octal, and Hexadecimal Representations

The decimal number 59, written in binary, is 111011_{two}, which requires six digits. For humans, writing the base-2 representations of large decimal numbers can be quite cumbersome. However, the base-8 and base-16 representations of a number are closely related to the base-2 representation, to the extent that they can be considered abbreviations of it.

Consider the binary number 11111110_{two}. If we group the digits in the binary number by threes, starting at the right, we obtain

$$(011) \quad (111) \quad (110)$$

where we have added a zero on the left to round out the third group. Now write the number in expanded form and group the terms of the expansion similarly; it then looks like

$$(0 \times 2^8 + 1 \times 2^7 + 1 \times 2^6)$$
$$+ (1 \times 2^5 + 1 \times 2^4 + 1 \times 2^3)$$
$$+ (1 \times 2^2 + 1 \times 2^1 + 0 \times 2^0)$$

From each group of three terms, factor out the highest power of 2 contained in each of its terms. The result is

$$(0 \times 2^2 + 1 \times 2^1 + 1 \times 2^0) \times 2^6$$

— 3

$$+(1 \times 2^2 + 1 \times 2^1 + 1 \times 2^0) \times 2^3$$

— 7

$$+(1 \times 2^2 + 1 \times 2^1 + 0 \times 2^0) \times 2^0$$

— 6

Since $2^0 = 8^0$, $2^3 = 8^1$, and $2^6 = (2^3)^2 = 8^2$, this binary number can be written as $3 \times 8^2 + 7 \times 8^1 + 6 \times 8^0 = 376_{\text{eight}}$. This illustrates the relation between the binary representation of a number and its octal representation: Each octal digit corresponds to three binary digits. Thus, to convert from binary to octal, group the binary digits by threes from the right, and for each group of three, write the corresponding octal digit.

EXAMPLE 1.4   Convert the binary number $1111001010101_{\text{two}}$ to octal.

$$\begin{array}{ccccc} 001 & 111 & 001 & 010 & 101 \\ \diagdown\diagup & \diagdown\diagup & \diagdown\diagup & \diagdown\diagup & \diagdown\diagup \\ 1 & 7 & 1 & 2 & 5 \end{array}$$

The answer is 17125_{eight}.   ○

EXAMPLE 1.5   Convert the octal number 6341_{eight} to binary.

To do this, we reverse the steps shown in Example 1.4. That is, for each octal digit, write the three binary digits representing it:

$$\begin{array}{cccc} 6 & 3 & 4 & 1 \\ \diagup\diagdown & \diagup\diagdown & \diagup\diagdown & \diagup\diagdown \\ 110 & 011 & 100 & 001 \end{array}$$

Thus, $6341_{\text{eight}} = 110011100001_{\text{two}}$.   ○

Notice that in order to make three binary digits for each octal digit, leading zeros are added to the binary representations of the octal digits 3 and 1.

The relation between bases 2 and 16 is similar: Each hexadecimal digit in the base-16 representation corresponds to four binary digits. To convert from binary to hexadecimal, group the binary number in groups of four, again starting at the right, then write the hexadecimal digit for each group of four. To convince yourself that this method works, remember that $2^4 = 16$. Reverse this procedure to convert from hexadecimal to binary.

EXERCISES 1.1

1. Convert to binary
   a. 17
   b. 1024
   c. 344_{eight}
   d. ABC_{hex}

2. Convert to binary
   a. 256
   b. 100
   c. 102_{eight}
   d. 256_{hex}

3. Convert to decimal
   a. 34567_{eight}
   b. 11111_{two}
   c. 123_{five}
   d. $DEAD_{hex}$

4. Convert to decimal
   a. 111111111_{two}
   b. 71_{eight}
   c. 208_{nine}
   d. ACE_{hex}

5. Convert the decimal number 124 to
   a. octal
   b. hex

6. Convert the decimal number 500 to
   a. octal
   b. hex

7. Convert the binary number 1010010001 to
   a. octal
   b. hex

8. Convert the binary number 11100001110000 to
   a. octal
   b. hex

9. Which of the following numbers are prime?
   a. 17_{eight}
   b. 17_{hex}
   c. 101010101_{two}
   d. 401_{five}

10. Which of the following numbers are prime?
    a. 65_{eight}
    b. 221_{three}
    c. CAB_{hex}
    d. 100111_{two}

1.2 • ARITHMETIC WITH NATURAL NUMBERS

The set **N** of natural numbers is equipped with two binary operations: addition and multiplication. A **binary operation** is a process that can be applied to two elements of a number system to yield a result that is an element of the same number system. The number system is then said to be **closed** under the operation. For example, **N** is closed under addition but not under subtraction. The addition and multiplication operations on **N** obey the following rules:

1. Commutative laws for addition and multiplication:

$$x + y = y + x \quad \text{and} \quad xy = yx$$

2. Associative laws for addition and multiplication:

$$x + (y + z) = (x + y) + z \quad \text{and} \quad x(yz) = (xy)z$$

3. Distributive law:

$$x(y + z) = xy + xz$$

Binary Arithmetic

The addition and multiplication tables in base 2 are

+	0	1
0	0	1
1	1	10

×	0	1
0	0	0
1	0	1

Notice that adding $1 + 1$ results in a carry into the next column position; that is, $1 + 1 = 10$. In Example 1.6, we omit the subscript "two" from the binary representation.

EXAMPLE 1.6   Add the binary numbers 111001 and 1101.

$$
\begin{array}{rl}
11 \quad 1 \quad & \longleftarrow \quad \text{Carries} \\
111001 & (57_{\text{ten}}) \\
+ \quad\underline{1101} & (13_{\text{ten}}) \\
1000110 & (70_{\text{ten}})
\end{array}
$$

○

Note that the number of digits in the sum of two positive numbers is equal to, or one greater than, the number of digits in the longer summand. This is true in every base. The size of the results of arithmetic operations is important in studying computer arithmetic, because the amount of memory allocated to a result may or may not be enough to contain it.

EXERCISES 1.2

1. Perform the following binary additions:
   a. $100111 + 1101$
   b. $11001 + 11001$
2. Perform the following binary additions:
   a. $10010011 + 1111111$
   b. $10101 + 11011$
3. Perform the following binary multiplications:
   a. 1110011×101
   b. 10001001×111
4. Perform the following binary multiplications:
   a. 11011×11011
   b. 110×1011
5. Subtraction in binary may require borrowing: When a 1 is to be subtracted from a 0, the 0 must be converted to 10_{two} (2_{ten}), and all 0's to its left up to the first 1 must be converted to 1's, while the 1 becomes a 0. For example, to

subtract 101 from 110000,

$$01112 \longleftarrow \text{Effect of borrowing}$$

$$110000$$
$$- \quad 101$$
$$\overline{101011}$$

Perform the binary subtraction $110111 - 10101$.

6. Perform the binary subtraction $10001101 - 1011010$.

7. Perform the following additions and subtractions in base 8:
   a. $2345 + 1777$ b. $73 - 25$

8. Perform the following additions and subtractions in base 8:
   a. $556 + 123$ b. $4321 - 1234$

9. Perform the following additions and subtractions in base 16:
   a. $12345 + 6789A$ b. $C6E18 - A5E2$

10. Perform the following additions and subtractions in base 16:
    a. $73C6E + A5EF2$ b. $FFFFFF - 12345$

11. In binary arithmetic, multiplication by a single-digit number never requires carrying. Is there any other base with this property? Explain.

12. State a rule relating the lengths (number of digits) of the quotient and remainder of a division to the lengths of the dividend and divisor.

13. The ancient Babylonians used the sexagesimal (base-60) number system. The base 60 uses 60 digits, which were built by the Babylonians from the symbols Y (for 1) and ◁ (for 10). For example, 11 was represented by ◁Y and 23 by ◁◁YYY. The Babylonians never invented a symbol for zero; a blank space was left instead. Consequently, they were extremely systematic about arranging their computations in neat columns, so that the blank spaces could be identified. Perform the following addition:

Y Y	◁◁ Y Y Y	◁◁◁◁ Y Y Y Y
+	◁◁◁◁ Y	Y Y Y Y Y

14. Perform the following sexagesimal addition:

◁ Y	◁◁◁◁◁◁ Y Y Y Y Y Y Y Y Y	◁◁◁
+ Y Y		◁◁◁

1.3 • SUMMATION AND PRODUCT NOTATION

The expression for the representation of a number n in base b,

$$n = a_k b^k + a_{k-1} b^{k-1} + \cdots + a_1 b + a_0$$

is the sum of several terms that follow a pattern. The pattern is more readily apparent if we rewrite the expression as

$$n = a_k b^k + a_{k-1} b^{k-1} + \cdots + a_1 b^1 + a_0 b^0 \tag{1}$$

Each term is of the form $a_i b^i$ for some value of the integer i. In the first term of the expression, i is k; in the second, it is $k - 1$. In the second to last term, i is 1, since $b^1 = b$. In the last term, i is 0, since $b^0 = 1$. Sums of this type are very common in mathematics. There is a standard "shorthand" notation for writing such expressions.

The Greek letter Σ (sigma) is used to indicate a sum. For example, the sum of the first five positive integers, which we write $1 + 2 + 3 + 4 + 5$, can be written in "shorthand" form as

$$\sum_{i=1}^{5} i$$

You can read the symbol Σ as "the sum of." In the expression above, i is called the **index of summation.** Although any variable can be used as the index of summation, i, j, and k are most common. The expression $\sum_{i=1}^{5} i$ means the same as $\sum_{k=1}^{5} k$. The numbers 1 and 5 are the limits of summation, with 1 being the lower limit and 5 the upper limit. You can read the expression $\sum_{i=1}^{5} i$ as "the sum of i as i goes from 1 to 5."

The following example illustrates the use of summation notation.

EXAMPLE 1.7   **a.** $\displaystyle\sum_{k=1}^{n} k = 1 + 2 + 3 + \cdots + n$

b. $\displaystyle\sum_{i=0}^{4} 2^i = 2^0 + 2^1 + 2^2 + 2^3 + 2^4 = 31$

c. $\displaystyle\sum_{j=1}^{10} 5 = 5 + 5 + 5 + 5 + 5 + 5 + 5 + 5 + 5 + 5 = 50$

d. $\displaystyle\sum_{i=2}^{5} 4i = 4(2) + 4(3) + 4(4) + 4(5) = 8 + 12 + 16 + 20 = 56$

e. $\displaystyle\sum_{k=0}^{3} k^2 = 0^2 + 1^2 + 2^2 + 3^2 = 0 + 1 + 4 + 9 = 14$

f. $\displaystyle\sum_{i=1}^{n} (x_i + y_i) = (x_1 + y_1) + (x_2 + y_2) + \cdots + (x_n + y_n)$

Now we can write the short form for Equation (1) as

$$n = \sum_{i=0}^{k} a_i b^i$$

○

Summation Rules

From the preceding examples, you can see that the following rules hold for expressions in which the Σ notation is used. They are consequences of the commutative, associative, and distributive laws. Here, c is any constant, x_i and y_i are real numbers, and i, k, and n are positive integers.

1. $\displaystyle\sum_{i=1}^{n} c = nc$

2. $\displaystyle\sum_{i=1}^{n} cx_i = c \sum_{i=1}^{n} x_i$

3. $\displaystyle\sum_{i=1}^{n} (x_i + y_i) = \sum_{i=1}^{n} x_i + \sum_{i=1}^{n} y_i$

4. $\displaystyle\sum_{i=1}^{n} x_i = \sum_{i=1}^{k} x_i + \sum_{i=k+1}^{n} x_i$ Where $1 \le k < n$

Product Notation

Although it is used much less frequently, there is a notation for products that is similar to the Σ notation for sums. The Greek letter Π (capital pi) is used to indicate a product of terms.

EXAMPLE 1.8   **a.** $\displaystyle\prod_{k=1}^{n} k = 1 \times 2 \times 3 \times \cdots \times n$

b. $\displaystyle\prod_{i=0}^{4} (x - 2i) = (x - 0)(x - 2)(x - 4)(x - 6)(x - 8)$

c. $\displaystyle\prod_{k=1}^{m} (y - a_k) = (y - a_1)(y - a_2) \cdots (y - a_m)$ ○

EXERCISES 1.3

1. Write out each of the following sums and evaluate if possible.

  a. $\displaystyle\sum_{i=8}^{20} x_i^2$ **b.** $\displaystyle\sum_{i=1}^{10} (x_i + y_i)^2$ **c.** $\displaystyle\sum_{i=1}^{10} x_i^2 + y_i^2$

  d. $\displaystyle\sum_{i=1}^{5} i^3$ **e.** $\displaystyle\sum_{j=1}^{4} j^2 + 9$ **f.** $\displaystyle\sum_{j=1}^{4} (j^2 + 9)$

2. Write out each of the following sums and evaluate if possible.

  a. $\displaystyle\sum_{k=1}^{10} k^2$ **b.** $\displaystyle\left(\sum_{k=1}^{10} k\right)^2$ **c.** $\displaystyle\sum_{i=1}^{8} 3i^2$

  d. $\displaystyle\sum_{k=1}^{100} 12$ **e.** $\displaystyle\sum_{k=1}^{6} (5 - k)$ **f.** $\displaystyle\sum_{j=1}^{6} 2^j$

3. Write the following sums using Σ notation.

  a. $5 + 6 + 7 + 8 + 9 + \cdots + 22$ **b.** $X_1 + X_2 + X_3 + \cdots + X_{20}$

4. Write the following sums using Σ notation.

  a. $0 + 2 + 4 + 6 + 8 + 10 + \cdots + 100$

  b. $x_1 y_1 + x_2 y_2 + \cdots + x_n y_n$

5. Use the summation rules to show that

$$\sum_{i=1}^{n} (x_i + 5) = \sum_{i=1}^{n} x_i + 5n$$

6. Write out each of the following products:

    a. $\displaystyle\prod_{i=8}^{15} y_i$ **b.** $\displaystyle\prod_{i=1}^{5} (x_i + y_i)$

7. Write out each of the following products:

    a. $\displaystyle\prod_{i=1}^{10} (x + 2i)^2$ **b.** $\displaystyle\prod_{i=1}^{4} (i - i^2)$

8. Rewrite each of the following summations so that the index of summation starts with the given different value:

    a. $\displaystyle\sum_{i=1}^{15} i^2 = \sum_{i=0}^{?} ?$ **b.** $\displaystyle\sum_{k=0}^{5} (k + 3)^3 = \sum_{k=3}^{?} ?$

9. Rewrite each of the following summations so that the index of summation starts with the given different value:

    a. $\displaystyle\sum_{i=0}^{9} \frac{1}{(1 + i)^2} = \sum_{i=1}^{?} ?$ **b.** $\displaystyle\sum_{k=5}^{10} (k - 1)^2 = \sum_{k=1}^{?} ?$

10. Evaluate $\displaystyle\sum_{i=1}^{5} \sum_{j=1}^{3} i^j$

11. Evaluate $\displaystyle\sum_{i=0}^{2} \prod_{j=0}^{3} (i + j)$

12. Evaluate $\displaystyle\prod_{i=k}^{n} c$

1.4 • INTEGERS AND RATIONAL NUMBERS

The operations of addition and multiplication on the natural numbers have corresponding inverse operations: subtraction and division. If a and b are natural numbers, to subtract b from a means to find a number x such that $b + x = a$; the number x, if it exists, is denoted by $a - b$. To divide a by b means to find a number y such that $by = a$; the number y, if it exists, is denoted by $\frac{a}{b}$. In the natural numbers, it is not always possible to find such numbers. For instance, $7 + x = 5$ has no solution in the natural numbers, because there is no natural number that can be denoted by $5 - 7$. Also, there is no natural number that can be denoted by $\frac{1}{2}$, because in the natural numbers there is no solution to the equation $2y = 1$. Other equations that have no solutions in the natural numbers are $x^2 = 2$ and $x^2 + 2 = 1$.

In these cases, we can construct a larger number system that contains solutions to the equations. To solve equations of the form $b + x = a$, we construct the integers. To solve equations of the form $by = a$, we construct the rational numbers.

The Integers

In order to be able to solve equations of the form $a + x = b$, we expand the natural number system to the **integers.** The set of integers consists of the natural numbers, the number 0, and new numbers that are the negatives of the natural numbers. Thus the integers consist of

$$\ldots, -3, -2, -1, 0, 1, 2, 3, \ldots$$

where the ellipsis indicates continuation indefinitely.

The set of integers is denoted by **Z**. Addition and multiplication are defined on the integers and satisfy the commutative, associative, and distributive laws. The integer 0 is an **identity element,** also called a **neutral element,** for the operation of addition; this means that $x + 0 = x$ for every integer x. The number 1 is an **identity element for multiplication:** $1x = x$ for every integer **x.**

In the set **Z**, every equation of the form $a + x = b$ has a solution, which is denoted by $b - a$. In particular, the equation $a + x = 0$ has a solution $0 - a$, which is called the **additive inverse** or **negative** of a and is written $-a$. For instance, the additive inverse of 5 is -5, and the additive inverse of -7 is $-(-7) = 7$.

The Rational Numbers

In the integers, it is usually not possible to find solutions of all equations of the form $ax = b$. Of course, there is no hope of finding a solution of $0x = b$ unless $b = 0$, because $0x = 0$ for every number x. Moreover, every number x is a solution of the equation $0x = 0$. Therefore, we exclude the case $a = 0$. In order to solve equations of the form $ax = b$ when $a \neq 0$, we introduce the **rational numbers.** The set of rational numbers is denoted by **Q**.

Rational numbers are represented by fractions of the form $\frac{b}{a}$ where a and b are integers and $a \neq 0$. Two fractions $\frac{b}{a}$ and $\frac{d}{c}$ represent the same rational number if and only if $bc = ad$.

If a and b are rational numbers and $a \neq 0$, then the equation $ax = b$ has a solution that is a rational number. To find it, suppose that a is represented by the fraction $\frac{c}{d}$ and b is represented by the fraction $\frac{e}{f}$, where c, d, e, and f are integers. Then the fraction $\frac{ed}{fc}$ represents the solution of the equation $ax = b$, since

$$ax = \frac{c}{d} \times \frac{ed}{fc} = \frac{e}{f} = b$$

In particular, if $a \neq 0$, the equation $ax = 1$ has a solution $\frac{1}{a}$, which is called the **multiplicative inverse** of a. The operations of addition and multiplication of rational numbers satisfy the commutative, associative, and distributive laws.

If the numerator and denominator of a fraction have any common factors greater than 1, the common factors can be canceled to obtain a representation for the same rational number having a smaller numerator and denominator. If there are no common factors, the fraction is said to be in **lowest terms.** For instance, $\frac{1}{3}$ is in lowest terms, but $\frac{2}{6}$ is not. If we impose the additional requirement that a fraction in lowest terms have a positive denominator, every rational number except 0 has a unique representation as a fraction in lowest terms.

Decimal-Fraction Representations of Rational Numbers

Writing a rational number as the quotient of two integers is just one way of representing it. Another way to represent a rational number is as a **decimal fraction.**

A decimal fraction is a sequence of decimal digits, followed by a decimal point, followed by another sequence of decimal digits. (A minus sign may be included to indicate a negative number.) The sequence of digits after the decimal point may be finite, or it may be infinite. Digit positions after the decimal point correspond to negative powers of ten.

To get the decimal-fraction representation of a rational number, do the long division. A sequence of remainders will be produced. There are two possible results: Either a remainder of zero will appear, and the division will stop; or no zero will ever appear as a remainder, and the division will go on forever. For example, $\frac{37}{20} = 1.85$, with a zero remainder occurring after three steps. On the other hand,

$$\frac{22}{7} = 3.142857142857142857\ldots$$

In fact, the pattern 142857 repeats indefinitely; the result is sometimes written $\frac{22}{7} = 3.\overline{142857}$.

Base-*b* Fractions

The concept of decimal fraction can be extended to any base b, in which case it is called a base-b fraction. If $a_k, a_{k-1}, \ldots, a_1, a_0$ and $c_1, c_2, \ldots, c_{m-1}, c_m$ are base-b digits, the number x represented by

$$a_k a_{k-1} \cdots a_1 a_0 . c_1 c_2 \cdots c_{m-1} c_m$$

is

$$x = \sum_{i=0}^{k} a_i b^i + \sum_{i=1}^{m} c_i b^{-i}$$

A base-b number with a fractional part can be converted to its base-10 representation by writing down the meaning of the number, remembering that digits to the right of the point are multiplied by negative powers of the base.

EXAMPLE 1.9   The binary number 110.1001_{two} means

$$1 \times 2^2 + 1 \times 2^1 + 0 \times 2^0 + 1 \times 2^{-1} + 0 \times 2^{-2} + 0 \times 2^{-3} + 1 \times 2^{-4}$$

$$= 1 \times 2^2 + 1 \times 2^1 + 0 \times 2^0 + 1 \times \left(\frac{1}{2}\right)^1 + 0 \times \left(\frac{1}{2}\right)^2 + 0 \times \left(\frac{1}{2}\right)^3 + 1 \times \left(\frac{1}{2}\right)^4$$

$$= 6.5625$$

○

There is an algorithm to convert a fractional number from base 10 to another base: First convert the integer part of the number using repeated division by the base, then convert the fractional part using repeated multiplication by the base. (Multiplying a base-b number by b moves the base-b point one place to the right.) The process stops when zero or a repeating pattern is reached.

EXAMPLE 1.10   Convert the decimal number 4.375 to binary.

$4_{\text{ten}} = 100_{\text{two}}$. To convert .375, multiply by 2:

$$
\begin{array}{r|c|l}
.375 \times 2 = & 0 & .750 \\
.750 \times 2 = & 1 & .500 \\
.500 \times 2 = & 1 & .000 \quad \longleftarrow \quad \text{Stop when zero}
\end{array}
$$

The fractional part of the binary number is written from left to right starting at the top of the box. Thus, 4.375 is written in binary as 100.011.   ○

EXAMPLE 1.11   Convert 7.35 to binary.

$7_{\text{ten}} = 111_{\text{two}}$.

$$
\begin{array}{r|c|l}
.35 \times 2 = & 0 & .70 \\
.70 \times 2 = & 1 & .40 \\
.40 \times 2 = & 0 & .80 \quad \longleftarrow \\
.80 \times 2 = & 1 & .60 \\
.60 \times 2 = & 1 & .20 \\
.20 \times 2 = & 0 & .40 \\
.40 \times 2 = & 0 & .80 \quad \longleftarrow \quad \text{Pattern repeats} \\
.80 \times 2 = & 1 & .60 \\
.60 \times 2 = & 1 & .20
\end{array}
$$

The answer is $111.01011001100110\ldots = 111.01\overline{0110}$ ○

The preceding example shows that a number whose base-10 representation terminates may not terminate when written in base 2. Conversely, a number whose base-2 representation terminates may not terminate when written in base 10. This explains why computers, which commonly do arithmetic in base 2, sometimes appear to give wrong answers to trivial problems. For instance, $0.4 + 0.6$, done on some computers, may result in 0.99999999. Roundoff errors are introduced in converting the summands to base 2 and the result back to base 10.

Our observations about base-b fractional representations can be collected in the following theorem.

Theorem 1.1   Let b be a natural number greater than 1. Then every rational number x can be represented in base b with an integer and a fractional part that is either finite or contains a repeating pattern.

Proof   If $x = \frac{c}{d}$ is a rational number, long division in base b will produce a base-b representation of x. Carry out the division beyond the point where digits of the numerator are brought down. To show that there is a repeating pattern, it is necessary only to show that some remainder is repeated. But in dividing by d, there are only d possible remainders—namely, $0, 1, \ldots, d-1$. Therefore, as soon as $d+1$ consecutive steps are performed, a repetition will occur.   ●

The method of reasoning used in proving Theorem 1.1 is so common that it is given a name: the **Pigeonhole Principle.** This principle states that if $n+1$ (or more) objects are arranged in n groups (pigeonholes), at least one group contains more than one object. This principle will be discussed further in Chapter 3.

EXERCISES 1.4

1.   What is the 100th digit in the decimal expansion of 5/7?

2.   What is the 729th digit in the decimal expansion of 3/13?

3.   Convert the binary number 11110.1101 to base 10.

4.   Convert the binary number 10001.0111 to base 10.

5.   Convert 19.2 to binary.

6.   Convert 3.45 to binary.

7.   Devise an algorithm to convert a decimal fraction to octal. Test your algorithm with the following base-10 numbers:

   a.   3.5   **b.**   12.3   **c.**   0.25   **d.**   7.01

8.   Devise an algorithm to convert a binary fraction to octal without first converting to decimal. Test your algorithm with the following binary fractions:

   a.   0.1111000011111   **b.**   0.1111001110
   c.   0.00011010100011

9.   Convert the following binary fractions to hexadecimal without converting to base 10:

   a.   0.1111000011111   **b.**   0.1100000000001011
   c.   0.0001111111111000

10.   Show that 0.5 and 0.4999 . . . represent the same rational number.

1.5 ● THE REAL AND COMPLEX NUMBERS

We have seen that rational numbers can be represented by decimal fractions (or fractions relative to any other base) that either terminate or repeat. Decimal fractions that do not repeat represent **irrational numbers.** For instance, the infinite decimal fraction 0.101001000100001000001 . . . , where each group of zeros contains one more zero than the preceding group, represents an irrational number. (The term decimal "fraction" for an expression such as 0.101001000100001000001 . . . is something of a misnomer, since the number represented is not equal to any fraction.) Of course, the same statements apply in bases other than 10.

Another very well-known example of an irrational number is $\sqrt{2}$. To prove that $\sqrt{2}$ is irrational, assume that $\sqrt{2}$ is rational and deduce a contradiction. If $\sqrt{2}$ is rational, it can be represented by a fraction $\frac{p}{q}$ in lowest terms. So $(\frac{p}{q})^2 = 2$, and so $p^2 = 2q^2$. Thus, p^2 is an even number, so p must be an even number also. Suppose $p = 2r$. It follows that $(2r)^2 = 2q^2$, so $2r^2 = q^2$. Similar reasoning now shows that q is even. But the statement that p and q are both even contradicts the assumption that the fraction $\frac{p}{q}$ is in lowest terms. This shows that $\sqrt{2}$ is not a rational number.

The rational and irrational numbers together make up the set of **real numbers,** denoted by **R**. Every real number can be written in a unique way as a non-terminating decimal. For example, 4.5 can be written as 4.4999 . . . .

Complex Numbers

The set **R** of real numbers contains solutions to many polynomial equations with rational coefficients, including (as we have seen) the equation $x^2 = 2$. But it does not contain solutions of all polynomial equations. For instance, there is no real solution of the equation $x^2 = -1$.

In order to solve all polynomial equations with rational coefficients, we construct a larger number system called the set of **complex numbers.** The set of complex numbers is denoted by **C.** The construction is done by introducing the symbol i as a solution of the equation $x^2 = -1$; that is, $i^2 = -1$. A complex number is an expression of the form $a + bi$, where a and b are real numbers. The real numbers are contained in the complex numbers, since the real number a can be written in the form $a + 0i$. The complex number $a + bi$ may be identified with a point in the plane whose coordinates are (a, b). The plane is then called the **complex plane;** it is an extension of the real line. Figure 1.1 shows the complex plane and several complex numbers in it.

If $z = a + bi$ with a and b real numbers, then a is called the **real part** of z and b is called the **imaginary part** of z. The number i is frequently called the **imaginary unit,** and numbers of the form bi are called **imaginary numbers.** The name is misleading; these numbers are no more imaginary (or less real) than the integers, rational numbers, or real numbers.

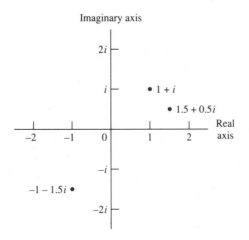

FIGURE 1.1   The complex plane

Arithmetic with Complex Numbers

Addition and multiplication are defined for complex numbers by the rules

$$(a + bi) + (c + di) = (a + c) + (b + d)i$$
$$(a + bi)(c + di) = ac + adi + bci + bdi^2$$
$$= ac + adi + bci + bd(-1)$$
$$= ac - bd + (ad + bc)i$$

The commutative, associative, and distributive laws hold for addition and multiplication of complex numbers. The number 0 is an identity for addition, and 1 is an identity for multiplication. Additive inverses exist, and all complex numbers except 0 have multiplicative inverses.

Absolute Value and Argument

If $z = a + bi$ is a complex number, the quantity $\sqrt{a^2 + b^2}$ is called the **absolute value** of z and is denoted by $|z|$. It is the distance from the point (a, b) to the origin $(0, 0)$ in the complex plane. The angle that the line from 0 to z makes with the positive real axis is called the **argument** of z and is denoted by Arg(z). Angles measured counterclockwise from the positive real axis are positive; angles measured clockwise are negative.

EXAMPLE 1.12   Figure 1.2 shows the complex numbers $z = 2 + 1.5i$ and $w = -1 - i$. The absolute value and argument of z are (approximately) $|z| = \sqrt{2^2 + 1.5^2} = 2.5$ and Arg(z) = 36.9 degrees. Note that Arg(w) = Arg($-1 - i$) may be given as either 225 degrees or as -135 degrees; both are names of the same angle.

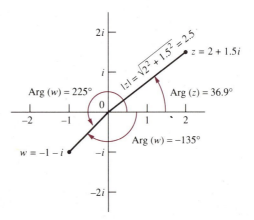

FIGURE 1.2   Absolute value and argument of complex numbers

Geometrical Interpretation of Complex Arithmetic

Addition of complex numbers has a simple geometrical interpretation. If z and w are complex numbers, the sum $z + w$ may be found by drawing lines from z and w to 0 and then completing a parallelogram with these lines as sides. Then $z + w$ is the vertex of the parallelogram opposite 0.

EXAMPLE 1.13   Figure 1.3 shows the sum of the complex numbers $z = 2 + 0.5i$ and $w = 1 + 1.7i$.

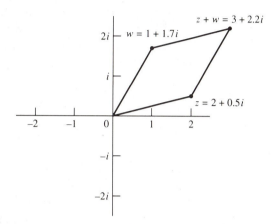

FIGURE 1.3   Sum of complex numbers

If z and w are complex numbers, then $\text{Arg}(zw) = \text{Arg}(z) + \text{Arg}(w)$. (The proof of this requires some trigonometry and is left as an exercise.) Also, $|zw| = |z||w|$. These facts provide a geometrical interpretation for multiplication of complex numbers. The product zw can be found by measuring off the distance $|z||w|$ from 0 and adding the angles $\text{Arg}(z)$ and $\text{Arg}(w)$.

EXAMPLE 1.14   In Figure 1.4, z is the complex number of absolute value 1.5 and argument 20 degrees, and w is the complex number of absolute value 2 and argument 45 degrees. Approximately, $z = 1.41 + 0.51i$ and $w = 1.41 + 1.41i$. The product zw has absolute value $1.5 \times 2 = 3$ and argument $20 + 45 = 65$ degrees.

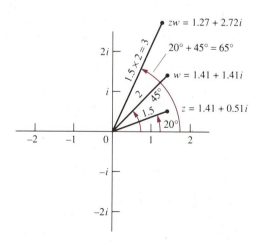

FIGURE 1.4   Product of complex numbers

Multiplication of complex numbers of absolute value 1 is particularly simple: It is accomplished entirely by adding angles.

EXAMPLE 1.15   Consider the complex number z whose absolute value is 1 and whose argument is 40 degrees; approximately, $z = 0.776 + 0.643i$. This is shown in Figure 1.5. The numbers $z^2, z^3, \ldots, z^9$ have absolute value 1 and arguments of 80, 120, 160, $\ldots$, 320, and 360 degrees. Therefore, $z^9 = 1$, since z^9 has absolute value 1 and argument 360 degrees. The numbers $z, z^2, \ldots, z^9$ are shown in Figure 1.5. Now, since $(z^2)^9 = (z^9)^2 = 1, (z^3)^9 = (z^9)^3 = 1$, and so on, the numbers $1, z, z^2, \ldots, z^8$ are nine different ninth roots of 1.

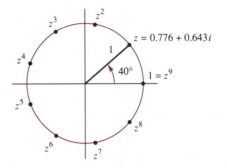

FIGURE 1.5   The nine ninth roots of 1

For any positive integer n, the complex numbers contain n different nth roots of 1, evenly spaced on the circle $|z| = 1$, at intervals of $\frac{360}{n}$ degrees. The number with absolute value 1 and argument $\frac{360}{n}$ degrees is called the **primitive nth root of unity.** All the other roots are powers of the primitive root.

The complex numbers were constructed by starting with the real numbers and adding a solution of the equation $x^2 = -1$. In the complex numbers, every polynomial equation with complex coefficients has a solution in the complex numbers. In fact, every polynomial can be factored into linear factors, that is, into polynomials of the form $z - a$. The proof of this fact requires advanced methods in the theory of complex numbers, so we only state it.

Theorem 1.2   (*Fundamental Theorem of Algebra*)   Let

$$f(x) = x^n + a_{n-1}x^{n-1} + \cdots + a_1x + a_0$$

be a polynomial of degree n with complex coefficients $a_{n-1}, \ldots, a_1, a_0$. Then $f(x)$ can be factored into linear factors; that is,

$$f(x) = \prod_{j=1}^{n} (x - r_j)$$

where the complex numbers $r_1, \ldots, r_n$ are all roots of $f(x)$. (The roots r_j need not be distinct.)

Comparison of Number Systems

The algebraic properties of the number systems we have discussed can be summarized in the following table.

	N	Z	Q	R	C
Addition is commutative.	Yes	Yes	Yes	Yes	Yes
Addition is associative.	Yes	Yes	Yes	Yes	Yes
Multiplication is commutative.	Yes	Yes	Yes	Yes	Yes
Multiplication is associative.	Yes	Yes	Yes	Yes	Yes
Distributive law holds.	Yes	Yes	Yes	Yes	Yes
Multiplicative identity exists.	Yes	Yes	Yes	Yes	Yes
Additive identity exists.	No	Yes	Yes	Yes	Yes
Additive inverses exist.	No	Yes	Yes	Yes	Yes
Multiplicative inverses exist.*	No	No	Yes	Yes	Yes
Solution for $x^2 = 2$?	No	No	No	Yes	Yes
Solutions for all polynomials?	No	No	No	No	Yes

*Except for zero

EXERCISES 1.5

1. Show that $\sqrt{3}$ is irrational.

2. The notation $\log_2 5$ denotes the solution x of the equation $2^x = 5$. (Approximately, $x = 2.321928\ldots$.) Show that $\log_2 5$ is irrational.

3. Plot each of the following complex numbers in the complex plane:

   a. 1.8 b. $2.2i$ c. $-1.2i$ d. -3.5

   e. $1 + 7i$ f. $3 - 5i$ g. $-3 + 2.5i$ h. $-2.5 - i$

4. Perform the indicated operations:

   a. $(17 + 34i) + (-4 + 9i)$ b. $(17.5 - 93.6i) - (-3.6 - i)$

   c. $(1 + i)(3 - 5i)$ d. $(4 + 3i)i$ e. i^3

5. Perform the indicated operations:

   a. $(-3.56 - 5.87i) + (3.62 - 6.28i)$

   b. $265i - (2 + 3i)$ c. $(-5 - 13i)(-6 + 4i)$

   d. $(-i)i$ e. i^4

6. Find the absolute value and argument of each of the following complex numbers. Approximate answers are acceptable. You may use a protractor or trigonometry.

   a. $3 + 5i$ b. $3 - 5i$ c. $-3 + 5i$

   d. $-3 - 5i$ e. -4 f. $-i$

7. Find the complex numbers having the following absolute values and arguments. Approximate answers are acceptable. You may use a protractor or trigonometry.

    a. $|z| = 1.3$, $\mathrm{Arg}(z) = 35$ degrees   **b.** $|z| = 2.5$, $\mathrm{Arg}(z) = 170$ degrees

    c. $|z| = 1$, $\mathrm{Arg}(z) = 270$ degrees   **d.** $|z| = 2$, $\mathrm{Arg}(z) = -200$ degrees

8. Find the additive and multiplicative inverses for each of the following complex numbers:

    a. $1 + i$ **b.** $1 - 2i$ **c.** $-5i$

    d. $-3 + i$ **e.** -4 **f.** $-2 - \sqrt{2i}$

9. Let a and b be real numbers, and let $z = a + bi$. Then the **complex conjugate** of z, denoted by $\bar{z}$, is $a - bi$. For each of the following complex numbers, find the complex conjugate and plot both the number and its conjugate in the complex plane.

    a. $2 + 3i$ **b.** $-6 - 7i$ **c.** 17 **d.** $2.2i$

10. Show that if z is a complex number, then

    a. $z\bar{z} = |z|^2$ **b.** real part $(z) = \dfrac{z + \bar{z}}{2}$

    c. imaginary part $(z) = \dfrac{z - \bar{z}}{2i}$

11. Perform each of the following divisions. (*Hint:* Multiply the numerator and denominator of the fraction by the complex conjugate of the denominator.)

    a. $\dfrac{2 + 3i}{3 + 4i}$ **b.** $\dfrac{i}{-2 + 4.5i}$ **c.** $\dfrac{-7 + 6i}{-7 - 6i}$ **d.** $\dfrac{1}{1 + i}$

12. Find the solutions of each of the following equations and plot the solutions in the complex plane. How are the solutions related to one another?

    a. $x^2 + x + 1 = 0$ **b.** $2x^2 - 7x + 7 = 0$ **c.** $x^2 + 9 = 0$

    d. $x^2 + x - 12 = 0$ **e.** $x^2 - 6x + 9 = 0$ **f.** $x^3 + x^2 + x = 0$

13. Prove that if p is prime, then $\sqrt{p}$ is irrational.

14. Prove that $|z||w| = |zw|$ for any complex numbers z and w.

15. (*Optional—requires trigonometry*) If $z = a + bi$ is a complex number, show that $a = |z| \cos(\mathrm{Arg}(z))$ and $b = |z| \sin(\mathrm{Arg}(z))$.

16. (*Optional—requires trigonometry*) Show that if z and w are complex numbers, then $\mathrm{Arg}(zw) = \mathrm{Arg}(z) + \mathrm{Arg}(w)$. (*Hint:* Use the results of the previous exercise and the formulas for the sine and cosine of a sum of angles.)

1.6 • ARITHMETIC IN THE INTEGERS MODULO *n*

When a computer does arithmetic, the operands and the result must be stored in the computer's memory. Computer memories are made up of two-state devices; that is, each unit of memory can be in any one of two states. This unit of memory is called a **bit** (**bi**nary digi**t**). A bit may be in one of two states, called YES and NO, or ON and OFF, or 1 and 0. Thus, computers are well suited for the use of the binary system.

Information stored in a computer's memory is made up of a large number of bits, which are arranged in groups called **words**. An m-bit word can be thought of as a string of bits, each of which is either 0 or 1.

Arithmetic in a computer is based on the representation of numbers by m-bit words. Since there are only 2^m bit strings of m bits, only 2^m different numbers can be represented. It follows that not all arithmetic computations can be done correctly. For instance, if X is the largest number that can be stored in a word, the computation of $X + X$, or even $X + 1$, cannot be done correctly. The issues are the same regardless of the word length. These considerations will lead us to a new class of number systems, called the integers modulo n and denoted by $\mathbf{Z_n}$.

Two's Complement Arithmetic

Most modern computers include an electronic circuit called a **two's complement adder.** A two's complement adder accepts as inputs two m-bit binary numbers (where m is the word length of the computer) and produces a result called the two's complement sum. The sum is obtained by doing ordinary binary addition and discarding any carry from the leftmost bit position. The consequences can be illustrated adequately by considering 8-bit words.

EXAMPLE 1.16   Add 01001111 and 00011000 using two's complement addition.

$$
\begin{array}{ll}
11 \qquad \longleftarrow \text{Carries} \\
01001111 & (79_{\text{ten}}) \\
\underline{00011000} & (24_{\text{ten}}) \\
01100111 & (103_{\text{ten}})
\end{array}
$$

There is no carry from the leftmost position, and this result is correct for integer arithmetic.   ○

EXAMPLE 1.17   Add 10101010 and 01010101 using two's complement addition.

$$
\begin{array}{ll}
10101010 & (170_{\text{ten}}) \\
\underline{01010101} & (85_{\text{ten}}) \\
11111111 & (255_{\text{ten}})
\end{array}
$$

There is no carry. The result appears to be correct for integer arithmetic, if we interpret 11111111 as denoting 255. However, the next example casts doubt on this interpretation.   ○

EXAMPLE 1.18   Add 00001111 and 11111111 using two's complement addition.

$$
\begin{array}{ll}
11111111 \qquad \longleftarrow \text{Carries} \\
00001111 & (15_{\text{ten}}) \\
\underline{11111111} & (255_{\text{ten}}) \\
\boxed{1}\,00001110 & (14_{\text{ten}}) \\
\longrightarrow \text{Carry discarded}
\end{array}
$$

Since $15 + 255 = 270$, the answer 14 is wrong by 256.   ○

In Example 1.18, if we interpret 11111111 as denoting -1 instead of 255, then 14 is the correct answer! Also, the fact that $11111111 + 00000001 = 00000000$ in two's complement arithmetic suggests that 11111111 should be interpreted as -1. Similarly, 11111110 can be interpreted as -2, because $11111110 + 00000010 = 00000000$, and so on through 10000001 (129 as a positive number), which can be interpreted as -127. This suggests that words that start with a 1-bit be considered as representing negative numbers. The negative number represented is 256 less than the positive number represented:

$$-1 = 255 - 256, \text{ and } -127 = 129 - 256$$

EXAMPLE 1.19   What negative number is represented by 10000100 in two's complement arithmetic?

$10000100_{two} = 132_{ten}$, so 10000100 represents $132 - 256 = -124$.   o

The next example shows that even with this interpretation, the computer cannot do all addition problems correctly.

EXAMPLE 1.20   Add 01100011 and 00100001 using two's complement addition.

$$
\begin{array}{ll}
11 \quad 11 \quad \longleftarrow & \text{Carries} \\
01100011 & (99_{ten}) \\
\underline{00100001} & (33_{ten}) \\
10000100 & (132_{ten} \text{ or } -124_{ten})
\end{array}
$$

The answer is correct if 10000100 is interpreted as 132, but wrong by 256 if 10000100 is interpreted as -124.   o

EXAMPLE 1.21   Add 10011101 and 11011111 using two's complement addition.

$$
\begin{array}{ll}
1 \quad 11111 \quad \longleftarrow & \text{Carries} \\
10011101 & (157_{ten} \text{ or } -99_{ten}) \\
\underline{11011111} & (223_{ten} \text{ or } -33_{ten}) \\
1 \,|\, 01111100 & (124_{ten}) \\
\quad \longrightarrow & \text{Carry discarded}
\end{array}
$$

The answer is incorrect by 256 in either interpretation.   o

Examples 1.16 through 1.21 show that there are two different interpretations of what a two's complement adder does:

1. *Unsigned-integer interpretation:* An m-bit two's complement adder does arithmetic with the integers $0, 1, 2, \ldots, 2^m - 1$.
    *   When the sum of two of these integers is greater than $2^m - 1$, the adder produces an incorrect result. The result produced is less than the correct answer by 2^m. (See Example 1.18.)
2. *Signed-integer interpretation:* An m-bit two's complement adder does arithmetic with integers in the range $-2^m/2$ through $2^m/2 - 1$.

- When the sum of two positive numbers exceeds $2^m/2 - 1$, the adder produces an incorrect result. The result produced is less than the correct answer by 2^m. (See Example 1.20.)
- When the sum of two negative numbers is less than $-2^m/2$, the adder produces an incorrect result. The result produced is greater than the correct answer by 2^m. (See Example 1.21.)

Both of these interpretations are useful in computer programming. In either case, when the adder produces a result that is incorrect for the interpretation, the adder is said to **overflow.** In signed-integer interpretation, overflow occurs when two positive numbers are added and the result is negative, or when two negative numbers are added and the result is positive.

Yet another view of the two's complement adder is that it does arithmetic not with integers, but with elements of a different number system: a finite number system with 2^m elements. Elements of such a number system are identified by integers, but two integers that differ by 2^m (or any multiple of 2^m) represent the same element of the finite number system. This number system is called the *system of integers modulo n*, where $n = 2^m$. For example, if $m = 4$ (so that $2^4 = 16$), the integers -3 and 13 can be taken to represent the same element of the integers modulo 16, since they differ by 16. Such number systems can be constructed with any number n of elements. It is not necessary that n be a power of 2.

The Number System Z_n

For any integer greater than 1, there is a number system of n elements called the **integers modulo *n*** and denoted by $\mathbf{Z}_n$. We will not say what the elements of $\mathbf{Z}_n$ are (although we shall do so in Chapter 3). Rather, we will describe how they are represented and how to do arithmetic with them.

If a is an integer, the element of $\mathbf{Z}_n$ corresponding to a is denoted by $[a]_n$, or just $[a]$ if there is no ambiguity. Two integers a and b represent the same element of $\mathbf{Z}_n$ if $b - a$ is a multiple of n; that is,

$$[a] = [b] \quad \text{means that} \quad b - a = nx \quad \text{for some integer } x$$

EXAMPLE 1.22   In $\mathbf{Z}_5$, $[2] = [17]$, since $17 - 2 = 15$ is a multiple of 5. Also, $[-30] = [0]$, and $[6] = [1]$.

In $\mathbf{Z}_{256}$, $[255] = [-1]$, since $(-1) - 255 = -256$ is a multiple of 256.   ○

For any integer a, it is possible to find an integer x such that $0 \le x < n$, and $[x]_n = [a]_n$. The integer x is the remainder when a is divided by n. For instance, in $\mathbf{Z}_8$, $[35] = [3]$, since $35/8$ gives a remainder of 3.

Arithmetic with elements of $\mathbf{Z}_n$ is accomplished by doing arithmetic with the integers representing them. That is,

$$[a] + [b] \quad \text{is defined as} \quad [a + b]$$
$$[a][b] \quad \text{is defined as} \quad [ab]$$

EXAMPLE 1.23   In Z_5, $[3] + [4] = [7]$ and $[3][4] = [12]$. Since $[7] = [2]$ and $[12] = [2]$, we can also write $[3] + [4] = [2]$ and $[3][4] = [2]$.

In Z_{256}, $[170] + [86] = [256]$ and $[170] + [-170] = [0]$; of course, $[256] = [0]$. Also in Z_{256}, $[16][16] = [256] = [0]$.   ○

There is one logical deficiency with the definition of addition and multiplication of elements of Z_n just given. Since the sum of $[a]$ and $[b]$ depends (at least in appearance) on the choice of a and b, it is conceivable that a different sum or product could be obtained if different integers were chosen to represent the elements. For instance, in Example 1.23, we computed $[3] + [4] = [7]$ in Z_5. But $[3] = [-2]$ and $[4] = [14]$. Therefore, the same sum could be computed as $[-2] + [14] = [12]$. Fortunately, $[7] = [12]$ in Z_5. The following theorem shows that definitions of addition and multiplication in Z_n are valid.

Theorem 1.3   Let $[a] = [a']$ and $[b] = [b']$. Then $[a + b] = [a' + b']$ and $[ab] = [a'b']$.

Proof   Since $[a] = [a']$ and $[b] = [b']$, we have $a - a' = nx$ and $b - b' = ny$ for some integers x and y. Therefore,

$$(a + b) - (a' + b') = (a - a') + (b - b')$$
$$= nx + ny$$
$$= n(x + y)$$

and this means that $[a + b] = [a' + b']$. The proof that $[ab] = [a'b']$ is similar and is left as an exercise.   ●

The same issue arises with the definition of addition and multiplication of rational numbers. That is, the sum (or product) of two rational numbers is defined in terms of fractions representing the numbers, but the result is the same if different fractions are used to represent the numbers. For instance, $\frac{1}{2} + \frac{2}{3} = \frac{4}{8} + \frac{6}{9}$.

In Z_n, the commutative and associative laws for addition and multiplication, as well as the distributive law, hold. There are identity elements for addition and multiplication (namely, $[0]$ and $[1]$, respectively). Additive inverses exist: The additive inverse of $[a]$ is $[-a]$. In the real numbers, if a product is zero, one of the factors must be zero. This rule does not necessarily apply in Z_n, as we saw in Example 1.23.

The existence of multiplicative inverses in Z_n is a more difficult question. For instance, in Z_8, the elements $[1]$, $[3]$, $[5]$, and $[7]$ have multiplicative inverses (each of these is its own multiplicative inverse), but the elements $[2]$, $[4]$, and $[6]$ do not. (This can be verified by checking all possibilities.)

One's Complement Arithmetic

Many older large-scale computers, and some current designs, have a circuit called a one's complement adder. One's complement addition differs from two's complement addition in that any carry that occurs at the leftmost bit position is brought around to the right and the addition is continued.

EXAMPLE 1.24   Add 01101000 and 11111111 using one's complement arithmetic.

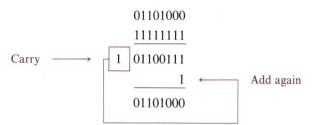

Since 01101000 is unchanged by the addition of 11111111, 11111111 must be a representation of zero. Thus, in one's complement arithmetic, there are two representations of zero—namely, 00000000 and 11111111. The representation consisting of all zero bits is called $+0$; the representation consisting of all one bits is called -0.

EXAMPLE 1.25   Add 01101000 (104_{ten}) and 10010111 using one's complement arithmetic.

$$\begin{array}{r} 01101000 \\ 10010111 \\ \hline 11111111 \end{array}$$

Since 11111111 is a representation of zero, 10010111 must represent the negative of 01101000 (that is, -104). In one's complement arithmetic, the representation of the negative of an integer is formed by complementing the representation of the integer—that is, by changing all the 1's to 0's and all the 0's to 1's. As with the signed-integer interpretation of two's complement arithmetic, words that begin with a 1 bit are considered to represent negative numbers. Since there are two different representations of zero, only $2^m - 1$ different integers can be represented in an *m*-bit word: the integers $-2^m/2 + 1$ through $2^m/2 - 1$. A one's complement adder does arithmetic in the number system $\mathbf{Z}_{2^m-1}$. As in two's complement arithmetic, overflow occurs when the sum of two positive numbers is negative or the sum of two negative numbers is positive.

EXERCISES 1.6

1. The following are binary representations of integers in an 8-bit machine with a two's complement adder. Find the integer represented (i) using the unsigned-integer interpretation, (ii) using the signed-integer interpretation.

    a.   00111011   **b.**   10110001   **c.**   11101111

2. Perform the following operations using two's complement arithmetic and the signed-integer interpretation. State whether a carry or overflow occurs.

    a.   11100011 + 00011111   **b.**   00111111 + 11000000

    c.   00101010 − 01100001   **d.**   11111111 − 10011001

3. In two's complement arithmetic with the signed-integer interpretation, what are the largest and smallest integers that can be stored
    a. in a 16-bit machine?
    b. in a 32-bit machine?

4. If the 8-bit binary representations of integers (using the two's complement signed-integer interpretation) are converted to octal,
    a. what digit(s) would the positive integers begin with?
    b. what digit(s) would the negative integers begin with?

5. Assuming two's complement signed integers, what integers are stored in the following 16-bit words?
    a. 0001111010111000
    b. 1110000000101111

6. Find a method for forming the negative of a number in two's complement signed-integer arithmetic.

7. Which of the following equations are true in Z_{64}?
    a. $[57] + [132] = [189]$
    b. $[35][83] = [-103]$
    c. $[16] - [65] = [-2]$

8. Which of the following equations are true in Z_{64}?
    a. $[35] + [58] = [29]$
    b. $[23][-1] = [37]$
    c. $[1]/[37] = [19]$

9. Which elements of Z_{16} have multiplicative inverses? Which elements of Z_7?

10. Which elements of Z_{15} have multiplicative inverses? Which elements of Z_5?

11. Prove the second part of Theorem 1.3. (*Hint:* $ab - a'b' = ab - a'b + a'b - a'b'$.)

12. Find the integers represented by the following 8-bit binary words in one's complement arithmetic.
    a. 00011110
    b. 11110001
    c. 10000110

13. Perform the following operations using one's complement arithmetic. State whether a carry and/or overflow occurs.
    a. $10001000 + 11110000$
    b. $11111111 + 00011110$
    c. $00110011 + 11100001$
    d. $00111111 + 01111110$

14. Show that in one's complement addition of any two numbers, at most one "end around carry" is necessary.

15. What is the relation between the two's complement representation of a negative integer and the one's complement representation of it? (*Hint:* Write down the one's and two's complement representations for several negative integers and compare them.)

16. Let n be an integer greater than 1, and let z be the primitive nth root of unity; that is, z is the complex number with absolute value 1 and argument $360/n$ degrees. Let $X = \{1, z, z^2, \ldots, z^{n-1}\}$. Show that the product of any two complex numbers in X is an element of X.

Computer Exercises for Chapter 1

1. Find out what are the largest and smallest integers that can be stored in your computer. (The answer may depend on what programming language you use.) What happens if you add 1 to the largest integer? What kind of adder does your computer have?

2. Write a program that calculates the number of bits set (equal to 1) in the internal representation of an integer. The program should accept input integers in the allowable range for your machine. For example, if input is the integer 5, output should be 2. If input is -1 and your machine is a 32-bit machine with a two's complement adder, the output should be 32.

3. Write a program that prints the internal binary representation for any integer (in the allowable range for your machine) entered by the user.

4. Write a menu-driven program that allows the user to choose from among the following options:
    a. Convert an integer from base 10 to base 8.
    b. Convert an integer from base 8 to base 10.
    c. Quit the program.

5. Write a menu-driven program that allows the user the following choices:
    a. Convert an integer from base 10 to base 16.
    b. Convert an integer from base 16 to base 10.
    c. Quit the program.

6. Write a program to convert an integer to an arbitrary base entered by the user.

7. Write a program to solve a quadratic equation and print the solutions. If the solutions are complex, the program should print them in the form $a + bi$.

8. Write a program to draw the x- and y-axes on the screen and plot the numbers $z, z^2, z^3, \ldots, z^{100}$ on the same set of axes, where z is input by the user. Run this program with several different values of z. (The most interesting results occur when $|z|$ is close to 1 and $\text{Arg}(z)$ is a small angle.)

• • • • CHAPTER REVIEW EXERCISES

1. Convert 57 to
    a. binary   b. octal   c. hex

2. Convert 32 to
    a. binary   b. octal   c. hex

3. Convert to decimal:
    a. 77_{eight} b. 10010_{two} c. 413_{five}

4. Convert to decimal:
   a. 256_{eight}
   b. 11110_{two}
   c. $F00D_{hex}$

5. Convert the binary number 11100010100 to:
   a. octal
   b. hex

6. Convert to binary:
   a. 527_{eight}
   b. $1CB_{hex}$

7. Convert 22102_{three} to base 9 without converting to base 10.

8. Convert 825_{nine} to base 3 without converting to base 10.

9. Which of the following are prime?
   a. 1212_{three}
   b. BAD_{hex}
   c. 776_{eight}
   d. 1111111_{two}

10. Perform the indicated binary arithmetic:
    a. $100001 + 10111$
    b. $1110100 - 1011$
    c. 10110×101

11. Show that $\sqrt{5}$ is irrational.

12. Plot each of the following complex numbers in the complex plane:
    a. 1.2
    b. $-1.2i$
    c. $1 - 5i$
    d. $2 + 5i$
    e. $-3 - 5i$
    f. $2.5 - 3i$

13. Perform the indicated operations:
    a. $(7 + 4i) + (-4 + 6i)$
    b. $25i - (2 - 3i)$

14. Perform the following multiplications:
    a. $(1 + 3i)(2 - 5i)$
    b. $(-5 - 3i)(-6 + 2i)$
    c. $(2 + 5i)i$
    d. $(-2i)i$
    e. i^7
    f. i^6

15. Find the absolute value and argument of each of the following complex numbers. Approximate answers are acceptable. You may use a protractor or trigonometry.
    a. $3 + 2i$
    b. $2 - 5i$
    c. $-4 + i$
    d. $-2 - 3i$
    e. -5
    f. $-2i$

16. Find the complex numbers having the following absolute values and arguments. Approximate answers are acceptable. You may use a protractor or trigonometry.
    a. $|z| = 1.2$, Arg(z) = 30 degrees
    b. $|z| = 0.5$, Arg(z) = 17 degrees
    c. $|z| = 2$, Arg(z) = 300 degrees
    d. $|z| = 5$, Arg(z) = -300 degrees

17. Find the additive and multiplicative inverses for each of the following complex numbers:
    a. $1 + 2i$
    b. $2 - 2i$
    c. $-3i$
    d. $-4 + i$
    e. -8
    f. $-1 + \sqrt{2}i$

18. How many different numbers can be stored in a 16-bit word?

19. Show how the integer -1572 would be represented in a 16-bit machine that has a
    a. one's complement adder
    b. two's complement adder

20. The following are binary representations for integers stored in an 8-bit machine with a one's complement adder. Find the decimal representation.
    a. 00111101
    b. 11111100

21. Repeat Exercise 20 if a two's complement adder is used.

22. Perform the following operations using one's complement arithmetic. State whether carry and/or overflow occurs.

    a. 10001001 + 11100000   **b.** 11111110 + 00001111   **c.** 00111110 + 01111101

    d. 10101010 − 00000111   **e.** 11111010 − 10000011

23. Repeat Exercise 22 using two's complement arithmetic.

24. Convert from binary to decimal.

    a. 11.101   **b.** 111.0011

25. Convert from decimal to binary:

    a. 12.4   **b.** 7.09

26. Which of the following equations are true in $\mathbf{Z}_{16}$?

    a. $[52] + [13] = [65]$ **b.** $[12] + [8] = [5]$ **c.** $[5][83] = [-103]$

    d. $[3][-1] = [7]$ **e.** $[16] - [65] = [-2]$ **f.** $\dfrac{[1]}{[17]} = [2]$

27. Which elements of $\mathbf{Z}_8$ have multiplicative inverses? Which elements of $\mathbf{Z}_{10}$?

28. Evaluate $\displaystyle\prod_{k=3}^{4} \sum_{i=5}^{6} (k - 2i)$

29. Evaluate $\displaystyle\prod_{i=1}^{3} \prod_{k=1}^{3} k^i$

2
SETS

The term **set** is used to mean any collection of objects. That is, if x is any object and A is a set, the statement "x is an element of A," written $x \in A$, is either true or false. The definition of a set is necessarily vague; set is the most fundamental mathematical concept and cannot be defined in terms of simpler concepts.

In the previous chapter, we looked at some familiar sets of numbers. This chapter will consider sets in general, notation for sets, and set algebra.

The branch of mathematics called *set theory* is, for the most part, the creation of one man, Georg Cantor (1845–1918). Cantor's work on set theory began in 1870 and occupied him for most of the remainder of his life. His discoveries, along with those of his collaborators and successors, have revolutionized mathematics—so much so that the concept of a set is now introduced in elementary schools. However, Cantor did not invent sets, set algebra, or the notation we use today. The algebra of sets was developed around 1850 by Augustus DeMorgan (1806–1871) and George Boole (1815–1864), and our modern set notation was developed in Italy around 1900.

The language and notations of set theory have now become the language of mathematics. In their modern presentations, all mathematical concepts are described in terms of sets. Sets are also useful for computer programming. Several modern programming languages, such as Pascal, provide operations to be performed on sets as well as on numbers.

Every variable used in a computer program is associated with a set of values that the variable can take on. Such a set is called a **data type.** The most common data types are **real** and **integer.** The data type **real** is a finite subset of the real numbers—namely, the set of rational numbers that can be represented in the computer. This set depends on the design of the computer and on the method of representing real numbers. Thus, the **real** data type for the VAX computer is not the same set as the **real** data type for the IBM 370. The size of a data type (that is, the number of values that a variable of that type may take on) depends on the amount of storage allocated to the variable.

2.1 ● SETS AND SUBSETS

A set is a collection of objects, called the **elements** of the set. Two sets are **equal** if they have the same elements. In mathematical writing, it is customary to use uppercase letters of the alphabet to denote sets and lowercase letters to denote elements of a set. (Programming languages follow different conventions.)

The sets **R**, **N**, and **Z** are all infinite sets. That is, they contain infinitely many elements. Most sets of interest in abstract mathematics are infinite sets. Of course, all sets processed by computer programs are finite sets.

Describing Sets

A finite set can be described by listing the elements in it and enclosing the list in braces. For instance, if the set A consists of all the integers between -3 and 2 inclusive, we write $A = \{-3, -2, -1, 0, 1, 2\}$.

It is impossible to list the elements of an infinite set, and often it is inconvenient to list all the elements of a finite set. In such cases, we can describe a set by specifying a property common to its elements. For example, the set of real numbers can be described in this way:

$\mathbf{R} = \{x \mid x \text{ is a real number}\}$.
The set of even integers can be described by $E = \{n \mid n \text{ is an even integer}\}$.

It is possible to describe the set of integers by listing some of its elements and using the ellipsis, as follows:

$\mathbf{Z} = \{\ldots, -3, -2, -1, 0, 1, 2, 3, \ldots\}$

To indicate that a is an element of the set A, we write $a \in A$, where the symbol $\in$ is read "is an element of." We write $a \notin A$ to indicate that a is not an element of the set A.

The Empty Set

A set may have no elements, such as the set of all major league teams to win ten consecutive world series. If we were to list this set, we would write $\{\ \}$, indicating that it has no elements. The set with no elements is called the **empty set** or the **null set** and is denoted by the symbol $\varnothing$.

Subsets

Definition 2.1

Given two sets A and B, A is a **subset** of B if every element of A is also an element of B. The set A is a **proper subset** of B if A is a subset of B and A is not equal to B.

If A is subset of B, we write $A \subset B$, which is read "A is a subset of B" or "A is contained in B." There is no standard symbol for the proper subset relation.

Here are some examples of sets and subsets.

1. If $A = \{1, 3, 5\}$ and $B = \{1, 2, 3, 4, 5\}$, then $A \subset B$. In fact, A is a proper subset of B.
2. If A is the set of hexadecimal digits and B is the set of octal digits, then $B \subset A$.

If $A = \{0, 2, 4, 6\}$ and $B = \{1, 2, 3, 4, 5\}$, then A is not a subset of B (why?), and we write $A \not\subset B$.

From Definition 2.1, it follows that a set is a subset of itself. That is, $A \subset A$. Also, for every set A, $\emptyset \subset A$.

Sums over Sets

In Chapter 1, we used the Σ notation to write the sum of a set of numbers. The set of numbers was indexed by a set of integers, such as $\{1, 2, 3\}$ or $\{0, 1, 2, \ldots, n\}$. In each case, we had a set X of consecutive integers and a rule assigning to each integer in X a number in **R**. This rule is a function defined on the set X. The Σ notation denoted the sum of the values of the function—namely, the numbers in **R**.

This suggests that the Σ notation can be extended to apply to the sum of the values of a function on any finite set, not just a set of consecutive integers. If X is a finite set, the sum of all of the values of a function f can be written

$$\sum_{x \in X} f(x)$$

where $f(x)$ represents a real number in **R**.

Instead of writing $x \in X$ below the Σ symbol, you can write any convenient description of the set X.

EXAMPLE 2.1   Write an expression for the sum of the cubes of all odd integers between 0 and 100. You can write this as

$$\sum_{\substack{0 < n < 100 \\ n \text{ odd}}} n^3$$

○

EXAMPLE 2.2   Write an expression for the sum of the reciprocals of the positive divisors of a positive integer n. You can write this as

$$\sum_{\substack{d \in \mathbf{N} \\ d \text{ divides } n}} \frac{1}{d}$$

○

EXERCISES 2.1

1. Given the sets $A = \{c, o, m, p, u, t, e, r\}$, $B = \{c, o, r, e\}$, and $C = \{t, e, r, m\}$, insert one of the symbols $\in, \notin, \subset, \not\subset$ between each of the following pairs:

    a. $B \quad C$ **b.** r $\quad C$ **c.** m $\quad B$

    d. $C \quad A$ **e.** $\emptyset \quad B$ **f.** $B \quad B$

2. Let $A = \{0, 1, 2, 3, 4, 5\}$, $B = \{2, 4, 6\}$, and $C = \{1, 3, 5\}$. Answer True or False:

    a. $A \subset B$ **b.** $B \subset A$ **c.** $C \subset A$

    d. $B = C$ **e.** $\emptyset \in B$ **f.** $\emptyset \not\subset C$

3. List the elements in each of the following sets. You may use ellipsis notation for infinite sets.

a. the set of natural numbers not greater than 10

b. $\{n \mid n \in \mathbf{N} \text{ and } n \text{ is a multiple of } 3\}$

c. $\{n \mid n \in \mathbf{Z} \text{ and } -2 \leq n < 3\}$

d. $\left\{\frac{1}{n} \,\middle|\, n \in \mathbf{N}\right\}$ **e.** $\{2^n \mid n \in \mathbf{N}\}$

f. $\{n \mid n \in \mathbf{Z} \text{ and } n^2 = 16\}$ **g.** $\{(-1)^n \mid n \in \mathbf{N}\}$

4. Write each of the following sets without using ellipses:

a. $\{1, 4, 9, 16, \ldots\}$

b. the set of integers greater than 5

c. $\{2, 4, 6, 8, \ldots\}$ **d.** $\{2, 3, 5, 7, 11, 13, 17, 19\}$

5. Determine which of the following sets are subsets of which:

$$A = \{p \mid p \in \mathbf{N} \text{ and } p \text{ is prime}\}$$
$$B = \{n \mid n \in \mathbf{N} \text{ and } n \text{ is odd}\}$$
$$C = \{n \mid n \in \mathbf{N} \text{ and } n \text{ is a multiple of } 10\}$$
$$D = \{n \mid n \in \mathbf{N} \text{ and } n \text{ is even}\}$$

6. Determine which of the following sets are equal:

$$A = \{1, 2, 3\}$$
$$B = \{n \mid n \in \mathbf{N} \text{ and } n^2 < 10\}$$
$$C = \{n \mid n \in \mathbf{N} \text{ and } n^2 < 1\}$$
$$\varnothing$$

7. Show that if $A \subset B$ and $B \subset C$, then $A \subset C$.

8. If $A \subset C$ and $B \subset C$, what can you conclude about sets A and B?

9. Let A be the **integer** data type for the Apple II, and let B be the **integer** data type on the VAX. (Apple II stores integers in 16 bits of memory; VAX stores integers in 32 bits.) Is $A \subset B$? Is $B \subset A$?

10. Using Σ notation, write an expression for the sum of the squares of the even integers between -101 and $+101$.

11. Using Σ notation, write an expression for the sum of the scores of the students in your discrete math class on the last quiz.

12. Using Σ notation, write an expression for the sum of the weights of the professors in the mathematics department.

2.2 • OPERATIONS ON SETS AND SET ALGEBRA

In this section, we will look at ways of combining sets to form new sets. The common operations we consider are union, intersection, difference, and complement of sets.

Definition 2.2

The **union** of A and B, written $A \cup B$, is the set of elements in A or B or both. The **intersection** of A and B, written $A \cap B$, is the set of elements in both A and B. When $A \cap B = \emptyset$, we say that A and B are **disjoint** sets. The **difference** of A and B, written $A - B$, is the set of elements in A and not in B.

EXAMPLE 2.3   If $A = \{1, 2, 3, 4, 5\}$, $B = \{4, 5, 6\}$, and $C = \{3, 4\}$, then

$$A \cup B = \{1, 2, 3, 4, 5, 6\}$$
$$A \cap B = \{4, 5\} \qquad A \cap C = \{3, 4\} = C$$
$$A - B = \{1, 2, 3\} \qquad A - C = \{1, 2, 5\}$$
$$B - A = \{6\} \qquad C - A = \emptyset$$

○

The concepts of union, intersection, and difference can be illustrated graphically by **Venn diagrams,** as shown in Figure 2.1. The shaded areas of Figure 2.1(a), (b), and (c) represent $A \cup B$, $A \cap B$, and $A - B$, respectively.

Complements of Sets

In Example 2.3, the set C is a subset of the set A. When this is the case, the difference set $A - C$ is called the **complement** of C with respect to A.

If, for some problem, we are considering the sets A and B to be subsets of a set X, then X is called the **universe of discourse** (or simply the **universe**) for that problem. In this case, the complement of A will be written A', meaning $X - A$. Similarly, B', the complement of B, means $X - B$. It follows from the definitions of union and intersection that $A \cup A' = X$ and $A \cap A' = \emptyset$.

Figure 2.2 illustrates the complement of a set. In Figure 2.2, the universe is represented by the rectangle.

Power Set of X

If, for any set X, we form the set of all subsets of X, we get what is called the *power set* of X.

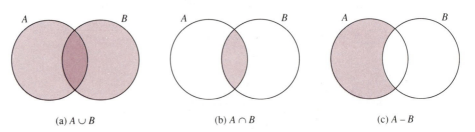

(a) $A \cup B$ (b) $A \cap B$ (c) $A - B$

FIGURE 2.1   Venn diagrams for (a) union, (b) intersection, and (c) difference

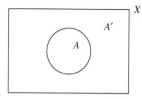

FIGURE 2.2   The complement of A

Definition 2.3

Let X be any set. Then the **power set** of X, written $P(X)$, is the set of all subsets of X.

Some examples of power sets are as follows:

$$P(\varnothing) = \{\varnothing\}$$
$$P(\{\varnothing\}) = \{\varnothing, \{\varnothing\}\}$$

If $A = \{1, 2, 3\}$, then

$$P(A) = \{\varnothing, \{1\}, \{2\}, \{3\}, \{1, 2\}, \{1, 3\}, \{2, 3\}, \{1, 2, 3\}\}$$

In a later section, we will show that if a set A has n elements, then the power set $P(A)$ has 2^n elements.

Rules of Set Algebra

The operations of union and intersection obey the following laws, some of which are analogous to the algebraic rules of arithmetic for real numbers (mentioned in Chapter 1). Here A, B, and C are subsets of a set X.

1. The Commutative Laws
   $$A \cup B = B \cup A$$
   $$A \cap B = B \cap A$$
2. The Associative Laws
   $$A \cup (B \cup C) = (A \cup B) \cup C$$
   $$A \cap (B \cap C) = (A \cap B) \cap C$$
3. The Distributive Laws
   $$A \cup (B \cap C) = (A \cup B) \cap (A \cup C)$$
   $$A \cap (B \cup C) = (A \cap B) \cup (A \cap C)$$
4. Identities for union and intersection
   $$A \cup \varnothing = A$$
   $$A \cap X = A$$
5. Properties of the complement
   $$A \cup A' = X$$
   $$A \cap A' = \varnothing$$
   $$(A')' = A$$

6.   Idempotent Laws

$A \cup A = A$

$A \cap A = A$

Notice that the union of A with its complement A' produces the identity for intersection. Similarly, the intersection of A with its complement A' produces the identity for union.

In addition to these five laws, union and intersection are related to taking complements by the following two laws, called **DeMorgan's Laws:**

$$(A \cup B)' = A' \cap B'$$

$$(A \cap B)' = A' \cup B'$$

To prove any of the laws stated previously, you need to remember that two sets A and B are equal if they have the same elements—that is, every element of A is an element of $B (A \subset B)$, and every element of B is an element of $A (B \subset A)$. You should study carefully the technique in the following proof of the first of DeMorgan's Laws.

Theorem 2.1 $(A \cup B)' = A' \cap B'$

Proof   We need to show that if x is an element of $(A \cup B)'$, then x is also an element of $A' \cap B'$; and conversely, if x is an element of $A' \cap B'$, then x is an element of $(A \cup B)'$.

Suppose $x \in (A \cup B)'$. Then $x \notin A \cup B$, so x is neither in A nor in B. Thus, $x \in A'$ and $x \in B'$. This means that $x \in A' \cap B'$ and, therefore, $(A \cup B)' \subset A' \cap B'$. Conversely, suppose $x \in A' \cap B'$. Then $x \in A'$, so $x \notin A$, and $x \in B'$, so $x \notin B$. It follows that $x \notin A \cup B$; therefore, $x \in (A \cup B)'$. This shows that $A' \cap B' \subset (A \cup B)'$, and the theorem is proved.   ●

Proving the second of DeMorgan's Laws and the Commutative, Associative, and Distributive Laws are left as exercises.

EXERCISES 2.2

1.   Let $A = \{c, o, m, p, u, t, e, r\}$, $B = \{c, o, r, e\}$, and $C = \{t, e, r, m\}$. Find

a. $A \cap B$ b. $B \cap C$ c. $B \cup C$

d. $B \cap (B \cup C)$ e. $A - B$ f. $C - B$

g. $B - A$ h. $B - C$ i. $A - (B \cup C)$

j. $(A - B) \cup C$ k. $A - (B \cap C)$ l. $(A - C) \cap B$

2.   Let the universe $A = \{0, 1, 2, 3, 4, 5, 6, 7\}$, and let $B = \{0, 2, 4, 6\}$, $C = \{1, 3, 5, 7\}$, and $D = \{2, 3, 5, 7\}$. Find

a. $C - D$ b. $C \cap D'$ c. $B' \cup C$ d. $D' \cap B'$

e. $(B \cap D)'$ f. $(C \cup D)'$ g. $(B - D)'$ h. $B' - D'$

i. $(B \cup C) \cap D'$ j. $(D')'$

3.   A roulette wheel has 36 slots, numbered 1–36. The numbers 19 to 36 inclusive

are high. Let H = the set of high numbers and E = the set of even numbers. Find

a. $H \cap E$ **b.** $H - E$ **c.** $E - H$ **d.** $E \cup H$

4. Let N be the universe and let $A = \{2, 4, 6, 8, \ldots\}$, $B = \{1, 3, 5, 7, \ldots\}$, $C = \{1, 2, 4, 8, \ldots\}$, and $D = \{5, 10, 15, \ldots\}$. Find

     a. $A \cap B$ **b.** $A \cap D$ **c.** $A \cap C \cap D$

     d. $C \cap D$ **e.** $A' \cap C$ **f.** $C - A$

5. Let $A = \{a, \{a\}, \{a, b\}\}$, $B = \{a\}$, $C = \{a, b\}$, and $D = \{a, \{a\}\}$. Find

     a. $B \cap D$ **b.** $B \cap C$ **c.** $A \cup C$

     d. $A \cap C$ **e.** $A - C$ **f.** $A - D$

6. Give an example of three sets, A, B, and C, such that $A \cap B$, $B \cap C$, and $A \cap C$ are nonempty but $A \cap B \cap C = \varnothing$.

7. Is it possible to find three sets, A, B, and C, such that $A \cap B \cap C \neq \varnothing$ but at least one of $A \cap B$, $B \cap C$, or $A \cap C$ is empty? If yes, give an example. If no, why not?

8. Give an example of three sets, A, B, and C, such that $A \in B$ and $B \in C$ but $A \notin C$.

9. If $A = \{a, b, c\}$, list the elements in $P(A)$, the power set of A.

10. If $A = \{a, \{a\}\}$, list the elements in $P(A)$.

11. If $A = \{a\}$, list the elements of $P(P(A))$.

12. If $A = \{a\}$, find

     a. $A \cup P(A)$ **b.** $A \cap P(A)$ **c.** $\{A\} \cup P(A)$

     d. $\{A\} \cap P(A)$ **e.** $P(A) - A$ **f.** $P(A) - \{A\}$

13. The symmetric difference of two sets, written $A \oplus B$, is the set of elements in either A or B but not both. That is, $A \oplus B = (A \cup B) - (A \cap B)$. If $A = \{0, 1, 2, 3, 4, 5, 6, 7\}$, $B = \{0, 3, 6, 9\}$, and $C = \{2, 3, 5, 7\}$, find

     a. $B \oplus C$ **b.** $A \oplus B$ **c.** $A \cap (B \oplus C)$

     d. $(A \oplus B) \oplus C$ **e.** $A \oplus \varnothing$ **f.** $A \oplus A$

14. Find the following sets:

     a. $Z \oplus N$ **b.** $R \oplus Q$

15. If A and B are subsets of a universe X, illustrate each of the following properties with a Venn diagram.

     a. $A - B = A \cap B'$ **b.** $A \oplus B = (A - B) \cup (B - A)$

16. If A, B, and C are subsets of a universe X, prove each of the following statements. Use the method of Theorem 2.1.

     a. $(A \cap B)' = A' \cup B'$ **b.** $(A')' = A$

     c. $A \cap (B \cup C) = (A \cap B) \cup (A \cap C)$

17. If A and B are subsets of a universe X, prove the following:

     a. $(A - B)' = A' \cup B$ **b.** $A \cap (A \cup B) = A$

18. If A and B are subsets of a universe X and if $A' = B$, show that $B' = A$.

19. If A, B, and C are subsets of a universe X, prove the following. ($A \oplus B$ is defined in Exercise 13.)

   a. $A \oplus B \subset (A \oplus C) \cup (B \oplus C)$

   b. $A = B$ if and only if $A \oplus B = \varnothing$

   c. $A \oplus B = A' \oplus B'$

   d. $A \cap (B \oplus C) = (A \cap B) \oplus (A \cap C)$

20. Let A, B, and C be subsets of a universe X. For each of the following statements, either prove the statement is true (using the technique of Theorem 2.1) or find an example for which the statement is false. (Such an example is called a **counterexample.**)

   a. $(A - B)' = A' - B'$ **b.** $(A \oplus B)' = A' \oplus B'$

   c. $P(A \cap B) = P(A) \cap P(B)$ **d.** $P(A \cup B) = P(A) \cup P(B)$

21. Prove or find a counterexample:

   a. If $A \cup B = A \cup C$, then $B = C$.

   b. If $A \cap B = A \cap C$, then $B = C$.

   c. If $A \cup B \subset A \cap B$, then $A = B$.

   d. If $A \subset B$ and $C \subset D$, then $A \cup C \subset B \cup D$.

2.3 • SETS OF REAL NUMBERS

In a computer, only a finite subset of the set of real numbers can be represented exactly. In engineering and scientific applications, however, numbers represent physical quantities, such as speed or weight, that can take on any real value. Thus, a real number stored in a computer is an approximation to a physical value. Frequently, the accuracy of the approximation is important. If, for instance, the number stored is 1.234, the actual value of the physical quantity represented may be in some range of numbers containing 1.234—for example, the range from 1.232 to 1.236. Such ranges of real numbers are called **intervals.** Analysis of the accuracy of approximate computations requires the use of set operations on intervals.

Definition 2.4

If a and b are real numbers, then

1. $[a, b] = \{x \mid x \in \mathbf{R} \text{ and } a \leq x \leq b\}$ is the **closed interval** from a to b.
2. $(a, b) = \{x \mid x \in \mathbf{R} \text{ and } a < x < b\}$ is the **open interval** from a to b.
3. $[a, b) = \{x \mid x \in \mathbf{R} \text{ and } a \leq x < b\}$ and
   $(a, b] = \{x \mid x \in \mathbf{R} \text{ and } a < x \leq b\}$
   are the **half-open intervals** from a to b.
4. $(-\infty, a) = \{x \mid x \in \mathbf{R} \text{ and } x < a\}$.

5. $(-\infty, a] = \{x \mid x \in \mathbf{R} \text{ and } x \le a\}$.
6. $(a, \infty) = \{x \mid x \in \mathbf{R} \text{ and } x > a\}$.
7. $[a, \infty) = \{x \mid x \in \mathbf{R} \text{ and } x \ge a\}$.
8. $(-\infty, \infty) = \mathbf{R}$.

The numbers a and b are called **endpoints** of the interval. Notice that open intervals do not include their endpoints, closed intervals include their endpoints, and half-open intervals include one endpoint but not the other. Intervals of types 4 through 7 are called **rays** or **half-lines.**

It is important to note that the symbol ∞ in the interval notation has no independent meaning; it is merely a part of the name of a set.

EXAMPLE 2.4 $[-2, 5] = \{x \in \mathbf{R} \mid -2 \le x \le 5\}$
$(.5, 1] = \{x \in \mathbf{R} \mid .5 < x \le 1\}$
$(3, \infty) = \{x \in \mathbf{R} \mid x > 3\}$ ○

Note that the empty set $\varnothing$ is an interval, because $\varnothing = (a, a)$ for any real number a. Also, a set containing a single element is an interval, because $\{a\} = [a, a]$ for any real number a.

Since intervals are sets of numbers, we can perform set operations on them.

EXAMPLE 2.5 $(-\infty, 2] \cup (0, 4) = (-\infty, 4)$
$(0, 1) \cap [0, 1] = (0, 1)$
$[0, 8] - (1, 4] = [0, 1] \cup (4, 8]$
$(-3, 3]' = (-\infty, -3] \cup (3, \infty)$ where the complement is taken with respect to $\mathbf{R}$.
                                                                                               ○

EXERCISES 2.3

1. Give an example of a set of real numbers that is *not* an interval.
2. Write in interval notation:
    a. $\{x \mid x \in \mathbf{R} \text{ and } -3 \le x \le -1\}$ b. $\{x \mid x \in \mathbf{R} \text{ and } x < -2\}$
    c. $\{x \mid x \in \mathbf{R} \text{ and } x \ge 7\}$
3. Write using interval notation:
    a. The set of all nonnegative real numbers
    b. The set of all numbers whose squares are between 4 and 9
4. Answer True or False:
    a. $\{0, 1\} \subset (0, 1)$ b. $\{0, 1\} \subset [0, 1]$
    c. $(1, 2] \subset \mathbf{N}$ d. $\{1, 2\} \subset \mathbf{N}$

5. Answer True or False:
    a. $[1, 2] \subset \mathbf{Q}$ b. $(0, 1) \subset [1, 2]$
    c. $(0, 1) \subset \{0, 1\}$ d. $\{1\} \in [0, 1]$

6. Perform the indicated operations on the given intervals.
    a. $(0, 1) \cup (-2, 0]$ b. $(-5, 5) \cap (3, 8)$
    c. $[-1, 2] \cup (0, 4)$ d. $(-\infty, 0) \cap (-5, 5)$
    e. $[7, \infty) \cap (0, 1)$

7. Perform the indicated operations on the given intervals.
    a. $[0, 1] \cap [1, 3]$ b. $(-\infty, 2] - (1, 2]$ c. $(-\infty, 2] - (1, 2)$
    d. $((-3, 3) \cap (1, 4)) \cup ((0, 1) \cap (-\infty, 3])$
    e. $((1, \infty) \cup (0, 1)) \cap ((1, 2) \cup [2, 4])$

8. If the universe is $\mathbf{R}$, find
    a. $(-\infty, 4]'$ b. $[0, 1]'$

9. If the universe is $\mathbf{R}$, find
    a. $(0, 1)'$ b. $(-5, \infty)'$

2.4 ● CARTESIAN PRODUCTS OF SETS

In addition to elementary data types such as **integer** and **real,** programming languages provide means for constructing larger data types, called **compound data types**, from the elementary ones. The simplest of these is the **array.** The value of a variable of type "array of two **real** numbers" is a list of two values of type **real,** and the value of a variable of type "array of three **integers**" is a list of three values of type **integer.** Variables of these types may be declared in BASIC by the statement

        100 DIM A(1), B%(2)

and in Pascal by the statements

        var A: array [0 . . 1] of real;

        var B: array [0 . . 2] of integer;

Compound data types can also be built of elementary types that are not all the same; these types are called **records.** Record types are supported in Pascal and COBOL but not in BASIC or FORTRAN. The construction of compound data types involves a method of combining sets to form new ones. This method is called forming the **Cartesian product** of sets.

Definition 2.5

The **Cartesian product** of two sets A and B, written $A \times B$, is the set of ordered pairs (a, b) where $a \in A$ and $b \in B$. The Cartesian product of n sets $A_1, \ldots, A_n$ is the set of all ordered n-tuples $(a_1, \ldots, a_n)$ where $a_1 \in A_1, a_2 \in A_2, \ldots, a_n \in A_n$.

We can write $A \times B = \{(a, b) \mid a \in A \text{ and } b \in B\}$. The term *ordered* indicates that the position of the elements in the pair is important; the pair (a, b) is not the same as the pair (b, a). Here are some examples of sets and their Cartesian products.

EXAMPLE 2.6   If $A = \{a, b, c\}$ and $B = \{x, y\}$, then

$$A \times B = \{(a, x), (a, y), (b, x), (b, y), (c, x), (c, y)\}$$
$$B \times B = \{(x, x), (x, y), (y, x), (y, y)\}$$
$$A \times \varnothing = \varnothing$$

○

EXAMPLE 2.7   If **R** is the set of real numbers, then **R** $\times$ **R** is the set of ordered pairs of real numbers. The set **R** $\times$ **R** can be represented graphically by the plane, sometimes called the *Cartesian plane*. Each point in the plane correspónds to some ordered pair of real numbers; conversely, each ordered pair of real numbers corresponds to a point in the plane.

○

Properties of Cartesian Products

Notice the following things about Cartesian products:

1. If $A \neq B$, the set $A \times B$ is not the same as the set $B \times A$. That is, the commutative law does not hold.
2. You may take the Cartesian product of a set with itself. The set $A \times A$ is sometimes written A^2. You will frequently see the plane **R** $\times$ **R** referred to as $\mathbf{R}^2$.
3. If A and B are finite, the number of pairs in the set $A \times B$ is the product of the number of elements in A times the number of elements in B.

Although we use the same notation for ordered pairs and for open intervals on the real line, it will always be clear from the context which we mean.

EXAMPLE 2.8   Let R be the data type **real** (for some specific computer) and let I be the data type **integer** for the same computer. (Note that R is a finite subset of **R** and I is a finite subset of **Z**.) Then the Pascal declaration

```
var X: record
      realpart: real;
      intpart: integer
   end;
```

defines X to be a variable of a compound data type having two components: a real and an integer. The data type is the set $R \times I$. The names realpart and intpart have no mathematical significance; they are chosen by the programmer and used to refer to the parts of the record.

○

EXERCISES 2.4

1. Let $A = \{1, 2, 3, 4\}$ and $B = \{0, 1\}$. List the elements of
    a. $A \times B$ b. $B \times A$ c. $B \times B$ d. $P(B \times B)$

2. List the elements of the set
$$A = \{(a, b) \mid a \in \mathbf{N}, b \in \mathbf{N}, \text{ and } a + b < 5\}$$

3. Let $A = \{1, 2, 3\}$ and $B = \{2, 4, 6, 8\}$. Let C be the subset of $A \times B$ defined as follows:
$$C = \{(a, b) \mid a \in A, b \in B, \text{ and } a + b \leq 5\}$$
List the elements of C.

4. Let $A = \{1, 2, 3\}$, $B = \{0\}$, and $C = \{a, b\}$.
   a. List the elements in $A \times (B \times C)$.
   b. List the elements in $(A \times B) \times C$.

5. Let $A = \{0, 1\}$ and $B = \{a\}$. List the elements of
   a. $P(A \times B)$ b. $P(A) \times P(B)$

6. Let $A = \{1, 2, 3, \ldots, 25\}$ and $B = \{a, b, c, d\}$. How many pairs are in the following sets?
   a. $A \times B$ b. $P(A \times B)$ c. $P(A) \times P(B)$

7. Given the interval $(0, 1)$, how would you interpret the set $(0, 1) \times (0, 1)$?

8. If $A \times B = \varnothing$, what can you say about the sets A and B?

9. If $(A \times B) \cap (B \times A) = \varnothing$, what can you conclude about the sets A and B?

10. If $A \times B = B \times A$, what can you conclude about A and B?

11. Prove the following using the method of Theorem 2.1: $A \times (B \cap C) = (A \times B) \cap (A \times C)$.

12. Prove that $(A \cap B) \times (C \cap D) = (A \times C) \cap (B \times D)$.

13. Prove that if $A \subset B$ and $C \subset D$, then $A \times C \subset B \times D$.

14. Prove the converse of the statement in the previous exercise—that is, if $A \times C \subset B \times D$, then $A \subset B$ and $C \subset D$.

15. Let R be the data type **real** and let I be the data type **integer**. Write as Cartesian products the data types occurring in each of the following Pascal declarations:

   a.   var X: record
                     part1: array [0 .. 3] of integer;
                     part2: array [0 .. 1] of real
              end;
   b.   var Y: record
                     part1: real;
                     part2: record
                                    part2a: integer;
                                    part2b: real
                            end
              end;

2.5 • FINITE AND INFINITE SETS

We have seen that while computers can deal only with finite sets, infinite sets are important in mathematics, including the mathematical analysis of computer programs. So we need to understand the differences. If A is a finite set, then there is an integer n such that A has exactly n elements. Any proper subset of A has fewer than n elements. However, consider the relationship between the set $\mathbf{N}$ of natural numbers and the set (call it E) of even natural numbers. Although E is a proper subset of $\mathbf{N}$, it has just as many elements as $\mathbf{N}$. To see this, associate with each x in $\mathbf{N}$ the number $2x$ in E. Then every element of $\mathbf{N}$ corresponds to exactly one element of E, and every element of E corresponds to exactly one element of $\mathbf{N}$. Therefore, E and $\mathbf{N}$ must have equally many elements!

One-to-One Correspondence

The association of each x in $\mathbf{N}$ with $2x$ in E is an example of a **one-to-one correspondence**. A one-to-one correspondence is a particular type of function. Functions in general will be considered in the next chapter.

Definition 2.6

> A **one-to-one correspondence** f from a set A to a set B, written $f: A \rightarrow B$, is a rule that associates a unique element of B with each element of A, in such a way that each element of B is associated with one and only one element of A. If $a \in A$, the element of B associated with a is written $f(a)$.

EXAMPLE 2.9   Let $A = \{1, 2, 3\}$ and $B = \{a, b, c\}$. Then the rule $f(1) = c$, $f(2) = a$, $f(3) = b$ is a one-to-one correspondence from A to B. The correspondence is illustrated in Figure 2.3.

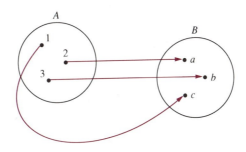

FIGURE 2.3   One-to-one correspondence from A to B

Definition 2.7

> An **infinite** set is a set that can be put into one-to-one correspondence with a proper subset of itself. A **finite** set is a set that is not infinite.

EXAMPLE 2.10   The set $\{1\}$ is finite, because the only proper subset of $\{1\}$ is $\varnothing$, and $\varnothing$ has no element that can be made to correspond to 1. The set $\mathbf{N}$ is infinite because, as shown earlier, the rule $f(x) = 2x$ establishes a one-to-one correspondence between $\mathbf{N}$ and its proper subset E of even numbers.

Two more examples of one-to-one correspondences follow.

EXAMPLE 2.11   Let $X = [0, 1)$ and $Y = (2, 3]$. Let $f(x) = 3 - x$. Then f is a one-to-one correspondence from X to Y.   ○

EXAMPLE 2.12   Let $X = [0, 2]$ and $Y = [0, 1]$, and let $g(x) = \frac{x}{2}$. Then g is a one-to-one correspondence from X to a proper subset of X (namely, Y), showing that X is an infinite set.   ○

Definition 2.8   A set is **countable** or **countably infinite** (or **denumerable**) if it can be put into one-to-one correspondence with the set **N**. A set that is neither finite nor countable is called **uncountable** (or **nondenumerable**).

The set E in Example 2.10 is countable, since there is a one-to-one correspondence between E and **N**.

EXAMPLE 2.13   The set **Z** of integers is countable. Simply write the set **Z** as 0, 1, -1, 2, -2, 3, $-3, \ldots$. Then the one-to-one correspondence is as follows:

$$
\begin{array}{ccccccccccc}
\mathbf{Z}: & 0 & 1 & -1 & 2 & -2 & 3 & -3 & 4 & -4 & \ldots \\
& \downarrow & \downarrow & \downarrow & \downarrow & \downarrow & \downarrow & \downarrow & \downarrow & \downarrow & \\
\mathbf{N}: & 1 & 2 & 3 & 4 & 5 & 6 & 7 & 8 & 9 & \ldots
\end{array}
$$
    ○

It can be shown that any infinite subset of **Z** is countable. (The proof of this is left as an exercise.) Therefore, to show that a set is countable, it is sufficient to show there is a one-to-one correspondence between that set and an appropriate subset of **Z**.

EXAMPLE 2.14   The set of positive rational numbers is countable. To see this, note that every positive rational number can be written uniquely as $\frac{a}{b}$, where a and b are integers with no common factor other than 1, $a > 0$, and $b > 0$. The fraction is said to be in lowest terms. Assume that a and b are written in the customary base-10 notation. Then we may also consider $\frac{a}{b}$ as representing a unique integer in base 11, where the fraction bar (—) serves as the digit for 10. For instance, $\frac{7}{12}$ corresponds to

$$7(11^3) + 10(11^2) + 1(11^1) + 2(11^0) = 10{,}540$$

This scheme gives us a one-to-one correspondence from the positive rational numbers to a subset of integers, showing that the positive rational numbers are countable.   ○

Not all infinite sets are countable. The next example proves the existence of uncountable sets by showing that the set **R** is uncountable.

EXAMPLE 2.15   The set of real numbers **R** is not countable. To show this, it is enough to show that the real numbers in the interval $(0, 1]$ are not countable. As noted in Chapter 1, every number in $(0, 1]$ can be written as a nonterminating decimal (for example,

$\frac{1}{2} = 0.49999 \ldots$). If the numbers in $(0, 1]$ were countable, they could be written as a sequence of nonterminating decimals:

$$x_1 = 0.a_{11}a_{12}a_{13}a_{14} \ldots$$
$$x_2 = 0.a_{21}a_{22}a_{23}a_{24} \ldots$$
$$x_3 = 0.a_{31}a_{32}a_{33}a_{34} \ldots$$
$$x_4 = 0.a_{41}a_{42}a_{43}a_{44} \ldots$$

.
.
.

Now write down a number $x = 0.b_1b_2b_3b_4 \ldots$ according to the following rule: b_1 may be any digit except 0 and a_{11}, b_2 may be any digit except 0 and a_{22}, and so on. Then x is not equal to x_1, because it differs from x_1 in the first decimal place. It is not equal to x_2, because it differs from x_2 in the second decimal place. Continuing this reasoning, we see that x cannot be any of the numbers in the list. It is also nonterminating, since we did not allow the digit zero. Therefore, the list was not a complete list of the real numbers in the interval $(0, 1]$, contradicting the original assumption.   ○

We have shown that, in a sense, the set of real numbers is "bigger" than the set of natural numbers. It can be shown that for any set A there exists a "bigger" set; in fact, the power set of A is bigger than A. This leads to a hierarchy of infinite sets that is a principle subject of Cantor's set theory.

EXERCISES 2.5

1. Let $A = \{1, 2, 3, 4, 5\}$ and $B = \{a, b, c, d, e\}$. Exhibit a one-to-one correspondence between A and B.

2. If $A = \{0, 1, 2, 3, \ldots\}$ and $B = \{1, 2, 3, \ldots\}$, exhibit a one-to-one correspondence between A and B.

3. Exhibit a one-to-one correspondence between $\mathbf{N}$ and the subset of perfect squares in $\mathbf{N}$.

4. If $A = \{0, 1, 2, 3, \ldots\}$ and $B = \{0, 5, 10, 15, \ldots\}$, exhibit a one-to-one correspondence between
    a. A and B b. A and $B - \{0, 5, 10\}$

5. Let $A = \{1, 2, 3, \ldots, 100\}$ and $B = \{3, 6, 9, \ldots, 300\}$. Give a formula for a one-to-one correspondence $f: B \to A$.

6. Which of the following are one-to-one correspondences from $\mathbf{R}$ to $\mathbf{R}$?
    a. $f(x) = x + 1$ b. $f(x) = 1 - x^2$
    c. $f(x) = |x|$ d. $f(x) = 2^x$

7. Define a one-to-one correspondence $f: (0, 1) \to (a, b)$, where $a, b \in \mathbf{R}$ and $a < b$.

8. Let $A = \{1, 2, 3, \ldots, 32\}$, and let B be the set of all possible bit strings in a 32-bit word. Find a one-to-one correspondence between $P(A)$ and B.

9. Show that the set of all integral powers of 2 is countable.

10. Show that the set $\mathbf{N} \times \mathbf{N}$ is countable. (*Hint:* Start with $(1, 1)$, then list the pairs of sum 3, then the pairs of sum 4, and so on.)

11. Prove that any infinite subset of $\mathbf{Z}$ is countable.

12. Find a one-to-one correspondence between $(0, 1] \times (0, 1]$ and some subset of $(0, 1]$.

2.6 • MATHEMATICAL INDUCTION

In this section, we will look at a very powerful proof technique called *proof by induction*. This method of proof can be used to prove theorems that are statements involving the positive integers. The technique is based on the principle of mathematical induction, which we state without proof in Theorem 2.2.

Theorem 2.2   (*Principle of Mathematical Induction*)   Suppose that S is a set of natural numbers such that

1. $1 \in S$, and
2. if k is a natural number such that all natural numbers less than k are in S, then k is in S.

Then S contains all the natural numbers.   ●

A proof using the Principle of Mathematical induction involves two steps:

1. Show that $1 \in S$.
2. Let $k > 1$, and assume that all natural numbers less than k are in S. (This assumption is called the **induction hypothesis.**) Then show that $k \in S$.

EXAMPLE 2.16   Prove that

$$\sum_{i=1}^{n} i = \frac{n(n + 1)}{2} \tag{1}$$

holds for all positive integers n. Let S be the set of all natural numbers for which this formula holds.

Step 1: Show that $1 \in S$. By replacing n with 1 in Equation (1), we get

$$\sum_{i=1}^{1} i = \frac{1(1 + 1)}{2}$$

and this is true because both sides of the equation are equal to 1.

Step 2: Let k be a natural number greater than 1, and assume that all natural numbers less than k are in S. (This is the induction hypothesis.) We have to show that $k \in S$—that is, that

$$\sum_{i=1}^{k} i = \frac{k(k + 1)}{2}$$

This equation is obtained by substituting k for n in Equation (1).

We can compute as follows:

$$\sum_{i=1}^{k} i = 1 + 2 + 3 + \cdots + (k - 1) + k$$

$$= \sum_{i=1}^{k-1} i + k \qquad (2)$$

Since $k - 1 < k$, the induction hypothesis tells us that $k - 1 \in S$, which means that

$$\sum_{i=1}^{k-1} i = \frac{(k-1)[(k-1)+1]}{2} \qquad \begin{array}{l}\text{This equation is obtained by substituting}\\ k-1 \text{ for } n \text{ in Equation (1).}\end{array}$$

$$= \frac{(k-1)k}{2} \qquad (3)$$

By substituting the right side of Equation (3) for the sum in Equation (2), we get

$$\sum_{i=1}^{k} i = \frac{(k-1)k}{2} + k$$

$$= \frac{k^2 - k + 2k}{2}$$

$$= \frac{k^2 + k}{2}$$

$$= \frac{k(k+1)}{2}$$

which shows that $k \in S$.

Since S satisfies Properties 1 and 2 of Theorem 2.2, it follows that $S = \mathbf{N}$. ○

EXAMPLE 2.17 Prove that any integer greater than 1 is prime or can be written as a product of primes.

Define S to consist of the integer 1 and also all integers greater than 1 that are prime or can be written as a product of primes.

Step 1: Show that $1 \in S$. This follows from the definition of S.

Step 2: Let k be a natural number greater than 1, and assume that all natural numbers less than k are in S. That is, assume that every integer m in the range $1 < m < k$ is prime or can be written as a product of primes. (This is the induction hypothesis.) Now show that $k \in S$. There are two cases:
  (i) If k is prime, we are done.
  (ii) If k is not prime, then $k = rs$, where $1 < r < k$ and $1 < s < k$. By the induction hypothesis, r is prime or can be written as a product of primes, and s is prime or can be written as a product of primes.
Therefore, k can be written as a product of primes, so $k \in S$. ○

EXAMPLE 2.18 Use induction to prove a generalized distributive law for intersection and union of sets: If $A_1, A_2, \ldots, A_n$, and B are subsets of a set X, then

$$(A_1 \cup A_2 \cup \cdots \cup A_n) \cap B = (A_1 \cap B) \cup (A_2 \cap B) \cup \cdots \cup (A_n \cap B)$$

This can be written

$$\left(\bigcup_{i=1}^{n} A_i\right) \cap B = \bigcup_{i=1}^{n} (A_i \cap B)$$

This is an extension of the notation for sums and products introduced in Section 1.3.
    Let S be the set of natural numbers for which this statement is true.

Step 1: Show that $1 \in S$: For $n = 1$, $A_1 \cap B = A_1 \cap B$, which is true.

Step 2: Let k be a natural number greater than 1, and assume that S contains all natural numbers that are less than k. (This is the induction hypothesis.) Now $k - 1 < k$, so $k - 1 \in S$. That is,

$$(A_1 \cup A_2 \cup \cdots \cup A_{k-1}) \cap B = (A_1 \cap B) \cup (A_2 \cap B) \cup \cdots \cup (A_{k-1} \cap B)$$

We have to show that $k \in S$, so we compute

$$(A_1 \cup A_2 \cup \cdots \cup A_{k-1} \cup A_k) \cap B$$
$$= [(A_1 \cup A_2 \cup \cdots \cup A_{k-1}) \cup A_k] \cap B$$

    By the Associative Law

$$= [(A_1 \cup A_2 \cup \cdots \cup A_{k-1}) \cap B] \cup (A_k \cap B)$$

    By the Distributive Law

$$= [(A_1 \cap B) \cup (A_2 \cap B) \cup \cdots \cup (A_{k-1} \cap B)] \cup (A_k \cap B)$$

    By the induction hypothesis

$$= (A_1 \cap B) \cup (A_2 \cap B) \cup \cdots \cup (A_{k-1} \cap B) \cup (A_k \cap B)$$

    By the Associative Law

This proves that $k \in S$, so the statement is true for all natural numbers n.   ○

Other Forms of the Principle of Mathematical Induction

In Examples 2.16 and 2.18, although the inductive hypothesis stated that the formula was true for all $n < k$, we used only the case $n = k - 1$. This situation is sufficiently common that it is worth stating separately.

Theorem 2.3   (*Principle of Mathematical Induction—Weak Form*)   If S is a set of natural numbers that

1. contains the number 1 and
2. contains the number k whenever it contains $k - 1$,
    then S contains all the natural numbers.   ●

    Theorem 2.3 is called the *weak form* for the induction principle, because in carrying out the induction step, we start from the assumption that $k - 1 \in S$, which is a weaker assumption than the assumption that all natural numbers less than k are in S. (Theorem 2.2 is called the *strong form*.)
    Also, there is nothing special about the role of the number 1 in mathematical induction. For the starting point, we can choose 1, or 0, or any other integer n_0. For

instance, if a set S of integers contains 0, and contains k whenever it contains $k - 1$, then S contains all integers greater than or equal to 0.

Finally, in proofs by induction it is customary to omit explicit mention of the set S. To prove a statement about integers $n \geq n_0$, there are two steps:

Step 1: Prove that the statement is true for the case $n = n_0$.

Step 2: Prove that the statement is true for the case $n = k$ whenever it is true for the case $n = k - 1$.

EXAMPLE 2.19   Prove that if a set A has n elements, the power set $P(A)$ has 2^n elements.

Step 1: Show that the statement is true for $n = 0$.

If A has 0 elements, then $A = \emptyset$ and $P(\emptyset) = \{\emptyset\}$. Thus, $P(\emptyset)$ has 1 element, and $1 = 2^0$.

Step 2: Assume that the statement is true for the case $n = k - 1$; that is, if A has $k - 1$ elements, then $P(A)$ has 2^{k-1} elements. We have to show that the statement is true for the case $n = k$.

Let A be a set with k elements, and let a_0 be any element of A. Let $B = A - \{a_0\}$. Then B has $k - 1$ elements. Now consider the subsets of A. They are of two types: those that do not contain a_0 and those that do. Those that do not contain a_0 are subsets of B, so by the induction hypothesis, there are 2^{k-1} of them. Each subset that does contain a_0 consists of a_0 and a subset of B, so there are 2^{k-1} of these also. Thus the total number of subsets of A is $2^{k-1} + 2^{k-1} = 2(2^{k-1}) = 2^k$.   ○

The following result will be useful in later chapters.

EXAMPLE 2.20   A **finite geometric series** is a sum of the form

$$\sum_{i=0}^{n} a^i$$

or, in expanded form,

$$1 + a + a^2 + \cdots + a^n$$

Show that if $a \neq 1$, the sum of a finite geometric series is given by the formula

$$\sum_{i=0}^{n} a^i = \frac{a^{n+1} - 1}{a - 1}$$

Step 1: Show that the formula is true in the case $n = 0$. In this case, the left side of the formula is a^0, which is equal to 1, and the right side is

$$\frac{a^{0+1} - 1}{a - 1} = 1$$

Step 2: Assume that the formula holds for the case $n = k - 1$; that is,

$$\sum_{i=0}^{k-1} a^i = \frac{a^k - 1}{a - 1}$$

Then

$$\sum_{i=0}^{k} a^i = \sum_{i=1}^{k-1} a^i + a^k$$

$$= \frac{a^k - 1}{a - 1} + a^k$$

$$= \frac{a^k - 1 + a^k(a - 1)}{a - 1}$$

$$= \frac{a^{k+1} - 1}{a - 1}$$

as required.   ○

In the next example, induction is used to prove an inequality.

EXAMPLE 2.21   Prove that if $x \in \mathbf{R}$ and $x \geq -1$, then $(1 + x)^n \geq 1 + nx$ for $n \in \mathbf{N}$.

Step 1: If $n = 1, (1 + x)^1 \geq 1 + 1x$, so the inequality is true.

Step 2: Assume that the inequality holds for the case $n = k - 1$; that is, $(1 + x)^{k-1} \geq 1 + (k - 1)x$. Then

$$
\begin{aligned}
(1 + x)^k &= (1 + x)^{k-1}(1 + x) \\
&\geq [1 + (k - 1)x](1 + x) &&\text{Since } 1 + x \geq 0 \\
&= 1 + x + (k - 1)x + (k - 1)x^2 \\
&= 1 + x + kx - x + kx^2 - x^2 \\
&= 1 + kx + kx^2 - x^2 \\
&= 1 + kx + (k - 1)x^2 \\
&\geq 1 + kx &&\text{Since } (k - 1)x^2 \text{ is nonnegative}
\end{aligned}
$$

                                                                                         ○

EXERCISES 2.6

1. Prove by induction the following statements in which $n \in \mathbf{N}$:

   a. $1 + 3 + 5 + 7 + \cdots + (2n - 1) = n^2$

   b. $1 + 4 + 7 + 11 + \cdots + (3n - 2) = \dfrac{n(3n - 1)}{2}$

   c. $1^2 + 2^2 + 3^2 + \cdots + n^2 = \dfrac{n(n + 1)(2n + 1)}{6}$

2. Prove by induction the following statements in which $n \in \mathbf{N}$:

   a. $1^3 + 2^3 + 3^3 + \cdots + n^3 = \left(\dfrac{n(n + 1)}{2}\right)^2$

b. $\dfrac{1}{1(2)} + \dfrac{1}{2(3)} + \dfrac{1}{3(4)} + \cdots + \dfrac{1}{n(n+1)} = \dfrac{n}{n+1}$

c. $2^1 + 2^2 + 2^3 + \cdots + 2^n = 2^{n+1} - 2$

3. For $n \in \mathbf{N}$, prove that $2^n > n$.

4. For $n \in \mathbf{N}$ and $n \geq 5$, prove that $2^n > n^2$.

5. Prove that if $0 \leq a < b$, then $a^n < b^n$.

6. Prove that $1 + 2n \leq 3^n$.

7. Prove that for $n > 1$, $\dfrac{1}{\sqrt{1}} + \dfrac{1}{\sqrt{2}} + \cdots + \dfrac{1}{\sqrt{n}} > \sqrt{n}$.

8. Prove by induction that if P dollars is invested at a yearly rate r compounded annually, the amount earned after n years is $P(1 + r)^n$.

9. Prove the following (generalized) DeMorgan's Laws:

a. $\left(\bigcap_{i=1}^{n} A_i \right)' = \bigcup_{i=1}^{n} A_i'$ **b.** $\left(\bigcup_{i=1}^{n} A_i \right)' = \bigcap_{i=1}^{n} A_i'$

10. Prove the generalized Distributive Law:

$$\left(\bigcap_{i=1}^{n} A_i \right) \cup B = \bigcap_{i=1}^{n} (A_i \cup B)$$

11. Find a formula for the sum of the finite geometric series

$$\sum_{i=0}^{n} a^i$$

for the case $a = 1$, and prove it by mathematical induction.

12. Define $H_n = 1 + \frac{1}{2} + \frac{1}{3} + \cdots + \frac{1}{n}$. Prove that $H_{2^m} \geq 1 + \frac{m}{2}$. (This shows that the sums H_n can be made arbitrarily large by taking a sufficient number of terms.)

13. What is wrong with the following proof that all math books have the same number of pages?

We will prove by induction on n that all sets of n math books have the same number of pages.

Step 1: $n = 1$. If X is a set of one math book, then all math books in X have the same number of pages.

Step 2: Assume that in every set of k math books, all the books have the same number of pages, and suppose X is a set of $k + 1$ math books. To show that all books in X have the same number of pages, it suffices to show that if $x_1 \in X$ and $x_2 \in X$, then x_1 has the same number of pages as x_2. Let $X_1 = X - x_1$, and let $X_2 = X - x_2$. Then X_1 and X_2 have k elements. By the inductive hypothesis, all books in X_1 have the same number of pages, and all books in X_2 have the same number of pages. So if $z \in X_1 \cap X_2$, x_1 has the same number of pages as z, and x_2 has the same number of pages as z. Therefore, x_1 has the same number of pages as x_2.

14. What is wrong with the following proof that $x^n = 1$ for all positive real numbers x and nonnegative integers n?

Step 1: $n = 0$. $x^0 = 1$ is true.

Step 2: Assume that the statement is true for all nonnegative integers less than $k + 1$. Then $x^{k+1} = x^k x^1 = 1 \times 1 = 1$.

15. Prove the following variation of the statement of the Principle of Mathematical Induction.

Let $n_0 \in \mathbf{Z}$, and let S be a subset of $\mathbf{Z}$ having the properties
(i) $n_0 \in S$ and
(ii) k is an element of S whenever $k - 1 \in S$.
Then $S = \{k \in \mathbf{Z} \mid k \geq n_0\}$.

16. The **Well-Ordering Principle** is the following statement about natural numbers: "Every nonempty set of natural numbers has a least element."

a. Deduce the Principle of Mathematical Induction from the Well-Ordering Principle. (*Hint:* Suppose S is a set of natural numbers such that $1 \in S$, and such that $k \in S$ whenever $k - 1 \in S$. Set $T = \mathbf{N} - S$. You must show that $T = \varnothing$. Assume that $T \neq \varnothing$, and apply the Well-Ordering Principle.)

b. Deduce the Well-Ordering Principle from the Principle of Mathematical Induction. (*Hint:* Suppose that T is a set of natural numbers without a least element, and set $S = \mathbf{N} - T$. Use mathematical induction to show that $S = \mathbf{N}$.)

2.7 • REPRESENTATION OF SETS IN COMPUTERS

In Chapter 1, we saw how integers are represented in a computer's memory. In this section, we will see how the subsets of a finite set can be represented in a computer's memory.

Suppose that set $X = \{x_1, x_2, x_3, \ldots, x_n\}$. In writing the set X in this way, we have defined an **order** for the set. To represent subsets of X in a computer, we must choose such an order. Different orders for the set will lead to different representations for the subsets of X. If $A \subset X$, we can represent A as a string of n bits. Each bit corresponds to one element of X. We write a 1 in the bit string if the corresponding element of X is in A, and 0 otherwise.

EXAMPLE 2.22   Suppose that

$$X = \{1, 2, 3, 4, 5, 6, 7, 8, 9, 10\}$$
$$A = \{2, 4, 6, 8, 10\}$$
$$B = \{1, 3, 5, 7, 9\}$$
$$C = \{2, 3, 5, 7\}$$

Then the set A is represented by the following ten bits:

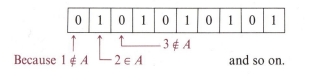

Similarly, the set C is represented by the bits

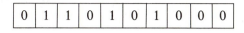

| 0 | 1 | 1 | 0 | 1 | 0 | 1 | 0 | 0 | 0 |

and the set X by

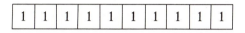

| 1 | 1 | 1 | 1 | 1 | 1 | 1 | 1 | 1 | 1 |

○

Bit strings offer economy in terms of memory space. In a file of student data, for instance, certain student attributes, such as male (0) or female (1) and noncitizen (0) or citizen (1), can be stored in a single bit rather than in an entire computer word, which may consist of from 8 to 32 bits or even more. With large data files, the resulting savings in memory space can be considerable.

Set Operations in Computers

We can perform the set operations of union, intersection, set difference, and complementation in a computer by bitwise (or logical) operations. In a bitwise operation, corresponding bits of the operands are combined according to some rule to yield a bit in the result. That is, the ith bit in the result depends only on the ith bit in the operands, not on anything else. Here are the rules for the set operations:

1. Union: The ith bit in $A \cup B$ is 0 if and only if the ith bit in A is 0 and the ith bit in B is 0. Otherwise, it is 1.
2. Intersection: The ith bit in $A \cap B$ is 1 if and only if the ith bit in A is 1 and the ith bit in B is 1. Otherwise, it is 0.
3. Complementation: A' is obtained by changing each 1 bit to 0 and each 0 bit to 1 in the representation of A.
4. Set difference: $A - B$ is obtained by computing $A \cap B'$.

Example 2.23 illustrates these operations using the sets X, A, and C of Example 2.22.

EXAMPLE 2.23 For the sets in Example 2.22, find $A \cap C$, $A \cup C$, and C'.

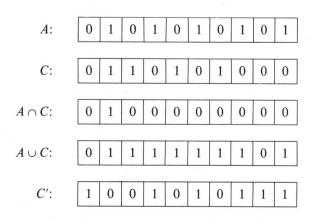

A: | 0 | 1 | 0 | 1 | 0 | 1 | 0 | 1 | 0 | 1 |

C: | 0 | 1 | 1 | 0 | 1 | 0 | 1 | 0 | 0 | 0 |

$A \cap C$: | 0 | 1 | 0 | 0 | 0 | 0 | 0 | 0 | 0 | 0 |

$A \cup C$: | 0 | 1 | 1 | 1 | 1 | 1 | 1 | 1 | 0 | 1 |

C': | 1 | 0 | 0 | 1 | 0 | 1 | 0 | 1 | 1 | 1 |

The results are interpreted as follows. Since the result of $A \cap C$ contains a 1 in the second bit only, the set $A \cap C$ contains only the second element of the set X (namely, the element 2). The set $A \cup C$ contains the second, third, fourth, fifth, sixth, seventh, eighth, and tenth elements of X. Thus,

$$A \cup C = \{2, 3, 4, 5, 6, 7, 8, 10\}$$

What are the elements in C'? In $A \cap B$?   ○

At least one computer programming language, Pascal, supports set variables as well as set operations. Sets in Pascal are stored in memory as bit strings, as shown in Example 2.22. The operations of union and intersection are performed as shown in Example 2.23, and the set difference $A - B$ is computed as $A \cap B'$.

The computer can determine whether two sets are equal by comparing their bit strings, and it can check whether A is a subset of B by computing $A \cap B'$. If $A \subset B$, then $A \cap B' = \emptyset$.

EXERCISES 2.7

1. Let $X = \{a, b, c, d, e, f, g, h\}$, $A = \{a, b, f\}$, $B = \{c, d, e, f, g\}$, and $C = \{c, e, g, h\}$. Represent each of the sets A, B, C, and X by bit strings.

2. For the sets in Exercise 1, find the bit string representation for

   a. $A \cup B$ **b.** $A \cap C$ **c.** $B - A$ **d.** C'

3. Cabin Creek College maintains a student file with the following information on each student:

sex:	$0 = $ male, $1 = $ female
marital status:	$0 = $ single, $1 = $ married
U.S. citizen:	$0 = $ no, $1 = $ yes
state resident:	$0 = $ no, $1 = $ yes
on financial aid:	$0 = $ no, $1 = $ yes
full-time student:	$0 = $ no, $1 = $ yes
employed:	$0 = $ no, $1 = $ yes
living on campus:	$0 = $ no, $1 = $ yes

   The file records for six students look like this:

student number	sex	mar st	US cit	st res	fin aid	f/t	emp	on cmp
1	1	0	1	1	0	1	0	1
2	1	0	1	0	1	1	0	0
3	0	1	1	1	1	0	1	0
4	1	1	0	0	0	1	0	0
5	0	0	0	0	1	0	0	1
6	1	0	1	1	0	1	1	1

According to the table, student #3 is a married male, a U.S. citizen, a state resident receiving financial aid, a part-time student, employed, and not living on campus.

    a. Find the bit string representing a single male French citizen who is an out-of-state resident receiving no financial aid, a full-time student, unemployed, and living on campus.

    b. Find all students in the table who are not U.S. citizens.

    c. Find all students who are state residents.

    d. Find all female students.

    e. Find all married students.

4. Refer to Exercise 3. Find

    a. all full-time married male students

    b. all out-of-state, single, employed students

    c. all non–U.S. citizens who are part-time students, employed, and living on campus

Computer Exercises for Chapter 2

1. Let the universe $X = \{a, b, c, \ldots, z\}$. Write a program to display the bit-string representation of an arbitrary subset of X. Use it to print the representations of the following sets: $A = \{a, d, f, h, m, x\}$, $B = \{m, n, p, r, t, v\}$, $C = \{h, k, q, s, u, y\}$.

2. Refer to the sets described in Exercise 1. Write a program to display both the bit-string representations and the actual elements in the following sets:

    a. $A \cup B$ **b.** $A \cap B$ **c.** $B \cap C$

    d. $A - B$ **e.** A' **f.** $(A \cup C)'$

    g. $A \oplus B$, where $\oplus$ is defined in Exercise 13 of Section 2.2.

3. Refer to the sets described in Exercise 1. Write a program to determine whether x is an element of A, where $x \in X$.

4. Refer to the sets described in Exercise 1. Write a program to find the number of elements in sets A, B, and C.

5. Refer to the sets in Exercise 1. Write a program to print the elements in $A \times B$.

● ● ● ● CHAPTER REVIEW EXERCISES

1. Let $A = \{1, 2, 3, 4\}$ and $B = \{0, 1, 2, 3\}$, and let $C \subset A \times B$ be defined by $C = \{(x, y) \mid x \in A, y \in B, \text{ and } x - y > 0\}$. List the elements of C.

2. If $A = \{1, 2, 3\}$, list the elements in $P(A)$.

3. Let $A = \{1, 2, 3, 4, 5, 7\}$ and $B = \{a, b, c, d, e, f\}$. Exhibit a one-to-one correspondence between A and B.

4. Exhibit a one-to-one correspondence between **N** and the subset of **N** consisting of multiples of 4.

5. Give an example of a one-to-one correspondence from **R** to **R**.

6. Write in interval notation:

    a. $\{x \mid x \in \mathbf{R} \text{ and } -5 \leq x < 3\}$ **b.** $\{x \mid x \in \mathbf{R} \text{ and } -2 < x < 2\}$

    c. $\{x \mid x \in \mathbf{R} \text{ and } x > -1\}$

7. Perform the indicated operations on the given intervals.

    a. $(1, 2) \cup (-1, 3)$ **b.** $[-1, 1] \cap (0, 2)$ **c.** $(-2, 4] \cup [0, 5]$

    d. $(-\infty, 2) - (0, 1)$ **e.** $[3, \infty) \cap (1, 5]$

8. If the universe is **R**, find

    a. $(-\infty, 1]'$ **b.** $(-1, 1)'$ **c.** $[2, 5]'$ **d.** $(-1, \infty)'$

9. If A, B, C are subsets of X, prove that $A - (B \cap C) = (A - B) \cup (A - C)$.

10. Prove the following by induction. Here, $n \in \mathbf{N}$.

    a. $2 + 4 + 6 + 8 + \cdots + (2n) = n^2 + n$ **b.** $1 + \dfrac{1}{2} + \dfrac{1}{4} + \dfrac{1}{8} + \cdots + \dfrac{1}{2^n} = 2 - \dfrac{1}{2^n}$

    c. If $n \geq 4$, then $n! > 2^n$.

11. Prove the following by induction. Here, $n \in \mathbf{N}$.

    a. $\left(1 - \dfrac{1}{2}\right)\left(1 - \dfrac{1}{3}\right)\left(1 - \dfrac{1}{4}\right)\cdots\left(1 - \dfrac{1}{n+1}\right) = \dfrac{1}{n+1}$

    b. $1 + 3 + 3^2 + 3^3 + \cdots + 3^{n-1} = \dfrac{3^n - 1}{2}$ **c.** $2n \leq 2^n$

12. Let $X = \{a, b, c, d, e, f\}$, $A = \{a, d, e\}$, $B = \{b, e, f\}$, and $C = \{c\}$. Represent each of the sets A, B, and C as bit strings.

13. Refer to the sets in Exercise 12. Find the bit-string representations for

    a. $A \cap B$ **b.** $A - B$ **c.** $(B \cup C)'$

    d. $B - A$ **e.** $(A \cap B)'$ **f.** $A - (B \cup C)$

    g. $(A \cup B) \cap C'$

3
FUNCTIONS AND RELATIONS

The concept of a function is fundamental to both mathematics and computing. Informally, a function is a rule for computing some value (or output) from an argument (or input). In algebra, many simple functions are described by equations. For example,

$$f(x) = 2x + 1 \qquad g(x) = \frac{x - 5}{2x^2 - 3x + 2} \qquad h(x) = \sqrt{x - 2}$$

In trigonometry, the so-called circular functions, such as sine and cosine, are studied. Most computer programming languages provide several built-in functions and allow the programmer to define many other functions.

Functions are very closely related to sets. The arguments, or inputs, to a function come from specific sets. The value, or output, of a function is an element of some set. When describing a function, it is essential to specify the sets that these arguments and values belong to.

Functions are a special case of a more general mathematical concept, the relation. The value of a function is "related" to the arguments by virtue of being computed from the arguments.

3.1 • FUNCTIONS

In this section, we introduce the formal definition of a function, the notations and terminology for functions, and some special functions.

Definition 3.1

A **function** consists of three things:

1. A set A called the **domain** of the function
2. A set B called the **range** of the function
3. A subset G of $A \times B$ called the **rule** or **graph** of the function and satisfying the following two conditions:

    (i) No two ordered pairs of G have the same first element
    (ii) Every element $a \in A$ is the first element of a pair in G

Functions are given names, usually lowercase letters such as f and g. The notation $f: A \to B$ indicates that f is a function from the set A (the domain) to the set B (the range). A function can be thought of as a rule that assigns a unique element of the range B to each element of the domain A.

EXAMPLE 3.1   Let $A = \{a, b, c, d\}$ and let $B = \{v, w, x, y, z\}$. Let $f = \{(a, v), (b, x), (c, x), (d, z)\}$. Then the domain A, the range B, and the graph f determine a function.   ○

Terminology and Notations for Functions

If $a \in A$, there is one and only one element $b \in B$ such that $(a, b) \in f$. Therefore, it is appropriate to give this element a name. It is called $f(a)$. In Example 3.1, we can write $f(a) = v$, $f(b) = x$, and so on. In this notation, the item in parentheses is called the **argument,** or **variable,** of the function.

In mathematical writing, a function is sometimes called a **functional,** a **map,** or a **mapping.** The range is sometimes called the **codomain.**

According to our definition, two functions are equal if and only if they have the same domain, same range, and same rule. This definition of a function, and our terminology, are not entirely standard. The term *range* is sometimes used to mean the set of values $b \in B$ that appear in the ordered pairs in the graph of the function. In this text, we will use the term **image** for this set.

Definition 3.2

Let $f: A \to B$ be a function. Then the **image** of f is the set

$$\{b \in B \mid (a, b) \in f \text{ for some } a \in A\}$$

The image of f is also called the image of A under f. It is denoted by $f[A]$. More generally, if $X \subset A$, the **image of X under f** is

$$\{b \in B \mid (a, b) \in f \text{ for some } a \in X\}$$

and is denoted by $f[X]$.

EXAMPLE 3.2   Let $f: \mathbf{R} \to \mathbf{R}$ be defined by $f(x) = x^2$. Then the image of f is $[0, \infty)$, the set of nonnegative reals. The image of the open interval $(2, 3)$ under f is $f[(2, 3)] = (4, 9)$. The image of the closed interval $[-3, 2]$ is $f[[-3, 2]] = [0, 9]$.   ○

Since the term *range* is used ambiguously in the mathematics literature, many current authors have dropped it in favor of the word *codomain.* Moreover, many authors of mathematics texts consider two functions equal if they have the same domain and the same rule, without regard to the range.

EXAMPLE 3.3   The greatest integer function, also called the **floor function,** has as its domain the set $\mathbf{R}$ of real numbers. The greatest integer in a real number x is the largest integer less than or equal to x. Traditionally, this has been denoted by $[x]$, but it is now more often written $\lfloor x \rfloor$. For instance, $\lfloor 5.25 \rfloor = 5$ and $\lfloor -3.75 \rfloor = -4$. To complete the definition of the greatest integer function, we must specify the range. Two possibilities are **Z** and **R**: We can define functions:

$$g_{\text{int}}: \mathbf{R} \to \mathbf{Z}$$

and

$$g_{\text{real}}: \mathbf{R} \to \mathbf{R}$$

by the same rule, $g_{\text{int}}(x) = g_{\text{real}}(x) = \lfloor x \rfloor$. These two functions differ only in their ranges. In the first instance, the value of the function is an integer, and in the second the value is a real number.   ○

Distinctions between ranges are frequently ignored in mathematical writing, but they are always explicit in computer programs. For instance, in the Pascal language, the two versions of the greatest integer function could be written

function gint(x:real): integer;

Range
Domain

and

function greal(x:real): real;

Note that the data type of the argument is also declared explicitly; that is, the domain of the function is a specified data type.

Some Special Functions

There are certain functions that occur with sufficient frequency that they are given names. Some of these functions are identified in the definitions that follow.

Definition 3.3

If X is a set, the function $f: X \to X$ defined by $f(x) = x$ is called the **identity** function on X. If X and Y are sets and $b \in Y$, then the function $f: X \to Y$ defined by $f(x) = b$ is a **constant** function.

Definition 3.4

If X is a set and $A \subset X$, then the function $f: X \to \mathbf{R}$ defined by

$$f_A(x) = \begin{cases} 1, & \text{if } x \in A \\ 0, & \text{if } x \notin A \end{cases}$$

is called the **characteristic function** of A.

In Example 2.22, we used characteristic functions to represent finite sets by bit strings.

Sequences

Definition 3.5

A **sequence** is a function whose domain is a set (possibly infinite) of consecutive integers.

EXAMPLE 3.4   Define $s: \mathbf{N} \to \mathbf{R}$ by $s(n) = n\pi$. Then $s(1) = \pi$, $s(2) = 2\pi$, $s(3) = 3\pi$, and so on.   ○

Since sequences are so common, special conventions are used to describe them. The values of the function are listed in order, like this:

$$s(1), s(2), s(3), \ldots$$

In addition, subscripts are used instead of the usual function notation; that is, s_1 instead of $s(1)$, s_2 instead of $s(2)$, and so on. Thus, a sequence is written:

$$s_1, s_2, s_3, \ldots$$

with the ellipsis denoting continuation indefinitely. The function s in Example 3.4 can be described as

$$\pi, 2\pi, 3\pi, \ldots$$

or by the formula $s_n = n\pi$ for $n = 1, 2, 3, \ldots$

EXAMPLE 3.5   Define the sequence $s: \mathbf{N} \cup \{0\} \to \mathbf{Z}$ by

$$s_0 = 0$$
$$s_1 = 1$$
$$s_n = s_{n-1} + s_{n-2} \qquad \text{if } n > 1$$

Then the first several elements of the sequence are

$$0, 1, 1, 2, 3, 5, 8, 13, 21, 34, \ldots$$

This sequence of numbers is called the **Fibonacci Sequence.** What is s_{12}?   ○

A definition that directly or indirectly references itself is said to be **recursive.** The sequence in Example 3.5 is defined recursively, and the next example also contains a recursive definition.

EXAMPLE 3.6   The factorial function, denoted by an exclamation point (!), has domain $\mathbf{N} \cup \{0\}$ and range $\mathbf{N}$ and is defined by the rule

$$0! = 1$$
$$n! = n \times (n - 1)!$$

Thus,

$$0! = 1$$
$$1! = 1$$

$$2! = 2 \times 1$$
$$3! = 3 \times 2 \times 1$$
$$4! = 4 \times 3 \times 2 \times 1$$

and so on.   ○

Real-Valued Functions

Most functions familiar from algebra are functions whose range is the set **R** of real numbers. If the domain X of a real-valued function is a subset of **R**, the graph of f is a subset of the plane.

EXAMPLE 3.7   Define $f: \mathbf{R} \to \mathbf{R}$ as follows:

$$f(x) = \begin{cases} x, & \text{if } x \geq 0 \\ -x, & \text{if } x < 0 \end{cases}$$

This function is called the **absolute value function** and is written $f(x) = |x|$.   ○

EXAMPLE 3.8   If $f: \mathbf{R} \to \mathbf{R}$ is defined by the formula

$$f(x) = 3x^4 - x^3 + 2x^2 + 5x - 12$$

then f belongs to a class of functions called polynomials. In general, a **polynomial** is a function in the form

$$f(x) = a_n x^n + a_{n-1} x^{n-1} + \cdots + a_2 x^2 + a_1 x + a_0$$

where n is a nonnegative integer, the a_i's are the coefficients, and the highest power of x with a nonzero coefficient is the **degree** of the polynomial. (The zero polynomial, $f(x) = 0$, does not have a degree, since it does not have any nonzero coefficient.)   ○

Functions of Several Variables

A function of several variables is a function whose domain is a subset of a Cartesian product of two or more sets. Suppose A and B are sets and D is a subset of $A \times B$. Then an element of D is an ordered pair (a, b) where $a \in A$ and $b \in B$. If f is a function with domain D, then the value of f at the point (a, b) could be written $f((a, b))$. This is a special case of the notation $f(x)$ where $x = (a, b)$. It is customary to omit the second set of parentheses; instead, we write simply $f(a, b)$. It is convenient, then, to think of f as a function of two variables (one from the set A and one from the set B), rather than as a function of one variable from the set $A \times B$.

EXAMPLE 3.9   Let S be the set of all students at Cabin Creek College, and let D be the set of all dates in the twentieth century. We can define a function $g: S \times D \to \mathbf{N}$ by the rule: $g(s, d)$ is the age of student s on date d (or zero if the student was born after date d). For instance, if Luisa Miller was born on 3/21/68, then

$$g(\text{Luisa Miller}, 5/16/87) = 19$$

                                                                    ○

EXAMPLE 3.10   We can define a function $p\colon \mathbf{R} \times \mathbf{N} \to \mathbf{R}$ by the rule $p(x, n) = x^n$. Then p is a function of two variables: a real variable x and a natural number variable n. What is $p(-\frac{1}{2}, 3)$?   ○

EXAMPLE 3.11   If $f(x, y) = \sqrt{1 - x^2 - y^2}$, the domain of f is a subset of $\mathbf{R} \times \mathbf{R}$—namely, the set of points where

$$1 - x^2 - y^2 \geq 0$$

since the square root of a negative number is not defined. The domain consists of the circle of radius 1 centered at the origin and all points inside it. The domain and the graph are shown in Figure 3.1. The graph is a surface in three-dimensional space.

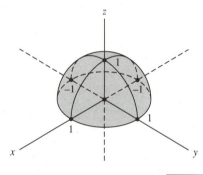

FIGURE 3.1   The graph of $f(x, y) = \sqrt{1 - x^2 - y^2}$ ○

In programs written in some computer languages, when a function of several variables is defined, the data type of each variable must be declared. In languages such as Pascal, this must always be done explicitly. For instance, a Pascal implementation of the function of Example 3.10 would be declared by

function p(x:real; n: integer):real;

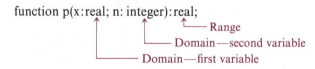

Range
Domain—second variable
Domain—first variable

The domain of this function is the set $R \times N$, where R is the data type real and N is the data type integer.

Binary Operations

One important class of functions of two variables consists of **binary operations.** A binary operation on a set A is a function whose domain is $A \times A$ and whose range is A. The most common binary operations are the ordinary arithmetic operations, such as addition and multiplication. It is customary to write the symbol for a binary operation between the two arguments rather than in front; that is, we write $x + y$ for the sum of two numbers, rather than $+(x, y)$. For instance, we write $3 + 5 = 8$ rather than $+(3, 5) = 8$. It is also customary to use the same symbol for similar operations on different sets. For instance, the binary operation of addition on the integers is a function

$$+: \mathbf{Z} \times \mathbf{Z} \to \mathbf{Z}$$

whereas addition of real numbers is a function

$$+: \mathbf{R} \times \mathbf{R} \to \mathbf{R}$$

Although the same symbol is used for these two functions, they are different, because they have different domains and different ranges.

EXERCISES 3.1

1. If $A = \{1, 2, 3, 4, 5\}$, $B = \{x, y, z\}$, and $f: A \to B$ is defined by $\{(1, x), (2, y), (3, x), (4, z), (5, x)\}$, find
   a. $f(1)$
   b. $f(3)$
   c. $f(4) + f(1)$
   d. $f(2 + 3)$

2. Define $f: \mathbf{N} \to \mathbf{N}$ to be $\{(n, n^2) \mid n \in \mathbf{N}\}$. Find
   a. $f(3)$
   b. $f(12)$

3. If $A = \{1, 2, 3\}$ and $B = \{u, v\}$, give an example of a subset of $A \times B$ that could *not* be the rule, or graph, of a function.

4. Determine the domain and image for each of the following functions whose graph is the specified set of ordered pairs.
   a. $\{(a, n), (b, d), (u, g), (n, d), (m, s)\}$
   b. $\{(1, 1), (2, \sqrt{2}), (3, \sqrt{3}), (4, 2)\}$

5. From the information in Exercise 4, can you determine the range of the functions?

6. Sketch the graph of the function $f: \mathbf{R} \to \mathbf{R}$ defined by

$$f(x) = \begin{cases} 1, & \text{if } x > 0 \\ 0, & \text{if } x = 0 \\ -1, & \text{if } x < 0 \end{cases}$$

This function is called the **signum** function. What is the image of f?

7. Let $f: \mathbf{N} \to \mathbf{N}$ be defined as follows: For $n \in \mathbf{N}$, $f(n)$ is the sum of the digits of n. Find
   a. $f(3)$
   b. $f(3333)$
   c. $f(1024)$
   d. $f(1123456)$

8. Define $f: \mathbf{N} \to \mathbf{N}$ as follows: For $n \in \mathbf{N}$, represent n in hexadecimal and add the digits. Thus,

$$f(57005_{\text{ten}}) = f(\text{DEAD}_{\text{hex}}) = D + E + A + D$$
$$= 32_{\text{hex}} = 50_{\text{ten}}$$

Find
   a. $f(1024)$
   b. $f(32767)$
   c. $f(15)$
   d. $f(65536)$

9. Let P be the set of Social Security numbers (SSN) of the students at Cabin Creek College. Define a function h as follows: $h(\text{SSN}) = (\text{remainder of } \frac{\text{SSN}}{5}) + 1$.

    a. Find $h(235347890)$ **b.** Find $h(841710234)$

10. Refer to Exercise 9. Suppose h is defined to be

$$h(\text{SSN}) = 10 \times \left(\text{remainder of } \frac{\text{SSN}}{5} \right) + 1$$

    a. Find $h(235347890)$ **b.** Find $h(841710234)$

11. If $f: [-2, 3] \to \mathbf{R}$ is defined by

$$f(x) = \begin{cases} x^3, & \text{if } x \in [-2, 1) \\ 3x - 1, & \text{if } x \in [1, 3] \end{cases}$$

find

    a. $f(-2)$ **b.** $f(0)$ **c.** $f(-\frac{1}{2})$ **d.** $f(\frac{3}{2})$

12. Sketch the graph of the identity function $f(x) = x$.

13. Sketch the graph of the absolute value function $f(x) = |x|$, defined in Example 3.7. What is the image of f?

14. Explain why a vertical line cannot be the graph of a function.

15. The function $f: \mathbf{R} \to \mathbf{R}$ defined by $f(x) =$ the smallest integer greater than or equal to x, written $\lceil x \rceil$, is sometimes called the **ceiling function.** For example, $f(1.235) = 2$, and $f(5) = 5$. Find

    a. $f(4.9362)$ **b.** $f(-0.41)$

    c. Show that $\lceil x \rceil \neq 1 + \lfloor x \rfloor$.

16. Given the following sequences, find a formula for the nth term, s_n.

    a. $1, 8, 27, 64, 125, \ldots$ **b.** $2, 4, 8, 16, 32, \ldots$

    c. $1, 2, 4, 8, 16, 32, \ldots$ **d.** $-1, 1, -1, 1, \ldots$

17. Given the following sequences, find a formula for the nth term, s_n.

    a. $1, -1, 1, -1, \ldots$ **b.** $1, \frac{1}{2}, \frac{1}{3}, \ldots$

    c. $\frac{1}{2}, \frac{1}{4}, \frac{1}{8}, \frac{1}{16}, \ldots$ **d.** $.1, .01, .001, .0001, \ldots$

18. Consider the sequences

$$1, \tfrac{1}{2}, \tfrac{1}{3}, \tfrac{1}{4}, \tfrac{1}{5}, \ldots \qquad \text{and}$$
$$1, \tfrac{1}{2}, 1, \tfrac{1}{3}, 1, \tfrac{1}{4}, 1, \tfrac{1}{5}, \ldots$$

    a. Determine the image for each.

    b. Find a formula for the nth term of each.

19. Consider the polynomial function $f: \mathbf{R} \to \mathbf{R}$ defined by $f(x) = 5x^3 - 3x^2 + 2x - 1$. A method of computing the value of f at a point x, which involves multiplication and addition only and no taking powers, is called **Horner's method** and is described as follows:

$$5x^3 - 3x^2 + 2x - 1 = x(5x^2 - 3x + 2) - 1$$
$$= x(x(5x - 3) + 2) - 1$$

Using this method, we find $f(4)$ by computing:

$$f(4) = 4(4(4(5) - 3) + 2) - 1$$
$$= 4(4(20 - 3) + 2) - 1$$
$$= 4(4(17) + 2) - 1$$
$$= 4(68 + 2) - 1$$
$$= 4(70) - 1 = 280 - 1 = 279$$

a. Find $f(3)$ using Horner's method.

b. Compare the number of multiplications performed in computing $f(3)$ using Horner's method with the number of multiplications performed in computing $f(3)$ by the usual method of substitution.

20. If $A = \{a, b\}$, list all functions from A to A.

21. Given $A = \{1, 2, 3\}$ and $B = \{u, v\}$, how many different functions are there from A to B? From B to A?

22. If $f: A \to B$ and if X and Y are subsets of A, show that $f[X \cup Y] = f[X] \cup f[Y]$.

23. Refer to Exercise 22. Show by an example that it is not necessarily true that $f[X \cap Y] = f[X] \cap f[Y]$.

24. Let P be the set of all U.S. presidents. For $p \in P$, define f as follows: $f(p) =$ the successor of p. Is f a function? Explain.

3.2 • OPERATIONS WITH FUNCTIONS

In Chapter 2, we introduced the concept of a one-to-one correspondence. We can now recognize a one-to-one correspondence as a special type of function: No two elements of the domain are assigned the same function value, and every element of the range occurs as a function value. These two properties are referred to so frequently that they are given names.

Definition 3.6

Let $f: A \to B$ be a function. The function f is said to be **one-to-one** (sometimes written **1-1**) if $f(x) \neq f(y)$ whenever x and y are different elements of A. The function f is said to be **onto** if the image of f is equal to the range B.

Another way of stating this definition is as follows. The function f is one-to-one if $f(x) = f(y)$ implies $x = y$. The function f is onto if, for every $b \in B$, there is an $a \in A$ such that $f(a) = b$.

EXAMPLE 3.12   Let $A = \{a, b, c, d\}$ and $B = \{v, w, x, y, z\}$. Then Figure 3.2 shows a function f from A to B that is one-to-one but not onto. Figure 3.3 shows a function g from B to A that is onto but not one-to-one.

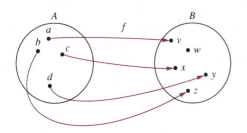

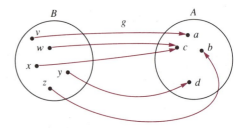

FIGURE 3.2   A function that is one-to-one but not onto   FIGURE 3.3   A function that is onto but not one-to-one   ○

EXAMPLE 3.13   Let $A = \{1, 2, 3\}$ and $B = \{a, b, c\}$. The function $f: A \to B$ defined in Example 2.9 is both one-to-one and onto. See Figure 2.3.   ○

We can now see that a one-to-one correspondence is just a function that is both one-to-one and onto. A function that is one-to-one is sometimes called **injective.** A function that is onto is sometimes called **surjective.** A function that is both one-to-one and onto (that is, a one-to-one correspondence) is sometimes called **bijective.**

EXAMPLE 3.14   Define $f: \mathbf{N} \to \mathbf{N}$ by the rule $f(x) = 2x$. Then f is one-to-one (if $2x = 2y$, then $x = y$). But f is not onto, because the image of f consists of only the even integers in $\mathbf{N}$.   ○

EXAMPLE 3.15   Let $f: \mathbf{R} \to \mathbf{R}$ be defined by the rule $f(x) = x^2$. Then f is not onto, since the image of f is the set of nonnegative real numbers, and f is not one-to-one, because $f(-2) = f(2)$ but $-2 \neq 2$.   ○

EXAMPLE 3.16   Define $g: \mathbf{N} \to \mathbf{N}$ by the rule $g(z) =$ the smallest integer greater than or equal to $\frac{z}{2}$. For instance, $g(14) = 7$ and $g(19) = 10$. Then g is onto, because if x is any element of $\mathbf{N}$, then $x = g(2x)$. However, g is not one-to-one, since $g(20) = g(19)$ but $20 \neq 19$.   ○

The Pigeonhole Principle

The following theorem is a formal statement of the Pigeonhole Principle, which was used in Theorem 1.1.

Theorem 3.1   Let A be a set of m elements, let B be a set of n elements, and suppose $m > n$. Let $f: A \to B$ be a function. Then f is not one-to-one.

Proof   The proof is by induction on n, the number of elements in the range B. First we show that the theorem is true for $n = 1$.

Suppose that B has one element—say, $B = \{b\}$. The theorem states that if A has more than one element, then f is not one-to-one. This is true because if a_1 and a_2 are distinct elements of A, then $f(a_1) = b$ and $f(a_2) = b$, but $a_1 \neq a_2$. Next assume that the theorem is true for $n \leq k$. That is, suppose that all functions from sets of more than k elements to sets of k elements are not one-to-one. We now show that the theorem is true for $n = k + 1$.

Assume that f is a function from A to B, B has $k + 1$ elements, and A has $m > k + 1$ elements. Let b_0 be an element of B. Then $B - \{b_0\}$ has k elements. We distinguish three cases:

1. b_0 is not assigned to any element of A.
2. b_0 is assigned to exactly one element of A.
3. b_0 is assigned to more than one element of A.

In case (3), f is clearly not one-to-one, and we are done. In case (1), we may discard b_0 from the range and obtain a function f' from a set of m elements to a set of k elements. By the induction hypothesis, f' is not one-to-one, so f is not one-to-one. In case (2), let a_0 be the unique element of A such that $f(a_0) = b_0$. Discard a_0 from the domain and discard b_0 from the range. We obtain a function f' from a set of $m - 1$ elements to a set of k elements. Since $m > k + 1$, we have $m - 1 > k$. By the induction hypothesis, f' is not one-to-one, so f is not one-to-one. ●

In Theorem 1.1, A is the set of $d + 1$ steps in the division operation, B is the set of d possible remainders, and f is the function that assigns to each step the remainder produced at that step.

Inverse Functions

Example 2.9 illustrates a function f from the set $A = \{1, 2, 3\}$ to the set $B = \{a, b, c\}$ that is both one-to-one and onto. Note that if we reverse each arrow in Figure 2.3, we define a function g from B to A that is also one-to-one and onto. This function is called the **inverse** of f and is shown in Figure 3.4.

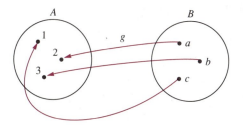

FIGURE 3.4   The inverse of the function of Figure 2.3

Definition 3.7

Let $f: A \to B$ and $g: B \to A$ be functions. Then f and g are **inverse** functions (to one another) if

$$g(f(a)) = a \quad \text{for all } a \in A, \quad \text{and}$$
$$f(g(b)) = b \quad \text{for all } b \in B$$

If f and g are inverse functions, we write $g = f^{-1}$ and $f = g^{-1}$.

EXAMPLE 3.17   In Example 2.11, we saw that the function

$$f: [0, 1) \to (2, 3] \text{ defined by } f(x) = 3 - x$$

is a one-to-one correspondence. Its inverse,

$$g = f^{-1}: (2, 3] \to [0, 1)$$

is also defined by the formula $g(x) = 3 - x$. Note that even though they are defined by the same formula, f and g are not the same function, since they have different domains and different ranges.   ○

EXAMPLE 3.18   In Example 2.12, we noted that

$$g: [0, 2] \to [0, 1] \text{ defined by } g(x) = \frac{x}{2}$$

is a one-to-one correspondence. The inverse g^{-1} is defined by the formula $g^{-1}(y) = 2y$.   ○

EXAMPLE 3.19   In Example 2.13, we showed that the set $\mathbf{Z}$ of integers is countable by writing the elements of $\mathbf{Z}$ in a sequence: $0, 1, -1, 2, -2, \ldots$. The sequence is a function (let us call it s) from $\mathbf{N}$ to $\mathbf{Z}$. Then s^{-1} is a function from $\mathbf{Z}$ to $\mathbf{N}$ that can be described by the rule

$$s^{-1}(x) = \begin{cases} \dfrac{x}{2} & \text{if } x \text{ is even} \\ -\dfrac{x-1}{2} & \text{if } x \text{ is odd} \end{cases}$$

(Verify this for several values of x.)   ○

Theorem 3.2   A function $f: A \to B$ has an inverse if and only if f is both one-to-one and onto.

Proof   Suppose $f: A \to B$ is both one-to-one and onto. Let $b \in B$. Since f is onto, there must be some $a \in A$ such that $f(a) = b$. There can be at most one such a, because otherwise the function f would not be one-to-one. So we can define $g(b)$ to be this a. This is true for each $b \in B$; therefore, g is the inverse of f. This proves the "if" part of the theorem: that if f is one-to-one and onto, then f has an inverse.

Now we prove the "only if" part or the converse—that if f has an inverse, then f is one-to-one and onto.

Suppose that f has the inverse $g: B \to A$. This means that $g(f(a)) = a$ for all $a \in A$ and that $f(g(b)) = b$ for all $b \in B$. We must show that f is one-to-one and onto. If $b \in B$, then $b = f(g(b))$. This shows that the image of f is all of B, so f is onto. To show that f is one-to-one, suppose $f(a_1) = f(a_2)$. Then $g(f(a_1)) = g(f(a_2))$; that is, $a_1 = a_2$.   ●

Composition of Functions

In defining the inverse g of a function $f: A \to B$, we encountered the expressions $f(g(b))$ and $g(f(b))$. These expressions illustrate a way of combining two functions to obtain a third: First apply one function, then the second. The resulting function is called the **composition** of the two functions.

Definition 3.8

Let $f: X \to Y$ and $g: Y \to Z$ be functions. Then the **composition** of g with f, written $g \circ f$, is the function with domain X, range Z, and rule

$$(g \circ f)(x) = g(f(x))$$

The relation of the sets X, Y, and Z and the three functions f, g, and $g \circ f$ is shown in Figure 3.5.

For both $f \circ g$ and $g \circ f$ to exist, it is necessary that the domain of f be the range of g and also that the domain of g be the range of f; that is, that $X = Z$.

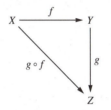

FIGURE 3.5   Composition of functions

EXAMPLE 3.20   Let $X = \{a, b, c\}$, $Y = \{v, w, x, y\}$, and $Z = \{1, 2, 3\}$. Suppose $f: X \to Y$ and $g: Y \to Z$ are as shown in Figure 3.6.

Then

$$(g \circ f)(a) = 3$$
$$(g \circ f)(b) = 1$$
$$(g \circ f)(c) = 2$$

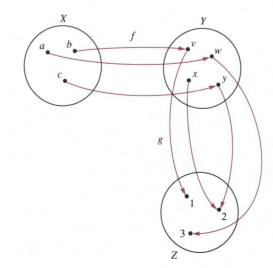

FIGURE 3.6   The function of Example 3.20

But $f(g(w)) = f(3)$, which does not exist, because 3 is not in the domain of f. Thus, $f \circ g$ does not exist.   ○

EXAMPLE 3.21   Let f and g be functions from $\mathbf{R}$ to $\mathbf{R}$, defined by the formulas

$$f(x) = 2x$$
$$g(x) = x - 1$$

Then $(g \circ f)(x) = g(f(x)) = g(2x) = 2x - 1$
while $(f \circ g)(x) = f(g(x)) = f(x - 1) = 2(x - 1) = 2x - 2$.   ○

Example 3.21 illustrates that $f \circ g$ and $g \circ f$ are usually not the same, even when both exist and have the same domain and range.

Inverse Functions and Composition

We can now restate the definition of inverse functions in terms of composition. Let $f: X \to Y$ be a one-to-one correspondence with inverse $g = f^{-1}$. Then $g \circ f$ is the identity function on X; that is, $(g \circ f)(x) = x$ for all $x \in X$, and $f \circ g$ is the identity function on Y; $(f \circ g)(y) = y$ for all $y \in Y$. The following example illustrates that both of these conditions are necessary in order that a pair of functions be inverses of one another.

EXAMPLE 3.22   Let $X = \{1, 2, 3, 4\}$, and let $Y = \{5, 6, 7\}$. Define $f: X \to Y$ and $g: Y \to X$ by

$$f(1) = 5 \qquad g(5) = 1$$
$$f(2) = 5$$
$$f(3) = 6 \qquad g(6) = 3$$
$$f(4) = 7 \qquad g(7) = 4$$

Then $f(g(y)) = y$, but $g(f(x)) \neq x$, because $g(f(2)) = 1$, not 2.   ○

In Example 3.22, note that f is onto but not one-to-one, whereas g is one-to-one but not onto.

EXERCISES 3.2

1. Which of the functions shown in Figure 3.7 are one-to-one? Which are onto? Which are both one-to-one and onto? Which are neither?

2. Given $A = \{1, 2, 3, 4, 5\}$, $B = \{a, b, c, d, e\}$, and $f: A \to B$, determine which of the following functions are one-to-one, onto, neither one-to-one nor onto, and both one-to-one and onto.

    a. $f = \{(1, d), (2, a), (3, c), (4, a), (5, b)\}$

    b. $f = \{(1, b), (2, c), (3, a), (4, d), (5, e)\}$

3. Which of the following functions from $\mathbf{N}$ to $\mathbf{N}$ are one-to-one, onto, neither one-to-one nor onto, and both one-to-one and onto?

    a. $f(n) = 2$ **b.** $f(n) = n + 1$ **c.** $f(n) = n^2 + 1$

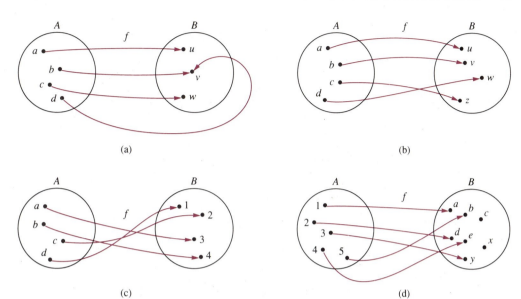

FIGURE 3.7

4. Which of the following functions from **R** to **R** are one-to-one, onto, neither one-to-one nor onto, and both one-to-one and onto?

   a. $f(x) = 2x + 1$

   b. $f(x) = \lfloor x \rfloor$

   c. $f(x) = \dfrac{1}{1 + x^2}$

   d. $f(x) = 2^x$

5. Let f be the factorial function defined in Example 3.6. Is f one-to-one? Is f onto? Explain.

6. Let s be the Fibonacci Sequence defined in Example 3.5. Is s one-to-one? Is it onto?

7. If f is the signum function defined in Exercise 6 of Section 3.1, is f one-to-one? Is it onto?

8. Let f be the function defined in Example 3.10. Is f one-to-one? Is it onto?

9. Define $f: \mathbf{N} \to \mathbf{Z}$ by $f(n) =$ remainder of $\frac{n}{7}$, where $n \in \mathbf{N}$. Is f one-to-one? Is it onto?

10. Let $A = \{a, b, c, d\}$ and $B = \{v, w, x, y\}$. Let the function f be defined by $f = \{(a, x), (b, y), (c, v), (d, w)\}$. Find the inverse of f.

11. Let $A = \{1, 2, 3, 4, 5\}$ and $B = \{a, b, c, d, e\}$. Let the function f be defined by $f = \{(1, d), (2, a), (3, d), (4, b), (5, c)\}$. Does f^{-1} exist? If so, find it.

12. Let $f: \mathbf{N} \to \mathbf{N}$ be defined by $f(n) = 5n$ for $n \in \mathbf{N}$. Does f^{-1} exist? If so, find it.

13. Find the inverse of the function $f(x) = 3x - 1$ from **R** to **R**.

14. Let f and g be functions from **R** to **R** defined by $f(x) = x^3$ and $g(x) = \sqrt[3]{x}$. Show that g is the inverse of f. (*Hint:* Show that $f(g(x)) = x$ and $g(f(x)) = x$.)

15. Let $A = \{1, 2, 3, 4\}$ and $B = \{a, b, c, d, e\}$. Let $f: A \to B$ be defined by $f = \{(1, b), (2, e), (3, d), (4, b)\}$ and $g: B \to A$ be defined by $g = \{(a, 2), (b, 1), (c, 3), (d, 1), (e, 4)\}$. List the ordered pairs in $f \circ g$ and $g \circ f$.

16. Let f and g be functions from $\mathbf{R}$ to $\mathbf{R}$ defined by $f(x) = x + 1$ and $g(x) = x^2$. Find

   a. $(g \circ f)(x)$ **b.** $(f \circ g)(x)$

17. Let f and g be functions from $\mathbf{N}$ to $\mathbf{N}$ defined by $f(n) = n^2 + 1$ and $g(n) = 2n + 5$. Find

   a. $f \circ g$ **b.** $f \circ f$

   c. $g \circ f$ **d.** $g \circ g$

18. Let $f: A \to B$ and $g: B \to C$ be functions that are one-to-one and onto. Show that $g \circ f$ is one-to-one and onto and that $(g \circ f)^{-1} = f^{-1} \circ g^{-1}$.

19. If f and g are functions with range $\mathbf{R}$ and the same domain, we can define arithmetic operations on these functions. If x is an element of the domain of f and g, then the sum $f + g$, the difference $f - g$, and the product fg are defined by $(f + g)(x) = f(x) + g(x)$, $(f - g)(x) = f(x) - g(x)$, and $(fg)(x) = f(x)g(x)$, respectively. If $f(x) = x^2 + 1$ and $g(x) = 1 - x$, find $f + g$, $f - g$, and fg.

20. Show that the function f from $\mathbf{R}$ to $\mathbf{R}$ defined by $f(x) = x^3 + 1$ is one-to-one.

21. Show that if $f: A \to B$ and $g: B \to C$ are both one-to-one, and if $g \circ f$ exists, then $g \circ f$ is one-to-one.

22. Show that if $f: A \to B$ and $g: B \to C$ are both onto, and if $g \circ f$ exists, then $g \circ f$ is onto.

23. Prove the following statement: If A and B are sets containing n elements each and f is a function from A to B, then f is one-to-one if and only if f is onto.

24. Suppose X is a set and $P(X)$ is the power set of X. Define $f: P(X) \to P(X)$ as follows: If A is a subset of X, then $f(A) = X - A$. Is f one-to-one? Is it onto?

25. Let $f: X \to Y$ be a function. Then the **inverse image** of f is a function from $P(Y)$ to $P(X)$ defined as follows: If B is a subset of Y, then the inverse image of B is

$$\{x \in X \mid f(x) \in B\}$$

The inverse image of B under f is denoted by $f^{-1}[B]$. Let $X = \{1, 2, 3, 4\}$ and $Y = \{a, b, c\}$, and let $f: X \to Y$ be defined by $f = \{(1, a), (2, c), (3, c), (4, b)\}$. Find

   a. $f^{-1}[\{a\}]$ **b.** $f^{-1}[\{b, c\}]$

   c. $f^{-1}[\varnothing]$ **d.** $f^{-1}[Y]$

26. Let $f: X \to Y$, and let $B \subset Y$. Show that $f^{-1}[B'] = (f^{-1}[B])'$.

27. Let $f: X \to Y$, and let $A \subset X$. Prove or disprove

$$f[A'] = (f[A])'.$$

28. Let $\mathbf{Q}^+$ be the set of positive rational numbers. Define the function $f: \mathbf{N} \to \mathbf{Q}^+$

by the rules

$$f(1) = 1$$
$$f(2n) = f(n) + 1$$
$$f(2n + 1) = \frac{1}{f(2n)}$$

a. Compute $f(n)$ for $n = 1, 2, 3, \ldots, 15$.

b. Show that f is one-to-one. (*Hint:* Use induction on n to prove the statement, "All of the numbers $f(1), f(2), \ldots, f(n)$ are different.")

c. Show that f is onto. (*Hint:* If $\frac{a}{b}$ is the representation of the rational number x in lowest terms, define $s(x) = a + b$. Use induction on $s(x)$.)

3.3 ● RELATIONS

We have seen that if $f \colon A \to B$ is a function, then the graph of f is a subset of the Cartesian product $A \times B$. Recall that a subset of $A \times B$ is the graph of a function if and only if it satisfies the condition that each element $a \in A$ occurs as the first element in exactly one ordered pair of f.

In general, any subset of $A \times B$ is called a **relation** on the sets A and B. A function, then, is a special type of relation. In the next several sections, we will study subsets of $A \times B$ satisfying various other conditions, but first we will consider relations in general and introduce some important properties of relations.

Definition 3.9

A **relation** consists of a list of sets $A_1, A_2, \ldots, A_n$ and a subset of the Cartesian product $A_1 \times A_2 \cdots \times A_n$. The subset of $A_1 \times A_2 \cdots \times A_n$ is called the **graph** of the relation.

The sets $A_1, \ldots, A_n$ are called the **domains** of the relation: A_1 is the first domain, A_2 is the second domain, and so on. It is customary to use the same symbol to denote both the relation and its graph, and to identify the domains verbally. Thus, we will say simply that R is a relation on sets $A_1, \ldots, A_n$. It is important to remember, however, that a relation is not completely defined unless its domains are specified.

EXAMPLE 3.23 Let $A = \{1, 2, 3\}$, and let $B = \{4, 5, 6\}$. Then the set of ordered pairs $\{(1, 4), (2, 5), (2, 4)\}$ is a relation with domains A and B. This set is not the graph of a function for two reasons: The element 2 occurs as the first element in two ordered pairs, and the element 3 does not occur as the first element in any pair.   ○

A relation on two sets is called a **binary** relation. A relation on three sets is called a **ternary** relation, and in general, a relation on n sets is called an **n-ary** relation. The relation of Example 3.23 is a binary relation on the sets A and B. A function is a special type of binary relation.

If R is a binary relation on the sets A and B, the second domain B is sometimes called the **range** of the relation, in agreement with the terminology for functions. We may also say that R is a relation **from A to B**. If R is a binary relation from A to A, we say simply that R is a binary relation **on A**.

If R is a binary relation on A, it is customary to write $a\ R\ b$ to indicate that $(a, b) \in R$. Examples of this notation are $x < y$, $z \geq w$, and $u = v$, where R is $<$, $\geq$, and $=$, respectively.

EXAMPLE 3.24   The following are binary relations on the set **Z**:

equality, $n = m$: $5 = 5,\ -2 = -2$
inequality, $n \neq m$: $2 \neq 7,\ -3 \neq 0$
is greater than, $n > m$: $18 > 17,\ -8 > -10$
is a divisor of, $n \mid m$: $2 \mid 6,\ 14 \mid -42$ ◦

Relations occur in everyday speech as well as in mathematics. When speaking of people, we use phrases such as "is the father of," "is a brother of," and "works for." These are relations on the set of all persons. In mathematics, the phrases "is greater than," "is less than," and "is equal to," describe relations on numbers. All of these are examples of binary relations. Here are some examples of ternary relations.

EXAMPLE 3.25   The relation of "betweenness" is a ternary relation on the real numbers. If x, y, and z are real numbers, then the ordered triple $(x, y, z) \in \mathbf{R} \times \mathbf{R} \times \mathbf{R}$ is an element of the "betweenness" relation if x is between y and z; that is, if $y < x < z$ or $z < x < y$. ◦

EXAMPLE 3.26   Let S be the set of students at Cabin Creek College, let C be the set of courses offered, and let F be the faculty. List all class enrollments for the current semester. The list may look (in part) like this:

Student	Course	Instructor
Chenier	Math 212	Bartolo
Miller	Math 212	Bartolo
Troyens	Math 212	Tell
Troyens	Music 301	Herring
Padilla	Poli Sci 101	Carlos

The complete list is a ternary relation on the sets S, C, F. The first line in this table indicates that (Chenier, Math 212, Bartolo) is an element of this relation. ◦

Example 3.26 illustrates that not all interesting relations are mathematical in origin. The efficient handling of large relations of the type shown in Example 3.26 is a primary concern of business data processing.

Some Important Properties of Binary Relations

We will now define some important properties that binary relations may possess.

Reflexive Property

Definition 3.10   A binary relation R on a set A is **reflexive** if $a\, R\, a$ for every $a \in A$.

EXAMPLE 3.27   The following relations on the set **Z** are reflexive:

equality, $=$ (for any integer n, $n = n$)
is greater than or equal to, $\geq$ (for any integer n, $n \geq n$)
is a divisor of, $|$ (for any integer n, $n\,|\,n$)   o

If B is the set of all possible bit strings in a 32-bit word and R is the relation "is the one's complement of," then R is not reflexive, because a bit string is not the one's complement of itself.

If R is a binary relation on the set **R** of real numbers, the graph of R is a set of points in the plane. The relation is reflexive if and only if the set of points contains the diagonal line $y = x$, where $x, y \in R$. The following example illustrates this.

EXAMPLE 3.28   Define a binary relation $\approx$ on **R** by the rule: $x \approx y$ if $|x - y| < 1$. We may think of $\approx$ as meaning "is within 1 of." Then $\approx$ is reflexive, because for any $x \in R$, $|x - x| = 0 < 1$. The graph of $\approx$ is shown in Figure 3.8.

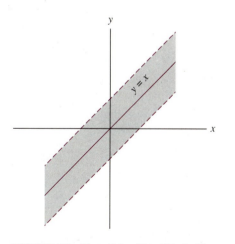

FIGURE 3.8   The relation "is within 1 of"   o

Symmetric Property

Definition 3.11   A binary relation R on a set A is **symmetric** if, whenever $a R b$, then $b R a$.

EXAMPLE 3.29   The following relations on **R** are symmetric:

equality: if $x = y$, then $y = x$
the "is within 1 of" relation of Example 3.28   ○

However, the relation $<$ is not symmetric: 2 is less than 3 but 3 is not less than 2. The relation $\geq$ is not symmetric either.

Geometrically, if R is a symmetric relation on the real numbers **R**, the graph of R remains the same if it is flipped about the line $y = x$. The graph is said to be *symmetric* about the line $y = x$. This is shown in Figure 3.9.

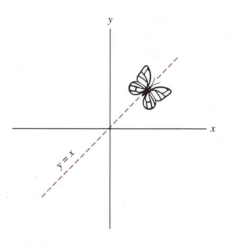

FIGURE 3.9   A symmetric relation on **R**

Transitive Property

Definition 3.12   A binary relation R on a set A is **transitive** if, whenever $x R y$ and $y R z$, then $x R z$.

EXAMPLE 3.30   The following relations of **R** are transitive:

equality   (if $x = y$ and $y = z$, then $x = z$)
the relations $<, >, \geq, \leq$ (if $x < y$ and $y < z$, then $x < z$)   ○

However, the "is within 1 of" relation of Example 3.28 is not transitive: $|1 - 1.5| < 1$ and $|1.5 - 2| < 1$, but $|1 - 2| = 1$, which is not less than 1.

Representations of Binary Relations

There are several ways of defining or describing relations, both on paper and in a computer. We have seen that a binary relation may be described by listing all of its ordered pairs as in Example 3.23. Example 3.26 shows a relation represented in tabular form.

If A and B are small finite sets, or if R is a simple relation on **R**, we may describe a binary relation from A to B by plotting its graph, as we would with a function. Figure 3.10 plots the graph of the relation of Example 3.23.

If R is a binary relation on a set A, we may describe R by means of a figure in which the elements of A are represented by points and the ordered pairs of R are represented by arrows. We draw an arrow from point a to point b to indicate that $(a, b) \in R$. Figure 3.11 illustrates a relation on the set $A = \{1, 2, 3, 4, 5, 6\}$ represented in this way.

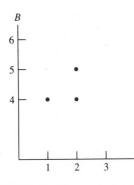

FIGURE 3.10   The graph of the relation of Example 3.23

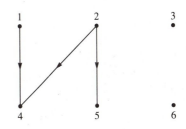

FIGURE 3.11   A relation described by a directed graph

This type of figure is called a **directed graph,** or **digraph.** The word *directed* is used because the arrow indicates a direction on each line connecting two points. The arrow from 1 to 4 represents the ordered pair (1, 4), and so on.

EXAMPLE 3.31   Let $A = \{a, b, c, d\}$, and let R be a binary relation on A. If $R = \{(a, a), (a, b), (b, a), (a, d), (d, c), (c, b)\}$, the directed graph for R is shown in Figure 3.12.

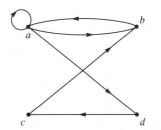

FIGURE 3.12   The relation of Example 3.31

Note that we have used the term **graph** with a new meaning. In all previous cases, a graph was a subset of a Cartesian product of sets. Here, a graph is a set of points, called **vertices,** or **nodes,** connected by lines, called **edges.** In Figures 3.11 and 3.12, the vertices correspond to the elements of the set A, and the edges correspond to the ordered pairs of the relation. The word *graph* is used with both meanings throughout the mathematics and computer science literature. It is necessary for the reader to determine from the context which meaning is intended. Properties of graphs will be studied in detail in Chapter 7.

In the next two sections, we will consider several classes of relations that arise in mathematics and computing. These classes of relations are characterized by the extent to which they do or do not have the reflexive, symmetric, and transitive properties.

EXERCISES 3.3

1. In Example 3.23, how many relations are there from A to B? How many of the relations are functions?

2. Each of the directed graphs in Figure 3.13 represents a binary relation R on the set $A = \{1, 2, 3, 4\}$. List the ordered pairs of R.

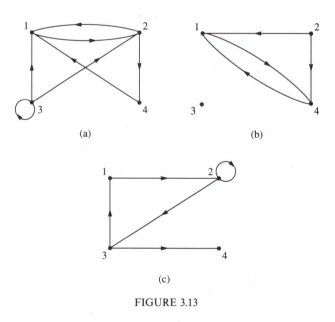

(a)

(b)

(c)

FIGURE 3.13

3. Let $A = \{1, 2, 3, 4, 5\}$. Describe the following relations on A using a directed graph.

   a.   >   **b.**   is a multiple of   **c.**   is within 1 of

4. Which of the following binary relations R are reflexive? Symmetric? Transitive?

   a. A = the set of all females
      R = "is the sister of"

   b. P = the set of lines in the plane
      R = "intersects"

   c. $\mathbf{N}$ = the set of natural numbers
      R = "is a multiple of"

5. Which of the following binary relations R are reflexive? Symmetric? Transitive?

   a. $\mathbf{N}$ = the natural numbers
      R = "have no common factor other than 1"

   b. $\mathbf{Z}$ = the integers
      R = "the difference is divisible by 5"

   c. B = the set of all possible bit strings in a 32-bit word
      R = "is the two's complement of"

6. Let $A = \{a, b, c, d, e\}$. Which of the following binary relations on A are reflexive? Symmetric? Transitive?

   a. $\{(a, b), (a, c), (a, d), (a, e)\}$

   b. $\{(a, b), (b, a), (b, d), (d, b), (a, e), (e, a)\}$

   c. $\{(a, a), (b, b), (c, c), (d, d)\}$

7. Draw each of the relations in the previous exercise as a directed graph.

8. Which of the binary relations described in Exercise 2 are reflexive? Symmetric? Transitive?

9. Give an example of a ternary relation on $\mathbf{R}$.

10. Let A be the set of ordered pairs of natural numbers. Define the relation $\approx$ to be: $(a, b) \approx (c, d)$ if $ad = bc$. Is $\approx$ reflexive? Symmetric? Transitive?

11. Let A be the set of ordered pairs of integers. Define the relation $\blacksquare$ to be: $(a, b) \blacksquare (c, d)$ if $a + d = b + c$. Is $\blacksquare$ reflexive? Symmetric? Transitive?

12. Let R be a relation on the set $\mathbf{Q} - \{0\}$ defined by $(x, y) \in R$ if $\frac{x}{y}$ is an integer. Is R reflexive? Symmetric? Transitive?

13. What is wrong with the following argument that if a binary relation R is symmetric and transitive, it is also reflexive?

   Since R is symmetric, if $a\ R\ b$, then $b\ R\ a$.
   Since R is transitive, if $a\ R\ b$ and $b\ R\ a$, then $a\ R\ a$.
   Thus, R is reflexive.

14. A relation R on a set A is **irreflexive** if for every $a \in A$, $(a, a) \notin R$. Which of the relations in Exercise 2 are irreflexive?

15. Which of the relations in Exercise 6 are irreflexive?

16. Give an example of a relation that is both symmetric and irreflexive.

17. Give an example of a relation that is neither reflexive nor irreflexive.

18. A relation R on a set A is **asymmetric** if whenever $(a, b) \in R$, then $(b, a) \notin R$. Which of the relations in Exercise 2 are asymmetric?

19. Which of the relations in Exercise 6 are asymmetric?

20. Give an example of a relation that is both symmetric and asymmetric.

21. Show that if a relation R is asymmetric, it is not reflexive.

22. Is the converse of Exercise 21 true?

3.4 • EQUIVALENCE RELATIONS

Classes of relations having certain combinations of the reflexive, symmetric, and transitive properties arise frequently. The simplest of these classes consists of relations having all three properties.

Definition 3.13

> An **equivalence relation** on a set A is a binary relation on A that is reflexive, symmetric, and transitive.

EXAMPLE 3.32   Let R be the relation on $\mathbf{Z} \times \mathbf{N}$ defined by

$$(a, b) \, R \, (c, d) \qquad \text{if } ad = bc$$

We will verify that R is an equivalence relation.

Reflexivity: $(a, b) \, R \, (a, b)$ means $ab = ba$, which is true, because multiplication is commutative.

Symmetry: If $(a, b) \, R \, (c, d)$, it must be true that $(c, d) \, R \, (a, b)$. The first statement means that $ad = bc$, and the second means that $cb = da$. Again, the statements are equivalent, because multiplication is commutative and equality is symmetric.

Transitivity: Suppose that $(a, b) \, R \, (c, d)$ and $(c, d) \, R \, (e, f)$. Then $ad = bc$ and $cf = de$. Multiplying the first equation by f, we obtain

$$adf = bcf$$

Since $cf = de$, we have

$$adf = bde$$

Dividing both sides by d (we can do this since $d \in \mathbf{N}$ and therefore cannot be zero) yields

$$af = be$$

which means that $(a, b) \, R \, (e, f)$.   ○

The significance of the relation R in Example 3.32 is as follows: Think of the pair (a, b) as the fraction $\frac{a}{b}$. Then $\frac{a}{b} \, R \, \frac{c}{d}$ if and only if the fractions $\frac{a}{b}$ and $\frac{c}{d}$ represent the same rational number.

EXAMPLE 3.33   Let $A = \{a, b, c, d, e\}$, and let $R = \{(a, a), (a, b), (b, a), (b, b), (c, c), (c, d), (d, c), (d, d), (e, e)\}$. Then it can be checked directly that R is reflexive, symmetric, and transitive. (Reflexive and symmetric are clear, and verification of transitivity is left as an exercise.)   ○

The equivalence relation R of Example 3.33 is represented by the directed graph in Figure 3.14.

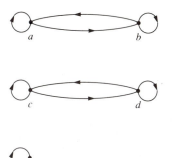

FIGURE 3.14   The relation of Example 3.33

Partitions

You can see from Figure 3.14 that R separates the set A into three subsets, or classes: $\{a, b\}$; $\{c, d\}$; and $\{e\}$. We say that the set A is **partitioned** into these subsets. This partitioning is a characteristic of equivalence relations.

Definition 3.14

Let A be a nonempty set. A **partition** of A is a set of subsets $A_1, A_2, \ldots, A_n$ of A with the following properties:

   (i)   Each A_i is nonempty
   (ii)   The intersection of any two different A_i's is empty
   (iii) $A_1 \cup A_2 \cup \cdots \cup A_n = A$

Figure 3.15 illustrates a partition of a set A.

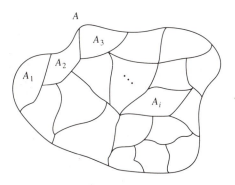

FIGURE 3.15   A partition of the set A

EXAMPLE 3.34   Let $A = \{1, 2, 3, 4, 5, 6, 7\}$. Then $\{\{1, 3, 4, 5\}, \{2\}, \{6, 7\}\}$ is a partition of A.   o

A partition $A_1, \ldots, A_n$ of a set A defines an equivalence relation on A in a very simple manner: Two elements of A are related (equivalent) if and only if they are in the same subset. Conversely, an equivalence relation on A defines a partition of A. This is proved in the following theorem.

Theorem 3.3   If R is an equivalence relation on the set A, then there is some partition $A_1, \ldots, A_n$ of A such that R is the equivalence relation defined by this partition.

Proof   Let R be an equivalence relation on A. For each $a \in A$, define $S_a \subset A$ by the rule

$$S_a = \{x \in A \,|\, (a, x) \in R\}$$

That is, S_a is the set of all elements related to a. We wish to show that the set of all sets S_a, $a \in A$, is a partition of A, so we must verify properties (i), (ii), and (iii) of Definition 3.14.

(i) Each set S_a is nonempty: Since R is reflexive, $(a, a) \in R$ and, therefore, $a \in S_a$. Therefore, S_a is nonempty.

(ii) If two of our sets are different (that is, $S_a \neq S_b$), then $S_a \cap S_b = \varnothing$. To see this, suppose that $S_a \cap S_b \neq \varnothing$. Then there is at least one element $x \in S_a \cap S_b$. This means that $(a, x) \in R$ and $(b, x) \in R$. Since R is symmetric and transitive, it follows that $(a, b) \in R$. Now let y be any element of S_a. Then $(a, y) \in R$. Again, since R is symmetric and transitive, $(b, y) \in R$, so $y \in S_b$. Thus, $S_a \subset S_b$. A similar argument shows that any $y \in S_b$ is also an element of S_a, so $S_b \subset S_a$. So S_a and S_b are equal, contradicting our assumption.

(iii) The union of the sets S_a is all of A. This is true because if z is any element of A, then $z \in S_z$.

It is now apparent from the definition of the sets S_a that R is the equivalence relation defined by the partition we have constructed.   ●

Equivalence Classes

Theorem 3.3 shows that every equivalence relation partitions a set into subsets; the subsets are called the **equivalence classes** defined by the relation. In Example 3.33, the equivalence classes are the sets $\{a, b\}$, $\{c, d\}$, and $\{e\}$.

EXAMPLE 3.35   Define a binary relation S on $\mathbf{Z} \times \mathbf{Z}$ by the rule

$$(x, y)\, S\, (z, w) \qquad \text{if } x + w = y + z$$

Then $(x, y)\, S\, (z, w)$ if and only if $x - y = z - w$. From this observation, it is easy to check that S is an equivalence relation. The equivalence classes of S are shown in Figure 3.16. An equivalence class of S consists of all the points with integer coordinates on one diagonal line.   o

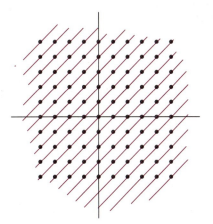

FIGURE 3.16   The equivalence classes of Example 3.35

EXAMPLE 3.36   Let m be an integer greater than 1. We will define a relation on **Z**, called **congruence modulo m.** We write

$$x \equiv y \,(\mathrm{mod}\ m)$$

and say "x is congruent to y mod m." This means that $x - y$ is a multiple of m. For instance,

$$7 \equiv 1 \,(\mathrm{mod}\ 3) \qquad \text{because } 7 - 1 = 6 \text{ is a multiple of } 3$$
$$-8 \equiv 12 \,(\mathrm{mod}\ 5) \qquad \text{because } (-8) - 12 = -20 \text{ is a multiple of } 5$$

It is easy to verify that congruence is an equivalence relation. It is also easy to see that two integers are congruent modulo m if and only if they have the same remainder on division by m. For instance, both 7 and 1 leave remainder 1 on division by 3, and both -8 and 12 leave remainder 2 on division by 5. The equivalence classes for the relation "congruent modulo 5" are shown in Figure 3.17, where the points on each horizontal line make up one equivalence class.

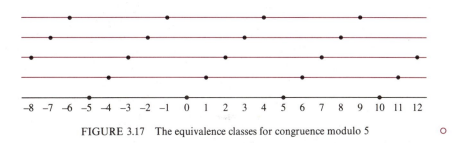

FIGURE 3.17   The equivalence classes for congruence modulo 5   ○

Note that the number system $\mathbf{Z}_m$ introduced in Chapter 1 can be defined as the set of equivalence classes of integers modulo m. The symbol $[a]_m$ (or just $[a]$, if there is no ambiguity) denotes the set of integers congruent to a modulo m. We can write out the members of the equivalence classes in $\mathbf{Z}_m$ as follows:

$$[0] = \{0, m, -m, 2m, -2m, 3m, -3m, \ldots\}$$
$$[1] = \{1, m + 1, -m + 1, 2m + 1, -2m + 1, \ldots\}$$
$$[2] = \{2, m + 2, -m + 2, 2m + 2, -2m + 2, \ldots\}$$
$$\vdots$$
$$[m - 1] = \{m - 1, m + m - 1, -m + m - 1, 2m + m - 1, -2m + m - 1, \ldots\}$$

This list contains all the elements of $\mathbf{Z}_m$. To see this, let $[k] \in \mathbf{Z}_m$, and let r be the remainder on division of k by m. Then $0 \le r < m$, and $[k] = [r]$.

EXAMPLE 3.37   Let $f: A \to B$ be a function. Define a relation R on A by the following rule: $a_1 \, R \, a_2$ if $f(a_1) = f(a_2)$. Then R is an equivalence relation. If $a \in A$ and $f(a) = b$, then the equivalence class containing a is the set $f^{-1}[b] = \{a \in A \mid f(a) = b\}$. The set $f^{-1}[b]$ is called the **fiber** of f **through** a, or the fiber **over** b. If f is a real-valued function of real variables, the fibers are also called **level sets.**   ○

Closure Operations

If R is a relation on a set A, we may ask whether R can be built up to an equivalence relation by adding more ordered pairs. Certainly there are equivalence relations on A that contain R; in fact, $A \times A$ itself is such an equivalence relation. Rather than choose $A \times A$, we would like to find the smallest equivalence relation on A that contains R. The following example shows how this can be done.

EXAMPLE 3.38   Let $X = \{a, b, c, d, e, f\}$, and let R be the relation defined by the directed graph in Figure 3.18(a). To build a reflexive relation, add an arrow from each element to itself. The result is the graph (b). To make the relation symmetric, add the reverse of each arrow; the result is graph (c). Finally, to make the relation transitive, add the arrows between points b and d. The final result is the equivalence relation in graph (d).   ○

The operation of finding the smallest set containing a given set and having some property is very common in mathematics. Such operations are frequently called **closure** operations. The three steps we have applied to R may be referred to as the **reflexive closure, symmetric closure,** and **transitive closure** operations. These closures can be formed for any relation. To form the reflexive closure of a binary relation R on A, add all pairs (a, a). To form the symmetric closure of R, add the reversed pair (b, a) for every pair $(a, b) \in R$. The transitive closure operation is more difficult to describe. If R is a relation on A, then (a, b) is in the transitive closure of R if there is a finite sequence $a_1, \ldots, a_n$ of elements of A such that $a_1 = a$, $a_n = b$, and each pair (a_i, a_{i+1}) of adjacent members of the sequence belongs to R. The transitive closure operation will also appear in the next section, where we consider relations that are not symmetric.

EXAMPLE 3.39   Let R be the relation on $\mathbf{Z}$ defined by the following rule: $x \, R \, y$ if $x - y = 3$. Then the reflexive closure of R is defined by the rule: $x - y = 3$ or $x - y = 0$. The symmetric

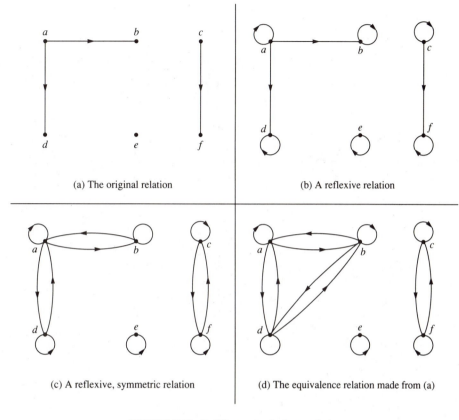

(a) The original relation

(b) A reflexive relation

(c) A reflexive, symmetric relation

(d) The equivalence relation made from (a)

FIGURE 3.18   Building an equivalence relation

closure is defined by the rule: $x - y = 3$ or $x - y = -3$. The transitive closure of R is defined by the rule: $x - y$ is a positive multiple of 3. The equivalence relation generated by R is congruence modulo 3.   ○

EXERCISES 3.4

1. List the 17 cases that must be checked to verify that the relation of Example 3.33 is transitive.

2. Define the relation & on $\mathbf{Z} \times \mathbf{Z}$ by the rule

    $$(x, y) \,\&\, (z, w) \qquad \text{if } xw = yz$$

    Determine whether & is an equivalence relation.

3. Verify in detail that the relation of Example 3.35 is an equivalence relation.

4. Verify algebraically that congruence (Example 3.36) is an equivalence relation.

5. Let $X = \{1, 2, 3, 4, 5, 6, 7\}$ be partitioned into the subsets $\{1, 3, 5\}$, $(2, 4, 7\}$, and $\{6\}$. Illustrate the corresponding equivalence relation by a directed graph.

6. Define a relation # on $\mathbf{R} \times (\mathbf{R} - \{0\})$ by the rule: $(x, y) \# (z, w)$ if $xw = yz$. Show that # is an equivalence relation. What are the equivalence classes? Describe them geometrically.

7. Describe the level sets of each of the following functions

   a. $f(x) = |x|$ b. $f(x) = \lfloor x \rfloor$

   c. $f(x) = \lceil x \rceil$ (See Exercise 15, Section 3.1, for the definition of $\lceil \ \ \rceil$.)

   d. $f(x, y) = 3x - 2y$ e. $f(x, y) = x^2 + y^2$

8. If $f: X \rightarrow Y$ is a one-to-one function, what are the fibers of f?

9. Illustrate the reflexive closure, symmetric closure, and transitive closure of each of the relations in Figure 3.19.

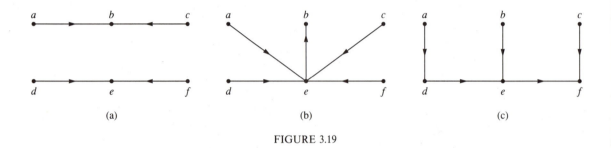

(a)   (b)   (c)

FIGURE 3.19

10. Illustrate the smallest equivalence relation containing each of the relations in Figure 3.20.

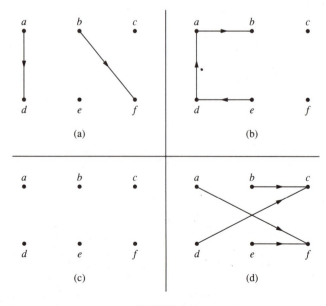

FIGURE 3.20

11. If A is the set of all persons and R means "is the child of," find the transitive closure of R.

12. If $A = \{a, b, c, d\}$ and $R = \{(a, b), (c, b), (a, c), (d, d)\}$, find the reflexive closure of R.

13. Find the symmetric closure of the relation R described in the previous exercise.

14. Find the transitive closure of the relation R described in Exercise 12.

15. Describe the equivalence classes for the relation "congruent modulo 0."

16. If B is the set of all bit strings in a 32-bit word and R means "is the one's complement of," what is the smallest equivalence relation containing R? Describe the equivalence classes of this equivalence relation.

3.5 • ORDER RELATIONS

In the last section, we studied relations that have the reflexive, symmetric, and transitive properties. Yet many interesting mathematical relations are characterized by a distinct lack of symmetry. The relations $<, \leq, >$, and $\geq$ on real numbers, and the relation $\subset$ on sets, are such relations. For instance, $x < y$ and $y < x$ are inconsistent statements. The statements $x \leq y$ and $y \leq x$ can be true only if $x = y$. Also, if A and B are sets with $A \subset B$ and $B \subset A$, then $A = B$. These facts suggest the following definition.

Definition 3.15

A relation R on a set A is **antisymmetric** if $a\ R\ b$ and $b\ R\ a$ imply $a = b$.

With this terminology, the relations $\geq$ and $\leq$ on real numbers and the relation $\subset$ on sets are antisymmetric.

EXAMPLE 3.40   On the set $\mathbf{N}$ of natural numbers, consider the divisibility relation $|$, defined as follows: $n \mid m$ if m is a multiple of n. Then $|$ is antisymmetric, because on $\mathbf{N}$, if $n \mid m$, then $n \leq m$. However, the divisibility relation on $\mathbf{Z}$ is not antisymmetric: $3 \mid -3$ and $-3 \mid 3$, but 3 is not equal to -3. ○

Partial and Total Order Relations

We will now introduce two classes of relations that possess the antisymmetric property.

Definition 3.16

A **partial order relation** on a set A is a relation on A that is reflexive, anti-symmetric, and transitive. If a partial order relation is defined on A, then A is said to be a **partially ordered set,** or **poset.**

The term **partial** is used to indicate that it is not necessary for any two elements of the set to be related in some way. For instance, in the set **N** with the divisibility relation, neither of the statements $2 \mid 3$ and $3 \mid 2$ is true.

Definition 3.17

A **total order** relation R on a set A (also called a **linear order** relation) is a partial order relation having this additional property: If a and b are elements of A, then either $a \, R \, b$ or $b \, R \, a$.

EXAMPLE 3.41   On the set **N**, the relations $\leq$ and $\geq$ are partial order relations that are also total order relations, but the relation $\mid$ is a partial order relation that is not a total order relation.

If X is a set, the relation $\subset$ on the subsets of X is a partial order relation that is not a total order.   ○

A total order relation on a finite set corresponds to writing the set as a finite sequence.

Note that the relation $<$ is not a partial order relation according to this definition, because it is not reflexive. However, this is not a real limitation. If R is a relation that we would like to think of as imposing some order on a set A, we can form the reflexive closure of R and work with it instead. In fact, the reflexive closure of $<$ is just $\leq$.

Graphic Representation of Order Relations

Let A be the set of positive divisors of 6 (that is, $A = \{1, 2, 3, 6\}$), and consider the divisibility relation $\mid$ on A. Figure 3.21 is the directed graph representing $\mid$. There are nine arrows, one for each ordered pair in the relation. Since we know that divisibility in A is a partial order relation, five of these arrows show no useful information: The four arrows pointing from a number to itself tell us only that the relation is reflexive, which we know, and the arrow from 1 to 6 is a consequence of the transitive property. So we may adequately illustrate the relation by omitting them. Furthermore, if we assume that all arrows point upward, we can omit the arrowheads. Using this convention, the result is Figure 3.22. The relation depicted in Figure 3.22 has only four ordered pairs, and its reflexive and transitive closure is the relation of Figure 3.21.

A diagram that contains all the information needed to define a partial order R, and that contains no unnecessary information, is called a **Hasse diagram** for R. The diagram in Figure 3.22 is a Hasse diagram for the relation of Figure 3.21. Figure 3.23 is a Hasse diagram for the divisibility relation on the divisors of 30. The following sequences define total order relations on the set $\{@, \#, \$, \%\}$:

   a. $@, \#, \$, \%$
   b. $\%, \$, \#, @$
   c. $\$, @, \#, \%$

The Hasse diagrams of these total order relations are shown in Figure 3.24.

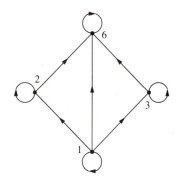

FIGURE 3.21   The divisors of 6

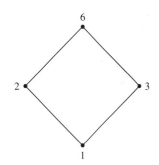

FIGURE 3.22   The divisors of 6, simplified

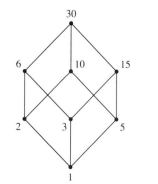

FIGURE 3.23   The divisors of 30, simplified

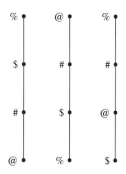

FIGURE 3.24   Three total orders on {@, #, $, %}

EXERCISES 3.5

1. Draw the complete directed graph of the divisibility relation on the divisors of 30. (Start with Figure 3.23 and draw arrows to form the reflexive and transitive closures.)

2. Draw the complete directed graph of the total order relation of Figure 3.24(c).

3. Which of the relations shown in Figure 3.25 are antisymmetric?

4. Consider the reflexive, transitive closures of the relations illustrated in Figure 3.26. Which are partial order relations? Which are total order relations?

5. Refer to Exercise 4 of Section 3.3 Which of the relations are partial order relations? Total order relations?

6. Let X be a set, and let R be the relation $X \times X$ on X. That is, $x\,R\,y$ for every x and y in X. For which sets X is R antisymmetric?

7. Show that the empty relation $\varnothing$ on a set X (that is, $x\,R\,y$ is false for every x and y in X) is antisymmetric.

8. Show that every subset of an antisymmetric relation is antisymmetric.

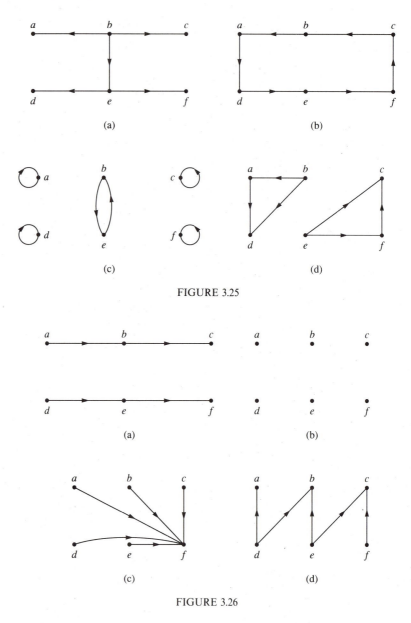

FIGURE 3.25

FIGURE 3.26

9. Explain why "antisymmetric closure" is not a meaningful concept.

10. Show that every asymmetric relation is antisymmetric. (See Exercise 18 of Section 3.3 for the definition of *asymmetric*.)

11. Give an example of a relation that is antisymmetric but not asymmetric.

12. Give an example of a relation that is both antisymmetric and irreflexive. (See Exercise 14 of Section 3.3 for the definition of *irreflexive*.)

13. Show that a relation that is both transitive and irreflexive is asymmetric.

3.6 • RELATIONAL DATA BASES

In Example 3.26, we looked at a relation on the sets of students, courses, and instructors at Cabin Creek College. The relation listed students, the courses being taken, and the instructors. Relations of this type are the subject matter of business data processing. A collection of relations having a common subject matter is often called a **data base.** Computer programs to handle data bases are called **database management systems.** The efficient organization of large volumes of information is critical to the design and use of such programs.

In this section, we will examine some of the issues related to management of large relations. In so doing, we will return to the concept of function with which we started this chapter.

Consider again the relation of Example 3.26. The relation may be thought of as a table, part of which follows:

Student	Course	Instructor
Chenier	Math 212	Bartolo
Miller	Math 212	Bartolo
Troyens	Math 212	Tell
Troyens	Music 301	Herring
Padilla	Poli Sci 101	Carlos

Each line in the table (that is, each n-tuple of the relation) is called a **record** in data processing terminology. The columns correspond to the domains of the relation. Each column is called a **field.** This table contains some (but not all) of the information that the registrar's office must maintain. Other items include course grades, classrooms, and students' addresses. Since different students may have the same name, each student is assigned an ID number. If all of this information is built into one relation, it could be arranged as a table with the following column heads:

ID Num.	Name	Address	Course	Room	Instr.	Grade

Unlike purely mathematical relations, business data processing relations change with time. Therefore, the registrar must add, delete, and update (change) records as circumstances change. Let us examine the work involved in doing this.

Change of instructor.   When Professor Tell became ill, his several mathematics classes, all of which met in the same room, were taken over by various colleagues. All occurrences of "Tell" in the instructor field must be changed, but not all in the same way. There is no practical way of determining what changes must be made—we do not have enough information. Thus, we need to add the section number to our relation, so that the column heads become

ID Num.	Name	Address	Course	Section	Room	Instr.	Grade

Now the change of instructor can be made, but the change must be made many times. For instance, if there are 37 students in Tell's section of Math 212, the change must be made in 37 different records. If an error occurs during the change, inconsistent data may be introduced into the relation. For example, two students taking the same section of the same course may have different instructors listed.

A student drops all classes.   All records having the student's ID number can be removed, but now all information about the student is lost, including the name and address.

A student adds a course.   A record must be added to the file, and information already in the file, such as the name of the instructor teaching the section, must be repeated. If this is not done correctly, inconsistent information may again be introduced.

The cause of these problems is the existence of dependencies among the fields of the relation. For instance, the name and address of a student depends only on the student's ID number, not on his course enrollments or grades. The instructor depends only on the course and section, not on the students who are enrolled. These dependencies are, of course, **functions.** It is also true that the instructor is a function of course, section, and room, but this is not useful information; room is a function of course and section, so course, section, and room are not independent.

Thus, the Cabin Creek College data can be efficiently organized as three relations:

Relation 1	Relation 2	Relation 3
*ID number	*Course	*ID number
Name	*Section	*Course
Address	Room	Section
	Instructor	Grade

where the asterisks denote key fields (that is, fields on which the other fields depend). More precisely, let

 I be the set of all ID numbers
 N be the set of all names
 A be the set of all addresses
 C be the set of all courses
 S be the set of all section numbers
 R be the set of all room numbers
 T be the set of all instructors
 G be the set of all grades

Then R_1, relation 1, is a subset of $I \times N \times A$ and is a function from I to $N \times A$. The relation R_2 is a subset of $C \times S \times R \times T$ and is a function from $C \times S$ to $R \times T$. Finally, the relation R_3 is a subset of $I \times C \times S \times G$ and is a function from $I \times C$ to $S \times G$.

The **relational model** is a theory of database management which holds that data is most efficiently represented as a set of relations that are, in fact, functions. Several database management systems based on the relational model are now on the market, such as dBase II + (Aston-Tate), RBase System V (MicroRim), and Accent-R (National Information Systems). Most older database management systems were based on other theories.

EXERCISES 3.6

1. An engineering services contractor provides services to several customers. There are several contracts with each customer, and there are several jobs being worked under each contract. The contractor must bill the customers on the basis of number of hours worked each week, on each job, by skill (engineer, technician, drafter). The following data items must be tracked in the data base:

   Employee SSN
   Employee name
   Employee skill
   Number of hours worked
   Week
   Job number
   Contract number
   Customer name
   Customer address

   Design a set of relations for this data. The relations should be functions.

2. Each rocket in the new interplanetary fleet contains a widget and a transmogrifier. Records on the configuration of each rocket must be kept. The data to be recorded includes:

   Rocket serial number
   Serial number of widget
   Serial number of transmogrifier
   Manufacturer of each widget and transmogrifier
   Manufacturer's address
   Contract number under which each was manufactured
   Test facility where widget or transmogrifier was tested
   Test facility address
   Test date

   Design a set of relations for this data.

3. A museum stores many archeological artifacts and keeps records on each item. The data in the museum's records includes:

   Location found
   Date found
   Culture with which the item is associated
   Period (start and end dates) of that culture

Finder

Finder's current address

Organization sponsoring dig at time of find

Organization's current address

Design a set of relations for this data.

Computer Exercises for Chapter 3

1. Write a program to compute the value of $n!$, where n is entered by the user. ($n!$ is defined in Example 3.6.)

    a. Use a nonrecursive function to compute $n!$.

    b. Use a recursive function to compute $n!$.

2. Write a program to print the first n numbers in the Fibonacci Sequence defined in Example 3.5. Let the user specify n.

3. Write a program to evaluate a polynomial function at a point x. Use Horner's method, described in Exercise 19 of Section 3.1.

4. A classic example of a function defined recursively is Ackermann's function: $A: \{0, 1, 2, \ldots\} \times \{0, 1, 2, \ldots\} \to \mathbf{N}$ defined by

$$A(m, n) = \begin{cases} n + 1 & \text{if } m = 0 \\ A(m - 1, 1) & \text{if } m \neq 0 \text{ and } n = 0 \\ A(m - 1, A(m, n - 1)) & \text{if } m \neq 0 \text{ and } n \neq 0 \end{cases}$$

    Write a program to print the value of Ackermann's function for nonnegative integers m and n.

5. Find out how your computer responds to exception conditions such as computing the square root of a negative number or division by zero. Does your computer provide a means for the programmer to specify what action is to be taken on an exception condition? If so, how does it work?

6. Write a program to read in a list of ordered pairs defining a binary relation on the set $X = \{1, 2, \ldots, 10\}$ and store the information as a 10×10 array.

7. Write a program to determine whether a relation on the set $X = \{1, 2, \ldots, 10\}$ is (a) reflexive, (b) symmetric, (c) antisymmetric, or (d) transitive. Begin by reading in a list of ordered pairs and using the program of the previous exercise.

8. Using a relational database management system available on your computer, implement the Cabin Creek College registrar's database (see Section 3.6). Invent more data records and add them to the database.

9. Use a relational database management system to implement the database of Exercise 1, Section 3.6. Invent data for each of the relations.

10. Use a relational database management system to implement the database of Exercise 2, Section 3.6. Invent data for each of the relations.

11. Use a relational database management system to implement the database of Exercise 3, Section 3.6. Invent data for each of the relations.

• • • • CHAPTER REVIEW EXERCISES

1. Given the domain $\{-1, 0, \frac{1}{3}, 2\}$ and the formula $f(x) = x^3 - 1$, list the ordered pairs of f.

2. Define $f: \mathbf{N} \to \mathbf{N}$ as follows: For $n \in \mathbf{N}$, represent n in octal and add the digits. Thus,
$$f(83) = f(123_{\text{eight}}) = 1 + 2 + 3 = 6_{\text{eight}} = 6$$
Find
   a. $f(1024)$
   b. $f(1234)$

3. What is the domain and image for the function f whose graph is $\{(1, 1), (2, 4), (3, 9), (5, 25)\}$?

4. Determine the image of the sequence $1, 4, 9, 16, 25, \ldots$. Give a formula for the nth term.

5. $A = \{1, 2, 3\}$ and $B = \{a, b, c, d\}$. Define a one-to-one function from A to B. Is there a function from A to B that is onto?

6. Let $A = \{1, 2, 3, 4, 5\}$ and $B = \{a, b, c, d, e\}$. Define a function f from A to B that is
   a. Neither one-to-one nor onto
   b. One-to-one
   c. Is there a one-to-one function from A to B that is not onto? Explain.

7. If $A = \{1, 2, 3, 4\}$ and $B = \{a, b, c\}$, determine whether the following functions from A to B are one-to-one, onto, neither one-to-one nor onto, or both one-to-one and onto.
   a. $\{(1, a), (2, a), (3, b), (4, b)\}$
   b. $\{(1, a), (2, c), (3, b), (4, a)\}$

8. If $A = \{0, 1, 2, 3, 4, \ldots\}$ define a one-to-one and onto function $f: A \to \mathbf{N}$.

9. Given $f: [0, 1] \to [0,1]$ defined by
$$f(x) = \begin{cases} x, & \text{if } x \text{ is an integer} \\ \frac{1}{2}, & \text{otherwise} \end{cases}$$
   a. Is f one-to-one?
   b. Is f onto?

10. If $A = \{a, b\}$ and $B = \{1, 2, 3\}$, how many different functions are there from A to B? How many are one-to-one? How many different functions are there from B to A? How many are one-to-one?

11. Find the inverses of each of the following one-to-one functions.
   a. $\{(a, u), (b, w), (c, z), (d, x)\}$
   b. $\{(-1, 1), (0, 0), (1, -1)\}$

12. If $f: \mathbf{R} \to \mathbf{R}$ is defined by $f(x) = 5x + 2$ and $g: \mathbf{R} \to \mathbf{R}$ is defined by $g(x) = \frac{x-2}{5}$, show that g is the inverse of f.

13. Given $f: \mathbf{N} \to \mathbf{N}$ defined by $f(n) = 2n$, does f have an inverse? If so, find it.

14. Show that $f: \mathbf{R} \to \mathbf{R}$ defined by $f(x) = 2x - 9$ is one-to-one.

15. Give an example of a relation that is not a function.

16. Give an example of a binary relation on a set A that is
   a. Not reflexive, not symmetric, and not transitive
   b. Reflexive but neither symmetric nor transitive
   c. Symmetric but neither reflexive nor transitive
   d. Transitive but neither reflexive nor symmetric

17. Give an example of a binary relation on a set A that is

   a. Reflexive and symmetric but not transitive

   b. Reflexive and transitive but not symmetric

   c. Symmetric and transitive but not reflexive

   d. An equivalence relation

18. Draw the directed graph to represent the relation R on A where $A = \{1, 2, 3, 4, 5\}$ and $R = \{(1, 2), (1, 3), (2, 2), (2, 4), (4, 5)\}$.

19. Define the relation R on $\mathbf{Z}$ to be: $a\ R\ b$ if $a - b$ is prime. Is R reflexive? Symmetric? Transitive?

20. Define the relation R on $\mathbf{Z} \times \mathbf{Z}$ to be: $(a, b)\ R\ (c, d)$ if $a - c = b - d$. Is R reflexive? Symmetric? Transitive?

21. Let $A = \{a, b, c, d, e, f\}$ be partitioned into the subsets $\{a, b\}$, $\{c, e, f\}$, and $\{d\}$. Illustrate the corresponding equivalence relation by a directed graph.

22. If $A = \{a, b, c, d\}$, find the transitive closure of the relation R defined to be

$$\{(a, a), (a, b), (a, d), (b, c), (c, c)\}$$

23. If B is the set of all possible bit strings in an 8-bit word and R means "is the two's complement of," what is the smallest equivalence relation containing R? Describe its equivalence classes.

24. Draw the complete directed graph for the divisibility relation on the divisors of 24.

25. Draw the simplified directed graph for the relation in the previous exercise.

26. Give an example of a

   a. Partially ordered set   **b.** Totally ordered set

27. Let X be a nonempty set, and let the binary relation R on X be the empty set. Is R reflexive? Symmetric? Transitive? Antisymmetric? Asymmetric? Irreflexive? What if X is the empty set?

28. In an urban school district with mandatory busing, each neighborhood is assigned a school. All students are bused. In each neighborhood, there are several pickup points. Each student is picked up at the pickup point in her neighborhood nearest her home. Each bus picks up the students from certain pickup points and goes to one school. There may be several buses on each route. The data items handled by the school administration include:

     Student name
     Neighborhood pickup point
     School name
     Route number
     Bus number

   Design a set of relations for this data.

29. Let $f: X \to Y$ and $g: Y \to X$ be one-to-one functions. Construct a one-to-one correspondence between X and Y. (*Hint:* Call a an nth order ancestor of b if b can be obtained from a by applying f and g alternately with n applications. For instance, if $a = f(g(f(b)))$, then a is a third-order ancestor of b. If the oldest ancestor of a is an ancestor of even order, then a is of even ancestry; similarly for odd ancestry. If a has infinitely many ancestors, then a is of infinite ancestry. Classify the elements of X and Y according to type of ancestry.)

30. Show that there exists a one-to-one correspondence between $(0, 1]$ and $(0, 1] \times (0, 1]$.

Part II

BASIC DISCRETE MATHEMATICS

4
LOGIC
AND PROOF

The purpose of mathematical logic is to provide a model of ordinary human reasoning, or "common sense." By *common sense* we mean the generally accepted standards of what constitutes a valid argument. These standards are the same in all languages and all cultures. Our goal in studying logic is to reduce the standards to precise rules. Ideally, we would be able to program a computer to test an argument and determine whether it is valid.

There are two levels of the study of logic. The first level, called *propositional logic,* deals with relationships among whole sentences. The second level, called *predicate logic,* considers the internal structure of sentences.

4.1 • PROPOSITIONS AND CONNECTIVES

The sentences that we study in this chapter are called *propositions.*

Definition 4.1   A **proposition** is a declarative sentence that is either true or false.

EXAMPLE 4.1   The following statements are propositions:

Tetanus is a disease.
$\frac{1}{2}$ is an integer.
There is intelligent life on Mars.
One plus one equals two.   ○

The first and last propositions are true, the second is false. Although the truth or falsity of the third is unknown, it makes sense to say that it is either true or false. However, the statement "This is green," by itself, is not a proposition unless it is clear from some context what object is being discussed. Similarly, "x is a prime number" would be a proposition only if the value of x were specified.

We denote propositions by lowercase letters p, q, r, s, etc., called **propositional variables.**

Truth Values

In our study of propositions, we will take the view that the only interesting thing about a proposition is whether it is true or false. We will call the uppercase letters T and F **truth values.** If a proposition is true, we say that it has truth value T, and if it is false, we say that it has truth value F.

EXAMPLE 4.2   The truth values of the first two propositions of Example 4.1 are T and F, respectively.   ○

Connectives

Natural languages provide a number of means by which sentences may be combined or altered to form new sentences. These means are called **connectives** and are expressed by words such as "and," "or," and "not." The most common connectives are negation ("not"), conjunction ("and"), disjunction ("or"), implication ("if . . . then"), and equivalence ("if and only if"). Negation is a unary operation on the set of propositions—that is, a function from the set of propositions to itself. The other connectives named are binary operations on the set of propositions. Connectives are also called **logical operations.**

Negation, Conjunction, and Disjunction

Negation, usually accomplished by incorporating the word "not" into the sentence, is denoted by the symbol ¬. The conjunction of two propositions, formed in English by use of the word "and", is denoted by the symbol ∧. The disjunction of two propositions is formed in English with the word "or" and in logic with the symbol ∨.

We have said that for purposes of propositional logic, the only significant thing about a proposition is its truth value. We take the same attitude toward connectives: To characterize these connectives, we specify how the truth value of the result of using "and," "or," or "not" is obtained.

Definition 4.2   The connectives **negation, conjunction,** and **disjunction** are defined by the following tables:

Negation		Conjunction			Disjunction		
p	$\neg p$	p	q	$p \wedge q$	p	q	$p \vee q$
T	F	T	T	T	T	T	T
F	T	T	F	F	T	F	T
		F	T	F	F	T	T
		F	F	F	F	F	F

The tables in Definition 4.2 show how the truth values of $\neg p$, $p \wedge q$, and $p \vee q$ are obtained from the truth values of p and q. Such tables are called **truth tables.**

EXAMPLE 4.3   The negations of the first two propositions of Example 4.1 are "Tetanus is not a disease" and "$\frac{1}{2}$ is not an integer," respectively.   ○

EXAMPLE 4.4   The conjunction of the first two propositions of Example 4.1 is "Tetanus is a disease and $\frac{1}{2}$ is an integer."

The disjunction of the last two propositions is "There is intelligent life on Mars or one plus one equals two."   ○

In English, the word "or" has two different meanings:

1. **inclusive "or"**—meaning "one or the other or both" and sometimes written "and/or"
2. **exclusive "or"**—meaning "one or the other but not both"

The ambiguity is in the English language, not in logic. Latin has two words: *vel,* from which the symbol $\vee$ is derived, means inclusive "or," and *aut* means exclusive "or."

The agreement among mathematicians is that "or" as used in mathematics will always mean "inclusive 'or.'" Therefore, the disjunction of two true propositions is true.

The exclusive "or" operator is denoted by $\underline{\vee}$. Writing the truth table for $\underline{\vee}$ is left as an exercise.

Implication

The connective **implication** corresponds to a variety of English expressions, including:

p implies *q*	only if *q* is *p*
If *p*, then *q*	*q* is necessary for *p*
q if *p*	*p* is sufficient for *q*
p only if *q*	

Definition 4.3   Implication is denoted by the symbol $\Rightarrow$. A proposition in the form $p \Rightarrow q$ is called a **conditional proposition,** or a **conditional.** The expression $p \Rightarrow q$ is usually read "*p* implies *q*" or "If *p*, then *q*." The proposition *p* is called the **hypothesis** or **antecedent** or **sufficient condition,** and *q* is called the **conclusion** or **consequent** or **necessary condition.** Implication is defined by the following truth table:

p	q	$p \Rightarrow q$
T	T	T
T	F	F
F	T	T
F	F	T

To see the reasonableness of this definition of implication, consider the following examples.

EXAMPLE 4.5   Let p be "$2 + 1 = 3$" and q be "$(2 + 1) + 5 = 3 + 5$." The implication $p \Rightarrow q$ is "If $2 + 1 = 3$, then $(2 + 1) + 5 = 3 + 5$."

Both the hypothesis and conclusion are true, and we can see that $p \Rightarrow q$ is also true (by adding 5 to both sides of the equation).   ○

EXAMPLE 4.6   Let p be "$1 = 2$" and q be "$0 = 0$." Then $p \Rightarrow q$ is "If $1 = 2$, then $0 = 0$" and has truth value T. (To convince yourself of this, multiply both sides of $1 = 2$ by 0.)   ○

EXAMPLE 4.7   Let p be "$1 + 1 = 3$" and q be "$(1 + 1) + 5 = 3 + 5$." Here both p and q are false, but $p \Rightarrow q$, which is "If $1 + 1 = 3$, then $(1 + 1) + 5 = 3 + 5$," must be true (since 5 was added to both sides of p).   ○

Notice that in mathematical logic, the use of implication may differ from that in ordinary language. A statement in English, such as "If it rains, the softball game is canceled," indicates a cause-and-effect relationship between hypothesis and conclusion. However, in logic, the implication "If p, then q" may connect any two propositions p and q, even totally unrelated ones. Thus, according to the truth table, the implication "If 4 times 2 equals 8, then Chicago is the largest city in Illinois" is true. If this seems surprising, remember that we are not concerned with the meaning or structure of propositions—only with their truth values.

A more serious objection is that $p \Rightarrow q$ should be true only if q can be deduced from p by following some specified rules of deduction. However, to decide whether q can be deduced from p requires looking at the structure of p and q, not just the truth value. The idea of deducing one thing from another is an important one in mathematical logic (considered in Section 4.7), but it is to be distinguished from implication as defined by our truth table.

Converse, Contrapositive, and Inverse

From a conditional proposition $p \Rightarrow q$, it is possible to form related propositions called the *converse*, the *contrapositive*, and the *inverse* of the conditional.

Definition 4.4   If p and q are propositions, the **converse** of the proposition $p \Rightarrow q$ is the proposition $q \Rightarrow p$. The **contrapositive** of $p \Rightarrow q$ is the proposition $\neg q \Rightarrow \neg p$. The **inverse** of $p \Rightarrow q$ is the proposition $\neg p \Rightarrow \neg q$.

EXAMPLE 4.8   Consider the proposition "If a polygon is a square, then it is a rectangle." Its converse is "If a polygon is a rectangle, then it is a square." The contrapositive is "If a polygon is not a rectangle, then it is not a square." The inverse is "If a polygon is not a square, then it is not a rectangle."   ○

Equivalence

The connective equivalence is stated as "p if and only if q," or "p is necessary and sufficient for q." Equivalence is denoted by the symbol $\Leftrightarrow$; it is the conjunction of $p \Rightarrow q$ and $q \Rightarrow p$.

Definition 4.5

Equivalence is characterized by the following truth table:

p	q	$p \Rightarrow q$	$q \Rightarrow p$	$p \Leftrightarrow q$
T	T	T	T	T
T	F	F	T	F
F	T	T	F	F
F	F	T	T	T

In statements of the form "p if and only if q," $q \Rightarrow p$ is the "if" or "necessary" part, and $p \Rightarrow q$ is the "only if" or "sufficient" part.

EXAMPLE 4.9   Let p be "Triangle ABC has two equal sides" and q be "Triangle ABC has two equal angles." Then $p \Leftrightarrow q$ is a familiar theorem from geometry: "Triangle ABC has two equal sides if and only if triangle ABC has two equal angles."   o

NAND and NOR

Two other connectives are sufficiently important to have common names, although they do not have standard symbols. The connectives are NAND ("not and") and NOR ("not or").

Definition 4.6

The connectives **NAND** and **NOR** are defined as follows:

p	q	p NAND q	p NOR q
T	T	F	F
T	F	T	F
F	T	T	F
F	F	T	T

The expression p NAND q usually appears in sentences in the form "not both p and q." The expression p NOR q usually appears as "neither p nor q."

EXAMPLE 4.10   Suppose p is the proposition "Tetanus is a disease" and q is the proposition "Euclid is the author of *Proof*." Then p NAND q could be written, "It is not true that both tetanus is a disease and Euclid is the author of *Proof*." The proposition p NOR q could be written "Tetanus is not a disease, nor is Euclid the author of *Proof*."   o

The connectives NAND and NOR are important in electronic circuit design, discussed in Chapter 5.

How a Computer Does Logical Operations

In Chapter 1, we saw how the computer stores integers and handles some arithmetic operations. In this section, we will see how the computer stores truth values and performs the logical operations of negation, conjunction, and disjunction.

Most programming languages support a data type called **Boolean** (as in Pascal or C) or **logical** (as in FORTRAN or some dialects of BASIC). A **Boolean variable** can assume one of two values, True or False; that is, the data type Boolean is the set {T, F}. Thus, Boolean variables correspond to the propositional variables of logic.

How the values of True and False are stored in a computer depends on both the machine and the programming language used. In Pascal, one bit is used to store a truth value, with 0 representing False and 1 representing True. However, in most computers, operations on entire words are faster than operations on individual bits. So in most implementations of programming languages other than Pascal, an entire word is used to store a truth value. In VAX BASIC, for instance, the value False is stored as a word with 32 0's, and True is stored as a word of 32 1's.

The logical operations in the computer are bitwise operations, as described in Section 2.7. For example, in a computer with 8-bit words,

$$01010110 \text{ AND } 11010011 = 01010010$$
$$01010110 \text{ OR } \quad 11010011 = 11010111$$
$$01010110 \text{ XOR } 11010011 = 10000101$$
$$\text{NOT } 11010011 = 00101100$$

where, using programming practice, we have written AND, OR, XOR, and NOT for $\land$, $\lor$, $\veebar$, and $\neg$, respectively.

EXERCISES 4.1

1. Which of the following sentences are propositions?
   a. There are dinosaurs in the Bronx Zoo.
   b. Do not feed the animals.
   c. π is a rational number or 4 is prime.
   d. Who wrote *Middlemarch?*   e. $1101_{two} = 13_{ten}$
2. Form the negation of the following propositions:
   a. The set of integers is finite.
   b. F. D. Roosevelt was elected president in 1944.
   c. 6 is not a prime number.

3. Form the conjunction of the following pairs of propositions:

   a. 5 is prime.
      $\frac{1}{3}$ is rational.

   b. Nixon was elected president in 1968 and 1972.
      Nixon did not complete two terms as president.

4. Form the disjunction of the propositions "π is a natural number" and "Euclid is the author of *Proof*."

5. If p is the proposition "Euclid was born in Sardinia" and q is the proposition "Euclid was six feet tall," form each of the following propositions:

   a. $p \Rightarrow \neg q$ b. $p \Leftrightarrow q$

   c. $\neg p$ NAND q d. p NOR q

6. If p is the proposition "The material is porous" and q is the proposition "The material is lightweight," write the following propositions using logic symbols.

   a. A material is porous only if it is lightweight.

   b. Being lightweight is necessary for porosity.

   c. The material is lightweight but it is also porous.

7. Refer to Exercise 6. Write the following using logic symbols.

   a. In order for a material to be porous, it is sufficient for it to be lightweight.

   b. A material is not both lightweight and nonporous.

   c. Heaviness is sufficient for porosity.

8. State the converse and contrapositive of the following propositions:

   a. If f and g are one-to-one functions, then $f \circ g$ is one-to-one.

   b. If a polygon is a square, then it is a parallelogram.

9. Construct the truth table for $p \veebar q$.

10. Construct truth tables for the following connectives:

    a. p is implied by q

    b. p is not implied by q

    c. p does not imply q

11. For each of the connectives defined in Exercise 10, write an expression for it using the connectives defined in Section 4.1.

12. Given the bit strings $x = 01001101$ and $y = 11010111$, find x AND y, x OR y, x XOR y, and NOT x.

13. A **ternary connective** is a function from $\{T, F\} \times \{T, F\} \times \{T, F\}$ to $\{T, F\}$. How many ternary connectives exist?

14. A **multivalued logic** is a system in which the set of truth values for propositional variables is larger than $\{T, F\}$. For instance, the set of truth values may be $\{T, U, F\}$, where U means "unknown." Propose reasonable definitions of the connectives "and," "or," "not," and "implies" for this set of truth values.

4.2 • TAUTOLOGIES AND THE ALGEBRA OF PROPOSITIONS

In the previous section, we used truth tables to define each of the logical connectives. We can construct a truth table for any expression, no matter how many variables and connectives are involved. In each of the following examples, the truth value for the expression is in the last column on the right.

EXAMPLE 4.11   Construct the truth table for $\neg(p \wedge q)$.

p	q	$p \wedge q$	$\neg(p \wedge q)$
T	T	T	F
T	F	F	T
F	T	F	T
F	F	F	T

Observe that the truth table for $\neg(p \wedge q)$ is identical to that of p NAND q.   ○

EXAMPLE 4.12   Construct the truth table for $p \vee q \Rightarrow r \vee q$. Here we have three variables and eight (2^3) possible combinations of their truth values.

p	q	r	$p \vee q$	$r \vee q$	$p \vee q \Rightarrow r \vee q$
T	T	T	T	T	T
T	T	F	T	T	T
T	F	T	T	T	T
T	F	F	T	F	F
F	T	T	T	T	T
F	T	F	T	T	T
F	F	T	F	T	T
F	F	F	F	F	T

○

Order of Operations

In Example 4.12, the expression is written without any parentheses. The question that can arise is "Which operation is performed first?" Among the ways we can evaluate this expression are: $(p \vee q) \Rightarrow (r \vee q)$; $p \vee (q \Rightarrow r) \vee q$; and $p \vee (q \Rightarrow (r \vee q))$. To avoid ambiguity when complicated expressions are used without parentheses, the operations are always performed in the following order:

negation
conjunction (from left to right)
disjunction (from left to right)
implication
equivalence

Thus, the correct interpretation of the expression of Example 4.12 is $(p \vee q) \Rightarrow (r \vee q)$. Of course, parentheses override the specified order, as in ordinary algebra.

Since there are no rules stating places for **NAND** and **NOR** in the operator hierarchy, adequate parentheses should be used with these connectives. In fact, it is good practice to use parentheses to make complicated expressions more readable.

Tautologies

The class of expressions that have only T's in their truth table are important in analyzing arguments. These expressions are called *tautologies*.

Definition 4.7

A **tautology** is an expression that is true for all possible values of its propositional variables.

EXAMPLE 4.13   Verify by a truth table that $p \Rightarrow (q \Rightarrow p \wedge q)$ is a tautology.

p	q	$p \wedge q$	$q \Rightarrow p \wedge q$	$p \Rightarrow (q \Rightarrow p \wedge q)$
T	T	T	T	T
T	F	F	T	T
F	T	F	F	T
F	F	F	T	T

This is a tautology, since all entries in the rightmost column are T.   ○

Tautologies of the form $A \Leftrightarrow B$, where A and B are expressions, play a role in logical expressions akin to equality of numerical expressions. Such a tautology expresses the fact that A has the same meaning as B and can therefore be substituted for B in expressions. The following examples illustrate some useful equivalences of this form.

EXAMPLE 4.14   The expression $\neg\neg p \Leftrightarrow p$ is a tautology, as can be seen from the following truth table.

p	$\neg p$	$\neg\neg p$	$\neg\neg p \Leftrightarrow p$
T	F	T	T
F	T	F	T

○

In the following examples, the truth table analyses are left as exercises.

EXAMPLE 4.15   The expression $(p \Rightarrow q) \Leftrightarrow (\neg q \Rightarrow \neg p)$ is a tautology; it expresses the fact that a conditional proposition and its contrapositive are equivalent.   ○

EXAMPLE 4.16   The expression $(p \Rightarrow q) \Leftrightarrow (\neg p \vee q)$ is a tautology; it expresses the fact that $\Rightarrow$ can be defined in terms of the connectives $\neg$ and $\vee$.   ○

EXAMPLE 4.17   The expression $\neg p \Leftrightarrow (p \text{ NAND } p)$ is a tautology; it shows that $\neg$ can be defined in terms of NAND.   ○

EXAMPLE 4.18   The connective $\wedge$ can be defined in terms of NAND as follows:

$p \wedge q \Leftrightarrow \neg\neg(p \wedge q)$	By Example 4.14
$\Leftrightarrow \neg(p \text{ NAND } q)$	By Example 4.11
$\Leftrightarrow (p \text{ NAND } q) \text{ NAND } (p \text{ NAND } q)$	By Example 4.17

We can check the accuracy of this reasoning by verifying that

$$p \wedge q \Leftrightarrow [(p \text{ NAND } q) \text{ NAND } (p \text{ NAND } q)]$$

is a tautology.   ○

Algebraic Rules for Propositions

In Chapter 2, we looked at the laws for set algebra. Here is a summary of the analogous laws for logical operations. The laws are formulated by tautologies of the form $A \Leftrightarrow B$, where A and B are expressions.

1. The Commutative Laws
   $p \wedge q \Leftrightarrow q \wedge p$
   $p \vee q \Leftrightarrow q \vee p$
2. The Associative Laws
   $p \wedge (q \wedge r) \Leftrightarrow (p \wedge q) \wedge r$
   $p \vee (q \vee r) \Leftrightarrow (p \vee q) \vee r$
3. The Distributive Laws
   $p \vee (q \wedge r) \Leftrightarrow (p \vee q) \wedge (p \vee r)$
   $p \wedge (q \vee r) \Leftrightarrow (p \wedge q) \vee (p \wedge r)$
4. Identities for Disjunction and Conjunction
   The constant F is the identity for $\vee$: $p \vee \text{F} \Leftrightarrow p$
   The constant T is the identity for $\wedge$: $p \wedge \text{T} \Leftrightarrow p$
5. Properties of Negation
   $p \wedge \neg p \Leftrightarrow \text{F}$
   $p \vee \neg p \Leftrightarrow \text{T}$
   $\neg(\neg p) \Leftrightarrow p$
6. Idempotent Laws
   $p \vee p \Leftrightarrow p$
   $p \wedge p \Leftrightarrow p$
7. DeMorgan's Laws
   $\neg(p \wedge q) \Leftrightarrow \neg p \vee \neg q$
   $\neg(p \vee q) \Leftrightarrow \neg p \wedge \neg q$

EXERCISES 4.2

1. Construct the truth table for

   a. $\neg p \wedge q$ **b.** $p \vee \neg q$ **c.** $\neg p \Rightarrow q$ **d.** $\neg(p \Rightarrow q)$

2. Construct the truth table for

    a. $p \veebar \neg q$ **b.** $\neg(p \veebar q)$

    c. $\neg p \veebar q$ **d.** $\neg p \Rightarrow (p \Rightarrow q)$

3. Verify that the following expressions are tautologies.

    a. $p \Rightarrow p$ **b.** $p \Rightarrow (q \Rightarrow p)$

    c. $(p \Rightarrow q) \Leftrightarrow (\neg p \vee q)$ **d.** $(p \wedge q) \wedge r \Leftrightarrow p \wedge (q \wedge r)$

4. Verify that the following expressions are tautologies.

    a. $p \wedge q \Leftrightarrow \neg(\neg p \vee \neg q)$

    b. $p \wedge q \Leftrightarrow [(p \text{ NAND } q) \text{ NAND } (p \text{ NAND } q)]$

    c. $(p \Rightarrow q) \Leftrightarrow (\neg q \Rightarrow \neg p)$ **d.** $\neg p \Leftrightarrow (p \text{ NAND } p)$

5. Determine whether the following are tautologies.

    a. $(p \Rightarrow q) \wedge p \Rightarrow q$ **b.** $(p \Rightarrow q) \wedge \neg q \Rightarrow \neg p$

    c. $(p \Rightarrow q) \wedge q \Rightarrow p$

6. Determine whether the following are tautologies.

    a. $(p \Rightarrow q) \wedge \neg p \Rightarrow \neg q$

    b. $(p \Rightarrow q) \wedge (q \Rightarrow r) \Rightarrow (p \vee q \Rightarrow r)$

    c. $(q \Rightarrow p) \wedge (q \Rightarrow \neg p) \Rightarrow \neg q$

7. Verify DeMorgan's Laws.

8. Verify that $p \vee q \Leftrightarrow \neg(\neg p \wedge \neg q)$ is a tautology.

9. Show that NAND is not associative.

10. A **contradiction** is an expression that is false for all truth values of its propositional variables. Verify that the following are contradictions.

    a. $p \wedge \neg p$ **b.** $p \wedge q \wedge \neg p$

    c. $p \wedge q \wedge \neg(p \vee q)$ **d.** $\neg(p \vee \neg p)$

11. Show that $p \Rightarrow q$ and its inverse are *not* equivalent.

12. Show that $p \Rightarrow q$ and its converse are *not* equivalent.

13. Refer to the multivalued logic of Exercise 14, Section 4.1. Are the operations of "and" and "or" commutative? Associative?

14. In the multivalued logic of Exercise 14 of Section 4.1, perform the truth table analysis for the following expressions:

    a. $p \wedge (q \vee r)$ **b.** $p \Rightarrow (q \vee \neg p)$

15. Does $\veebar$ obey the Commutative Law? The Associative Law?

16. Does $\veebar$ obey any Distributive Laws with $\wedge$?

17. Is there an identity element for $\veebar$?

18. Write $p \Leftrightarrow q$ in terms of $\neg$ and $\wedge$ alone.

19. Write $p \vee q$ in terms of NAND alone.

20. Write $p \Rightarrow q$ in terms of NAND alone.

21. A **primitive set of connectives** is a set of connectives with the property that every connective can be written in terms of the connectives in the set. Show that $\{\neg, \Rightarrow\}$ is a primitive set of connectives.

22. Show that $\{\text{NOR}\}$ is a primitive set of connectives.

23. Write $p \wedge q$ in terms of NOR alone.

24. Write $p \Rightarrow q$ in terms of NOR alone.

25. Use induction to prove the generalized DeMorgan's Laws:

a. $\neg(p_1 \wedge p_2 \wedge \cdots \wedge p_n) = \neg p_1 \vee \neg p_2 \vee \cdots \vee \neg p_n$

b. $\neg(p_1 \vee p_2 \vee \cdots \vee p_n) = \neg p_1 \wedge \neg p_2 \wedge \cdots \wedge \neg p_n$

4.3 • METHODS OF MATHEMATICAL PROOF

In this section, we consider how propositions can be combined to form valid arguments. By an argument, in ordinary language, we mean a list of propositions, called **premises** or **hypotheses,** together with another proposition, the **conclusion.** Informally, we say that an argument is valid if the conclusion is a logical consequence of the premises. The validity of an argument is based solely on the form of the argument, not on the truth or falsity of the propositions involved.

Argument Forms

Until the time of DeMorgan and Boole, logic as a subject consisted of collecting and classifying argument forms generally accepted as valid. These forms are called **syllogisms;** some examples are shown here.

EXAMPLE 4.19   If Springfield is the capital of Illinois, then Pascal invented a computer.
Springfield is the capital of Illinois.
Therefore, Pascal invented a computer.   ○

EXAMPLE 4.20   If Springfield is the capital of Illinois, then Euclid is the author of *Proof*.
Springfield is the capital of Illinois.
Therefore, Euclid is the author of *Proof*.   ○

EXAMPLE 4.21   If there is intelligent life on Mars, then Pythagoras had blue eyes.
There is intelligent life on Mars.
Therefore, Pythagoras had blue eyes.   ○

In Example 4.19, we reasoned from true hypotheses to a true conclusion; in Example 4.20, one of the premises is false, and so is the conclusion. In Example 4.21, we do not know the truth value of the hypotheses and conclusion. Yet the argument is **formally** correct; its form is the same as that in the two previous arguments. We can describe the form of these arguments by the sequence of expressions:

$$p \Rightarrow q$$

$$p$$

Therefore, q

This form of argument is known as **modus ponens.**

Definition 4.8 An **argument form** is a sequence of expressions $A_1, A_2, \ldots, A_n$ (the hypotheses) and another expression B (the conclusion).

For instance, in modus ponens,

A_1 is the expression $p \Rightarrow q$

A_2 is the expression p, and

B is the expression q

Valid Arguments

By relating an argument form to a tautology, we can use truth tables to analyze the validity of the argument. The next definition shows how to do this.

Definition 4.9 An argument of the form

$A_1, A_2, \ldots, A_n$, therefore B

is **valid** if $A_1 \wedge A_2 \wedge \cdots \wedge A_n \Rightarrow B$ is a tautology.

EXAMPLE 4.22 The argument form modus ponens is valid because $A_1 \wedge A_2 \Rightarrow B$ is the expression $[(p \Rightarrow q) \wedge p] \Rightarrow q$, which is a tautology. ○

EXAMPLE 4.23 Consider the argument
If there is intelligent life on Mars, then Pythagoras had blue eyes.
Pythagoras did not have blue eyes.
Therefore, there is not intelligent life on Mars.
The form of this argument is

$$p \Rightarrow q$$
$$\neg q$$

Therefore, $\neg p$

This form is known as **modus tollens;** it is valid because $[(p \Rightarrow q) \wedge \neg q] \Rightarrow \neg p$ is a tautology. ○

EXAMPLE 4.24 Consider the argument
If Springfield is the capital of Illinois, then Pythagoras was a mathematician.
Pythagoras was a mathematician.
Therefore, Springfield is the capital of Illinois.
The form of this argument is

$$p \Rightarrow q$$
$$q$$

Therefore, p

It is not valid, since $[(p \Rightarrow q) \wedge q] \Rightarrow p$ is not a tautology. ○

Proof Techniques

In previous math courses, you may have seen a variety of mathematical theorems and their proofs. Several methods of reasoning are commonly used in mathematics to prove theorems. The following examples will illustrate some proof techniques and relate them to the concept of valid argument given in Definition 4.9.

A **direct proof** in mathematics is an argument leading directly to the conclusion that the theorem is true. The reasoning must draw upon previously known mathematical facts.

EXAMPLE 4.25   Prove that if $x \mid a$ and $x \mid b$, then $x \mid (b - a)$.

The hypothesis that $x \mid a$ and $x \mid b$ means that there are integers u and v such that $xu = a$ and $xv = b$. Subtracting these equations yields

$$b - a = xv - xu = x(v - u)$$

which shows that $x \mid (b - a)$.   ○

In Example 4.25, we used known mathematics (subtracting equations, the Distributive Law) to reason directly from the hypothesis to the conclusion. We did not appeal explicitly to propositional logic at any point.

Very often, it is not practical to give a direct proof of a theorem. Instead, an **indirect proof** is given. An indirect proof consists of two steps: First, we prove some statement other than the statement of the theorem; second, we use logic to show that the theorem follows from the statement actually proved. There are two common methods of indirect proof: proof by contradiction and proof by contraposition.

Proof by Contradiction

To prove a statement p, by the method of contradiction, assume $\neg p$ and deduce a contradiction—that is, a statement of the form $r \wedge \neg r$. The truth of p follows, because $[\neg p \Rightarrow (r \wedge \neg r)] \Rightarrow p$ is a tautology.

EXAMPLE 4.26   Prove by contradiction that there are infinitely many primes.

Suppose there were only finitely many primes; call them $a_1, a_2, \ldots, a_n$. Let

$$a = 1 + a_1 a_2 \cdots a_n$$

Then a must be divisible by some prime b, which must therefore be one of the primes $a_1, a_2, \ldots, a_n$. Thus, $b \mid a$ and $b \mid a_1 a_2 \cdots a_n$, so it follows from Example 4.25 that $b \mid 1$. Therefore, b is not prime, since the smallest prime is 2. In this proof, the proposition p is "There are infinitely many primes," and r is "The number b is prime."
                                        ○

Another form of proof by contradiction is the following: To prove $p \Rightarrow q$ by contradiction, we start from the hypothesis and the negation of the conclusion, and we use mathematical facts to reason to a contradiction. Then we apply logic. The argument form

$$p \wedge \neg q \Rightarrow r \wedge \neg r$$

Therefore, $p \Rightarrow q$

is valid because the expression

$$[(p \wedge \neg q) \Rightarrow (r \wedge \neg r)] \Rightarrow (p \Rightarrow q)$$

is a tautology. It then follows that $p \Rightarrow q$.

EXAMPLE 4.27   The proof that $\sqrt{2}$ is irrational (see Section 1.5) is a proof by contradiction. The hypothesis is the proposition p, "x is a number such that $x^2 = 2$." The conclusion is the proposition q, "x is irrational." To prove the theorem, we started from the assumption $p \wedge \neg q$ (that is, $x^2 = 2$ and x is rational). Since x is rational, x can be written as a fraction: $x = \frac{a}{b}$. We let r be the proposition "the fraction $\frac{a}{b}$ is in lowest terms." By reasoning about divisibility, we arrived at the conclusion $r \wedge \neg r$.   ○

Proof by Contraposition

To prove $p \Rightarrow q$ by contraposition, we prove instead that $\neg q \Rightarrow \neg p$. The argument form used is

$$\neg q \Rightarrow \neg p$$

Therefore, $p \Rightarrow q$

which is valid because

$$(\neg q \Rightarrow \neg p) \Rightarrow (p \Rightarrow q)$$

is a tautology.

EXAMPLE 4.28   A positive integer is said to be *perfect* if it is the sum of its proper divisors; for instance, $6 = 1 + 2 + 3$. Prove that a perfect number is not prime.

We will do this by proving the contrapositive—that is, if a number is prime, then it is not perfect. Suppose p is prime. Then the only proper divisor of p is 1, so the sum of the proper divisors is 1, which is not equal to p.   ○

EXAMPLE 4.29   A binary relation R on a set X is said to be asymmetric if, for every x and y in X, if $(x, y) \in R$, then $(y, x) \notin R$. Prove that if R is asymmetric, then R is antisymmetric.

Suppose that R is not antisymmetric. Then there exist x and y in X such that $x\,R\,y$, $y\,R\,x$, but $x \neq y$. Therefore, R is not asymmetric.   ○

EXAMPLE 4.30   A binary relation R on a set X is said to be irreflexive if, for every $x \in X$, $(x, x) \notin R$. Prove that if R is transitive and irreflexive, then R is asymmetric.

Suppose R is not asymmetric; then there exist x and y in X such that $(x, y) \in R$ and $(y, x) \in R$. By transitivity, $(x, x) \in R$, which contradicts the assumption that R is irreflexive.   ○

Strictly speaking, the form of the argument used in Example 4.30 is

$$p \wedge \neg r \Rightarrow \neg q$$
Therefore, $p \wedge q \Rightarrow r$

where p is the statement, "R is transitive."

EXERCISES 4.3

In Exercises 1–4, determine whether the argument is valid.

1. If Euclid had red hair, then he was myopic. Euclid did not have red hair. Therefore, Euclid was not myopic.

2. If Pascal invented the transistor, then he was a poet. If Pascal did not invent the transistor, then he was a poet. Therefore, Pascal was a poet.

3. If Archimedes was tall, then Euclid was short. Euclid was not short unless Pythagoras was fat. Pythagoras was fat only if he ate too much. Pythagoras did not eat too much. Therefore, Archimedes was not tall.

4. If Archimedes lived in Syracuse, then he was killed by the Romans. Archimedes lived in Syracuse, but he died of natural causes. If Archimedes died of natural causes, he was not killed by the Romans. Therefore, Plato was Egyptian.

5. What proof technique is used in the following?
    Theorem: If f is a nonconstant linear function (that is, a function in the form $f(x) = mx + b, m \neq 0$), then f is one-to-one.
    Proof: Suppose $mx_1 + b = mx_2 + b$. Then

$mx_1 = mx_2$	Subtracting b from both sides
$x_1 = x_2$	Dividing both sides by m

6. Prove directly that if p and q are positive real numbers and $p = q$, then $\sqrt{pq} = \frac{p+q}{2}$.

7. Prove directly that the square of any even integer is even.

8. Prove by contraposition that if n is an integer whose square is even, then n is even.

9. Prove that the product of two odd integers is odd.

10. Prove by contraposition that if n is an odd integer, then the equation $x^2 + x - n = 0$ has no odd-integer solutions.

11. Prove that if $f: \mathbf{R} \to \mathbf{R}$ is a nonconstant linear function, then f is one-to-one.

12. A relation R on a set X is said to be a **quasi order** on X if R is transitive and irreflexive. (See Example 4.30 for the definition of *irreflexive*.) Show that a quasi order is antisymmetric.

13. Give a direct proof of the statement of Example 4.29.

14. Look at a math book of your choice. Locate some theorems with proofs and determine whether the proofs are direct or indirect. If indirect, identify the argument form used.

4.4 • PROPOSITIONAL FUNCTIONS

In the preceding sections, we gave a method for analyzing certain types of arguments and determining whether they are valid. The method is based on truth-table analysis. However, the method is not powerful enough to analyze some of the simplest types of arguments. Consider the argument

All prime numbers are odd.

17 is a prime number.

Therefore, 17 is odd.

The reasoning is certainly valid, even though the first hypothesis is not true. What is the argument form? It is not the modus ponens form,

$$p \Rightarrow q$$

$$p$$

Therefore, q

because the first hypothesis does not begin, "If 17 is a prime number, then . . . ."

To understand the argument, we must examine the structure of the first hypothesis, not merely its truth value. The first hypothesis makes a statement about the set $\mathbf{N}$ of natural numbers. For each element $x \in \mathbf{N}$, consider the proposition "If x is prime, then x is odd." This generates a set of propositions, one for each $x \in \mathbf{N}$. For example, for $x = 1$ we get the proposition "If 1 is prime, then 1 is odd." In effect, we have a function from the set $\mathbf{N}$ of natural numbers to a set of propositions. The statement "All prime numbers are odd" asserts that all propositions in the image of this function are true.

Functions whose values are propositions are the subject of predicate logic. By studying such functions, we will arrive at a class of expressions, called *valid formulas,* that generalize the concept of tautology.

Definition 4.10

A **propositional function** on the set X is a function whose domain is X and whose range is a set of propositions. Propositional functions are also called **predicates.**

You can think of a propositional function as a rule that assigns a proposition $P(x)$ to each element $x \in X$. The term *predicate* is derived from grammar; the predicate part of a sentence consists of the verb and object and provides some information about the subject of the sentence. This usage is the origin of the term *predicate logic.*

EXAMPLE 4.31   Let $X = \{1, 2, 3\}$. Define the propositional function P on X as follows:

x	$P(x)$
1	"Chicago is the capital of Illinois."
2	"$1 + 1 = 2$"
3	"$\frac{1}{2}$ is an integer."

Then $P(1)$ is the statement "Chicago is the capital of Illinois," $P(2)$ is the statement "$1 + 1 = 2$," and $P(3)$ is the statement "$\frac{1}{2}$ is an integer."   ○

EXAMPLE 4.32   On the set **Z**, define the propositional function Q by the rule

$$Q(x) = \text{"}2x - 3 = 7\text{"}$$

Then $Q(15)$ is the statement "$2(15) - 3 = 7$," which is false, and $Q(5)$ is the statement "$2(5) - 3 = 7$," which is true.   ○

EXAMPLE 4.33   On the set **R**, define the propositional function P by the rule $P(x) =$ "$(x > 2) \wedge (x < 6)$." Then $P(5)$ is the statement "$(5 > 2) \wedge (5 < 6)$," which is true, and $P(6)$ is the statement "$(6 > 2) \wedge (6 < 6)$," which is false.   ○

We can also define propositional functions of several variables. If $X_1, \ldots, X_n$ are sets, a propositional function whose domain is the Cartesian product $X_1 \times \cdots \times X_n$ can be thought of as a function of n variables. The next example is a propositional function of two variables whose domain is **R** $\times$ **R**.

EXAMPLE 4.34   On **R** $\times$ **R**, define the propositional function Q by the rule $Q(x, y) =$ "$x^2 + y^2 = 4$." Then $Q(1, 3)$ is the statement "$1^2 + 3^2 = 4$," which is false, whereas $Q(0, 2)$ is true.   ○

We use lowercase letters (x, y, z, and so on) to denote elements of the domain of a propositional function. These letters are called **individual variables.** Similarly, we use uppercase letters (P, Q, R, and so on) to denote propositional functions. These letters are called **predicate variables.**

In each of the previous examples, we used an individual variable to help describe the functions. Any other individual variable could have been used. For instance, the rule of Example 4.32 could have been written $Q(y) =$ "$2y - 3 = 7$." When a variable is used in this way, it is called a **bound variable** or a **dummy variable.**

Truth Sets

Every proposition has a truth value, either T or F. For propositional functions, the corresponding concept is the truth set.

Definition 4.11

> The **truth set** of a propositional function is the set of elements in its domain that are assigned true propositions by the function.

EXAMPLE 4.35 The truth set of the propositional function P of Example 4.31 is the set $\{2\}$. ○

EXAMPLE 4.36 The truth set of the propositional function Q of Example 4.32 is the set $\{5\}$. ○

EXAMPLE 4.37 The truth set of the propositional function P of Example 4.34 is the set of numbers both greater than 2 and less than 6. This is the open interval $(2, 6)$. ○

Notice that the truth set of a propositional function can be finite, as in Examples 4.36 and 4.37, or infinite, as in Example 4.37.

EXAMPLE 4.38 On the set $\mathbf{R} \times \mathbf{R}$ of real numbers, define $P(x, y) = $ "$x < y$." The truth set of P is equal to the relation "$<$" and consists of all points below the line $y = x$. ○

Combining Propositional Functions

Just as propositions can be combined, using connectives, to form new propositions, so can propositional functions be combined to form new propositional functions. For instance, the propositional functions P and Q defined on $\mathbf{Z}$ by the rules $P(x) = $ "$x > 1$" and $Q(x) = $ "$x < 5$" can be combined into the propositional function S defined by the rule $S(x) = $ "$x > 1 \vee x < 5$." The new propositional function S is a compound propositional function on $\mathbf{Z}$.

Definition 4.12

If P and Q are propositional functions on a set X, then $\neg P$, $P \wedge Q$, and $P \vee Q$ are defined by the formulas

$$(\neg P)(x) = \neg P(x)$$
$$(P \wedge Q)(x) = P(x) \wedge Q(x)$$
$$(P \vee Q)(x) = P(x) \vee Q(x)$$

The following theorem explains how combining propositional functions affects their truth sets.

Theorem 4.1

Let P and Q be propositional functions on a set X. Then

1. The truth set of $\neg P$ is the complement of the truth set of P.
2. The truth set of $P \wedge Q$ is the intersection of the truth sets of P and Q.
3. The truth set of $P \vee Q$ is the union of the truth sets of P and Q.

Proof
1. If x belongs to the truth set of $\neg P$, then $\neg P(x)$ has truth value T, so $P(x)$ has truth value F. Thus, x is not an element of the truth set of P. Conversely, if x belongs to the truth set of P, then $P(x)$ has truth value T and $\neg P(x)$ has truth value F. Thus, x is not in the truth set of $\neg P$.
2. If x is an element of the truth set of $P \wedge Q$, then $(P \wedge Q)(x)$ is true; that is, $P(x) \wedge Q(x)$ is true. This means that $P(x)$ and $Q(x)$ are both true, so x is an element of the intersection of the truth sets of P and Q. The converse is similar and is left as an exercise.
3. The proof of the third part of the theorem is left as an exercise. ●

The next examples show how to use Theorem 4.1 to find the truth sets of compound propositions.

EXAMPLE 4.39   If P and Q are propositional functions on $\mathbf{Z}$ defined by $P(x) = $ "$x < 0$" and $Q(x) = $ "$x < 5$," find the truth set of the propositional function $\neg P \wedge Q$.

The truth set of P is the set of negative integers, so the truth set of $\neg P$ is $\{0, 1, 2, 3, 4, 5, \ldots\}$. The truth set of Q is $\{\ldots, -1, 0, 1, 2, 3, 4\}$. Thus, the truth set of $\neg P \wedge Q$ is the intersection of these sets: $\{0, 1, 2, 3, 4\}$.   ○

EXAMPLE 4.40   If P and Q are propositional functions on $\mathbf{R}$ defined by $P(x) = $ "$x \geq 1$" and $Q(x) = $ "$x > 3$," find the truth set of $P \Rightarrow Q$.

Recall that $P \Rightarrow Q$ can be expressed as $\neg P \vee Q$. Now, the truth set of P is the interval $[1, \infty)$, so the truth set of $\neg P$ is the interval $(-\infty, 1)$. The truth set of Q is the interval $(3, \infty)$. Thus, the truth set of $\neg P \vee Q$ is the union of the intervals $(-\infty, 1)$ and $(3, \infty)$, as shown in Figure 4.1.

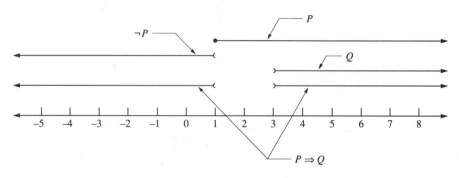

FIGURE 4.1   The truth sets of Example 4.40   ○

EXERCISES 4.4

1.   If L is the set of letters in the word "newelpost" and P is the propositional function on L defined by $P(x) = $ "x is a vowel," find the truth set of P.

2.   If X is the set of all U.S. presidents and P is the propositional function defined by $P(x) = $ "x was assassinated," find the truth set of P.

3.   If P is the propositional function on $\mathbf{Z}$ defined by $P(x) = $ "x is even," find the truth set of P.

4.   Find the truth set of the following propositional functions on $\mathbf{R}$.
     a. $P(x) = $ "$x + 1 = 5$"   b. $Q(x) = $ "$x^2 - x - 6 = 0$"

5.   Complete the proof of Theorem 4.1.

6.   Find the truth set of the following propositional functions on $\mathbf{R}$.
     a. $P(x) = $ "$(x < 4) \vee (x > -7)$"   b. $P(x) = $ "$x < 1 \Rightarrow x < -1$"

7.   Find the truth set of the following propositional functions on **R**.

   a. $P(x) = "x^2 > 0 \Rightarrow x = x + 1"$

   b. $Q(z) = "(z = 5 \vee z \geq 7) \Leftrightarrow (z \geq 1 \wedge z \leq 8)"$

8.   Find the truth set of the propositional function P on $\mathbf{R} \times \mathbf{R}$ defined by $P(x, y) = "x^2 + y^2 = -1"$

9.   Sketch the truth set of each of the following propositional functions on $\mathbf{R} \times \mathbf{R}$.

   a. $P(x, y) = "x < y"$ **b.** $P(x, y) = "(x = 2) \vee (y = -3)"$

   c. $P(x, y) = "x < 1 \Rightarrow y \geq 1"$

10.   Give a rule for finding the truth set of $P \Leftrightarrow Q$ from those of P and Q.

11.   Give a rule for finding the truth set of P NAND Q from those of P and Q.

12.   Give a rule for finding the truth set of P NOR Q from those of P and Q.

13.   Let P and Q be propositional functions defined on the same set X. Then P and Q are **equivalent** if for each $x \in X$, the propositions $P(x)$ and $Q(x)$ have the same truth value. Which of the following pairs of propositional functions are equivalent?

   a. $X = \{1, 2, 3\}$, P is the propositional function defined in Example 4.31, and Q is the propositional function defined by $Q(x) = "x = 2."$

   b.   On the set $\mathbf{Z}$, define the propositional functions P and Q as follows: $P(x) = "x^2 - 1 = 0"$ and $Q(x) = "x^6 - 1 = 0."$

   c.   On $\mathbf{Z}$, define P and Q to be: $P(x) = "x < 2 \vee x = 5"$ and $Q(x) = "x > 0 \wedge x < 6."$

14.   Given a set X and a subset A of X, find a propositional function P whose truth set is A.

4.5 • QUANTIFIERS

In the introduction to Section 4.4, we considered the argument,

All prime numbers are odd.

17 is a prime number.

Therefore, 17 is odd.

The first hypothesis contains two propositional functions on the set $\mathbf{N}$ of natural numbers: $P(x) = "x$ is prime" and $Q(x) = "x$ is odd." The compound propositional function $P \Rightarrow Q$ has truth set $\mathbf{N} - \{2\}$. The first hypothesis is a statement about $P \Rightarrow Q$; it states (incorrectly) that the truth set of $P \Rightarrow Q$ is all of $\mathbf{N}$.

   The process of forming a single proposition (in this case, "All prime numbers are odd") from a propositional function (in this case, $P \Rightarrow Q$) is called *quantification*.

Definition 4.13   A **quantifier** is a rule that assigns a single proposition to a propositional function.

There are two quantifiers that play a major role in mathematical logic: the existential quantifier and the universal quantifier.

Definition 4.14

The **existential quantifier,** denoted by $\exists$, assigns to a propositional function P the proposition, "The truth set of P is not empty." The result of applying the existential quantifier $\exists$ to P is written $\exists x P(x)$ and is read "There exists x such that $P(x)$."

EXAMPLE 4.41   If P is a propositional function on $\mathbf{Z}$ defined by $P(x) = $ "$x > 5$," the proposition obtained by applying the existential quantifier $\exists$ to P is written $\exists x(x > 5)$ and is read "There exists an x such that $x > 5$." The resulting proposition is true, since 6 (among other integers) belongs to the truth set of P. Thus, the truth set of P is not empty.   ○

Definition 4.15

The **universal quantifier,** denoted by $\forall$, assigns to a propositional function P the proposition, "The truth set of P is equal to its domain." The result of applying the universal quantifier $\forall$ to P is written $\forall x P(x)$. This is read, "For all x, $P(x)$" or "For every x, $P(x)$."

EXAMPLE 4.42   If P is the propositional function of Example 4.41, the proposition obtained by applying the universal quantifier to P is written $\forall x(x > 5)$ and is read "For all x, $x > 5$" or "For each x, $x > 5$." This proposition is false, because the truth set of P is the set $\{6, 7, 8, \ldots\}$, which is not equal to its domain $\mathbf{Z}$.   ○

EXAMPLE 4.43   If P is the propositional function on $\mathbf{Z}$ defined by $P(x) = $ "$x + 1 = 3$," then $\exists x(x + 1 = 3)$ is true, since the truth set of P is $\{2\}$, which is not empty. But the proposition $\forall x(x + 1 = 3)$ is false, since $\{2\} \neq \mathbf{Z}$.   ○

EXAMPLE 4.44   If P is the propositional function on $\mathbf{R}$ defined by $P(x) = $ "$x^2 = -1$," the proposition obtained by applying $\exists$ to P is $\exists x(x^2 = -1)$. This is false, since the truth set of P is the empty set $\varnothing$. The proposition $\forall x(x^2 = -1)$ is also false, since the truth set of P (namely, $\varnothing$) is not equal to the domain $\mathbf{R}$.   ○

EXAMPLE 4.45   Let P be the propositional function on $\mathbf{R}$ defined by $P(x) = $ "$(x + 1)^2 = x^2 + 2x + 1$." By applying $\forall$ to P, we get $\forall x[(x + 1)^2 = x^2 + 2x + 1]$, which is true because the truth set of P is $\mathbf{R}$. Applying $\exists$, we get $\exists x[(x + 1)^2 = x^2 + 2x + 1]$, which is also true, because the truth set of P is not empty.   ○

Quantifiers are found in almost every useful mathematical statement. But while the existential quantifier is almost always explicit in phrases such as "there exists" or "there is," the universal quantifier must often be inferred from the context. The theorem "If a function f is one-to-one and onto, then f has an inverse" has an implicit quantifier, $\forall f$. The statement "An asymmetric relation is antisymmetric" has an implicit quantifier, $\forall R$, where R refers to relations.

EXAMPLE 4.46   Write the statement "Some prime numbers are even" in symbolic form. Let P and Q be defined on $\mathbf{N}$ by $P(n) = $ "n is prime" and $Q(n) = $ "n is even." The symbolic form is $\exists x(P(x) \land Q(x))$. ○

EXERCISES 4.5

1. Identify all of the quantifiers in each of the following statements and write in symbolic form.

   a. There is an even perfect number.

   b. Every polynomial of odd degree has a root.

   c. Every even number greater than 2 is the sum of two primes.

   d. A function that has an inverse is one-to-one.

2. Identify all of the quantifiers in each of the following and write in symbolic form.

   a. There are transitive relations that are not reflexive.

   b. A sum of the form $1 + \frac{1}{2} + \frac{1}{3} + \cdots + \frac{1}{n}$ can be made arbitrarily large by choosing N large enough.

   c. There are numbers that differ by less than 0.005.

   d. For every positive number ε, there is a positive number δ such that if $|x - y| < \delta$, then $|2x - 2y| < \varepsilon$.

3. Write each of the following statements about the integers in succinct English with a minimum of symbols. Which are true?

   a. $\forall x(x = x)$

   b. $\exists x(x^2 - x - 6 = 0)$

   c. $\forall x(x \le 0 \lor x \ge 0)$

   d. $\forall x(x < 0 \Rightarrow x < 1)$

4. Write each of the following statements about the integers in succinct English with a minimum of symbols. Which are true?

   a. $\exists x(x = 2x)$

   b. $\forall x(x^2 - x - 6 = 0)$

   c. $\forall x(x > 5 \Rightarrow x > 7)$

   d. $\forall x(x > 2 \Rightarrow \exists r \exists s(r < x \land s < x \land rs = x)$

5. Prove that two propositional functions P and Q on the same set X are equivalent if and only if the proposition $\forall x[P(x) \Leftrightarrow Q(x)]$ is true. (See Exercise 13 of Section 4.4 for the definition of equivalence.)

6. Find the truth sets of each of the following propositional functions on $\mathbf{Z}$.

   a. $P(y) = $ "$\forall x(x > y)$"

   b. $Q(y) = $ "$\exists x(x > y)$"

   c. $S(x) = $ "$\exists y(x = y^2) \Rightarrow (x \ge 0)$"

7. Find the truth sets of each of the following propositional functions on $\mathbf{R}$.

   a. $R(x) = $ "$\forall y(y^2 = 1)$"

   b. $T(z) = $ "$\forall w(zw = 0) \Rightarrow (z = 0)$"

   c. $U(x) = $ "$\exists y(-x = 2^y) \land \forall y(|x| < y^2 + 1)$"

8. Which of the following statements about the real numbers are true?

   a. $\exists x \exists y(x^2 + y^2 = 0)$ b. $\forall x \exists y(y^2 = x)$

   c. $\exists x \forall y(y > x)$ d. $\forall x(x > 0 \Rightarrow \exists y(x = y^2))$

   e. $\exists y(y^2 - 5y = -17)$

9. Which of the following statements about the real numbers are true?

   a. $\forall x \forall y((x + y)^2 = x^2 + y^2)$ b. $\exists x \forall y((x + y)^2 = x^2 + y^2)$

   c. $\forall z(\exists w(w^2 = z) \Rightarrow z > 0)$ d. $\exists z(\exists w(w^2 = z) \Rightarrow z > 0)$

   e. $\forall w[(w^2 = -1) \Rightarrow \exists x(x < 0 \land x > 0)]$

10. Let $P(x, y)$ be the propositional function on $\mathbf{R} \times \mathbf{R}$ defined by "$xy \leq 1$." Find the truth set of P.

11. Let Q be defined on $\mathbf{R} \times \mathbf{R}$ by $Q(x, y) = $ "$y \leq 2^x$." Find the truth set of Q.

12. Suppose that P is a propositional function on the plane $\mathbf{R} \times \mathbf{R}$. Give a geometrical interpretation, in terms of the truth set of P, of each of the following expressions:

    a. $\forall x \forall y P(x, y)$ b. $\exists x \exists y P(x, y)$

    c. $\exists y \forall x P(x, y) \land \exists x \forall y P(x, y)$

13. Suppose that P is a propositional function on the plane $\mathbf{R} \times \mathbf{R}$. Give a geometrical interpretation, in terms of the truth set of P, of each of the following expressions:

    a. $\exists y \forall x P(x, y) \Rightarrow \exists x \forall y P(x, y)$ b. $\forall x P(x, x)$

    c. $\forall x \forall y(P(x, y) \Rightarrow P(y, x))$

14. For each of the expressions in Exercise 12, give an example of a propositional function P on $\mathbf{R} \times \mathbf{R}$ for which the expression is false.

15. For each of the expressions in Exercise 13, give an example of a propositional function P on $\mathbf{R} \times \mathbf{R}$ for which the expression is false.

4.6 • WORKING WITH FORMULAS CONTAINING QUANTIFIERS

Formulas involving quantifiers are very useful for writing mathematical statements precisely. In this section, we discuss some techniques for manipulating statements with quantifiers.

Free and Bound Variables

Definition 4.16

Let A be a formula of predicate logic. The **scope** of a quantifier in A is the portion of the formula to which the quantifier applies. Let x be an individual variable occurring in A. Then an occurrence of x in A is **bound** if it is located within the scope of a quantifier, $\forall x$ or $\exists x$. An occurrence of x is **free** if it is not bound. The variable x is a **free variable of** A if it has at least one free occurrence in A.

This definition merely formalizes the usage that we have tacitly assumed up to this point. It is the individual occurrence of a variable x in a formula that is free or bound, rather than the variable itself. In fact, a variable can have both free and bound occurrences in the same formula.

EXAMPLE 4.47 Identify the free and bound occurrences of individual variables in the formula

$$\forall x \exists y (P(x) \Rightarrow Q(y, z)) \Rightarrow R(z)$$

The scope of the quantifier $\exists y$ is $(P(x) \Rightarrow Q(y, z))$. The scope of the quantifier $\forall x$ is $\exists y(P(x) \Rightarrow Q(y, z))$. The occurrence of x in $P(x)$ is bound by the quantifier $\forall x$. The one occurrence of y is bound by the quantifier $\exists y$. Both occurrences of z are free. Thus, z is a free variable of the formula, but x and y are not. ○

EXAMPLE 4.48 Identify the free and bound occurrences of individual variables in the formula

$$\forall x(P(x) \Rightarrow Q(x)) \wedge \forall x(Q(x) \Rightarrow R(x)) \wedge P(x) \Rightarrow R(x)$$

The scope of the first $\forall x$ is $(P(x) \Rightarrow Q(x))$, and the scope of the second $\forall x$ is $(Q(x) \Rightarrow R(x))$. The occurrences of x within the two scopes are bound, as indicated by the arrows. The remaining two occurrences of x are free, so x is a free variable of the formula. ○

Change of Bound Variable

The formulas $\forall y P(x, y)$ and $\forall z P(x, z)$ can be obtained, one from the other, by a change of bound variable. The following rule defines precisely what changes of bound variable are allowed.

Rule for change of bound variable. The formulas A and B are obtained from one another by change of bound variable if all the following conditions are satisfied:

1. A and B have the same length—that is, they contain the same number of symbols.
2. Each symbol of A that is not an individual variable is the same as the corresponding symbol of B.
3. A and B have the same free individual variables, and they occur in the same positions in both formulas.
4. Corresponding bound individual variables of A and B are bound by quantifiers in corresponding positions.

EXAMPLE 4.49 Each of the following pairs of formulas satisfies the rule for changing bound variables:

a. $\forall x P(x) \Rightarrow \forall x P(x)$ and $\forall x P(x) \Rightarrow \forall y P(y)$
b. $P(x) \Rightarrow \forall x(Q(x) \Rightarrow \forall x R(x))$ and $P(x) \Rightarrow \forall y(Q(y) \Rightarrow \forall x R(x))$
c. $P(x) \Rightarrow \forall x(Q(x) \Rightarrow \forall x R(x))$ and $P(x) \Rightarrow \forall y(Q(y) \Rightarrow \forall z R(z))$ ○

EXAMPLE 4.50   The formula $\forall x P(x, x)$ is not obtained from $\forall y P(x, y)$ by change of bound variable, because condition 3 is violated: x is a free variable of $\forall y P(x, y)$ but not of $\forall x P(x, x)$.

EXAMPLE 4.51 $\exists x \forall x P(x, x)$ is not obtained from $\exists x \forall y P(x, y)$, because condition 4 of the rule is violated.

From now on, we will freely change bound variables whenever it will make a formula easier to understand.

Qualified Quantifiers

In order to write symbolic sentences that closely approximate ordinary English, we use **qualified quantifiers.** That is, we build into the quantifier itself some information about the individual variable in the quantifier. You can think of a formula involving a qualified quantifier as an abbreviation for a formula using only ordinary quantifiers. In the following examples, A is some subset of the domain of the propositional functions involved.

Qualified Quantifier	Meaning
$(\forall x > 0)(\cdots)$	$\forall x(x > 0 \Rightarrow \cdots)$
$(\forall x \in A)(\cdots)$	$\forall x(x \in A \Rightarrow \cdots)$
$(\exists x \geq 0)(\cdots)$	$\exists x(x \geq 0 \wedge \cdots)$
$(\exists x \in A)(\cdots)$	$\exists x(x \in A \wedge \cdots)$

EXAMPLE 4.52   The statement "For every positive number x, there is a positive number y such that $x = y^2$" can be written

$$(\forall x > 0)(\exists y > 0)(x = y^2)$$

Without the qualified quantifiers, this would be written

$$\forall x[(x > 0) \Rightarrow \exists y[(y > 0) \wedge x = y^2]]$$

EXAMPLE 4.53   The statement "Every real number has an nth root in the complex numbers for every positive integer n" can be written

$$(\forall n \in \mathbf{N})(\forall x \in \mathbf{R})(\exists y \in \mathbf{C})(x = y^n)$$

EXAMPLE 4.54   In calculus, a real-valued function f is said to be uniformly continuous on a set X if, for every positive number ε, there is a positive number δ such that $|f(x) - f(y)| < \varepsilon$ whenever x and y are in X and $|x - y| < \delta$. Using qualified quantifiers, the definition of uniform continuity can be written

$$(\forall \varepsilon > 0)(\exists \delta > 0)(\forall x \in X)(\forall y \in X)(|x - y| < \delta \Rightarrow |f(x) - f(y)| < \varepsilon)$$

EXAMPLE 4.55   Write

$$(\forall w > 0)(\exists z > 0)(\forall x \in A)(\forall y \in A)(|x - y| < z \Rightarrow |x^2 - y^2| < w)$$

without using qualified quantifiers. The result is

$$\forall w[w > 0 \Rightarrow \exists z(z > 0 \land \forall x(x \in A \Rightarrow \forall y(y \in A \Rightarrow (|x - y| < z \Rightarrow |x^2 - y^2| < w))))]$$

○

Negations of Formulas with Quantifiers

The quantifiers $\forall$ and $\exists$ introduced in the last section are closely related. To determine the relationship, let us consider the proposition $\exists x P(x)$ about a propositional function P. If $\exists x P(x)$ is false, the truth set of P is $\varnothing$. But if the truth set of P is $\varnothing$, then the truth set of $\neg P$ is all of the domain of P. That is, $\forall x \neg P(x)$ is true. Conversely, if $\forall x \neg P(x)$ is true, the truth set of P is $\varnothing$, so $\exists x P(x)$ is false. Thus, we see that for any propositional function P, the propositions $\exists x P(x)$ and $\neg \forall x \neg P(x)$ have the same truth value.

These facts, plus tautologies from propositional logic, provide five rules for transforming statements beginning with negation into more easily understood forms.

	Formula	Negation	Result
(1)	$\exists x P(x)$	$\neg \exists x P(x)$	$\forall x \neg P(x)$
(2)	$\forall x P(x)$	$\neg \forall x P(x)$	$\exists x \neg P(x)$
(3)	$(p \land q)$	$\neg(p \land q)$	$\neg p \lor \neg q$
(4)	$(p \lor q)$	$\neg(p \lor q)$	$\neg p \land \neg q$
(5)	$(p \Rightarrow q)$	$\neg(p \Rightarrow q)$	$p \land \neg q$

Rules (3) and (4) are, of course, DeMorgan's Laws.

EXAMPLE 4.56   The negation of "All odd numbers are prime" is "There is an odd number that is not prime." Rule (2) was applied here. The negation of "Some perfect numbers are prime" is "All perfect numbers are composite." Rule (1) was used in this case.   ○

EXAMPLE 4.57   Find the negation of $\forall x(x = 1)$.

The negation is $\neg \forall x(x = 1)$, which is equivalent to $\exists x \neg(x = 1)$, by Formula (2). This is more commonly written $\exists x(x \neq 1)$.   ○

EXAMPLE 4.58   Find the negation of $\exists x(x^2 < 0)$.

The negation is $\forall x \neg(x^2 < 0)$, by Formula (1). This is usually written $\forall x(x^2 \geq 0)$.   ○

EXAMPLE 4.59   Find the negation of $\forall x \exists y(x < y)$.

The negation is

$$\neg \forall x \exists y(x < y)$$

which is equivalent to

$$\exists x \neg \exists y (x < y) \qquad \text{By Rule (2)}$$

This is equivalent to

$$\exists x \forall y \neg (x < y) \qquad \text{By Rule (1)}$$

which is usually written

$$\exists x \forall y (x \geq y) \hspace{4cm} \circ$$

EXAMPLE 4.60 Find the negation of

$$\forall x [x > 0 \Rightarrow \exists y (0 < y \wedge y < x)]$$

We can write the negation in the following ways:

$$\exists x \neg [x > 0 \Rightarrow \exists y (0 < y \wedge y < x)] \qquad \text{By Rule (2)}$$
$$\exists x [x > 0 \wedge \neg \exists y (0 < y \wedge y < x)] \qquad \text{By Rule (5)}$$
$$\exists x [x > 0 \wedge \forall y \neg (0 < y \wedge y < x)] \qquad \text{By Rule (1)}$$
$$\exists x [x > 0 \wedge \forall y (0 \geq y \vee y \geq x)] \qquad \text{By Rule (3)} \hspace{1cm} \circ$$

The negation of a qualified quantifier is formed just as easily as that of an ordinary quantifier. For instance, $\neg (\forall x > 0) P(x)$ is equivalent to $(\exists x > 0) \neg P(x)$, and $\neg (\exists x > 0) P(x)$ is equivalent to $(\forall x > 0) \neg P(x)$. To verify this, note that

$$\neg \forall x (P(x) \Rightarrow Q(x))$$

is equivalent to

$$\exists x (P(x) \wedge \neg Q(x)) \qquad \text{By Rules (2) and (5)}$$

Similarly,

$$\neg \exists x (P(x) \wedge Q(x))$$

is equivalent to

$$\forall x (\neg P(x) \vee \neg Q(x)) \qquad \text{By Rules (1) and (3)}$$

which may be written

$$\forall x (P(x) \Rightarrow \neg Q(x))$$

EXAMPLE 4.61 Find the negation of

$$(\forall x > 0)(\exists y > 0)(y^2 = x)$$

The negation is

$$(\exists x > 0)(\forall y > 0)(y^2 \neq x) \hspace{4cm} \circ$$

EXAMPLE 4.62 In Example 4.54, we wrote the definition of uniform continuity of a function $f: X \to \mathbf{R}$. The statement "f is *not* uniformly continuous" can be written

$$(\exists \varepsilon > 0)(\forall \delta > 0)(\exists x \in X)(\exists y \in X)(|x - y| < \delta \wedge |f(x) - f(y)| \geq \varepsilon) \hspace{1cm} \circ$$

EXERCISES 4.6

1. In each of the following formulas, indicate the scope of each quantifier. Determine which occurrences of individual variables are bound and which are free. For each bound occurrence of a variable, identify the quantifier that binds it.

   a. $\forall x P(x, y) \Rightarrow \exists y P(y, z)$

   b. $\forall x (\exists y Q(x, y) \Rightarrow \forall x Q(y, x))$

   c. $\exists y \exists x (Q(x, y, w) \Leftrightarrow \forall z P(x, y, z))$

2. Do the same for the following:

   a. $\exists w \forall x (P(w) \wedge Q(x) \wedge \exists x R(x, x))$ **b.** $P(x) \Rightarrow \exists x P(x)$

3. For each of the following pairs of formulas, determine whether the formulas can be obtained, one from the other, by change of bound variable. If not, state which condition of the rule for change of bound variable is violated.

   a. $\forall x \forall y (Q(y) \Rightarrow P(x))$ and $\forall x \forall x (Q(y) \Rightarrow P(x))$

   b. $\forall x \forall y (Q(y) \Rightarrow P(x))$ and $\forall y \forall x (Q(x) \Rightarrow P(y))$

   c. $\forall x \forall y Q(y)$ and $\forall x \forall x Q(x)$

4. Do the same for the following:

   a. $\forall x (\forall y Q(y) \Rightarrow \forall y P(y))$ and $\forall x (\forall x Q(x) \Rightarrow \forall y P(y))$

   b. $Q(y) \Rightarrow \forall z (R(y, z) \Rightarrow \forall w P(y, z, w))$ and
   $Q(y) \Rightarrow \forall y (R(y, y) \Rightarrow \forall w P(y, y, w))$

   c. $Q(y) \Rightarrow \forall z (R(y, z) \Rightarrow \forall w P(y, z, w))$ and
   $Q(y) \Rightarrow \forall w (R(y, w) \Rightarrow \forall w P(y, w, w))$

5. Write each of the following using qualified quantifiers.

   a. $\forall x [x \in A \Rightarrow \exists y (y \in B \wedge y^2 = x)]$

   b. $\exists x [x < 5 \wedge \forall y (y > x \Rightarrow 2y > 1)]$

   c. $\exists x \exists y (x < 0 \wedge y > 0 \wedge xy = 5)$

6. Write each of the following without qualified quantifiers.

   a. $(\exists x > 0)(\exists y > 0)(\forall z > 0)(x - y < z)$

   b. $(\forall x \in \mathbf{Q})(\exists y \in \mathbf{Z})(\exists z \in \mathbf{Z})\left(x = \dfrac{y}{z}\right)$

   c. $(\forall n \in \mathbf{Z})[\neg (\exists m \in \mathbf{Z})(m^2 = n) \Rightarrow \neg (\exists p \in \mathbf{Z})(\exists q \in \mathbf{Z})(p^2 = nq^2)]$

7. For each of the following statements about the real numbers, write a form of the negation of the statement in which the symbol $\neg$ does not appear.

   a. $(x > y) \Rightarrow (x^2 > z)$ **b.** $(x < y) \wedge (x + y \neq 5)$

   c. $(x \geq 7) \vee (x < 3)$ **d.** $\forall x (x \geq 7)$

   e. $\exists x (x = -1)$

8. Do the same for the following:

   a. $\forall x \exists y (xy = x)$ **b.** $\exists y \forall x (xy = x)$

   c. $\exists x \forall y [(y < 0) \Rightarrow (y^2 = x)]$ **d.** $\forall x \forall y \exists x [(x < z) \wedge (z < y)]$

     e. $(\forall x > 0)(\exists z > -1)[(z + x = 5) \Rightarrow (z < x)]$

     f. $(\forall x \in A)(\exists y \in B)(f(y) = x)$

9. Write each of the following statements using the symbolism of predicate logic. (Refer to Chapter 3 for definitions.)

     a. The relation R is not reflexive.   **b.** The relation R is not symmetric.

     c. The relation R is not transitive.

4.7 • VALID ARGUMENTS OF PREDICATE LOGIC

In Section 4.3, we determined whether an argument of the form "$A_1, A_2, \ldots, A_n$, therefore B" was valid by checking whether $A_1 \wedge A_2 \wedge \cdots \wedge A_n \Rightarrow B$ was a tautology. We want to do something similar for predicate logic. However, the situation is more complicated, since the formulas involve individual and predicate variables as well as propositional variables.

Definition 4.17

> A formula A of predicate logic is **valid** if, for all choices of a set X, and for all possible values for the propositional variables of A, and for all possible values (from the set X) for the free individual variables of A, and for all possible values for the predicate variables of A, the formula takes on the truth value T.

EXAMPLE 4.63   Show that the formula $\forall x P(x) \Rightarrow P(y)$ is valid.

     To find truth values for this formula, we need a nonempty set X, a propositional function P on X, and an element y of X. We consider two cases.

Case 1: If P is the constant function T on X, then $\forall x P(x)$ is T and so is $P(y)$. Therefore, from the truth table for implication, the value assigned to the formula $\forall x P(x) \Rightarrow P(y)$ is T.

Case 2: Suppose that P is not the constant function T. Then $\forall x P(x)$ is F, and therefore the value assigned to the formula is T.   ○

EXAMPLE 4.64   Show that the formula A given by

$$\forall x(p \Rightarrow P(x)) \Rightarrow (p \Rightarrow \forall x P(x))$$

is valid. To find truth values, we need a truth value for the proposition p, a nonempty set X, and a propositional function P on X. Again, we consider two cases.

Case 1: p has truth value F. Then $p \Rightarrow \forall x P(x)$ is T, and A is T.

Case 2: p has truth value T. Then if B is any formula, $p \Rightarrow B$ always has the same truth value as B. So in this case, the formula A reduces to the formula

$$\forall x P(x) \Rightarrow \forall x P(x)$$

which has value T.   ○

EXAMPLE 4.65   Show that $P(y) \Rightarrow \forall x P(x)$ is not valid.

If we choose

$$X = \{1, 2\}$$
$$P(z) = \text{“}z = 1\text{”}$$
$$y = 1$$

then the formula $P(y) \Rightarrow \forall x P(x)$ becomes

$$1 = 1 \Rightarrow \forall x(x = 1)$$

which is false. This choice of X, P, and y is called a **counterexample** to the formula $P(y) \Rightarrow \forall x P(x)$. ○

Note that if A is a formula of propositional logic, then A is valid if and only if A is a tautology. You can think of a valid formula as a formula that is always true, no matter what values are assigned to the variables in the formula. Conceptually, this is the same as in propositional logic. However, in predicate logic, there is no obvious, systematic way to determine whether a formula is valid.

EXAMPLE 4.66   Show that the argument

All prime numbers are odd.

17 is a prime number.

Therefore, 17 is odd.

is a valid argument. The argument is of the form

$$\forall x(P(x) \Rightarrow Q(x))$$
$$P(y)$$

Therefore, $Q(y)$

The reasoning pattern is valid because

$$[\forall x(P(x) \Rightarrow Q(x)) \wedge P(y)] \Rightarrow Q(y)$$

is a valid formula of predicate logic. The verification is left as an exercise. ○

Undecidability and Completeness

Beginning in the late nineteenth century, it became a goal of mathematical logicians to reduce standards of correct mathematical reasoning to precise rules. The objective was to be able, at least in principle, to decide the validity of an argument form by rote computation. In modern terms, we would want to be able to program a computer to determine whether a proposed proof is correct. One approach would be to find an algorithm that determines whether a formula of predicate logic is valid. However, in 1936 the logician Alonzo Church (b. 1902) showed that no such algorithm exists. Church's Theorem is known as the **undecidability** of predicate logic.

Another approach, developed before Church's result appeared, was based on the observation that certain operations on valid formulas produce new valid formulas.

In fact,

1. If A and $A \Rightarrow B$ are valid formulas, then so is B (This is the modus ponens rule.)
2. If A is a valid formula and if B is obtained from A by substituting expressions for variables, then B is valid.
3. If A is a valid formula and if B is obtained from A by change of bound variable, then B is valid.

In 1930, Kurt Gödel (1906–1978) showed that using these rules, every valid formula can be derived from tautologies and the two valid formulas

$$\forall x P(x) \Rightarrow P(y)$$
$$\forall x(p \Rightarrow Q(x)) \Rightarrow (p \Rightarrow \forall x Q(x))$$

This result is known as the **completeness of predicate logic.** It shows that an adequate definition of mathematical proof can be formulated by requiring that the logical principles used be the tautologies, the two formulas above, and further valid formulas derived from them.

EXERCISES 4.7

In Exercises 1 and 2, find the truth values assigned to the given formulas.

1. **a.** $P(x) \Rightarrow \forall x q$
    $q\colon T$
    set: $\{1, 2\}$
    $P\colon P(y) = $ "$y = 1$"
    $x\colon 2$

    b. $\forall x(P(x) \Rightarrow Q(x))$
    set: $\{1, 2, 3\}$
    $P\colon P(y) = $ "$y = 2$"
    $Q\colon Q(y) = $ "$y = 3$"

2. **a.** $\forall x(P(x) \Rightarrow \forall x P(x))$
    set: $\{1, 2\}$
    $P\colon P(y) = $ "$y = 1$"

    b. $\forall x(P(x) \Rightarrow \forall y Q(x, y))$
    set: $\{1, 2, 3\}$

P:

x	$P(x)$
1	T
2	F
3	T

Q:

x \ y	1	2	3
1	T	T	T
2	F	T	T
3	T	F	T

3. Which of the following formulas are valid? For each formula that is not valid, give a counterexample.

   a. p **b.** $(p \Rightarrow q) \land q \Rightarrow p$

   c. $P(x)$ **d.** $P(x) \Rightarrow P(y)$

4. Which of the following formulas are valid? For each formula that is not valid, give a counterexample.

   a. $P(x) \Rightarrow Q(x)$ **b.** $\neg(P(x) \Rightarrow Q(x))$

   c. $\forall x(P(x) \Rightarrow P(x)) \Rightarrow (P(x) \Rightarrow \forall x P(x))$

   d. $(Q(y) \Rightarrow \forall x P(x)) \Rightarrow \forall x(Q(y) \Rightarrow P(x))$

5. Which of the following formulas are valid? For each formula that is not valid, give a counterexample.

   a. $(\exists x P(x) \land \exists x Q(x)) \Rightarrow \exists x(P(x) \land Q(x))$

   b. $\forall x \exists y P(x, y) \Rightarrow \exists y \forall x P(x, y)$

   c. $\neg \forall x P(x) \Rightarrow \forall x \neg P(x)$ **d.** $\exists x \neg P(x) \Rightarrow \neg \exists x P(x)$

6. Do the same for the following:

   a. $\forall x(P(x) \land q) \Leftrightarrow (\forall x P(x) \land q)$ **b.** $\exists x(P(x) \land q) \Leftrightarrow (\exists x P(x) \land q)$

   c. $\forall x(P(x) \lor q) \Leftrightarrow (\forall x P(x) \lor q)$ **d.** $\exists x(P(x) \lor q) \Leftrightarrow (\exists x P(x) \lor q)$

   e. $\forall x(P(x) \land Q(x)) \Leftrightarrow \forall x P(x) \land \forall x Q(x)$

7. Verify that the formula in Example 4.66 is valid.

8. Find the error in the following reasoning. Consider the statement "A positive integer that is both even and prime is equal to 2." Let $p =$ "The integer is even," $q =$ "The integer is prime," and $r =$ "The integer is equal to 2." Then the statement is of the form $(p \land q) \Rightarrow r$. Now

$$[(p \land q) \Rightarrow r] \Rightarrow [(p \Rightarrow r) \lor (q \Rightarrow r)]$$

is a tautology, so from the given statement it follows that $(p \Rightarrow r) \lor (q \Rightarrow r)$, which says, "A positive integer that is even is equal to 2, or a positive integer that is odd is equal to 2."

Computer Exercises for Chapter 4

1. For each of the following expressions, write a sequence of statements in your computer programming language that declare suitable variables and perform the truth-value computation indicated by the expression. (*Example:* If the expression were $(p \land q) \Rightarrow r$ and you were writing in Pascal, you would write: Boolean p, q, r, x, y; $x := p$ and q; $y :=$ not x or r.)

   a. $\neg p \lor q$ **b.** $(\neg p \land q) \lor r$

   c. $(p \land \neg q) \lor \neg p \lor q$ **d.** $p \land q \lor \neg r \land s$

2. For each of the expressions of Exercise 1, write a program that prints out the truth table of the expression. (*Hint:* Put some of the statements from Exercise 1 in a loop.)

3. For each of the expressions of Exercise 1, write a program to determine whether the expression is a tautology.

4. Some programming languages, including Pascal, define an order relation on the set of truth values by declaring that F is less than T. In such languages, it is permitted to use the order operators of the language, <, <=, >, >=, in expressions containing Boolean variables. Express the connectives $\wedge$, $\vee$, and $\Rightarrow$ in terms of the order operators and NOT.

• • • •   CHAPTER REVIEW EXERCISES

1. If p is the proposition "6 is a perfect number" and q is "6 is prime," express in English each of the following propositions.

    a. $\neg p \wedge q$ **b.** $\neg p$ **c.** q NOR p **d.** $p \Rightarrow q$

2. If p is the proposition "The patient is hyperactive," and q is the proposition "The patient is ambidextrous," write the following propositions using logic symbols.

    a. Ambidexterity is necessary for hyperactivity.

    b. The patient is hyperactive only if she is ambidextrous.

    c. Hyperactivity is sufficient for nonambidexterity.

3. Construct the truth table for $p \wedge (p \Rightarrow q) \vee q$.

4. Determine which of the following are tautologies, contradictions (defined in Exercise 10 of Section 4.2), or neither.

    a. $\neg p \vee (p \wedge \neg q) \vee q$ **b.** $q \vee (p \wedge \neg q) \vee \neg p$

    c. $p \wedge (p \Rightarrow q) \Rightarrow q$ **d.** $p \Rightarrow (p \Rightarrow q) \Rightarrow q$

5. Determine which of the following are tautologies, contradictions, or neither.

    a. $\neg q \wedge (p \Rightarrow q) \Rightarrow \neg p$ **b.** $\neg q \Rightarrow (p \Rightarrow q) \Rightarrow \neg p$

    c. $p \wedge (p \Rightarrow q) \Leftrightarrow q$ **d.** $[(p \wedge \neg q) \Rightarrow \neg p] \Leftrightarrow (p \Rightarrow q)$

6. Express in terms of $\vee$ and $\neg$ only.

    a. $(p \vee p) \Rightarrow p$ **b.** $(q \Rightarrow r) \Rightarrow [(p \vee q) \Rightarrow (p \vee r)]$

    c. $[p \vee (q \vee r)] \Rightarrow [q \vee (p \vee r)]$

7. Express in terms of $\vee$ and $\neg$ only.

    a. $(p \Rightarrow q) \wedge r$ **b.** $(p \wedge q) \vee r$ **c.** $\neg p \Leftrightarrow (p \wedge \neg q)$

8. Express $(p \vee q) \Rightarrow p$ in terms of:

    a. NAND only   **b.** NOR only

9. Write the converse, inverse, and contrapositive of:

    a. $q \Rightarrow p$ **b.** $\neg p \Rightarrow \neg q$ **c.** $\neg q \Rightarrow \neg p$ **d.** $p \Rightarrow \neg q$

10. Show that the converse of the inverse of an implication is the contrapositive of the implication.

11. Show that NOR is not associative.

12. Determine whether the following argument is valid: Pascal wrote *Proof* or he did not invent the

transistor. If Euclid wrote *Proof*, then Pascal invented the transistor. Pascal did not both invent the transistor and write *Proof*. Therefore, Euclid did not write *Proof*.

13. What proof technique is used in the following?
    Theorem: If n and m are even integers, then $n + m$ is even.
    Proof:

$n = 2k$ for some $k \in \mathbf{Z}$	Definition of even
$m = 2j$ for some $j \in \mathbf{Z}$	
$n + m = 2k + 2j = 2(k + j)$	Factoring 2 from each term
$= 2i$ for some $i \in \mathbf{Z}$	Integers are closed under +

14. Prove that the sum of two odd integers is an even integer.

15. Prove that the sum of two rational numbers is a rational number.

16. If A is the alphabet and P is the propositional function on A defined by $P(x) =$ "x is followed by a vowel in 'newelpost,'" find the truth set of P.

17. If P is the propositional function on $\mathbf{Z}$ defined by $P(x) =$ "x is congruent to 2 or 3 mod 5," find the truth set of P.

18. Find the truth set of the following propositional functions on $\mathbf{R}$.
    a. $S(y) =$ "$(y < 8 \wedge y > -2) \vee (y^2 < 100)$"   **b.** $P(x) =$ "$x^3 - 2x^2 + 7x = 0$"

19. Find the truth set of the following propositional functions on $\mathbf{R}$.
    a. $P(x) =$ "$x < 1 \Leftrightarrow x < -1$"   **b.** $P(z) =$ "$z^2 > 1 \Rightarrow z = z - 7$"

20. Sketch the truth set of each of the following propositional functions on $\mathbf{R} \times \mathbf{R}$.
    a. $P(x, y) =$ "$(x = -3) \vee (x = 3)$"   **b.** $P(x, y) =$ "$x > 2 \Rightarrow y \le 3$"

21. Write the following definitions using the symbolism of predicate logic.
    a. [Definition of $x \mid y$] There is a z with $xz = y$.
    b. [Definition of a symmetric relation] If $a \, R \, b$, then $b \, R \, a$.
    c. [Definition of $\lim_{x \to a} f(x) = b$] For every positive number ε, there is a positive number δ such that $|f(x) - b| < \varepsilon$ whenever $0 < |x - a| < \delta$.

22. Which of the following statements about integers are true?
    a. $\forall x(x > x)$ **b.** $\exists x(x^2 - 3x + 15 = 0)$
    c. $\forall x(100x^2 - 230x + 132 > 0)$ **d.** $\forall x(x^2 > 2 \vee x^2 < 2)$

23. Write each of the following statements using the symbolism of predicate logic. (Refer to Chapter 3 for definitions.)
    a. The relation R is not irreflexive.   **b.** The relation R is not antisymmetric.
    c. The relation R is not asymmetric.

24. Which of the following statements about the real numbers are true?
    a. $\forall x \forall y(x^2 + y^2 \ge 0)$ **b.** $\forall x \exists y(x + 2y + 1 = 5)$
    c. $\exists x \forall y(|x| < |y|)$ **d.** $\forall x(x \le 0 \vee \exists y(x = y^4))$

25. Which of the following statements about the real numbers are true?
    a. $\forall x \forall y((x - y)(x + y) = x^2 - y^2)$ **b.** $\forall x \exists y((x + y)^2 = x^2 + y^2)$
    c. $\exists w(w > 51 \Rightarrow w + 1 = 3)$

26. In each of the following formulas, indicate the scope of each quantifier. Determine which occurrences of individual variables are bound and which are free. For each bound occurrence of a variable, identify the quantifier that binds it.

  a. $\forall x(P(x, y) \Rightarrow \exists y P(y, z))$

  b. $\exists x(\exists y Q(x, y) \Rightarrow \exists x Q(y, x))$

  c. $\exists y \exists x(\forall x \forall y \forall w Q(x, y, w) \Leftrightarrow P(x, y, z))$

27. Do the same for the following:

  a. $\forall x(\exists w P(w) \wedge Q(x) \wedge \exists w R(x, x))$ **b.** $\forall x P(x) \Rightarrow \exists x P(x)$

28. For each of the following pairs of formulas, determine whether the formulas can be obtained, one from the other, by change of bound variable. If not, state which condition of the rule for change of bound variable is violated.

  a. $\forall x(\forall y Q(y) \Rightarrow P(x))$ and $\forall x(\forall x Q(y) \Rightarrow P(x))$

  b. $\forall x(\forall y Q(y) \Rightarrow P(x))$ and $\forall y \forall x(\forall x Q(x) \Rightarrow P(y))$

29. Do the same for the following:

  a. $\forall x \forall y \exists y Q(y)$ and $\forall x \exists y \forall x Q(x)$

  b. $\forall x(\exists y Q(y) \wedge \forall y P(y))$ and $\forall x(\exists x Q(x) \wedge \forall y P(y))$

30. Write each of the following using qualified quantifiers.

  a. $\forall x[x > 5 \Rightarrow \exists y(y < 2 \wedge y = x^2)]$ **b.** $\exists x[x \in C \wedge \forall y(y > x \Rightarrow 2y > 1)]$

  c. $\forall x \forall y(x < 0 \wedge y > 0 \Rightarrow x + y = -1)$

31. Write each of the following without qualified quantifiers.

  a. $(\exists x \in A)(\exists y \in B)(\forall z > 5)(x + y = z)$ **b.** $(\forall x \in \mathbf{Z})(\exists y \in \mathbf{N})(\exists z \in \mathbf{N})(x = y - z)$

32. For each of the following statements about the integers, write a form of the negation of the statement in which the symbol $\neg$ does not appear.

  a. $(x \leq y) \vee (x \geq z)$ **b.** $(x \neq y) \Rightarrow (x - y < 5)$ **c.** $(x \geq 7) \wedge (x < 3)$

33. Do the same for the following:

  a. $\forall y(y < 17)$ **b.** $\exists z(z^2 = -1)$

  c. $\exists x \forall y(x + y > y)$ **d.** $\forall y \exists x(x > y^2 + 1)$

34. Find the truth values assigned to the given formulas.

  a. $\neg P(x) \Rightarrow \exists x q$ **b.** $\exists x(P(x) \vee Q(x))$
    q: F   set: $\{1, 2, 3\}$
    set: $\{1, 2, 3\}$ P: $P(y) = $ "$y > 2$"
    P: $P(y) = $ "$y = 1$" Q: $Q(y) = $ "$y > 3$"
    x: 3

  c. $\forall x(P(x) \Rightarrow \forall y Q(x, y))$
    set: $\{1, 2, 3\}$

P: x	$P(x)$
1	F
2	F
3	T

Q:

x \ y	1	2	3
1	T	F	T
2	F	T	F
3	T	F	T

35. Which of the following formulas are valid? For each formula that is not valid, give a counterexample.

   a. $\neg P$
   b. $\neg(p \Rightarrow \neg q) \Rightarrow p$
   c. $\neg\neg\neg P(x)$

36. Which of the following formulas are valid? For each formula that is not valid, give a counterexample.

   a. $P(x) \Rightarrow \neg P(y)$
   b. $\neg P(x) \Rightarrow Q(x)$
   c. $\neg(P(x) \Rightarrow \neg P(x))$

5

BOOLEAN ALGEBRA AND LOGIC CIRCUITS

In Chapter 4, we saw that the truth values T and F, under the operations of ∧ ("and") and ∨ ("or"), satisfy many of the same algebraic rules that sets satisfy under the operations of ∩ (intersection) and ∪ (union). In fact, combining propositional functions using ∧ and ∨ was naturally related to combining their truth sets using ∩ and ∪. In this chapter, we study systems that obey these rules, and we apply the principles to electronic circuit design.

5.1 • BOOLEAN ALGEBRAS

Sets with two binary operations that satisfy the same laws as ∧ and ∨ (or ∩ and ∪) are called Boolean algebras. Boolean algebra is named after the English mathematician George Boole (1815–1864), who formulated the algebraic rules for logic. This was done independently, at about the same time, by Augustus DeMorgan.

Definition 5.1

A **Boolean algebra** is a set B with two binary operations ∨ and ∧, satisfying the following:

1. Each operation is commutative.
2. Each operation is associative.
3. Each operation is distributive over the other:

$$x \wedge (y \vee z) = (x \wedge y) \vee (x \wedge z)$$
$$x \vee (y \wedge z) = (x \vee y) \wedge (x \vee z)$$

4. B contains identity elements 0 and 1 for ∨ and ∧, respectively, such that

$$0 \vee x = x \quad \text{and} \quad 1 \wedge x = x$$

5. Each element x of B has a complement x' such that

$$x \vee x' = 1 \quad \text{and} \quad x \wedge x' = 0$$

Here are some examples of Boolean algebras.

EXAMPLE 5.1   The set of truth values $\{T, F\}$ under the operations of conjunction and disjunction forms a Boolean algebra. This follows from the tautologies listed in Chapter 4.   ○

EXAMPLE 5.2   The set $P(X)$ of all subsets of a set X forms a Boolean algebra under the operations of union and intersection. The identity for union is $\varnothing$ and the identity for intersection is X.   ○

EXAMPLE 5.3   Let D_{30} be the set of positive divisors of 30. Then D_{30} is a Boolean algebra under the operations of g.c.d. (greatest common divisor) and l.c.m. (least common multiple). The identity for g.c.d. is 30, since if d is any divisor of 30, the greatest common divisor of d and 30 is d. The identity for l.c.m. is 1, since the least common multiple of d and 1 is d. The complement of the divisor d under both g.c.d. and l.c.m. is $\frac{30}{d}$. The commutative, associative, and distributive laws can be checked directly by testing all possible cases. The distributive laws can be stated as follows:

$$g.c.d.(l.c.m.(a, b), l.c.m.(a, c)) = l.c.m.(a, g.c.d.(b, c))$$

and

$$l.c.m.(g.c.d.(a, b), g.c.d.(a, c)) = g.c.d.(a, l.c.m.(b, c))$$ ○

EXAMPLE 5.4   The set of positive divisors of 4 is *not* a Boolean algebra under the operations of l.c.m. and g.c.d. If it were, the divisor 2 would have a complement x satisfying both g.c.d.$(x, 2) = 1$ and l.c.m.$(x, 2) = 4$. You can easily check that neither 1 nor 2 nor 4 satisfies both these equations.   ○

EXAMPLE 5.5   Let $X = \{p_1, p_2, \ldots, p_n\}$ be a set of prime numbers, and let

$$N = \prod_{i=1}^{n} p_i$$

Then the set D_N of positive divisors of N is a Boolean algebra under l.c.m. and g.c.d. The complement of a divisor d is $\frac{N}{d}$. This is a generalization of Example 5.3.   ○

The algebra in Example 5.5 looks very different from the familiar one in Example 5.2. However, there is a sense in which these algebras are "alike." We say that the Boolean algebras D_N and $P(X)$ are *isomorphic*.

Definition 5.2   Let B and C be two Boolean algebras. Use the symbols $\wedge$, $\vee$, and $'$ to denote the operations for both algebras. Then B and C are **isomorphic** if there is one-to-one correspondence $f: B \to C$ such that

    a. $f(x \wedge y) = f(x) \wedge f(y)$
    b. $f(x \vee y) = f(x) \vee f(y)$
    c. $f(x') = f(x)'$

EXAMPLE 5.6   Let X and N be as in Example 5.5, and define $f: P(X) \to D_N$ by the following: If $A \subset X$, then $f(A)$ is the product of the prime numbers in A. Then f is a Boolean algebra isomorphism. The empty set $\varnothing \subset X$ corresponds to the divisor 1, and

$f(X) = N$. Furthermore,

$$f(A \cap B) = \text{g.c.d.}(f(A), f(B))$$
$$f(A \cup B) = \text{l.c.m.}(f(A), f(B))$$
$$f(A') = \frac{N}{f(A)}$$

○

Properties of Boolean Algebras

The algebraic rules that hold for union and intersection of sets and for conjunction and disjunction of propositions hold in all Boolean algebras. The rules follow from the statements in Definition 5.1.

Theorem 5.1   In any Boolean algebra B, the identity elements 0 and 1 are unique.

Proof   The statement means that if $1_a \wedge x = x$ for all $x \in B$ and $1_b \wedge x = x$ for all $x \in B$, then $1_a = 1_b$; and similarly for 0. To see that $1_a = 1_b$, compute

$1_a = 1_b \wedge 1_a$	Because 1_b is an identity for $\wedge$
$= 1_a \wedge 1_b$	By the commutative law
$= 1_b$	Since 1_a is an identity for $\wedge$

The uniqueness of 0 is proved similarly and is left as an exercise.   ●

Theorem 5.2   In any Boolean algebra B, the complement x' of each element x is unique.

Proof   Suppose that a and b are two complements for x; that is, $x \vee a = 1$, $x \wedge a = 0$, $x \vee b = 1$, $x \wedge b = 0$. We want to show that $a = b$. Compute as follows:

$a = a \wedge 1$	Since 1 is the identity for $\wedge$
$= a \wedge (x \vee b)$	Since $x \vee b = 1$
$= (a \wedge x) \vee (a \wedge b)$	Distributive law
$= 0 \vee (a \wedge b)$	Since $a \wedge x = x \wedge a = 0$
$= (x \wedge b) \vee (a \wedge b)$	Since $0 = x \wedge b$
$= (x \vee a) \wedge b$	Distributive law
$= 1 \wedge b$	Since $x \vee a = 1$
$= b$	Since 1 is the identity for $\wedge$

●

Theorem 5.3   In any Boolean algebra, the idempotent laws hold: $x \wedge x = x$ and $x \vee x = x$.

Proof

$$x \wedge x = (x \wedge x) \vee 0$$
$$= (x \wedge x) \vee (x \wedge x')$$
$$= x \wedge (x \vee x')$$
$$= x \wedge 1$$
$$= x$$

The proof of the second part is obtained from the above by interchanging $\wedge$ and $\vee$ and interchanging 0 and 1.   ●

Theorem 5.4   In any Boolean algebra, $x \wedge 0 = 0$ and $x \vee 1 = 1$.

Proof

$$x \wedge 0 = (x \wedge 0) \vee 0$$
$$= (x \wedge 0) \vee (x \wedge x')$$
$$= x \wedge (0 \vee x')$$
$$= x \wedge x'$$
$$= 0$$

Again, the proof of the second formula is similar and is left as an exercise.   ●

Theorem 5.5   DeMorgan's Laws hold in any Boolean algebra: $(x \wedge y)' = x' \vee y'$ and $(x \vee y)' = x' \wedge y'$

Proof   To prove the first formula, we must show that $x' \vee y'$ is the complement of $x \wedge y$; that is, that $(x \wedge y) \vee (x' \vee y') = 1$ and $(x \wedge y) \wedge (x' \vee y') = 0$. To verify the first of these statements, compute

$$(x \wedge y) \vee (x' \vee y') = [x \vee (x' \vee y')] \wedge [y \vee (x' \vee y')]$$

By the distributive law

$$= [(x \vee x') \vee y'] \wedge [(y \vee y') \vee x']$$

By the commutative and associative laws

$$= [1 \vee y'] \wedge [1 \vee x']$$

By the definition of complement

$$= 1 \wedge 1 \qquad \text{By Theorem 5.4}$$

$$= 1 \qquad \text{By Theorem 5.3}$$

The statement that $(x \wedge y) \wedge (x' \vee y') = 0$ and the two parts of the second DeMorgan Law are proved by similar computations, which are left as exercises.   ●

The proofs of the rules that $0' = 1$ and $1' = 0$ and the **absorption laws** $x = x \wedge (x \vee y)$ and $x = x \vee (x \wedge y)$ are left as exercises.

The Duality Principle˙

By now you have noticed that all theorems about Boolean algebras come in pairs. The statements in Definition 5.1 are paired: $\wedge$ is commutative and so is $\vee$; $\wedge$ has identity element 1 and $\vee$ has identity element 0; and so on. In fact, given any formula A of Boolean algebra, we can form its **dual** A' by interchanging $\wedge$ and $\vee$ and interchanging 1 and 0. We may state this as follows.

The Duality Principle for Boolean Algebra.   If A is a theorem of Boolean algebra (that is, a statement that can be deduced from the statements in Definition 5.1), then its dual A' is also a theorem.

For example, the commutative law for $\wedge$ and the commutative law for $\vee$ are dual, as are the two associative laws. The dual of the DeMorgan Law $(x \wedge y)' = x' \vee y'$ is the other DeMorgan Law, $(x \vee y)' = x' \wedge y'$.

Boolean Algebras and Partial Order Relations

The Boolean algebra $P(X)$ of subsets of a set X is a partially ordered set under the subset relation $\subset$. This relation can be characterized in terms of the intersection operation: If A and B are subsets of X, then $A \subset B$ if and only if $A = A \cap B$. This fact can be extended to define a partial order relation $\leq$ on any Boolean algebra by defining $x \leq y$ to mean $x = x \wedge y$.

Theorem 5.6 If B is a Boolean algebra, then the relation $\leq$ on B defined by $x \leq y$ if $x = x \wedge y$ is a partial order.

Proof We must show that $\leq$ is reflexive, transitive, and antisymmetric.
Reflexive: $x \leq x$ means, by definition, $x = x \wedge x$, which is true by Theorem 5.3.
Transitive: If $x \leq y$ and $y \leq z$, then $x = x \wedge y$ and $y = y \wedge z$. So $x = x \wedge y = x \wedge (y \wedge z) = (x \wedge y) \wedge z = x \wedge z$, which means that $x \leq z$.
Antisymmetric: If $x \leq y$ and $y \leq x$, then $x = x \wedge y$ and $y = y \wedge x$. These say that $x = y$.   ●

Greatest Lower Bound and Least Upper Bound

When is a partially ordered set a Boolean algebra? The first requirement is that we must be able to reconstruct the $\wedge$ and $\vee$ operations from the partial order relation. In the algebra $P(X)$, the union of two sets A and B is the smallest set containing both A and B, and their intersection is the largest set contained in both A and B. These statements characterize union and intersection in terms of the relation $\subset$. We can extend the concept to arbitrary partially ordered sets with the following definition.

Definition 5.3

Let a and b be elements of a set X with a partial order relation $\leq$, and let $c \in X$. Then c is a **least upper bound** of a and b, written lub(a, b), if

  a. $a \leq c$ and $b \leq c$; and
  b. If $a \leq d$ and $b \leq d$ and $d \leq c$, then $d = c$.

Similarly, c is a **greatest lower bound** of a and b, written glb(a, b), if

  a. $a \geq c$ and $b \geq c$; and
  b. If $a \geq d$ and $b \geq d$ and $d \geq c$, then $d = c$.

EXAMPLE 5.7 Figure 5.1 is a Hasse diagram for a partially ordered set. In this set, x and y are both least upper bounds for a and b. The element y is the only least upper bound for b and c. The vertex z is an upper bound for a and c but is not a least upper bound. The vertices x and y have no least upper bound. Also, a and b are both greatest lower bounds for x and z.

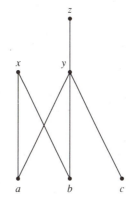

FIGURE 5.1   The graph of Example 5.7   ○

EXAMPLE 5.8   Let X be the set of all divisors (positive and negative) of 30, with the divisibility relation. Then 6 is a least upper bound of 2 and 3, and so is -6. Also, 3 is a greatest lower bound of 6 and -15, and so is -3. Note that in this context, the terms *greatest* and *least* are referenced to the divisibility relation on X, not to the standard order relation $\leq$ on X. With respect to the divisibility relation, 3 is "less than" -15, since $3 \mid -15$.   ○

Lattices

Definition 5.4   A partially ordered set L is called a **lattice** if each pair of elements of L has a unique least upper bound and a unique greatest lower bound.

If L is a lattice, we can define two binary operations $\wedge$ and $\vee$ on L by the rules:

$\quad x \wedge y$ is the greatest lower bound of x and y

$\quad x \vee y$ is the least upper bound of x and y

The operations $\wedge$ and $\vee$ so defined are often called the **meet** and **join** operations, respectively.

EXAMPLE 5.9   The Hasse diagram of Figure 5.2 is a lattice.

FIGURE 5.2   A lattice   ○

EXAMPLE 5.10   If N is a positive integer, the set D_N, with the "divides" relation, is a lattice; the join is the l.c.m. and the meet is the g.c.d., and these are unique in the set of positive divisors. But the set of *all* divisors of N is not a lattice; in this set, both l.c.m.(x, y) and $-$l.c.m.(x, y) are least upper bounds of x and y.   ○

Given a lattice L, we can ask whether it is a Boolean algebra under the operations of meet and join. If it is, it is called a **Boolean lattice.** In Example 5.5, if N is a product of distinct primes, then the set of positive divisors of N is a Boolean lattice. But if any prime factor of N occurs with exponent greater than 1, the divisors do not form a Boolean lattice. (See Example 5.4.)

EXAMPLE 5.11   Figure 5.3 shows another lattice that is not Boolean. In this lattice, the vertex a is the identity for $\wedge$, and the vertex d is the identity for $\vee$; that is, $a = 1$ and $d = 0$. The vertex b is a complement of e, since $e \vee b = a$ and $e \wedge b = d$. But c is also a complement of e! Since complements are unique in a Boolean algebra (Theorem 5.2), this lattice cannot be Boolean. It is left as an exercise to find which of the rules in the definition of Boolean algebra is violated.

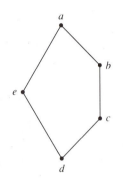

FIGURE 5.3   The lattice of Example 5.11   ○

Finite Boolean Algebras

You will have noticed that our examples of Boolean algebras lack variety: All algebras that we have seen are isomorphic to the algebra of subsets of a set X. In fact, all finite Boolean algebras are of this type. To see the correspondence, we need the following definition.

Definition 5.5   An element a of a Boolean algebra B is called an **atom** if a is a minimal nonzero element of B. That is, if $b \leq a$ and $b \neq 0$, then $b = a$.

EXAMPLE 5.12   In the algebra $P(X)$, the atoms are the singletons (sets with one element). In the algebra of positive divisors of a product of distinct primes, the atoms are the primes.   ○

Now let B be a finite Boolean algebra, and let A be the set of atoms of B. Define $f: B \to P(A)$ by the rule

$$f(x) = \{a \in A \mid a \le x\}$$

It can be shown that f is a Boolean algebra isomorphism.

EXERCISES 5.1

1. Does the set **R** of real numbers form a Boolean algebra under the operations $+$ and $\times$? Explain.

2. Show that the set of divisors of 12 is not a Boolean algebra under the operations of g.c.d. and l.c.m.

3. To verify directly the commutative, associative, and distributive laws for the Boolean algebra of Example 5.3, how many cases would you have to check?

4. In Example 5.5, what is the identity element for g.c.d.? For l.c.m.?

5. Show that in a Boolean algebra the identity 0 is unique.

6. Show that
   a. $0' = 1$
   b. $1' = 0$

7. Write the duals of the following expressions:
   a. $(x \wedge 1) \vee (y \vee 0)$
   b. $[(x \wedge y) \vee z]'$

8. Write the duals of the following expressions:
   a. $(x \wedge 0) \wedge (x \wedge 1)$
   b. $(x \wedge y \wedge 1)' \vee (z \vee 0)'$

9. Show that in a Boolean algebra, $x \vee x = x$.

10. Show that in a Boolean algebra, $x \vee 1 = 1$.

11. Complete the proof of the first DeMorgan Law (Theorem 5.5) by showing that $(x \wedge y) \wedge (x' \vee y') = 0$.

12. Prove the DeMorgan Law $(x \vee y)' = x' \wedge y'$.

13. Prove the absorption laws
   a. $x = x \wedge (x \vee y)$
   b. $x = x \vee (x \wedge y)$

14. In the Hasse diagram of Figure 5.4, find the greatest lower bounds and least upper bounds, if any, of each of the following pairs of vertices:
   a. g and h **b.** f and l **c.** c and d **d.** d and k

15. In the Hasse diagram of Figure 5.5, find the greatest lower bounds and least upper bounds, if any, of each of the following pairs of vertices:
   a. a and j **b.** h and g **c.** b and c **d.** h and j

16. Let X be the set of positive divisors of 60 with the divisibility relation. Find, if any, the greatest lower bound and least upper bound of the following elements:
   a. 2 and 5   **b.** 1 and 15   **c.** 5 and 6   **d.** 6 and 20

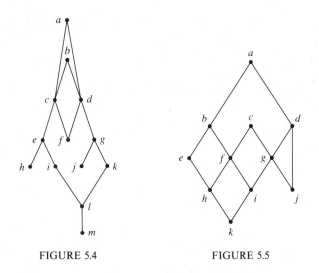

FIGURE 5.4   FIGURE 5.5

17. Let X be the set of positive divisors of 210 with the divisibility relation. Find, if any, the greatest lower bound and least upper bound of the following elements:

a.   2 and 5   **b.**   3 and 6   **c.**   5 and 7   **d.**   6 and 7

18. Let $P(X)$ be the set of subsets of X with the subset relation. If A and B are subsets of X, what is the least upper bound of A and B? The greatest lower bound?

19. Is the set of Exercise 14 a lattice? Explain.

20. Is the set of Exercise 15 a lattice? Explain.

21. Is the set **N** with the divisibility relation a lattice?

22. Which of the rules in the definition of a Boolean algebra is violated in the lattice of Example 5.11 (Figure 5.3)?

23. Find the atoms in the set of Example 5.3.

24. Find the atoms in the lattice of Figure 5.2.

25. Find the atoms in the lattice of Figure 5.6.

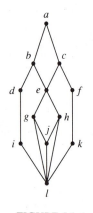

FIGURE 5.6

26. Let X be a set with a partial order relation $\leq$. An element $a \in X$ is a **universal lower bound** if $a \leq x$ for all $x \in X$; similarly, a is a **universal upper bound** if $a \geq x$ for all $x \in X$. Find the universal lower bounds and universal upper bounds, if any, in each of the following:

    a. Figure 5.5   **b.** $\mathbf{Z}$ with the usual order relation

27. Find the universal lower bounds and universal upper bounds, if any, in each of the following:

    a. Figure 5.6   **b.** $\mathbf{N}$ with the divisibility relation

28. **a.** Show that if a lattice contains a universal lower bound, it is unique.

    b. Show that if a lattice contains a universal upper bound, it is unique.

29. Let B be a Boolean algebra. Define the relation $\#$ on B to be: $x \# y$ if and only if $y = y \vee x$. Show that $\#$ is the same relation as that of Theorem 5.6.

30. A subset M of a lattice L is a **sublattice** if $x, y \in M$ implies that $x \vee y \in M$ and $x \wedge y \in M$. Which of the following subsets of the set of Figure 5.6 are sublattices?

    a. $\{l, i, k, a\}$ **b.** $\{i, h, d, b\}$ **c.** $\{a, c, k, l\}$

31. Which of the following subsets of the lattice of positive divisors of 210 are sublattices?

    a. $\{6, 35, 70, 1, 2\}$ **b.** $\{1, 5, 7, 35\}$

    c. $\{2, 6, 10, 14, 30, 42, 70, 210\}$

32. Prove or give a counterexample: If L is a lattice and $M \subset L$ is also a lattice, then M is a sublattice of L.

5.2 ● REPRESENTING BOOLEAN FUNCTIONS BY EXPRESSIONS

Now let us return to the simplest of all Boolean algebras, the algebra $B = \{F, T\}$ of truth values. The operations are conjunction ($\wedge$) and disjunction ($\vee$), and complementation is negation ($\neg$). A **Boolean function** of n variables is a function from B^n to B, where B^n is the Cartesian product of B with itself n times. Such a function is completely defined by a truth table with 2^n rows and a column for the function value. If $x_1, x_2, \ldots, x_n$ are n variables, an expression involving the x_i's and the operations of $\wedge$, $\vee$, and $\neg$ defines a Boolean function. We say that the expression represents the function. For example, the function f defined by the truth table

x	y	$f(x, y)$
T	T	T
T	F	F
F	T	T
F	F	T

is represented by the expression $\neg x \vee y$.

An expression that consists of a single variable or its negation, such as x or $\neg y$, is called a **literal.** An expression, such as $x \wedge y \wedge \neg z$, that is a conjunction of literals is called a **simple term.** An expression that is a disjunction of simple terms is called a **disjunctive form.** For example,

$$(x \wedge z) \vee (y \wedge \neg z)$$

is a disjunctive form.

Given a Boolean function, we want to write a disjunctive expression that represents the function. We will describe two methods, the disjunctive normal form (DNF) and Karnaugh maps. The DNF is highly systematic but often leads to long expressions. The Karnaugh map method is less systematic but produces shorter expressions. Both are important tools in circuit design, which we consider in the next section.

The Disjunctive Normal Form

The disjunctive normal form of a Boolean function is constructed directly from the truth table of the function. It contains one term for each T in the truth table. The method is best illustrated by examples.

EXAMPLE 5.13   Find the disjunctive normal form of $x \Rightarrow y$.

The truth table for $x \Rightarrow y$ is

x	y	$x \Rightarrow y$
T	T	T
T	F	F
F	T	T
F	F	T

From the truth table, we can see that $x \Rightarrow y$ is true in three cases:

a. x is T and y is T (that is, $x \wedge y$)
b. x is F and y is T (that is, $\neg x \wedge y$)
c. x is F and y is F (that is, $\neg x \wedge \neg y$)

By writing these three cases, joined by "or," we obtain the disjunctive normal form

$$(x \wedge y) \vee (\neg x \wedge y) \vee (\neg x \wedge \neg y) \qquad\qquad \circ$$

In a disjunctive normal form, each variable (or its complement) occurs in every term. A term containing every variable is called a **minterm,** so the disjunctive normal form is sometimes called the **minterm canonical form.**

EXAMPLE 5.14   Find the disjunctive normal form of the Boolean function defined by $(x \Leftrightarrow y) \Leftrightarrow z$. The truth table is

x	y	z	$(x \Leftrightarrow y) \Leftrightarrow z$
T	T	T	T
T	T	F	F
T	F	T	F
T	F	F	T
F	T	T	F
F	T	F	T
F	F	T	T
F	F	F	F

There are four cases having value T, so the disjunctive normal form will contain four terms. It is:

$$(x \wedge y \wedge z) \vee (x \wedge \neg y \wedge \neg z) \vee (\neg x \wedge y \wedge \neg z) \vee (\neg x \wedge \neg y \wedge z) \qquad \circ$$

Because disjunctive normal forms can be so long, the expressions are often abbreviated by omitting the $\wedge$ and indicating negation by drawing a bar over the variable. In this way, the DNF for Example 5.13 is written

$$xy \vee \bar{x}y \vee \bar{x}\bar{y}$$

and the DNF in Example 5.14 is

$$xyz \vee x\bar{y}\bar{z} \vee \bar{x}y\bar{z} \vee \bar{x}\bar{y}z$$

In a disjunctive normal form, there is one minterm for each T in the truth table of the expression. Moreover, each minterm of the expression involves each variable, either as the variable itself or as its negation. Thus, the expression $x \vee y \vee z$ is not in disjunctive normal form, because each term of the disjunction contains only one variable, not all three. If the truth table for an expression does not contain a T, then the disjunctive normal form is the empty expression (the expression containing no symbols).

Karnaugh Maps

Example 5.13 illustrates that the disjunctive normal form may be a very inefficient way to write the function as an expression involving only $\wedge$, $\vee$, and $\neg$. The short way to represent $x \Rightarrow y$ as a disjunctive form is, of course, $\neg x \vee y$. To see an even worse case, consider the function of three variables defined by $f(F, F, F) = F$ and $f(x, y, z) = T$ otherwise. The DNF for this function will contain seven terms! Karnaugh maps are a graphical device for finding simple disjunctive forms for Boolean functions in two to four variables. In fact, a Karnaugh map can be used to find the disjunctive form having the smallest number of terms.

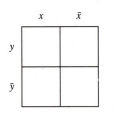

FIGURE 5.7   The two-variable Karnaugh map

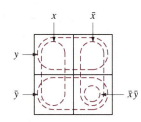

FIGURE 5.8   Some regions in the two-variable Karnaugh map

The Karnaugh map for two variables is the two-by-two square shown in Figure 5.7. Terms involving x, y, $\bar{x}$, and $\bar{y}$ correspond to regions of the square. For example, the region corresponding to the term x is the left column, and the region corresponding to the term $\bar{x}\bar{y}$ is the lower right box. Figure 5.8 shows these regions, as well as the regions corresponding to $\bar{x}$, y, and $\bar{y}$.

EXAMPLE 5.15   Again consider $x \Rightarrow y$, as in Example 5.13. The three T's in the truth table correspond to the three squares marked $+$ in Figure 5.9. The two regions outlined by dashed lines correspond to the terms $\bar{x}$ and y. The expression represented by the map is obtained by taking the disjunction of the terms corresponding to the outlined regions: $\bar{x} \lor y$.

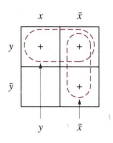

FIGURE 5.9   The Karnaugh map for $x \Rightarrow y$ ○

The three-variable Karnaugh map is shown in Figure 5.10. It is a two-by-four rectangle in which the left and right sides are considered as sewn together, forming a

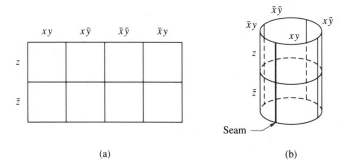

(a)   (b)

FIGURE 5.10   The three-variable Karnaugh map

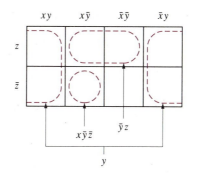

FIGURE 5.11   Some regions in the three-variable Karnaugh map

cylinder. The region corresponding to a term is permitted to overlap the seam. Figure 5.11 shows the regions corresponding to several terms. The region corresponding to the term y overlaps the seam and is drawn in two parts.

The order of xy terms across the top of the squares is chosen so that each term differs from its neighbors in only one variable. This order is different from the usual order in truth tables.

EXAMPLE 5.16   Consider the function f defined by the truth table

x	y	z	$f(x, y, z)$
T	T	T	T
T	T	F	F
T	F	T	F
T	F	F	F
F	T	T	T
F	T	F	T
F	F	T	F
F	F	F	T

Figure 5.12 shows the Karnaugh map for this function. Again, the squares marked + correspond to the T's in the truth table. Three one-by-two regions corresponding to the expressions yz, $\bar{x}\bar{z}$, and $\bar{x}y$, together include all the marked squares. The region

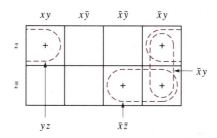

FIGURE 5.12   The Karnaugh map of Example 5.16

corresponding to the term $\bar{x}y$ can be ignored, since all squares in this region are also included in other regions. The function is represented by the expression $yz \lor \bar{x}\bar{z}$.

○

EXAMPLE 5.17   The Karnaugh map for the function f defined by $f(F, F, F) = F$ and $f(x, y, z) = T$ otherwise, is shown in Figure 5.13. Two two-by-two regions and one one-by-four region are required to include all seven marked squares. The expression is $x \lor y \lor z$.

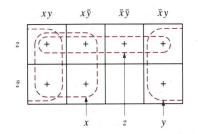

FIGURE 5.13   The Karnaugh map for Example 5.17

○

EXAMPLE 5.18   Figure 5.14 shows the Karnaugh map for the function f defined by

x	y	z	$f(x, y, z)$
T	T	T	F
T	T	F	T
T	F	T	T
T	F	F	T
F	T	T	T
F	T	F	T
F	F	T	T
F	F	F	F

Six one-by-two regions are shown, but the marked squares can be covered by either of two sets of three regions. The corresponding expressions are

$$\bar{y}z \lor \bar{x}y \lor x\bar{z} \quad \text{and} \quad \bar{x}z \lor y\bar{z} \lor x\bar{y}$$

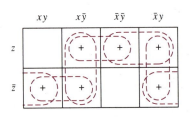

FIGURE 5.14   The Karnaugh map for Example 5.18

○

The four-variable Karnaugh map is shown in Figure 5.15. The two sides are considered as sewn together, and the top and bottom are also joined, forming a

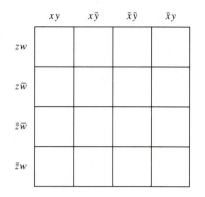

FIGURE 5.15   The four-variable Karnaugh map

torus. One-variable terms correspond to two-by-four regions, and two-variable terms correspond to regions that are two-by-two or four-by-one. A region can overlap the horizontal seam or the vertical seam or both.

EXAMPLE 5.19   Figure 5.16 shows the Karnaugh map for the function f defined by the table

x	y	z	w	$f(x, y, z, w)$	x	y	z	w	$f(x, y, z, w)$
T	T	T	T	F	F	T	T	T	F
T	T	T	F	T	F	T	T	F	T
T	T	F	T	F	F	T	F	T	F
T	T	F	F	T	F	T	F	F	T
T	F	T	T	T	F	F	T	T	F
T	F	T	F	F	F	F	T	F	T
T	F	F	T	T	F	F	F	T	F
T	F	F	F	F	F	F	F	F	F

Three regions are required to cover the marked squares. The expression is $y\bar{w} \vee x\bar{y}w \vee \bar{x}z\bar{w}$.

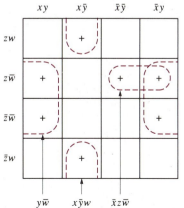

FIGURE 5.16   The Karnaugh map of Example 5.19

EXAMPLE 5.20   Figure 5.17 shows a Karnaugh map in which four one-by-two regions are required to cover the marked squares. All of these regions must be included in the expression for the function, since each covers a square that is not covered by any other region. The one-by-four region corresponding to the term $x\bar{y}$ is also shown, but it need not be included in the expression because all of the squares it includes are also included in the one-by-two regions. The expression is

$$zw\bar{y} \lor z\bar{w}x \lor \bar{z}\bar{w}\bar{y} \lor \bar{z}wx$$

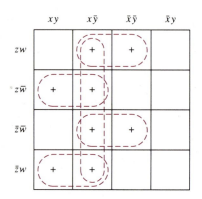

FIGURE 5.17   The Karnaugh map of Example 5.20   o

EXAMPLE 5.21   Figure 5.18 shows the worst case for a Karnaugh map in four variables. The regions containing only marked squares are one-by-one, and eight of them are required. In this case, the expression resulting from the Karnaugh map is the same as the DNF for the function.

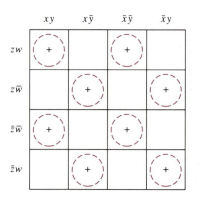

FIGURE 5.18   The Karnaugh map of Example 5.21   o

Finding Minimal Expressions

In general, the following steps yield the minimal expression for a Boolean function— that is, the expression that contains the smallest number of terms.

  1. List all terms corresponding to regions that are made up of marked squares and that are not subsets of any other such regions.

2. List the essential regions—that is, those that contain squares not contained in any other region. The essential regions will be included in the final expression.
3. By checking all possibilities, select the smallest set of other regions that suffices to cover the remaining marked squares completely.
4. The desired expression is the disjunction of the terms corresponding to the essential regions (step 2) and the selected other regions (step 3).

EXERCISES 5.2

1. Which of the following expressions are disjunctive normal forms?

    a. $x\bar{y} \vee x\bar{z} \vee x\bar{w}$ **b.** $y\bar{z}\bar{w} \vee z\bar{w}\bar{y} \vee w\bar{y}\bar{z}$

2. Which of the following are disjunctive normal forms?

    a. $xy\bar{z} \vee xyz \vee \bar{x}yz \vee xyz$ **b.** $\bar{x}y \vee yz \vee zw \vee w\bar{x}$

3. Find the disjunctive normal forms for the functions f and g defined by the following table:

x	y	$f(x, y)$	$g(x, y)$
T	T	F	F
T	F	T	T
F	T	F	T
F	F	T	F

4. Find the disjunctive normal forms for the functions defined by the expressions $x \Leftarrow y$ and $x \Leftrightarrow y$.

5. Find the disjunctive normal form for the following function:

x	y	z	$f(x, y, z)$
T	T	T	F
T	T	F	F
T	F	T	F
T	F	F	T
F	T	T	T
F	T	F	T
F	F	T	F
F	F	F	T

6. Find the disjunctive normal form for the following function:

x	y	z	$f(x, y, z)$
T	T	T	T
T	T	F	T
T	F	T	F
T	F	F	F
F	T	T	T
F	T	F	F
F	F	T	F
F	F	F	T

7. Draw the Karnaugh maps for the following functions and find the corresponding minimal expressions.

x	y	z	$f(x, y, z)$	$g(x, y, z)$	$h(x, y, z)$
T	T	T	T	T	F
T	T	F	T	F	T
T	F	T	F	F	F
T	F	F	T	T	F
F	T	T	T	T	T
F	T	F	T	T	T
F	F	T	T	T	T
F	F	F	F	F	T

8. Draw the Karnaugh maps for the following functions and find the corresponding minimal expressions.

x	y	z	$f(x, y, z)$	$g(x, y, z)$	$h(x, y, z)$
T	T	T	F	F	F
T	T	F	F	T	T
T	F	T	T	F	T
T	F	F	T	T	F
F	T	T	F	T	F
F	T	F	T	T	T
F	F	T	T	T	T
F	F	F	F	T	F

9. The correspondence between the squares of the three-variable Karnaugh map and the rows of a three-variable truth table in the usual order is described by the following diagram:

1	3	7	5
2	4	8	6

where the number in a square indicates the number of the corresponding row in the truth table. Make a similar diagram for the four-variable case.

10. Draw the Karnaugh maps for the following functions and find the corresponding minimal expressions.

x	y	z	w	$f(x, y, z, w)$	$g(x, y, z, w)$
T	T	T	T	F	F
T	T	T	F	F	T
T	T	F	T	F	F
T	T	F	F	T	F

x	y	z	w	$f(x, y, z, w)$	$g(x, y, z, w)$
T	F	T	T	T	T
T	F	T	F	F	T
T	F	F	T	T	T
T	F	F	F	F	F
F	T	T	T	F	F
F	T	T	F	T	T
F	T	F	T	F	F
F	T	F	F	T	F
F	F	T	T	F	T
F	F	T	F	F	T
F	F	F	T	F	T
F	F	F	F	F	F

11. Draw the Karnaugh maps for the following functions and find the corresponding minimal expressions.

x	y	z	w	$f(x, y, z, w)$	$g(x, y, z, w)$
T	T	T	T	F	T
T	T	T	F	F	T
T	T	F	T	T	T
T	T	F	F	T	F
T	F	T	T	F	F
T	F	T	F	F	T
T	F	F	T	T	T
T	F	F	F	T	F
F	T	T	T	F	T
F	T	T	F	T	F
F	T	F	T	T	T
F	T	F	F	T	F
F	F	T	T	T	F
F	F	T	F	F	T
F	F	F	T	T	T
F	F	F	F	T	T

12. Write the expression for the Karnaugh map of Figure 5.18.

5.3 • LOGIC CIRCUITS

We have seen how mathematical objects, such as numbers, sets, and functions, can be represented in a computer. These representations are imperfect models of the objects they represent. For instance, the data type "real" in a computer is only a finite subset of the set **R** of all real numbers. We now look at an entirely different representation of mathematical logic in a computer: representation of propositional logic in the electronic circuits of the computer itself.

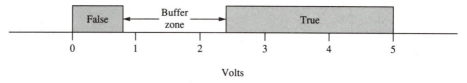

FIGURE 5.19   TTL logic levels

In an electronic circuit, the truth values T and F are represented by ranges of voltages that may appear on a signal line or wire. The ranges chosen depend on the types of electronic components used. If the voltage used to represent T is larger (more positive) than the voltage for F, the circuit is said to use **positive logic.** If the voltage used to represent F is larger, the circuit is said to use **negative logic.**

One very common convention using positive logic is:

T is represented by the range $+2.4$ volts to $+5.0$ volts;
F is represented by the range 0.0 volts to $+0.8$ volts.

These ranges are known as **TTL (transistor–transistor logic)** levels and are shown in Figure 5.19. Circuits using this convention use a single 5-volt power supply, so voltages above 5 and below 0 volts are irrelevant. Note, however, that there is a 1.6-volt-wide "buffer zone" between T and F. This is required because physical quantities such as voltage can never be measured perfectly. If there were no buffer zone and a voltage were close to the T–F boundary, it might be impossible to tell whether it represented T or F.

In circuit design, the truth values T and F are often called 1 and 0 or "high" and "low," the latter terms corresponding to the voltage levels representing the truth values.

Gates

A logical connective, such as "not," "and," or "or," is implemented by an electronic circuit called a **gate.** A gate that represents "not" and has one input and one output is called an **inverter.** Figure 5.20 shows the performance of a TTL inverter. Similarly, a gate that represents "and" has two inputs, x and y, and one output, $x \wedge y$.

If the input x to an inverter is T, the output y is F. If x changes from T to F, the output must change from F to T. But no physical quantity can change instantaneously; the output moves gradually out of the F range, through the buffer zone and into the T range. During this period, called the **gate delay,** the output of the gate is wrong or indeterminate. The circuit designer must be careful not to use the output of a gate until the gate delay has elapsed.

The speed of a computer is limited by the gate delays in its circuits. In early vacuum-tube computers, gate delays were measured in microseconds. Delays of 10 nanoseconds were achieved by the early 1970s; the fastest computers today have delays measured in picoseconds.

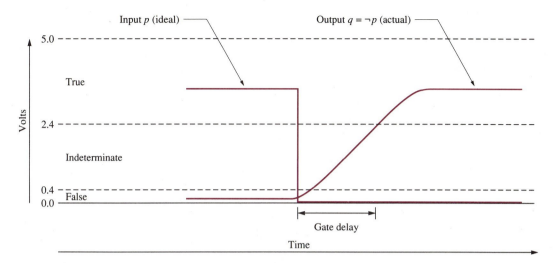

FIGURE 5.20 Actual performance of an inverter circuit

Implementing Boolean Expressions

Complex logic circuits are built by interconnecting gates. Circuits are represented by diagrams in which the gates are represented by symbols. Figure 5.21 shows the standard symbols for the six most common gates. Note the small circle on the right of the inverter, NAND, and NOR symbols. The circle indicates negation. The symbol for the NAND gate is formed by adding the circle to the symbol for the AND gate, and similarly for the NOR symbol.

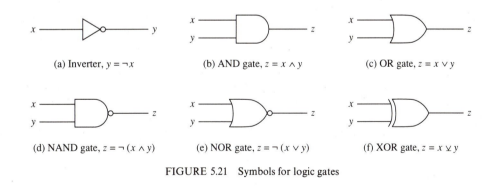

FIGURE 5.21 Symbols for logic gates

EXAMPLE 5.22 The expression $(x \wedge y) \vee (z \wedge w)$ contains three connectives. The results of the "and" operations are the inputs to the "or" operation. A circuit to implement this expression can be made from two AND gates and an OR gate. The circuit is shown in Figure 5.22.

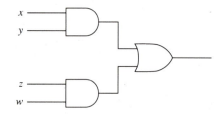

FIGURE 5.22   A logic circuit for $(x \wedge y) \vee (z \wedge w)$ o

EXAMPLE 5.23   The expression $(\neg x \vee y) \wedge (x \vee \neg z)$ can be implemented by the circuit of Figure 5.23.

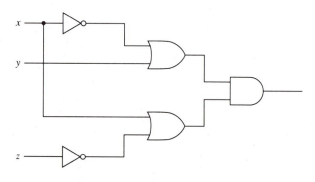

FIGURE 5.23   A logic circuit for $(\neg x \vee y) \wedge (x \vee \neg z)$ o

In electronic circuit design, it is possible to build AND and OR gates (as well as NAND and NOR gates) with more than two inputs. Also, a single signal line can be connected to inputs of several gates. Both concepts are illustrated in the next example.

EXAMPLE 5.24   The expression $\neg[\neg(x \wedge y) \wedge (\neg y \vee r \vee \neg t) \wedge (s \veebar \neg t)]$ contains a three-term AND, and one of the terms is a three-term OR. Both y and $\neg t$ are used in two places. The circuit for this expression is shown in Figure 5.24.

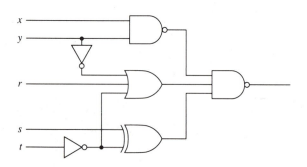

FIGURE 5.24   A logic circuit for $\neg(\neg(x \wedge y) \wedge (\neg y \vee r \vee \neg t) \wedge (s \veebar \neg t))$ o

EXAMPLE 5.25   Consider the task of designing a two's complement adder (see Section 1.6). The first step is to add the rightmost bits of the two summands, producing a sum bit and a carry bit. The circuit, called a **half adder,** has two inputs and two outputs; its required performance is defined by the table

x	y	Carry	Sum
1	1	1	0
1	0	0	1
0	1	0	1
0	0	0	0

which is nothing more than the base-2 addition table. The bit values 1 and 0 correspond to the truth values T and F, respectively. The carry column is easily recognized as the truth table for conjunction, while the sum column is the truth table for exclusive "or." Figure 5.25 shows two circuits that implement a half adder, the second of which uses only NAND gates and inverters. Although the first looks simpler, both are of equal complexity, since in actual practice, AND and OR gates are built out of NAND gates and inverters. At the other bit positions, there are three inputs to the addition: the bits in the two summands and the carry bit from the previous position. Again, two outputs are produced: a sum bit and a carry bit. The circuit that performs this function is called a **full adder.** The truth table and corresponding circuit are left as exercises.

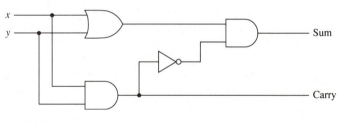

(a)

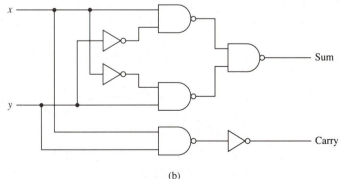

(b)

FIGURE 5.25   Two implementations of a half adder

Circuits for DNF and Karnaugh Expressions

Circuits corresponding to expressions obtained by the DNF and Karnaugh map methods are highly regular. They consist of three stages: an inversion stage, generating the negations of the inputs; an AND stage, with one gate for each term; and a final OR gate. Figure 5.26 shows the circuit for the DNF of $x \Rightarrow y$, from Example 5.13.

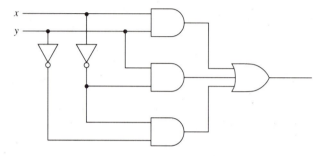

FIGURE 5.26   Circuit for the DNF of $x \Rightarrow y$

In most methods of gate design, the simplest and fastest gate is a NAND gate (or a NOR gate, depending on the type of transistors used). The negation operator $\neg$ can be implemented using a NAND gate because $\neg x = x$ NAND x; and $\wedge$ can be implemented using NAND gates because $x \wedge y = \neg(x$ NAND $y)$. The connective $\vee$ can also be written in terms of NAND alone by using the DeMorgan Law $x \vee y = \neg(\neg x \wedge \neg y) = (\neg x$ NAND $\neg y)$. It follows that every logical expression can be implemented by a single type of gate.

DNF expressions can be implemented in NAND gates alone by applying DeMorgan's Laws. Multiple-input gates are required for this technique. Suppose that the DNF for a Boolean function is

$$A_1 \vee A_2 \vee \cdots \vee A_n$$

where the A_i's are minterms. By DeMorgan's Laws, this expression is equivalent to

$$\neg(\neg A_1 \wedge \neg A_2 \wedge \cdots \wedge \neg A_n)$$

which is written

$$\text{NAND}(\neg A_1, \neg A_2, \ldots, \neg A_n)$$

Note that each argument $\neg A_i$ is itself a NAND, since $\neg A$ is equivalent to A NAND A.

EXAMPLE 5.26   Applying this transformation to Figure 5.26 (the circuit for the DNF $xy \vee \bar{x}y \vee \bar{x}\bar{y}$), we get

$$\neg[\neg(xy) \wedge \neg(\bar{x}y) \wedge \neg(\bar{x}\bar{y})]$$
$$= \text{NAND}\,[\neg(xy), \neg(\bar{x}y), \neg(\bar{x}\bar{y})]$$
$$= \text{NAND}\,(x\ \text{NAND}\ y, \bar{x}\ \text{NAND}\ y, \bar{x}\ \text{NAND}\ \bar{y})$$

This yields the circuit in Figure 5.27.

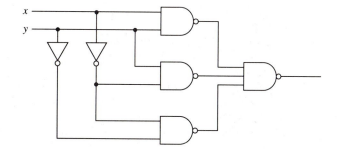

FIGURE 5.27   A NAND gate circuit equivalent to Figure 5.26   o

Figure 5.28 shows a NAND gate implementation of the DNF in Example 5.14. The same transformation can be applied to expressions derived from Karnaugh maps. In Example 5.15, we saw that the Karnaugh map method leads to the expression $\neg x \vee y$ for $x \Rightarrow y$. The direct implementation of this expression is shown in Figure 5.29(a); this can be transformed to the NAND gate shown in Figure 5.29(b). This is the simplest implementation of $x \Rightarrow y$.

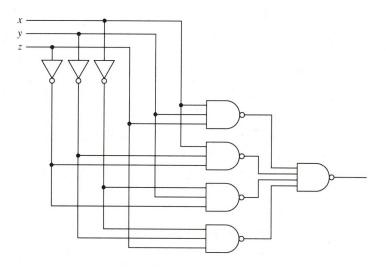

FIGURE 5.28   A NAND gate circuit for Example 5.14

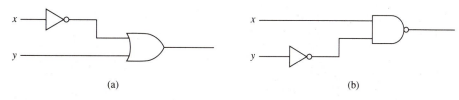

(a)   (b)

FIGURE 5.29   Implementation of $x \Rightarrow y$

The Minimization Problem

In this section, we have seen the basic relationship between Boolean algebra and digital circuit design. Boolean algebra is related to how circuit components are interconnected. Another mathematical problem in circuit design is the *minimization problem*. Each component of a circuit has a cost, which may be measured in dollars but is usually measured as the area (on a board or an integrated circuit) required by the component. The problem is to find the circuit of least cost from a given library of available components.

Karnaugh maps provide the simplest example of a circuit minimization method. However, the method cannot be implemented easily by a computer program. Other methods have been devised that handle many variables and can be programmed for a computer. Such methods are studied in advanced courses in digital circuit design.

EXERCISES 5.3

1. Refer to the gate symbols in Figure 5.30.

   a. Explain the meaning of symbol (a). How is it related to symbol (b)?

   b. Explain the meaning of symbol (c). How is it related to symbol (d)?

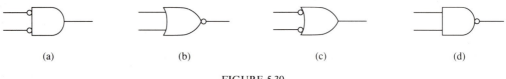

(a)   (b)   (c)   (d)

FIGURE 5.30

2. Write expressions corresponding to the circuits in Figure 5.31.

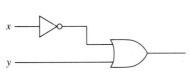

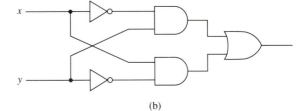

(a)   (b)

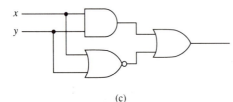

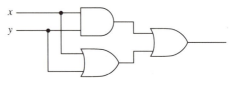

(c)   (d)

FIGURE 5.31

3. Draw circuit diagrams corresponding to each of the following expressions:

    a. $\neg\,\neg x$ **b.** $x \wedge \neg y$

    c. $(\neg x \vee y) \vee (x \vee \neg y)$

4. Draw circuit diagrams corresponding to each of the following expressions:

    a. $x \vee (y \wedge \neg z)$ **b.** $\neg(x \vee y) \wedge \neg(z \vee \neg w)$

    c. $(x \wedge y \wedge \neg z) \veebar (\neg x \wedge y \wedge z)$

5. Write expressions corresponding to the circuits in Figure 5.32.

6. Write expressions corresponding to the circuits in Figure 5.33.

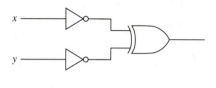

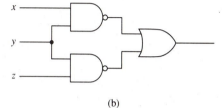

             (a)   (b)

FIGURE 5.32

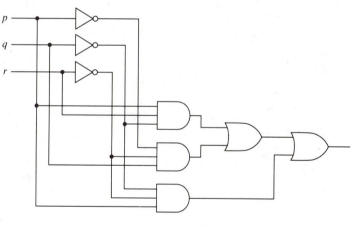

(a)

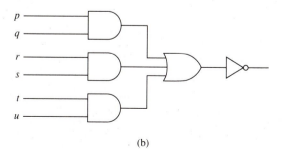

FIGURE 5.33

7. Using only inverters and (multiple-input) NAND gates, draw a circuit for the function of Example 5.16.

8. Do the same for the function of Example 5.17.

9. Do the same for the function of Example 5.18.

10. Do the same for the function of Example 5.19.

11. Construct the truth table that describes the performance of a full adder (see Example 5.25).

12. Design a circuit to implement a full adder.

5.4 • FEEDBACK AND MEMORY CIRCUITS

Up to this point, we have interconnected gates only to form circuits for logical expressions. In the circuits we have drawn, the inputs are on the left, the output is on the right, and signal flow is always from left to right. To make more sophisticated circuits, we can use feedback: The output of a circuit is "fed back" into the input. In a circuit with feedback, there is a loop in the signal flow path.

The simplest useful circuit with feedback is the **set–reset flip-flop** (or S–R flip-flop), shown in Figure 5.34. In this circuit, the output of each gate is fed back to one input of the other gate. The loop in the signal path is a figure-eight. To analyze this circuit, we must first write some equations describing it. There is one equation for each gate. The equations are

$$q = \neg(s \wedge q')$$
$$q' = \neg(r \wedge q)$$

We may look at this as a system of two equations in four unknowns. From our experience with algebraic equations, we may expect that there are several solutions. This is indeed the case. To find them, we can list all possible combinations of truth values for s, r, q, and q' and compute the truth values of $q \Leftrightarrow \neg(s \wedge q')$ and $q' \Leftrightarrow \neg(r \wedge q)$ in each case. The computations are left to you as an exercise. The solutions are shown in the following table.

s	r	q	q'	
T	T	T	F	(a)
T	T	F	T	(b)
F	T	T	F	(c)
T	F	F	T	(d)
F	F	T	T	(e)

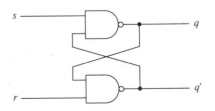

FIGURE 5.34   The S–R flip-flop

We can see from this table that if s or r is false (cases (c), (d), and (e)), the inputs s and r determine the outputs q and q'. But if s and r are both true (cases (a) and (b)), the circuit can be in either of two states. The circuit is stable in both of the states (a) and (b): The outputs will not change unless one of the inputs changes. Thus, when the inputs are both T, the circuit is a one-bit memory. It "remembers" what state it has been in.

The use of the flip-flop is illustrated in Figure 5.35. The figure plots the truth values of $s, r, q,$ and q' as functions of time. This type of diagram is called a **timing diagram.** At time T_1, the circuit is in state (b). It stays in this state until the input s changes to F. When s changes to F, the output of the top gate in Figure 5.34 changes to T, which causes the output of the lower gate to change to F. The circuit changes to state (c) at time T_2. Now when the input s is returned to the T state, neither gate changes its output. The circuit changes to state (a) and remains in this state.

The change in output state was caused by the **pulse** to F on the s input. A pulse is a short change in state of a signal line, after which the line returns to the previous state. The pulse on the s input is said to have **set** the q output to T. Similarly, a pulse on the r input will **reset** the q output to F. Throughout the sequence of events in Figure 5.35, the q' output is equal to the negation of the q output.

Notice that in the events of Figure 5.35, solution (e) of the equations plays no role. In fact, in state (e) it is not true that q' is the negation of q. This state is avoided in actual applications.

We have now seen how propositional logic contributes to the design of the two fundamental parts of a computer—the processor and the memory. The processor

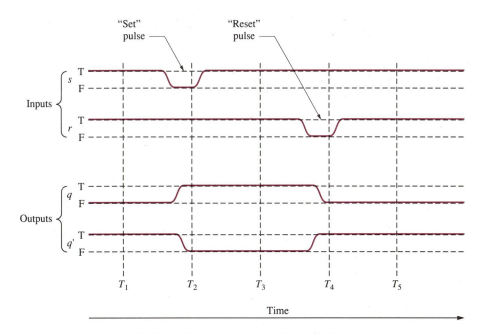

FIGURE 5.35   Setting and resetting a flip-flop

can be built of logic circuits, with the design based on formulas of Boolean algebra. Circuits with feedback can function as the computer's memory. Thus, it is possible in principle to build an entire computer out of NAND gates.

EXERCISES 5.4

1. Complete the analysis of the S–R flip-flop by computing the values of the expressions $q = \neg(s \wedge q')$ and $q' = \neg(r \wedge q)$ for all combinations of truth values of the variables.

2. **a.** Analyze the circuit in Figure 5.36(a) by writing a logical equation and finding all solutions.

    b. Analyze the circuit in Figure 5.36(b) by writing a logical equation and finding all solutions.

    c. What happens if the input x to the circuit in Figure 5.36(b) is T?

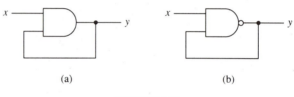

(a)   (b)

FIGURE 5.36

3. Figure 5.37 illustrates a more sophisticated memory circuit, the **D-type flip-flop,** or **latch.** It consists of an S–R flip-flop whose inputs are connected to some simple logic circuits. In this circuit, the d input is called the *data input* (hence the term *D-type flip-flop*), and the c input is called the clock input. (In this circuit, the q output equals the d input when c is true and remembers the previous d when c is false.)

    a. Write the equations describing the circuit.

    b. Find the solutions of the equations.

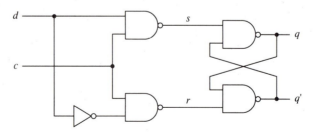

FIGURE 5.37   D-type flip-flop

Computer Exercises for Chapter 5

1. Write a program to read a table of values of a three-variable Boolean function and list the terms that appear in its DNF.

2. Do the same for functions of four variables.

3. Write a program to read a table of values of a three-variable Boolean function and print the corresponding Karnaugh map.

4. Do the same for functions of four variables.

• • • • CHAPTER REVIEW EXERCISES

1. Write the duals of the following expressions:
   a. $(x \wedge y') \vee (z \wedge 1)'$
   b. $\neg(p \vee F) \wedge (T \vee q)$
   c. $(A \cup B)' \cap C$ (where A, B, and C are subsets of X)

2. Write the duals of the following formulas:
   a. $\neg(\neg(x \vee y)) = \neg(\neg x \wedge \neg y)$
   b. $p \wedge F = F$
   c. $A \cup \emptyset = A$ (where A is a subset of X)

3. When is a totally ordered set a lattice?

4. When is a totally ordered set a Boolean lattice?

5. Define a partial order $\leq$ on the complex plane $\mathbf{C}$ by: $z \leq w$ whenever $\text{Re}(z) \leq \text{Re}(w)$ and $\text{Im}(z) \leq \text{Im}(w)$.
   a. Is this a lattice?
   b. If so, what are $\text{lub}(z, w)$ and $\text{glb}(z, w)$?
   c. If so, is it a Boolean lattice?

6. Let $X = \{1, 2, 3, 4, 5, 6, 7, 8, 9\}$, and let Y be the set of subsets of X having an even number of elements. Y is partially ordered by the inclusion relation $\subset$.
   a. Is this a lattice?
   b. If so, what are $\text{lub}(z, w)$ and $\text{glb}(z, w)$?
   c. If so, is it a Boolean lattice?

7. Let B be a Boolean algebra, and define $\leq$ on B by the rule $x \leq y \Leftrightarrow x = x \wedge y$. Show that, with respect to the partial order $\leq$, B is a lattice, and that $\text{lub}(x, y) = x \vee y$ and $\text{glb}(x, y) = x \wedge y$.

8. Which of the following expressions are in disjunctive normal form?
   a. $(\neg p \wedge r) \vee (p \wedge \neg r)$
   b. $(A \cap B \cap C') \cup (C \cap B' \cap A') \cup (B' \cap C' \cap A')$
   c. $\bar{x}\bar{y}\bar{z}w \vee x\bar{y}\bar{z}\bar{w} \vee \bar{x}yz\bar{w} \vee \bar{x}\bar{z}\bar{w}$

9. Find the disjunctive normal forms of the Boolean functions defined by the following expressions:
   a. $(x \vee y) \Rightarrow \neg z$
   b. $p \Rightarrow (q \wedge r)$

10. Find minimal disjunctive forms for the Boolean functions of Exercise 9.

11. Write expressions corresponding to the circuit diagrams of Figure 5.38.

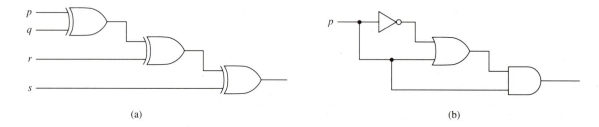

(a)

(b)

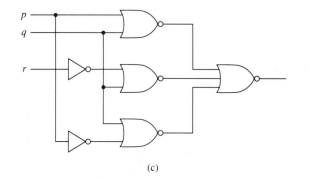

(c)

FIGURE 5.38

12. Draw circuits for the following expressions:

  a. $x \lor y \lor \neg r \lor s$ **b.** $(x \land \neg x) \veebar (x \lor \neg x)$ **c.** $r \land [\neg y \lor (s \veebar r)]$

13. Analyze the circuit of Figure 5.39. What does it do?

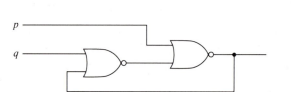

FIGURE 5.39

14. Analyze the circuit of Figure 5.40. (*Hint:* Consider the effects of gate delay.)

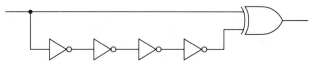

FIGURE 5.40

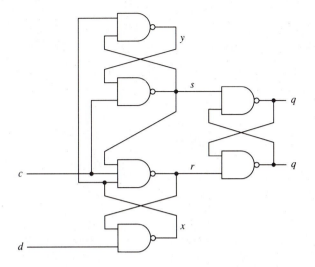

FIGURE 5.41   Edge-triggered D-type flip-flop

15. Analyze the edge-triggered flip-flop circuit in Figure 5.41 by writing a logical equation and finding all solutions. (This circuit remembers the value of the d input as of the time that the c input changed from F to T.)

16. Let B be the set of all subsets X of $\mathbf{N}$ such that either X or $\mathbf{N} - X$ is finite.

    a. Show that B is a Boolean algebra.

    b. Show that B is not isomorphic to the algebra $P(Y)$ for any set Y.

6

COMBINATORICS AND DISCRETE PROBABILITY

Quite often it is necessary to determine the number of elements in a set. For instance, in Chapter 2, we counted the number of subsets of a set. In this chapter, we introduce some counting techniques that apply to many situations. The branch of mathematics that deals with counting problems is called *combinatorics*.

One of the principal applications of combinatorics is probability problems. To determine the probability of winning a game of chance, it is usually necessary to count the possible outcomes of a play and also to count the number of outcomes on which you win.

6.1 • COUNTING TECHNIQUES

Your local pizzeria offers three sizes of pizza—small, medium, and large—and two flavors—plain and sausage. How many different pizzas can you order? One way to answer this question is to list all the possibilities: small plain, small sausage, medium plain, medium sausage, large plain, and large sausage. Thus, you have a choice of six different pizzas.

A second way to solve this problem is to draw a tree diagram as shown in Figure 6.1. The first set of edges of the diagram represents the different sizes possible, and the second set of edges represents the flavors. The number of possible pizzas is

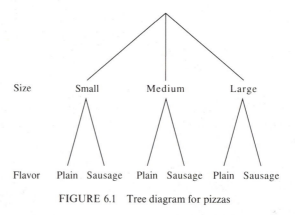

FIGURE 6.1   Tree diagram for pizzas

the total number of paths from the top of the diagram (called the **root** of the tree) to the bottom of the diagram (called the **leaves** of the tree). Of course, this number is also equal to the number of leaves.

EXAMPLE 6.1   As an opponent of all-digit dialing, you want to change your telephone exchange from 52 to an exchange beginning with two letters. How many ways can you do this?

Since the 5 corresponds to the letters J, K, and L and the 2 to the letters A, B, and C, the possibilities are:

JA	JB	JC
KA	KB	KC
LA	LB	LC

Thus, there are nine possible names for the exchange. The tree diagram solution is shown in Figure 6.2.

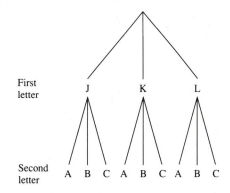

FIGURE 6.2   Tree diagram for telephone exchanges beginning with 52   o

EXAMPLE 6.2   The Math Department at Cabin Creek College offers three sections of calculus during the daytime and two sections of discrete math in the evening. If you want to take at least one math course, how many ways can you arrange your schedule? Let C1 denote the first section of calculus, D2 the second section of discrete math, etc. Then the possibilities are:

One math course: C1, C2, C3, D1, D2
Two math courses: C1 + D1, C2 + D1, C3 + D1, C1 + D2,
                    C2 + D2, C3 + D2

There are 11 possible schedules. They can be organized into the tree diagram of Figure 6.3.   o

Of course, it is not always convenient to list all possible choices or to draw a tree. However, there are two basic principles of counting that can be applied to most problems: the multiplication rule and the addition rule.

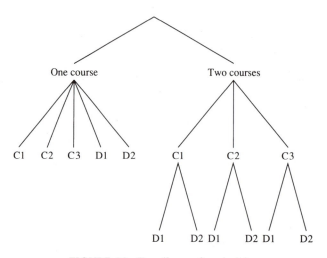

FIGURE 6.3   Tree diagram for schedules

Multiplication rule.   If a first task can be performed in n ways and a second task can be performed in m ways, the number of ways to perform the first task and then the second task is the product n × m. The rule extends to more than two tasks. In general, if a first task can be performed in n_1 ways, a second task in n_2 ways, . . . , a kth task in n_k ways, then the number of ways to perform the first task followed by the second, . . . , followed by the kth task is the product $n_1 \times n_2 \times \cdots \times n_k$

Addition rule.   If task A and task B are **mutually exclusive** (that is, if one is performed, then the other cannot be performed), and task A can be performed in m ways and task B in n ways, then exactly one of the tasks can be performed in $m + n$ ways. The rule extends to more than two tasks. If $A_1, A_2, \ldots, A_k$ are k mutually exclusive tasks, and A_1 can be performed in n_1 ways, A_2 in n_2 ways, and so on, then exactly one of the tasks can be performed in $n_1 + n_2 + \cdots + n_k$ ways.

In the pizza problem that began this discussion, the multiplication rule applies: The number of pizzas is $3 \times 2 = 6$. In Example 6.1, the number of telephone exchanges is $3 \times 3 = 9$, by the multiplication rule. In Example 6.2, both rules are used. There are $3 \times 2 = 6$ ways to arrange your schedule if you take both courses; $3 + 2 = 5$ ways to take exactly one course; and $6 + 5 = 11$ possible schedules.

EXAMPLE 6.3   How many ways can you fill an 8-bit byte? The first task is to fill the first bit. This can be done in two ways—with a 0 or a 1. The second task is to fill the second bit, and so on. Since each of the eight bits can be filled in two ways, the byte can be filled in

$$2 \times 2 \times 2 \times 2 \times 2 \times 2 \times 2 \times 2 = 2^8 = 256$$

ways.   ○

The tree diagram for Example 6.3 would be quite large. However, it is easy to understand the structure of the tree: At each stage, each branch splits in two. A tree diagram is often a useful way to think about a problem, even if drawing it is impractical.

EXAMPLE 6.4   How many nine-digit numbers can be formed from the digits 0, 1, 2, 3, 4, 5, 6, 7, 8, 9 if a number cannot begin with the digit 0 and no digit can be used twice?

Consider each of the nine positions and how many ways you can fill it.

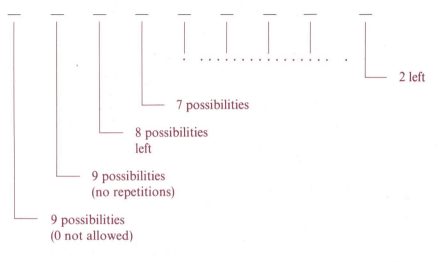

2 left

7 possibilities

8 possibilities left

9 possibilities (no repetitions)

9 possibilities (0 not allowed)

Using the multiplication rule, there are $9 \times 9 \times 8 \times 7 \times 6 \times 5 \times 4 \times 3 \times 2 = 3{,}265{,}920$ nine-digit numbers possible.   ○

EXAMPLE 6.5   Suppose your computer password may contain not more than four characters, must begin with a letter, and may contain letters and digits only. (Assume that the computer does not distinguish between upper- and lowercase letters in passwords.) How many different passwords can you create? Consider all one-character passwords, all two-character passwords, and so on.

One-character passwords. Since the password must begin with a letter, there are 26 possible.
Two-character passwords. There are 26 choices for the first character and 36 choices (26 letters + 10 digits) for the second, for a total of $26 \times 36 = 936$.
Three-character passwords. Using the same reasoning, there are $26 \times 36 \times 36 = 33{,}696$ possibilities.
Four-character passwords. There are $26 \times 36 \times 36 \times 36 = 1{,}213{,}056$ of these.
Now apply the addition rule to get $26 + 936 + 33{,}696 + 1{,}213{,}056 = 1{,}247{,}714$ passwords.   ○

The Inclusion–Exclusion Principle

In this section, we look at a technique for counting the number of elements in the union of sets. The following examples illustrate the principle of **inclusion–exclusion.**

EXAMPLE 6.6   You listen to 28 of your favorite operatic arias and find that 19 are about love, 14 are about death, and 8 are about both. How many arias are about love only? How many are about neither? Let L be the set of arias about love, and D be the set of arias about death. The sets L and D are illustrated in the Venn diagram in Figure 6.4.

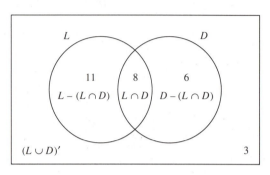

FIGURE 6.4   Venn diagram for Example 6.6

The number of arias about love only is the number of elements in the set $L - (L \cap D)$—namely, $19 - 8 = 11$. The number of arias about neither is equal to the number of elements in the set $(L \cup D)'$. To compute this, note that the number of arias about death only is the number of elements in the set $D - (L \cap D)$, which is $14 - 8 = 6$. So the number of elements in $L \cup D$ is the sum $11 + 8 + 6 = 25$, and the number of elements in $(L \cup D)'$ is $28 - 25 = 3$.   ○

In the next example, there are three sets involved. Although there are formulas for working out the answers, it is easier to deal with Venn diagrams.

EXAMPLE 6.7   In a survey of 80 college students, it was found that 35 play bridge, 54 play frisbee, 23 play go, 8 play all three games, 17 play both bridge and go, 23 play both bridge and frisbee, and 10 play both frisbee and go. How many play go only? How many play none of the three games?

Let B be the set of students who play bridge, F the set who play frisbee, and G the set who play go. We begin from the inside out—that is, by filling in the number of elements in the set $B \cap F \cap G$, obtaining the diagram in Figure 6.5(a). Next, we compute the number of students who play bridge and go only. This is the number in the set $(B \cap G) - (B \cap G \cap F)$—namely, $17 - 8 = 9$. We now have Figure 6.5(b).

Continuing to work from the inside out, we compute the number of students who play bridge and frisbee only, then frisbee and go only. The results thus far are shown in Figure 6.5(c).

Now compute the number of students who play bridge only. Since 35 students play bridge and $9 + 8 + 15 = 32$ are already accounted for, that leaves $35 - 32 = 3$ who play only bridge. Similarly, we can compute the number of students who play frisbee only and the number who play go only. Figure 6.5(d) shows the results so far.

Finally, the number of students who play none of the games is the number of elements in the set $(B \cup F \cup G)'$. To compute this, add all the numbers in the Venn

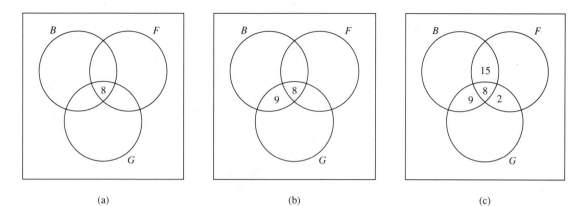

FIGURE 6.5   The diagram of Example 6.7

diagram so far and subtract this sum from the total number surveyed. This leaves 10 students who play none of the games. The final results are given in Figure 6.5(e).

In order to write down the inclusion–exclusion formulas, we need some additional notation for the number of elements in a set. We will denote the number of elements in a set X by $n(X)$.

Inclusion–exclusion formula for two sets.

$$n(A \cup B) = n(A) + n(B) - n(A \cap B)$$

Inclusion–exclusion formula for three sets.

$n(A \cup B \cup C)$
$$= n(A) + n(B) + n(C) - n(A \cap B) - n(B \cap C) - n(A \cap C) + n(A \cap B \cap C)$$

To understand the latter formula, we will draw a Venn diagram and label each region. The diagram is Figure 6.6. By adding the elements in A, B, and C, we count the items in regions R_2, R_3, and R_4 twice, and the items in region R_1 are counted

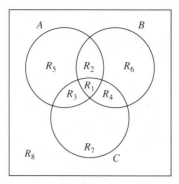

FIGURE 6.6   Venn diagram for three sets

three times. However, by subtracting the number of elements in the three pairwise intersections, we subtract the number of elements in $A \cap B \cap C$ (region R_1) three times so that the items in that region are not counted at all. Therefore, we must add back the number of elements in $A \cap B \cap C$.

General inclusion–exclusion formula.   If $A_1, A_2, \ldots, A_k$ are k sets, then

$$
\begin{aligned}
n(A_1 \cup A_2 \cup \cdots \cup A_k) = {} & n(A_1) + n(A_2) + \cdots + n(A_k) \\
& - n(A_1 \cap A_2) - n(A_1 \cap A_3) - \cdots + n(A_1 \cap A_2 \cap A_3) \\
& + n(A_1 \cap A_2 \cap A_4) + \cdots - n(A_1 \cap A_2 \cap A_3 \cap A_4) \\
& - \cdots + (-1)^{n+1} n(A_1 \cap A_2 \cap \cdots \cap A_k)
\end{aligned}
$$

The general formula states that to find the number of elements in the union of k sets, you first add the number of elements in each of the sets, then subtract the number of elements in each of the pairwise intersections, then add the number of elements in each intersection of three sets, then subtract the number of elements in each intersection of four sets, and so on. Proving that this works is left as an exercise.

EXERCISES 6.1

1.   How many ways are there to fill a 16-bit word?

2.   If $A = \{a, b, c\}$, how many functions are there from A to A?

3.   How many ways can you answer a 20-question multiple-choice test if there are 5 choices for each question?

4.   How many k-letter words can be formed from n letters, if any letter can be used any number of times?

5.   The population of Olympia is approximately 18,273. Show that at least two people in Olympia have the same initials. (Note that some people do not have middle names.)

6.   A burglar alarm key pad contains nine digits. The disarm code consists of four digits.

a. How many disarm codes are possible?

b. If a burglar can hit 20 codes in one minute, how long would it take to try every possible code?

c. How many disarm codes are possible if no two consecutive numbers can be the same?

7. How many ways can you pick five numbers for a lottery ticket if each number must be in the range 0–99?

8. A type of combination lock consists of a dial with 40 numbers from 1 through 40. How many three-number combinations are possible if the second number must differ from the first and third?

9. The licensing department issues plates with three letters followed by three digits for cars and two letters followed by four digits for trucks—except that there are 137 "objectional" three-letter prefixes that cannot be used. How many vehicles can be licensed?

10. If you live in the state of Washington, your driver's license number consists of the first five letters of your last name, your middle initial, four digits, and one letter. If you don't have a middle initial, or have a short name, asterisks are substituted for letters. How many license numbers are possible? (Assume that all last names have at least two letters).

11. A deck of cards contains 52 cards. How many ways can you fill a 5-card poker hand?

12. A PTA group has 40 members. How many ways can the slate of president, vice-president, secretary, and treasurer be chosen?

13. Use the multiplication rule to verify that a set of n elements has 2^n subsets.

14. In a study of 45 female mathematicians, it was found that 26 were from Protestant backgrounds, and 7 of the 26 were rated high in creativity. Of the women from non-Protestant backgrounds, 13 were rated high in creativity. (Fox, Brody, and Tobin (eds.), *Women and the Mathematical Mystique*, Johns Hopkins University Press, 1980.) How many women from non-Protestant backgrounds were rated low in creativity?

15. Of 20 high school students surveyed, 12 were in favor of mandatory busing, 10 were in favor of school-based clinics, and 6 were in favor of both.

a. How many were in favor of neither?

b. How many were in favor of busing only?

16. A survey of 100 college students showed that 30 played soccer, 60 played violin, and 25 played neither.

a. How many students played both?

b. How many played soccer only?

c. How many played violin only?

17. A survey of movie lovers produced the following results: 30 like war movies, 42 like love stories, 55 like comedies, 4 like war movies and love stories, 20 like war movies and comedies, 28 like love stories and comedies, 2 like all three, and 22 like none of the three.

   a. How many people were surveyed?

   b. How many like war movies only?

   c. How many like comedies only?

   d. How many like love stories only?

18. State the inclusion–exclusion formula for four sets.

19. If A is a set of n elements, how many reflexive relations on A are there?

20. If A is a set of n elements, how many symmetric relations on A are there?

21. Prove by induction the general inclusion–exclusion principle.

6.2 • PERMUTATIONS

It is often necessary to count the number of arrangements or orderings of the elements of a set. For instance, suppose we want to count the number of arrangements of the set of letters in the word *bit*. One way to do this is to list all the possibilities. There are six:

   bit   bti   ibt   itb   tbi   tib

Another way to do this is to draw a tree diagram as shown in Figure 6.7 and count the number of leaves.

A third way to approach this problem is to consider it as a sequence of three tasks, where each task is to choose a letter. There are three ways to choose the first letter, two ways to choose the second, and one way to choose the third. By the multiplication rule, the answer is $3 \times 2 \times 1 = 6$.

An arrangement of the elements of a set in a particular order is called a **permutation** of the set. In a permutation, repetition of the elements is not allowed. The

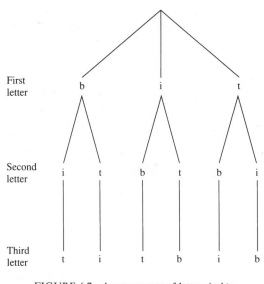

FIGURE 6.7   Arrangements of letters in *bit*

multiplication rule for counting leads to an important formula for the number of permutations of a set.

Theorem 6.1   The number of permutations of a set having n elements is $n!$.

Proof   There are n ways to choose the first object in the arrangement, leaving $(n - 1)$ ways to choose the second, $(n - 2)$ ways to choose the third, and so on. By the multiplication rule, the number of permutations is

$$n(n - 1)(n - 2) \cdots (3)(2)(1) = n!$$ ●

EXAMPLE 6.8   How many nine-digit numbers can be formed from the digits 1, 2, 3, 4, 5, 6, 7, 8, 9 if no digit can be used twice?

This is a permutation of nine objects, so there are $9! = 362,880$ numbers that can be formed.   ○

Permutations of n Objects r at a Time

In this discussion, we consider the orderings of r elements of a set of size n, where $r \leq n$. Such orderings are called *permutations of n objects r at a time*. The number of such permutations is denoted by $P(n, r)$.

EXAMPLE 6.9   How many four-letter words can be made from the alphabet if no letter is repeated? There are 26 ways to choose the first letter, 25 ways for the second letter, 24 for the third, and 23 for the fourth. Therefore, there are $26 \times 25 \times 24 \times 23 = 358,800$ words. That is, $P(26, 4) = 358,800$.   ○

The reasoning of Example 6.9 leads to a general formula for $P(n, r)$.

Theorem 6.2   The number $P(n, r)$ of permutations of n objects taken r at a time is given by

$$P(n, r) = \prod_{i=n-r+1}^{n} i = \frac{n!}{(n - r)!}$$

Proof   By the multiplication rule,

$$P(n, r) = n(n - 1) \cdots (n - r + 1) = \prod_{i=n-r+1}^{n} i$$

where the product stops at $(n - r + 1)$, because there are r factors. The formula $\frac{n!}{(n-r)!}$ is equal to this product because all of the factors in the denominator—that is, the numbers 1 through $(n - r)$—cancel with factors in the numerator.   ●

The second formula in Theorem 6.2 is more useful in studying the properties of $P(n, r)$, but the first is more efficient for computation. Note also that

$$P(n, n) = \frac{n!}{(n - n)!} = \frac{n!}{0!} = \frac{n!}{1} = n!$$

EXAMPLE 6.10   If eight soccer teams compete in a tournament, in how many ways can the teams finish? The number of ways is given by $P(8, 8) = 8! = 40{,}320$. In how many ways can the top three positions be filled? This is $P(8, 3) = 8(7)(6) = 336$.   ○

Permutations of Objects That Are Not All Distinct

In the previous examples, we counted the number of orderings of distinct objects. Suppose we want to count the number of permutations of the letters in the word *loop*. Since the two o's are not distinguishable, a rearrangement that merely switches them has no effect. Since there are 2! ways to permute two objects, we divide the number of permutations of four objects by 2! to get the correct count. Thus, the number of permutations of 4 objects in which 2 are the same is given by $\frac{4!}{2!} = \frac{24}{2} = 12$.

EXAMPLE 6.11   How many ways can we arrange the letters in the word *mollycoddle*? In eleven letters there are two o's, two d's, and three l's. So we twice divide out the number of permutations of two objects, and once divide out the number of permutations of three objects. The answer is

$$\frac{11!}{2!2!3!} = \frac{11 \times 10 \times 9 \times 8 \times 7 \times 6 \times 5 \times 4 \times 3 \times 2 \times 1}{2 \times 1 \times 2 \times 1 \times 3 \times 2 \times 1}$$
$$= 1{,}663{,}200$$ ○

The general formula for the number of permutations of n objects of which r_1 are alike, another r_2 are alike, ... , another r_j are alike is given by

$$\frac{n!}{r_1!r_2!r_3!\ldots r_j!}$$

EXERCISES 6.2

1. Use the formula for $P(n, r)$ to compute the value of
    a. $P(4, 1)$ **b.** $P(5, 2)$ **c.** $P(3, 0)$ **d.** $P(9, 9)$

2. If you rent six James Bond movies, in how many different orders can you watch them?

3. How many five-digit numbers can be formed from the digits 0 through 9, with no digit repeated? How many if the first digit cannot be 0?

4. How many k-letter words can be made from m distinct letters without repetition?

5. How many different license plates are possible consisting of four letters followed by two digits, if no letter can be repeated? If no letter and no digit can be repeated?

6. A committee of five instructors is selected to choose the winner in a writing competition. If eight papers are submitted, and each instructor ranks the papers from 1 (best) to 8 (worst), how many different rankings are possible?

7. The Tornadoes soccer team has 18 players. How many ways can they be lined up for a team picture? How many ways if the two goalies cannot be adjacent?

8. How many ways can seven people be arranged in a circle?

9. How many ways can you arrange the letters in the word *newelpost*? In *loblolly*?

10. How many seven-digit telephone numbers are possible if no digit can be repeated and the first digit may not be 0 or 1?

11. The Tornadoes soccer team has two goalies, five forwards, seven halfbacks, and four fullbacks. How many ways can they line up for a group picture if the goalies, forwards, halfbacks, and fullbacks must remain together?

12. In how many arrangements of the letters in the word *luscious* does the l immediately precede the c?

13. In how many arrangements of the letters in the word *luscious* does an s immediately precede the c?

14. A palindrome is a word that reads the same backwards as forwards. How many five-letter palindromes are there? How many six-letter palindromes?

15. How many *n*-letter palindromes are there?

6.3 ● COMBINATIONS

In the previous section, we counted the number of permutations or orderings of the elements of a set. In counting permutations, the order of the elements is important. For instance, *abc* and *cba* are two distinct arrangements of the elements of the set $\{a, b, c, d, e\}$. However, they represent the same subset.

EXAMPLE 6.12   How many three-element subsets can we form from the set $\{a, b, c, d, e\}$?

Divide $P(5, 3)$ by the number of permutations of three objects, $P(3, 3)$. The answer is 10.   ○

The number of *r*-element subsets in a set of *n* elements is called the **combinations** of *n* objects taken *r* at a time and is written $C(n, r)$. You can read $C(n, r)$ as "*n* choose *r*." Another notation used frequently is $\binom{n}{r}$.

Theorem 6.3   The formula for $C(n, r)$ is given by

$$C(n, r) = \frac{P(n, r)}{r!} = \frac{n!}{r!(n - r)!}$$

Proof   In the number $P(n, r)$ of permutations of *n* elements *r* at a time, each *r*-element subset is counted *r*! times. Therefore, $C(n, r)$ can be obtained from $P(n, r)$ by dividing by *r*!.   ●

It is clear from the formula that $C(n, r) = C(n, n - r)$.

EXAMPLE 6.13   How many different five-card poker hands can be dealt from a deck of 52 cards?

The answer is given by

$$C(52, 5) = \frac{52!}{5!(52 - 5)!} = \frac{52!}{5!47!} = \frac{48 \times 49 \times 50 \times 51 \times 52}{1 \times 2 \times 3 \times 4 \times 5}$$
$$= 2{,}598{,}960$$

○

EXAMPLE 6.14   How many 8-bit bytes contain exactly two 0's?

This problem can be thought of as the number of ways to choose two positions in which to put the 0's—that is, $C(8, 2)$. We compute

$$C(8, 2) = \frac{8!}{2!(8 - 2)!} = \frac{8!}{2!6!} = \frac{7 \times 8}{1 \times 2} = 28$$

○

EXAMPLE 6.15   The Tornadoes soccer team has 18 players, of whom 2 are goalies, 5 are forwards, 7 are halfbacks, and 4 are fullbacks. How many ways can the coach choose a starting lineup consisting of 1 goalie, 3 forwards, 4 halfbacks, and 3 fullbacks? There are $C(2, 1)$ ways to choose 1 goalie from a set of 2, $C(5, 3)$ ways to choose the forwards, $C(7, 4)$ ways to choose the halfbacks, and $C(4, 3)$ ways to choose the fullbacks. By the multiplication rule, the number of possible starting lineups is $C(2, 1) \times C(5, 3) \times C(7, 4) \times C(4, 3) = 2 \times 10 \times 35 \times 4 = 2800$.   ○

EXAMPLE 6.16   A typical punch card has 80 columns, each of which has 12 rows. The rows are numbered as shown here.

Column

Row	12	▪
	11	▪
	0	▪
	1	▪
	2	▪
	3	▪
	4	▪
	5	▪
	6	▪
	7	▪
	8	▪
	9	▪

According to data processing conventions, a column can be punched in one of three ways:

1. A single punch in any row
2. Two punches in any two rows
3. Three punches: one in row 12, 11, or 0, and two in rows 1 through 9.

How many different ways can a column be punched? There are $C(12, 1)$ ways to make a single punch in any row, and $C(12, 2)$ to make two punches in any row. To make three punches in a row, there are $C(3, 1)$ ways to make a punch in rows 12, 11, or 0, times $C(9, 2)$ ways to make two punches in the remaining nine rows. So the total number of ways to punch a column is

$$C(12, 1) + C(12, 2) + [C(3, 1) \times C(9, 2)] = 12 + 66 + 3 \times 36$$
$$= 12 + 66 + 108 = 186 \qquad \circ$$

Pascal's Triangle

If n is not too large, there is an easy way to compute the value of $C(n, r)$ by using a configuration of numbers known as *Pascal's Triangle*. Each row of the triangle is numbered, beginning with row 0, as shown here.

```
Row 0                                   1
Row 1                                1     1
Row 2                             1     2     1
Row 3                          1     3     3     1
Row 4                       1     4     6     4     1

Row 5                    1     5    10    10     5     1

Row 6                 1     6    15    20    15     6     1
Row 7              1     7    21    35    35    21     7     1
Row 8           1     8    28    56    70    56    28     8     1
Row 9        1     9    36    84   126   126    84    36     9     1
Row 10    1    10    45   120   210   252   210   120    45    10     1
```

Each row begins and ends with a 1. Notice that each of the other numbers is the sum of the two numbers directly above it. For instance, in row 5, $10 = 4 + 6$. The value of $C(7, 3)$ can be found in row 7 of the triangle:

```
1    7    21    35    35    21    7    1
```
$\longrightarrow C(7, 3)$

To see this, study the rows in Pascal's Triangle. You will notice that the numbers in row n are the values of $C(n, r)$, for $r = 0, 1, 2, 3, \ldots, n$. For instance, the numbers in row 7, from left to right, are $C(7, 0), C(7, 1), C(7, 2), C(7, 3), C(7, 4), C(7, 5), C(7, 6)$, and $C(7, 7)$.

We now prove a combinatorial identity, called *Pascal's Identity*, that gives the rule for computing the numbers in Pascal's Triangle.

Theorem 6.4   **(*Pascal's Identity*)**   If $1 \le r \le n$, then $C(n, r) = C(n - 1, r - 1) + C(n - 1, r)$

Proof   The proof using algebra is straightforward and is left as an exercise. Here we give a *combinatorial proof*, by showing that the expressions on both sides of the equals sign are a count of the same thing.

Suppose A is a set with $n - 1$ elements, and $a \notin A$. Then the set $A \cup \{a\}$ contains n elements and has $C(n, r)$ r-element subsets. Those r-element subsets are of two types: those containing a and those not containing a. Those not containing a are r-element subsets of the set A; there are $C(n - 1, r)$ of these. Those containing a have at most $r - 1$ elements of A, so there are $C(n - 1, r - 1)$ of these. Since the r-element subsets of $A \cup \{a\}$ either contain a or do not contain a, the total number of them (by the addition rule) is $C(n - r, r - 1) + C(n - 1, r)$, which must be equal to $C(n, r)$.   ●

Pascal's identity gives us the rule for computing the numbers in a row of Pascal's Triangle from the numbers in the previous row. For this reason, the formula in Theorem 6.4 is called a **recurrence relation**: an equation defining the nth term in a sequence of terms by relating it to preceding terms. In the case of Pascal's Triangle, a term is a row in the triangle. Example 3.6 defining the factorial of a number, and Example 3.5 defining the nth Fibonacci number are other examples of recurrence relations, which are the subject of Chapter 9.

EXERCISES 6.3

1. Compute the value of

    a. $C(3, 2)$ **b.** $C(8, 3)$ **c.** $C(9, 9)$ **d.** $C(6, 1)$

2. The following algorithm is an efficient way to calculate $C(n, r)$. In the formula, the factors in $(n - r)!$ cancel with the first $(n - r)$ factors in $n!$. The remaining factors in the numerator are the numbers $(n - r + 1)$ through n. That is,

    $$C(n, r) = \frac{(n - r + 1)(n - r + 2) \cdots n}{1 \times 2 \times \cdots \times r}$$

    Compute as follows: start with the first factor in the numerator, divide by the first factor in the denominator, multiply by the next factor in the numerator, divide by the next factor in the denominator, and so on. Use this algorithm to compute $C(52, 5)$.

3. Use the formula for $C(n, r)$ to show that $C(n, n - 1) = n$.

4. Show that $C(n, r) = C(n, n - r)$ without using the formula.

5. A soccer team has 18 members. How many 11-person starting lineups are possible?

6. If eight coins are tossed, how many ways can the outcome contain exactly two heads?

7. How many different 13-card bridge hands are possible from a 52-card deck?

8. A basketball team has three centers, four guards, and four forwards. How many ways can a starting lineup of one center, two guards, and two forwards be chosen?

9. Refer to Example 6.16. How many ways can a card be punched?

10. How many 8-bit bytes contain an even number of 0's?

11. A group of eight girls and four boys goes camping with three tents. How many ways can the tents be filled if

   a. Each tent must have the same number of people.

   b. Each tent must have the same number of people, and the boys must be together.

   c. Each tent must have the same number of people, and two of the boys refuse to occupy the same tent.

12. In Pascal's Triangle, what is the second number in row 314? What is the 314th number in this row?

13. Prove Theorem 6.4 using algebra.

14. Draw Pascal's Triangle as shown here:

1	1	1	1
1	2	3	4
1	3	6	10
1	4	10	20

   a. Write a formula for the entry in row r, column c.

   b. Show that each entry is equal to the sum of all the entries above and to the left of it, plus 1.

15. **a.** Show that $C(n, r) = \dfrac{nC(n-1, r-1)}{r}$.

   b. Show that, in the algorithm for computing $C(n, r)$ described in Exercise 2, all intermediate results are integers.

6.4 • BINOMIAL COEFFICIENTS

The word *binomial* refers to an algebraic expression that contains two terms, such as $x + y$, $2x + 3$, $5 - a$, and the like. In mathematics, it is often necessary to compute the terms in the expansion of products such as $(x + y)^3$ and $(3a - 5)^5$. Here is an example of a binomial expansion.

$$
\begin{aligned}
(x + y)^3 &= (x + y)(x + y)(x + y) \\
&= (x^2 + xy + yx + y^2)(x + y) \\
&= x^3 + x^2y + x^2y + xy^2 + yx^2 + y^2x + y^2x + y^3 \\
&= x^3 + 3x^2y + 3xy^2 + y^3
\end{aligned}
$$

In the expansion of the product, each term contains a product of the variables x and y in the form $x^{n-k}y^k$, where $0 \le k \le n$, and a numerical coefficient. These coefficients are called **binomial coefficients.** The coefficients in the expansion of $(x + y)^3$ are 1, 3, 3, and 1.

There is a systematic way to compute the binomial coefficients for the expansion of a binomial to any power. This method is given by the Binomial Theorem, which we prove here in two different ways: first using mathematical induction and second using a combinatorial argument.

Theorem 6.5 (*Binomial Theorem*) If x and y are any real numbers and n is a positive integer,

then

$(x + y)^n$

$$= C(n, 0)x^n + C(n, 1)x^{n-1}y + C(n, 2)x^{n-2}y^2 + \cdots + C(n, n-1)xy^{n-1} + C(n, n)y^n$$

$$= \sum_{i=0}^{n} C(n, i)x^{n-i}y^i$$

Proof   (by mathematical induction on n)

Step 1. For the case $n = 1$,

$$(x + y)^1 = x + y = C(1, 0)x^1y^0 + C(1, 1)x^{1-1}y^1$$

Step 2. The induction hypothesis ($n = k$) is

$$(x + y)^k = \sum_{i=0}^{k} C(k, i)x^{k-i}y^i$$

We have to show that the formula holds in case $n = k + 1$—that is, that

$$(x + y)^{k+1} = \sum_{i=0}^{k+1} C(k + 1, i)x^{k+1-i}y^i$$

We can show that this is true by the following computation:

$(x + y)^{k+1} = (x + y)^k(x + y)$

$\qquad = [C(k, 0)x^k + C(k, 1)x^{k-1}y + C(k, 2)x^{k-2}y^2 + \cdots +$

$\qquad\qquad C(k, k - 1)xy^{k-1} + C(k, k)y^k](x + y)$ By the induction hypothesis

$\qquad = C(k, 0)x^{k+1} + C(k, 1)x^{k-1}yx \leftarrow$

$\qquad\qquad + C(k, 2)x^{k-2}y^2x \leftarrow$

$\qquad\qquad + \cdots + C(k, k - 1)xy^{k-1}x$

$\qquad\qquad + C(k, k)xy^k$

$\qquad\qquad + C(k, 0)x^ky \leftarrow$

$\qquad\qquad + C(k, 1)x^{k-1}yy \leftarrow$

$\qquad\qquad + \cdots + C(k, k - 1)xy^{k-1}y \leftarrow$

$\qquad\qquad + C(k, k)y^{k+1}$ By the Distributive Law

Now, combine terms whose powers of x and y are alike (indicated by the arrows). We then get

$(x + y)^{k+1} = C(k, 0)x^{k+1} + x^ky[C(k, 1) + C(k, 0)]$

$\qquad\qquad\qquad\qquad\qquad\qquad\qquad\qquad\longrightarrow$ $C(k + 1, k)$
By Theorem
6.4, with
$n = k + 1$
and $r = 1$

$\qquad\qquad + x^{k-1}y[C(k, 1) + C(k, 2)]$

$\qquad\qquad\qquad\qquad\qquad\qquad\qquad\longrightarrow$ $C(k + 1, 2)$
By Theorem 6.4, with
$n = k + 1$ and $r = 2$

$$+ \cdots + xy^k[C(k, k) + C(k, k - 1)]$$

$\xrightarrow{\hspace{2cm}}$ $C(k + 1, k)$
By Theorem 6.4,
with $n = k + 1$
and $r = k$

$$+ C(k, k)y^{k+1}$$

Since $C(k, 0) = C(k + 1, 0) = C(k, k) = C(k + 1, k + 1) = 1$, we have

$$(x + y)^{k+1} = C(k + 1, 0)x^{k+1} + C(k + 1, 1)x^k y + C(k + 1, 2)x^{k-1}y^2$$
$$+ \cdots + C(k + 1, k)xy^k + C(k + 1, k + 1)y^{k+1}$$

which proves the theorem.

Proof (by a combinatorial argument)
We know that each term in the expansion of $(x + y)^n$ contains a coefficient times a product in the form $x^{n-k}y^k$. Now,

$$(x + y)^n = \underbrace{(x + y)(x + y) \cdots (x + y)}_{n \text{ times}}$$

Therefore, to get a term containing $x^{n-k}y^k$, we must choose $(n - k)$ x's from among the n factors and k y's from among the remaining k factors. There are $C(n, n - k)$ ways to choose $(n - k)$ x's and $C(k, k)$ ways to choose the k y's. By the multiplication rule, there are $C(n, n - k)C(k, k)$ ways to choose factors of the form $x^{n-k}y^k$. But

$$C(n, n - k)C(k, k) = C(n, k)(1) = C(n, k)$$

Adding the terms $C(n, k)x^{n-k}y^k$ for $k = 0, 1, 2, \ldots, n$ gives the desired result. ●

EXAMPLE 6.17   Use the Binomial Theorem to expand $(x + 2)^5$.

$$(x + 2)^5 = \sum_{k=0}^{5} C(5, k)x^{5-k}2^k$$
$$= x^5 + 5x^4(2) + 10x^3(4) + 10x^2(8) + 5x(16) + 32$$
$$= x^5 + 10x^4 + 40x^3 + 80x^2 + 80x + 32$$ ○

EXAMPLE 6.18   Find the coefficient of the $x^6 y^4$ term in the expansion of $(x + y)^{10}$.

Here $n = 10$, $n - k = 6$, and $k = 4$. By the Binomial Theorem, this coefficient is $C(10, 4) = 210$.   ○

EXAMPLE 6.19   Find the sixth term in the expansion of $(x + 3)^{18}$.

The sixth term has the factors x^{13} and 3^5, and coefficient $C(18, 5)$. The answer is

$$8568x^{13}(3)^5 = 8568x^{13}(243) = 2,082,024x^{13}$$ ○

Notice that the coefficients in the expansion of $(x + y)^n$ are the numbers in row n of Pascal's Triangle.

The next theorem is another example of a recurrence relation.

Theorem 6.6 $C(n + r, r) = 1 + C(n, 1) + C(n + 1, 2) + \cdots + C(n + r - 1, r)$

Proof   We use induction on r, keeping n fixed.

Step 1. For the case $r = 1$,

$$C(n + 1, 1) = \frac{(n + 1)!}{1![(n + 1) - 1]!} = \frac{(n + 1)n!}{n!}$$

$$= n + 1 = 1 + C(n, 1)$$

Step 2. The induction hypothesis $(r = k)$ is

$$C(n + k, k) = 1 + C(n, 1) + C(n + 1, 2) + \cdots + C(n + k - 1, k)$$

We have to show that the formula holds in case $r = k + 1$; that is, that

$$C(n + k + 1, k + 1) = 1 + C(n, 1) + C(n + 1, 2) + \cdots + C(n + k - 1, k)$$
$$+ C(n + k, k + 1)$$

By Theorem 6.4,

$$C(n + k + 1, k + 1)$$
$$= C(n + k + 1 - 1, k + 1 - 1) + C(n + k + 1 - 1, k + 1)$$
$$= C(n + k, k) + C(n + k, k + 1)$$

$\longrightarrow 1 + C(n, 1) + C(n + 1, 2) + \cdots + C(n + k - 1, k)$

By the induction hypothesis

Therefore,

$$C(n + k + 1, k + 1) = 1 + C(n, 1) + C(n + 1, 2) + \cdots + C(n + k - 1, k)$$
$$+ C(n + k, k + 1)$$

and the theorem is proved.   ●

Theorem 6.7 $\displaystyle\sum_{r=0}^{n} C(n, r) = 2^n$

Proof

$$2^n = (1 + 1)^n$$

$$= C(n, 0)1^n + C(n, 1)1^{n-1}(1) + C(n, 2)1^{n-2}1^2$$
$$+ \cdots + C(n, n - 1)(1)1^{n-1} + C(n, n)1^n$$

By the Binomial Theorem

$$= C(n, 0) + C(n, 1) + C(n, 2) + \cdots + C(n, n)$$ ●

Notice that in Theorem 6.7, the sum on the left side of the equation is the number of 0-element subsets plus 1-element subsets plus 2-element subsets ... plus n-element subsets of a set of size n. This sum represents the total number of subsets of a set of n elements, which we know to be 2^n (from Example 2.19).

EXERCISES 6.4

1. Use the Binomial Theorem to expand the following:
   a. $(x + 1)^6$ **b.** $(5 - 2a)^4$

2. What is the coefficient of the $x^4 y^3$ term in the expansion of $(x + y)^7$?

3. What is the coefficient of the $x^3 y^{19}$ term in the expansion of $(x + y)^{22}$?

4. What is the tenth term in the expansion of
   a. $(x + y)^{14}$ **b.** $(x + 1)^{12}$

5. What is the fourth term in the expansion of $(5 - 2x)^9$?

6. What is the fifth term in the expansion of $(2x - 3y)^8$?

7. Prove Theorem 6.7 using mathematical induction on n.

8. Use Theorem 6.4 to show that

$$\sum_{i=r}^{n} C(i, r) = C(n + 1, r + 1)$$

9. Use the result of Exercise 8 to find the sum of the first n positive integers. (*Hint:* Note that $C(n, 1) = n$ and choose an appropriate value for r.)

10. Show that

$$C(2n, n) = \sum_{r=0}^{n} [C(n, r)]^2$$

using a combinatorial argument.

11. Use the Binomial Theorem to show that
$$C(n, 0) - C(n, 1) + C(n, 2) - C(n, 3) + \cdots + (-1)^n C(n, n) = 0$$

12. If set A has n elements, how many subsets of A have an even number of elements? How many have an odd number?

13. Use algebra to show that

$$C(n, r + 1) = \frac{n - r}{r + 1} C(n, r)$$

14. Use the binomial theorem to show that

$$\sum_{r=0}^{n} 2^r C(n, r) = 3^n$$

15. Show that $C(n, r)C(r, k) = C(n, k)C(n - k, r - k)$.

16. Show that

$$C(m + n, r) = \sum_{k=0}^{r} C(m, r - k)C(n, k)$$

This identity is known as **Vandermonde's identity.**

6.5 • DISCRETE PROBABILITY

The term *probability* refers to the likelihood of chance, or random, occurrences. A random occurrence is the result of an experiment whose outcome cannot be predicted with certainty. Questions involving chance occurrences include the following. What is the probability that it will rain tomorrow? That the Cubs will win the pennant? That you will get an "A" in Discrete Math? That a die will come up 6 when tossed? That a five-card poker hand will contain four aces? That a new drug will cure a particular disease? The probability of each of these occurrences is a real number that indicates how likely the occurrence is. Historically, the development of the counting techniques discussed in the previous sections was motivated by the need to compute probabilities, primarily for gambling games.

In this section, we will look at methods of assigning probabilities to outcomes of experiments. First, here are some examples of experiments and their outcomes.

EXAMPLE 6.20   Toss a coin and observe the side that shows up. The possible outcomes are heads and tails. For this experiment, you may ask what the probability is that the outcome is heads (or tails).   ○

EXAMPLE 6.21   Toss two dice and observe the result. There are two ways to view this experiment: The numbers on the two dice may be noted, or only the sum may be noted. In the first view, the possible outcomes are the ordered pairs of integers in the range 1 to 6; that is,

$$(1, 1), (1, 2), (1, 3), (1, 4), (1, 5), (1, 6),$$
$$(2, 1), (2, 2), (2, 3), (2, 4), (2, 5), (2, 6),$$
$$(3, 1), (3, 2), (3, 3), (3, 4), (3, 5), (3, 6),$$
$$(4, 1), (4, 2), (4, 3), (4, 4), (4, 5), (4, 6),$$
$$(5, 1), (5, 2), (5, 3), (5, 4), (5, 5), (5, 6),$$
$$(6, 1), (6, 2), (6, 3), (6, 4), (6, 5), (6, 6)$$

In the second view, the possible outcomes are the sums of the numbers in these pairs: the integers from two to twelve. In this experiment, you may ask what the probability is that the result is a double (that is, one of the outcomes $(1, 1)$, $(2, 2)$, $(3, 3)$, $(4, 4)$, $(5, 5)$, or $(6, 6)$), or that the sum is 11, or that the sum is greater than 5.   ○

EXAMPLE 6.22   Put a pan of beer in your garden; the next morning, count the number of dead slugs in the pan. The possible outcomes are integers greater than or equal to zero. In this experiment, you may ask what the probability is that there is at least one slug, or fewer than 100 slugs, or an even number of slugs.   ○

Sample Spaces and Events

In each case, the possible outcomes of the experiment form a set. As we saw in Example 6.21, there may be more than one way to choose the set. The choice of set depends on the kinds of questions about probabilities that we are going to ask. These considerations motivate the following definitions.

Definition 6.1   A **sample space** is a set of all possible outcomes to an experiment. An **event** is any subset of a sample space.

If B is an event in a sample space S, the elements of B are sometimes said to be **favorable** outcomes of the experiment, relative to the event B.

In Example 6.20 (tossing a coin), the sample space is {Heads (H), Tails (T)}. In Example 6.21 (rolling two dice), we saw two possibilities for the choice of sample space: the set $X \times X$, where $X = \{1, \ldots, 6\}$, and the set $\{2, \ldots, 12\}$. All of these sample spaces are finite. In Example 6.22, the sample space is $\{n \in \mathbf{Z} \mid n \geq 0\}$, which is countably infinite. In the next example, the sample space is uncountably infinite.

EXAMPLE 6.23   You measure, in inches, the yearly rainfall in the Olympic rain forest. The possible outcomes for this experiment are the nonnegative real numbers. Thus, the sample space is $[0, \infty)$. ○

Probability Measures

In this text, we will consider only sample spaces that are finite; hence the term *discrete probability*. In a typical problem in discrete probability, we start with some assignment of probabilities to the elements of the sample space. Then we compute the probability of any event.

Definition 6.2

Let S be a finite sample space—that is, a finite nonempty set. A **probability measure** on S is a function $p: S \to \mathbf{R}$ such that

1. $p(x) \geq 0$ for each $x \in S$, and
2. $\sum_{x \in S} p(x) = 1$

If $A \subset S$ (that is, if A is an event), the **probability** of A, denoted $p(A)$, is given by

$$p(A) = \sum_{x \in A} p(x)$$

Note that if p is a probability measure on S, the probability of S is 1, and the probability of the empty set $\varnothing$ is zero. Furthermore, for any event A, $0 \leq P(A) \leq 1$.

EXAMPLE 6.24   Consider an experiment in which one die is rolled. The sample space is $S = \{1, \ldots, 6\}$. Define $p: S \to \mathbf{R}$ by $p(x) = \frac{1}{6}$ for every $x \in S$. Then p is a probability measure on S. If A is the event that the outcome is at least 5, then $A = \{5, 6\}$ and $p(A) = \frac{1}{6} + \frac{1}{6} = \frac{1}{3}$. ○

EXAMPLE 6.25   With the same experiment as in Example 6.24, define $q: S \to \mathbf{R}$ as follows:

x	$q(x)$
1	0.14
2	0.15
3	0.16
4	0.17
5	0.18
6	0.20

Then q is a probability measure on S. (The die is loaded.) If A is the event that an even number is rolled, then

$$q(A) = q(2) + q(4) + q(6) = 0.15 + 0.17 + 0.20 = 0.52$$

If B is the event that the result is at least 5, then

$$q(B) = q(5) + q(6) = 0.18 + 0.20 = 0.38$$

EXAMPLE 6.26 In Example 6.21 (rolling two dice), choose the sample space to be the set of ordered pairs $X \times X$ of elements of $X = \{1, \ldots, 6\}$. There are 36 such pairs. If all of the elements of the sample space are equally probable (and we may judge that they are, unless we know that the dice are loaded), then the probability of any particular outcome is $\frac{1}{36}$. We may then ask what the probability is of rolling a double; there are six doubles, so the probability is $\frac{6}{36} = \frac{1}{6}$.

EXAMPLE 6.27 Suppose that in the two-dice experiment, we choose the sample space Y to be the set of all possible totals of spots on the dice; that is,

$$Y = \{2, 3, 4, 5, 6, 7, 8, 9, 10, 11, 12\}$$

Then each outcome in Y is an event in the sample space $X \times X$ of Example 6.21. For instance, the outcome 4 in Y corresponds to the event

$$\{(1, 3), (2, 2), (3, 1)\}$$

in $X \times X$. This gives us a way of assigning a probability measure to Y: Assign to each outcome in Y the probability of the corresponding event in $X \times X$. The following table shows all the ways of getting each outcome in Y.

	1	2	3	4	5	6
1	2	3	4	5	6	7
2	3	4	5	6	7	8
3	4	5	6	7	8	9
4	5	6	7	8	9	10
5	6	7	8	9	10	11
6	7	8	9	10	11	12

The five ways to get a sum of 6

Since the probability of each outcome in $X \times X$ is $\frac{1}{36}$, you can see from the table that the probabilities of the outcomes of Y are

Outcome	2	3	4	5	6	7	8	9	10	11	12
Probability	$\frac{1}{36}$	$\frac{2}{36}$	$\frac{3}{36}$	$\frac{4}{36}$	$\frac{5}{36}$	$\frac{6}{36}$	$\frac{5}{36}$	$\frac{4}{36}$	$\frac{3}{36}$	$\frac{2}{36}$	$\frac{1}{36}$

EXAMPLE 6.28 Consider the experiment consisting of dealing a hand of five cards from a 52-card deck. The sample space S contains $C(52, 5) = 2,598,960$ elements. We can define a probability measure p on S by setting $p(x) = \frac{1}{2598960}$ for each $x \in S$. What is the

probability that a hand contains four aces? A hand with four aces contains one of the 48 other cards, so the event "four aces" contains 48 elements. To compute the probability of this event, we add up 48 numbers each equal to $\frac{1}{2598960}$. The probability of this event is then $\frac{48}{2598960} = \frac{1}{54145} = 0.0000184$.   ○

Choosing a Probability Measure

We have noted that the task of probability theory is to compute probabilities of events, given a sample space and a probability measure. This is different from the problem of choosing a probability measure. Consider Example 6.22, in which the experiment was leaving a pan of beer in the garden overnight and counting the dead slugs in the morning. If we choose a large enough number N (greater than the number of slugs in the neighborhood), we can choose the set $X = \{0, 1, \ldots, N\}$ as the sample space. What is the probability measure? If 100 neighbors leave pans of beer in their gardens and all count the slugs in the morning, the results may be as follows.

No. of Slugs	No. of Neighbors	Fraction of Neighbors
0	15	0.15
1	17	0.17
2	27	0.27
3	29	0.29
4	8	0.08
5	3	0.03
6	1	0.01
> 6	0	0.00

The third column defines a probability measure on X. Is this the "true" probability measure for this experiment? Not necessarily. If the experiment with 100 neighbors were done again, the results would surely be different. The "true" probability measure is unknown. We may take the third column in the table as an estimate of the probability measure. Better yet, we could perform the experiment 50 times and average the results. This would produce a better estimate, but it would still be an estimate. The science of estimating probabilities from results of experiments is called *statistics*.

Unions, Intersections, and Complements of Events

Since events are subsets of a sample space, it is reasonable to ask how probabilities of events are related to the set operations of union, intersection, and complement. The relationship is stated in the following theorem.

Theorem 6.8   Let S be a sample space with probability measure p, and let A and B be events. Then

$$p(A') = 1 - p(A)$$

and

$$p(A \cup B) = p(A) + p(B) - p(A \cap B)$$

Proof   Since each element of S is in either A or in A' but not both,

$$p(A) + p(A') = \sum_{x \in A} p(x) + \sum_{x \notin A} p(x)$$
$$= \sum_{x \in S} p(x)$$
$$= 1$$

because p is a probability measure. This proves the first part of the theorem. The proof of the second part is an application of the inclusion–exclusion principle and is left as an exercise.   ●

EXAMPLE 6.29   Again consider the experiment leaving a pan of beer in the garden overnight, and assume that the probabilities of finding different numbers of slugs in the morning are given by

No. of Slugs	Probability
0	0.15
1	0.17
2	0.27
3	0.29
4	0.08
5	0.03
6	0.01
> 6	0.00

Let A be the event that the number of slugs is even and let B be the event that the number of slugs is greater than 2. By adding the probabilities of the outcomes in each of the events, you can see that

$$p(A) = p(0) + p(2) + p(4) + p(6) = 0.51$$
$$p(A') = p(1) + p(3) + p(5) = 0.49 = 1 - 0.51$$
$$p(B) = p(3) + p(4) + p(5) + p(6) = 0.41$$
$$p(B') = p(0) + p(1) + p(2) = 0.59 = 1 - 0.41$$
$$p(A \cup B) = p(0) + p(2) + p(3) + p(4) + p(5) + p(6) = 0.83$$
$$p(A \cap B) = p(4) + p(6) = 0.09$$
$$p(A) + p(B) - p(A \cap B) = 0.51 + 0.41 - 0.09$$
$$= 0.83$$
$$= p(A \cup B)$$ ○

Note that if $A \cap B = \varnothing$, then $p(A \cup B) = p(A) + p(B)$. In this case, A and B are said to be **mutually exclusive** events.

Equiprobable Outcomes

In many problems, including almost all problems involving games of chance, the sample space is such that all outcomes are equally probable. This includes tossing a coin, rolling dice, dealing cards, etc.

Definition 6.3

> Let S be a nonempty finite set, and let N be the number of elements in S. Then the **equiprobable measure** on S is the probability measure that assigns the probability $\frac{1}{N}$ to each element of S.

If S is a sample space with the equiprobable measure, then the probability of any event $A \subset S$ is

$$p(A) = \frac{n(A)}{n(S)} = \frac{\text{Number of favorable outcomes}}{\text{Total number of outcomes}}$$

Thus, the problem of computing the probability of A is reduced to a counting problem.

EXAMPLE 6.30   If the designation of a telephone exchange numbered 322 is changed to begin with two letters, what is the probability that the two letters contain a vowel? Since 3 corresponds to D, E, F and 2 to A, B, C, the sample space is

$$S = \{\text{DA, DB, DC, EA, EB, EC, FA, FB, FC}\}$$

The event A is

$$A = \{\text{DA, EA, EB, EC, FA}\}$$

Then

$$p(A) = \frac{n(A)}{n(S)} = \frac{5}{9}$$

○

EXAMPLE 6.31   A person guesses the order in which three cards (red, blue, and yellow) have been placed on a table. What is the probability that the guess is correct? The sample space consists of the $3! = 6$ permutations of red, blue, and yellow:

RBY   BYR   YRB   RYB   YBR   BRY

If these are equiprobable and only one is correct, the probability of a correct guess is $\frac{1}{6}$.

○

EXAMPLE 6.32   The Tornadoes soccer team has two goalies, five forwards, seven halfbacks, and four fullbacks. If three players are chosen at random, what is the probability that all three are halfbacks? The team has 18 members. The sample space is the set of 3-element subsets of the team, so it has $C(18, 3) = 816$ elements. The number of favorable outcomes is the number of ways to choose three halfbacks from the seven (that is, $C(7, 3) = 35$). Thus, the desired probability is $\frac{35}{816} = 0.043$.

○

EXAMPLE 6.33   The Tornadoes line up in one row for a team picture. What is the probability that the two goalies are together? The sample space consists of the 18! ways for the team to line up. The first goalie may be in any one of 17 positions. Either of the two goalies may be first (two possibilities). There are 16! arrangements for the rest of the team. So the number of permutations in which the goalies are together is $17 \times 2 \times 16!$. Thus, the desired probability is

$$\frac{17 \times 2 \times 16!}{18!} = \frac{2}{18} = \frac{1}{9}$$

○

EXERCISES 6.5

1. A box contains five red, one green, and four yellow balls. Two balls are drawn from the box. List all possible outcomes of this experiment.

2. Two dice are rolled. Event A is the event that the sum showing is even. List the outcomes in A.

3. Which of the following functions are probability measures on the set $\{1, 2, 3, 4, 5\}$?

x	$p_1(x)$	$p_2(x)$	$p_3(x)$	$p_4(x)$	$p_5(x)$
1	0.25	0.17	0.05	0.31	0.37
2	0.25	0.22	0.22	0.33	0.36
3	0	0.04	0.37	0.14	0.25
4	0.35	0.38	-0.23	0.27	0.01
5	0.15	0.18	0.59	0.02	0.01

4. Define the probability measure p on the set $\{1, 2, 3, 4\}$ by the rule $p(1) = 0.07$, $p(2) = 0.31$, $p(3) = 0.37$, $p(4) = 0.25$. Find the probabilities of each of the following events:

   a. $\{1\}$ **b.** $\{2, 3\}$

   c. $\{1, 3, 4\}$ **d.** $\{1, 4\}$

5. A die is rolled. Let A be the event that the number showing is odd. What is the probability of A?

6. Two coins are tossed. Find the probability that the sides showing are the same.

7. A five-card poker hand is dealt. What is the probability that the hand contains four of a kind (that is, four aces, or four kings, etc.).

8. Find the probability that a thirteen-card bridge hand contains

   a. Thirteen cards of one suit

   b. Exactly seven spades

   c. A void in diamonds (that is, no diamonds)

9. Find the probability that an 8-bit byte contains exactly three zeros.

10. Find the probability that an 8-bit byte contains at least three zeros.

11. Refer to Example 6.30. What is the probability that the designation chosen has no vowels?

12. Refer to Example 6.7. If a student is chosen at random from this group, what is the probability that she plays none of the three games?

13. Refer to Example 6.32. If six players are chosen at random, what is the probability that none of the six is a goalie?

14. Refer to Example 6.33. What is the probability that

    a. The five forwards are together?

    b. The two goalies are *not* together?

15. A burglar alarm key pad contains nine digits. The disarm code consists of four digits. If the alarm system allows 90 seconds for the system to be disarmed, and if a burglar can enter 20 codes in one minute, what is the probability that she will disarm the system?

16. To win your state lottery, you must correctly pick six numbers in the range 0–44 in the correct order. What is the probability of winning?

17. What is the probability that a random arrangement of the letters in the word *newelpost* does not begin with "e"?

18. What is the probability that a random arrangement of the letters in the word *newelpost* does not contain an "e"?

19. A committee of five instructors is selected to choose the winner of a writing competition. Five papers are submitted. Each instructor ranks the papers from 1 (best) to 5 (worst). If all papers are equally meritorious, what is the probability that all five rankings are the same?

20. A panel of 76 people was chosen to taste-test soda pop to determine whether caffeine is detectable in the taste of the soda. Each panelist was given three samples of orange soda (one of which was different) and asked to identify the different soda. ("Caffeine: How to consume less." *Consumer Reports* 46, October 1981, 597–599.)

    a. If a panelist is just guessing, what is the probability that she will correctly identify the different soda?

    b. If all 76 panelists are guessing, how many would you expect to identify the different soda correctly?

    c. In the actual test, 34 people identified the different soda correctly. Do you think this result is due to chance, or do you think it indicates that caffeine affects the flavor of the soda?

21. Eight coins are tossed. What is the probability that exactly two heads appear?

22. XYZ Microelectronics receives 35 circuit boards, of which 5 are defective. If they test 5 of the boards, what is the probability that

    a. There are no defective boards in the sample?

    b. There is exactly one defective board in the sample?

    c. There is at least one defective board in the sample?

23. What is the probability that a five-card poker hand is

   a. A full house (three cards of one rank and two of another)?

   b. A flush (all cards of the same suit)?

   c. A straight (five cards in sequence—for example, 7-8-9-10-J)?

   d. A straight flush?

   e. Two pair?

   f. One pair?

24. Prove the second part of Theorem 6.8.

6.6 • ADDITIONAL TOPICS IN PROBABILITY

Frequently, an experiment is composed of several parts, such as rolling two dice, tossing two coins, or noting both the air temperature and barometric pressure. If the parts are unrelated, or independent, the probability measure for the complete experiment can be derived from the probability measures for the parts. A probability measure can be modified if additional information on the results of the experiment is provided. For instance, some outcomes may be excluded, increasing the probabilities of the remaining outcomes. Or, you may need to compute an "average value" of a function on a sample space. These topics are considered in this section.

Independent Events

Consider an experiment consisting of two independent, or unrelated, parts. For instance, suppose that you toss a coin and roll a die. The first part has two possible outcomes and the second has six possible outcomes. In this case, it is reasonable to assign the equiprobable measure to each part. That is, the probabilities of heads and tails are each $\frac{1}{2}$, and the probability of each of the numbers 1 through 6 is $\frac{1}{6}$. For the complete experiment, there are $2 \times 6 = 12$ possible outcomes. With the equiprobable measure, the probability of each outcome of the complete experiment is $\frac{1}{12}$. Note that $\frac{1}{12} = \frac{1}{2} \times \frac{1}{6}$. That is, the probability of each outcome of the complete experiment is the product of the probabilities of the two parts.

This multiplication rule applies to any experiment made up of independent events, even if the probability measures are not the equiprobable measures. That is, if A and B are independent events, then

$$p(A \cap B) = p(A)p(B)$$

This rule can be used in two ways. If you want to describe a real-world problem by constructing a probability measure on a sample space, and if, in your judgment, the outcomes are made up of two independent parts, you can use the multiplication rule to find a probability measure. This is what we did in the coin-and-die example. On the other hand, if you are given a probability measure p on a sample space, then two events A and B are defined to be independent if the multiplication rule holds.

In constructing a probability measure, you must be careful in deciding whether two parts of an experiment are independent. In the case of tossing coins and rolling dice, you can judge that the outcome of one step does not influence the outcome of the other. But consider Example 6.22, in which you put a pan of beer in your garden and count the dead slugs in the morning. If you perform this experiment twice in a row, the probabilities will be different the second day because the neighborhood slug population is smaller.

Independence and Cartesian Products

When two independent experiments are performed, the sample space of the combined experiment is the Cartesian product of the sample spaces of the parts. The following theorem shows how a probability measure on the Cartesian product of two sets is constructed from probability measures on the two sets.

Theorem 6.9   Let p be a probability measure on the sample space S, and let q be a probability measure on the sample space T. Define a function $r: S \times T \to \mathbf{R}$ by the rule $r(s, t) = p(s)q(t)$. Then r is a probability measure on $S \times T$.

Proof   We have to show that

(1) $r(s, t) \geq 0$ for every $(s, t) \in S \times T$, and

(2) $\displaystyle\sum_{s \in S, t \in T} r(s, t) = 1.$

We can see that (1) is true because $r(s, t)$ is a product of two nonnegative numbers. To see that (2) is true, note that

$$\sum_{s \in S, t \in T} r(s, t) = \sum_{s \in S, t \in t} p(s)q(t)$$

$$= \sum_{s \in S}\left(\sum_{t \in T} q(t)\right)p(s) \qquad \text{Distributive Law}$$

$$= \sum_{s \in S} p(s) \qquad\qquad \text{Because } \sum_{t \in T} q(t) = 1$$

$$= 1 \qquad\qquad\qquad\qquad\qquad\qquad \bullet$$

EXAMPLE 6.34   According to a local bookmaker, the probabilities of the outcomes of the University of Washington vs. Washington State University football game are: Washington, 0.57; Washington State, 0.40; tie, 0.03. In the same bookmaker's judgment, the probabilities for the Purdue vs. Indiana game are: Purdue, 0.55; Indiana, 0.44; tie, 0.01. The games are being played on the same day. Let $X = \{UW, WSU, tie\}$ and let $Y = \{PU, IU, tie\}$. Use the formula of Theorem 6.9 to compute a probability measure on $X \times Y$:

	PU	IU	Tie
UW	0.57×0.55	0.57×0.44	0.57×0.01
WSU	0.40×0.55	0.40×0.44	0.40×0.01
Tie	0.03×0.55	0.03×0.44	0.03×0.01

Probability of outcome (WSU, IU)   ○

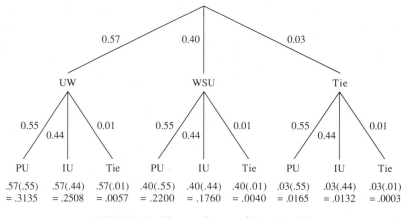

FIGURE 6.8   The tree diagram of Example 6.34

Another way to present this result is by using a tree diagram. In Figure 6.8, the first set of edges corresponds to the results of the UW–WSU game and the second set to the results of the PU–IU game. Each edge in the diagram is labeled with the corresponding probability. The probability associated with each leaf in the tree is the product of the probabilities along the path to the leaf.

EXAMPLE 6.35   Refer to the previous example. What is the probability of at least one tie? The probabilities of the various outcomes were given by the following table.

	PU	IU	Tie
UW	0.3135	0.2508	0.0057
WSU	0.2200	0.1760	0.0040
Tie	0.0165	0.0132	0.0003

— Ties

The probability of at least one tie is

$$0.0165 + 0.0132 + 0.0003 + 0.0040 + 0.0057 = 0.0397$$

This answer can be obtained from the tree diagram of Figure 6.8 by adding the products at each leaf that contains a tie.   ○

EXAMPLE 6.36   Cabin Creek College is playing soccer against Ivy Tech in a domed stadium. According to the bookmaker, the probability of a Cabin Creek College win is 0.37. According to the weather service, the probability of rain is 80%. If you regard both the bookmaker and the weatherman as reliable, what is the probability that it will rain and Cabin Creek College will win? Since the game is in a domed stadium, it is reasonable to judge that the experiments are independent. The probability is $0.37 \times 0.80 = 0.296$. This result is shown in Figure 6.9.   ○

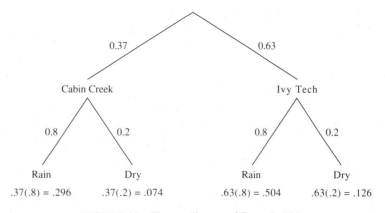

FIGURE 6.9   The tree diagram of Example 6.36

Determining Independence

The multiplication rule for independent events can be used to give a meaning to the concept of independent events, even if the sample space is not a Cartesian product.

Definition 6.4   Let S be a sample space with probability measure p, and let A and B be events. Then A and B are **independent events on S** if $p(A \cap B) = p(A)p(B)$.

EXAMPLE 6.37   Suppose that a single die is rolled. Let A be the event that an even number is rolled, and let B be the event that a number greater than 3 is rolled. Then $A \cap B$ is the event that an even number greater than 3 is rolled—that is, $A \cap B = \{4, 6\}$. Then

$$p(A) = \frac{1}{2}$$

$$p(B) = \frac{1}{2}$$

$$p(A \cap B) = \frac{1}{3}$$

Since $p(A)p(B) = \frac{1}{4} \neq p(A \cap B)$, events A and B are *not* independent.   ○

EXAMPLE 6.38   A coin is tossed 15 times. What is the probability that all results are heads? That exactly three of the results are heads?

The first part of the experiment is tossing one coin, with sample space $\{H, T\}$ and equiprobable measure (the probability of each result is $\frac{1}{2}$). The sample space for the complete experiment contains $2^{15} = 32{,}768$ outcomes, also with the equiprobable measure. Since the tosses are independent, the probability of 15 heads is $(\frac{1}{2})^{15}$. If A

is the event that exactly 3 heads occur, then

$$n(A) = C(15,3) = 455$$

so

$$p(A) = \frac{455}{32,768} = 0.0139$$

○

Conditional Probability

In posing probability questions, it is common to include some additional information. For instance, if two dice are rolled, what is the probability that the sum is at least 5, given that the roll is not a double? Two transistors are chosen from a batch containing 11 good transistors and 5 bad transistors. What is the probability that both transistors are bad if it is known that one is bad? In each case, we have a sample space and a probability measure. But the problem posed restricts our attention to some subset of the sample space. To answer such questions, we must define a probability measure on the subset. We can do this by "scaling up" the probabilities of the outcomes in the subset so that their sum becomes equal to 1.

Theorem 6.10   Let S be a sample space, and let p be a probability measure on S. Let $A \subset S$, and assume that $p(A)$ is not zero. Define a function $p_A \colon A \to \mathbf{R}$ by the rule

$$p_A(x) = \frac{p(x)}{p(A)}$$

Then p_A is a probability measure on A.

Proof   We have to show that

(1) $p_A(x) \geq 0$ for every $x \in S$, and
(2) $\displaystyle\sum_{x \in A} p_A(x) = 1$

Statement (1) is true, because $p_A(x)$ is equal to the quotient of a nonnegative number and a positive number.

To see that (2) is true, observe that

$$\sum_{x \in A} p_A(x) = \sum_{x \in A} \frac{p(x)}{p(A)}$$

$$= \frac{1}{p(A)} \sum_{x \in A} p(x) \qquad \text{Distributive Law}$$

$$= \frac{p(A)}{p(A)}$$

$$= 1$$

●

EXAMPLE 6.39   Suppose a die is loaded so that the probabilities of the outcomes are

x	$p(x)$
1	0.14
2	0.15
3	0.16
4	0.17
5	0.18
6	0.20

Find the probability of rolling an even number, given that a number greater than 2 is rolled. The condition "greater than 2" corresponds to the subset $A = \{3, 4, 5, 6\}$. Then

$$p(A) = 0.16 + 0.17 + 0.18 + 0.20 = 0.71$$

You can compute the probability measure p_A as follows:

	x	$p(x)$	$\frac{p(x)}{p(A)} = p_A(x)$
	1	0.14	
	2	0.15	
Subset A	3	0.16	$\frac{0.16}{0.71} = 0.225$
	4	0.17	$\frac{0.17}{0.71} = 0.239$
	5	0.18	$\frac{0.18}{0.71} = 0.254$
	6	0.20	$\frac{0.20}{0.71} = 0.282$

Note that $0.225 + 0.239 + 0.254 + 0.282 = 1$. The desired probability is then $p_A(4) + p_A(6) = 0.239 + 0.282 = 0.521$. ○

Probabilities assigned by the measure p_A are called **conditional probabilities.** If A and B are subsets of a sample space S, then the probability of event B occurring given that event A has occurred is written $p(B \mid A)$. This is read "the probability of B given A." It is the probability of the event $A \cap B$, computed from the probability measure p_A. The value of $p(B \mid A)$ is given by the formula

$$p(B \mid A) = \frac{p(A \cap B)}{p(A)}$$

Note that if events A and B are independent, then $p(A \cap B) = p(A)p(B)$ (see Definition 6.4), so $p(B \mid A) = p(B)$. Now we can state the following general formula for $p(A \cap B)$, which holds even if A and B are not independent:

$$p(A \cap B) = p(A)p(B \mid A)$$

EXAMPLE 6.40   Two dice are tossed. Compute the probability that the sum showing is 9, given that the sum is greater than 7. The sample space is

$$S = \{2, 3, 4, 5, 6, 7, 8, 9, 10, 11, 12\}$$

The probabilities of each of the possible sums were computed in Example 6.27. To apply the formula, let A be the event that the sum is 9 and B be the event that the sum is greater than 7. Then $A = \{9\}$, $B = \{8, 9, 10, 11, 12\}$, and $A \cap B = \{9\}$. The probability of A given B is computed as follows:

$$p(A \mid B) = \frac{p(A \cap B)}{p(B)}$$

$$= \frac{\frac{4}{36}}{\frac{5}{36} + \frac{4}{36} + \frac{3}{36} + \frac{2}{36} + \frac{1}{36}}$$

$$= \frac{4}{15}$$

○

EXAMPLE 6.41   Find the probability that an 8-bit byte contains exactly two zeros, given that it contains an even number of zeros. The sample space S consisting of all possible 8-bit bytes contains 2^8 elements. Let A be the set of bytes containing two zeros, and let B be the number of bytes containing an even number of zeros. In this case, $A \subset B$, so $A \cap B = A$. Then

$$n(A) = C(8, 2) = 28$$

and

$$n(B) = C(8, 0) + C(8, 2) + C(8, 4) + C(8, 6) + C(8, 8)$$
$$= 1 + 28 + 70 + 28 + 1$$
$$= 128$$

Thus,

$$p(A \mid B) = \frac{p(A \cap B)}{p(B)} = \frac{p(A)}{p(B)} = \frac{\dfrac{28}{2^8}}{\dfrac{128}{2^8}} = \frac{28}{128} = 0.219$$

○

The formula $p(A \cap B) = p(A)p(B \mid A)$ can be used to find the probability of the event $A \cap B$ when $p(A)$ and $p(A \mid B)$ are known.

EXAMPLE 6.42   A batch of transistors contains 12 good transistors and 2 defective ones. If two transistors are chosen at random, what is the probability that both transistors chosen are defective? Refer to the tree diagram in Figure 6.10.

When the first transistor is chosen, the probability that it is good $\frac{12}{14}$, and the probability that it is defective is $\frac{2}{14}$. If the first transistor is good, there is an $\frac{11}{13}$ probability that the second is good and a $\frac{2}{13}$ probability that the second is defective. These are conditional probabilities. Also, if the first transistor is defective, there is a

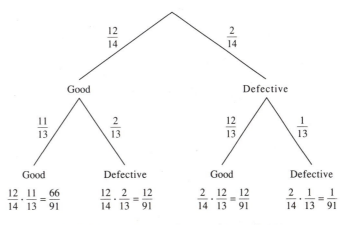

FIGURE 6.10   The tree diagram of Example 6.42

$\frac{12}{13}$ probability that the second is good and a $\frac{1}{13}$ probability that the second is defective. The edges of the tree are labeled with these probabilities. The probability of each of the outcomes is obtained by multiplying the probabilities along the path to the corresponding leaf. In particular, the probability that both are defective is $\frac{1}{91}$.

We can use the sample space constructed by the tree diagram of Figure 6.10 to find a different conditional probability, as shown in the next example.

EXAMPLE 6.43   Refer to Example 6.42. Find the probability that both transistors chosen are defective, given that at least one is defective. The sample space constructed in Example 6.42 is

First	Second	Probability	
good	good	$\frac{66}{91}$	
good	defective	$\frac{12}{91}$	$(\frac{12}{91})/(\frac{25}{91}) = \frac{12}{25}$
defective	good	$\frac{12}{91}$	$(\frac{12}{91})/(\frac{25}{91}) = \frac{12}{25}$
defective	defective	$\frac{1}{91}$	$(\frac{1}{91})/(\frac{25}{91}) = \frac{1}{25}$

The relevant subset of the sample space is indicated by the box, and the sum of the probabilities in the box is $\frac{25}{91}$. The answer is $(\frac{1}{91})/(\frac{25}{91}) = \frac{1}{25}$.

Expected Value

Let $S = \{s_1, s_2, \ldots, s_N\}$ be a sample space with a probability measure p, and suppose that $f: S \to \mathbf{R}$ is a real-valued function on S. If an outcome is chosen from the sample space, you may ask what value of f can be expected. For example, if a die is rolled, what number of spots can be expected? You can think of the answer to such a question as the "average" value of f if the experiment were to be performed a large number of times.

Now suppose that an experiment is performed M times, where M is some large number. Then you may expect that the outcome

s_1 occurs about $p(s_1)M$ times

s_2 occurs about $p(s_2)M$ times

$\cdots$

s_N occurs about $p(s_N)M$ times.

Consequently,

$f(s_1)$ occurs about $p(s_1)M$ times

$f(s_2)$ occurs about $p(s_2)M$ times

$\cdots$

$f(s_N)$ occurs about $p(s_N)M$ times.

Thus, the expected, or average, value of f is

$$\frac{f(s_1)p(s_1)M + f(s_2)p(s_2)M + \cdots + f(s_N)p(s_N)M}{M}$$

$$= f(s_1)p(s_1) + f(s_2)p(s_2) + \cdots + f(s_N)p(s_N)$$

$$= \sum_{i=1}^{N} f(s_i)p(s_i)$$

Since M canceled, the answer is independent of the number of times the experiment was performed.

Definition 6.5

Let S be a sample space with a probability measure p, and let $f: S \rightarrow \mathbf{R}$ be a function. Then the **expected value** of f is

$$\sum_{x \in S} f(x)p(x)$$

In practice, the function f of Definition 6.5 is usually a numerical value associated with an outcome.

EXAMPLE 6.44   Let $S = \{1, 2, 3\}$, and suppose that the probability measure p and the function f on S are given by

x	$f(x)$	$p(x)$
1	2.17	0.37
2	5.25	0.22
3	1.00	0.41

Then the expected value of f is

$$2.17 \times 0.37 + 5.25 \times 0.22 + 1.00 \times 0.41 = 2.3679$$

○

EXAMPLE 6.45   If a die is tossed, what is the expected number of spots? The sample space S is $\{1, 2, 3, 4, 5, 6\}$, and the function f on S is the function that assigns to an outcome x the number of spots showing; that is, $f(x) = x$. The probability of each outcome is $\frac{1}{6}$. This is summarized in the following table:

x	$f(x)$	$p(x)$
1	1	$\frac{1}{6}$
2	2	$\frac{1}{6}$
3	3	$\frac{1}{6}$
4	4	$\frac{1}{6}$
5	5	$\frac{1}{6}$
6	6	$\frac{1}{6}$

The expected number of spots is

$$1 \times \frac{1}{6} + 2 \times \frac{1}{6} + 3 \times \frac{1}{6} + 4 \times \frac{1}{6} + 5 \times \frac{1}{6} + 6 \times \frac{1}{6} = 3.5$$

○

EXAMPLE 6.46   Two coins are tossed. What is the expected number of heads? The sample space S is $\{HH, HT, TH, TT\}$, and the function f on S is the function that assigns to each outcome the number of heads in that outcome. Each outcome x has probability $\frac{1}{4}$. This is summarized in the following table:

x	$f(x)$	$p(x)$
HH	2	$\frac{1}{4}$
HT	1	$\frac{1}{4}$
TH	1	$\frac{1}{4}$
TT	0	$\frac{1}{4}$

The expected number of heads is

$$2 \times \frac{1}{4} + 1 \times \frac{1}{4} + 1 \times \frac{1}{4} + 0 \times \frac{1}{4} = 1$$

○

EXAMPLE 6.47   In blackjack, the values of the cards 2 through 10 are 2 through 10 respectively, the ace has value 1 or 11, and the value of a face card is 10. If a card is chosen at random from a deck, and if the card is valued at 1 if it is an ace, what is the expected value of the value of the card? The possible values, the cards having them, and the associated probabilities are shown in the following table.

Value	Cards	Probability
1	Ace	$\frac{1}{13}$
2	2	$\frac{1}{13}$
3	3	$\frac{1}{13}$
4	4	$\frac{1}{13}$
5	5	$\frac{1}{13}$
6	6	$\frac{1}{13}$
7	7	$\frac{1}{13}$
8	8	$\frac{1}{13}$
9	9	$\frac{1}{13}$
10	10, J, Q, K	$\frac{4}{13}$

So the expected value of the card value is

$$1\left(\frac{1}{13}\right) + 2\left(\frac{1}{13}\right) + 3\left(\frac{1}{13}\right) + 4\left(\frac{1}{13}\right) + 5\left(\frac{1}{13}\right) + 6\left(\frac{1}{13}\right) + 7\left(\frac{1}{13}\right)$$

$$+ 8\left(\frac{1}{13}\right) + 9\left(\frac{1}{13}\right) + 10\left(\frac{4}{13}\right) = \frac{85}{13} = 6.5385$$

EXAMPLE 6.48   In a five-card poker hand, what is the expected number of spades? The probabilities of the possible numbers of spades in the hand are as follows:

Number	Probability
0	$C(13, 0) \times \frac{C(39, 5)}{C(52, 5)} = 0.2215$
1	$C(13, 1) \times \frac{C(39, 4)}{C(52, 5)} = 0.4114$
2	$C(13, 2) \times \frac{C(39, 3)}{C(52, 5)} = 0.2744$
3	$C(13, 3) \times \frac{C(39, 2)}{C(52, 5)} = 0.0815$
4	$C(13, 4) \times \frac{C(39, 1)}{C(52, 5)} = 0.0107$
5	$C(13, 5) \times \frac{C(39, 0)}{C(52, 5)} = 0.0005$

So the expected number of spades is

$$0.2215(0) + 0.4114(1) + 0.2744(2) + 0.0815(3) + 0.0107(4)$$

$$+ 0.0005(5) = 1.25$$

EXERCISES 6.6

1. A researcher is investigating the results of using drug X to treat sesquipedalianism and drug Y to treat hypatia. In her judgment, the success rate of drug X is 37%, and the success rate of drug Y is 22%. There is no evidence

that the conditions or treatments are related. List the possible outcomes of administering both drugs to a patient suffering from both conditions, and the probabilities of each outcome.

2. The probability of rain in Akron is 15%. The probability of rain in Auckland is 45%. The probability of rain in Ahmedabad is 95%. List all possible outcomes of noting in which of these three cities it rains, and compute the probabilities of each outcome.

3. An experiment consists of dealing five cards from a deck of 52 cards and noting the number of spades. The sample space is {0, 1, 2, 3, 4, 5}. Derive a probability measure on this sample space from the equiprobable measure on the set of all five-card hands.

4. Two coins are tossed. If A is the event that the first coin is heads and B is the event that the second coin is heads, are A and B independent?

5. A die is tossed. If A is the event that the number showing is odd and B is the event that the number showing is greater than 4, are A and B independent?

6. Of the students at Cabin Creek College, 38% are members of minorities. If a random sample of 20 students is chosen, find the probability that at least one student in the sample is a member of a minority.

7. With the skin test for tuberculosis, 5% of those having a positive reaction to the test actually have the disease. Among a group of students given the skin test, 25% had positive reactions. If there are 1000 students in the school, what is the expected number of students having the disease?

8. About 5% of the population are carriers of cystic fibrosis. If two people are chosen at random, find the probability that

   a. Both are carriers.

   b. Exactly one is a carrier.

   c. Neither is a carrier.

9. You toss a pair of dice 20 times. What is the probability that the sum of 12 occurs exactly once?

10. You draw a card from a deck of 52 cards. Find the probability that

    a. The card is a picture card, given that the card is black.

    b. The card is a black card, given that it is a picture card.

    c. The card is a heart, given that it is red.

11. Refer to Exercise 14, Section 6.1. If a woman is chosen at random from the group of 45, find the probability that

    a. She is highly creative given that she is from a non-Protestant background.

    b. She is from a non-Protestant background given that she is highly creative.

12. Refer to Example 6.34 (two football games).

    a. What is the probability that Washington will defeat Washington State, given that the result is not a tie?

    b.   What is the probability that Purdue will defeat Indiana, given that the result is not a tie?

    c.   What is the probability of two ties, given that there is at least one tie?

13. Solve the problem of Example 6.30 using a tree diagram.

14. Refer to Example 6.42. What is the probability that both transistors chosen are good if at least one is good?

15. A coin is tossed four times. What is the probability that all results are heads, given that at least one is heads?

16. Two dice are rolled. What is the probability of rolling a double, given that the sum is 7 or larger?

17. A hand of five cards is dealt from a 52-card deck. What is the probability that it contains all spades, given that it is not void in spades?

18. If two dice are rolled, what is the expected value of the sum of the numbers showing?

19. A piggy bank contains seven pennies, three nickels, two dimes, and three quarters. Two coins are taken from the bank. What is the expected monetary value of the two coins?

20. You play a game in which you pay \$8, roll two dice, and receive a number of dollars equal to the number of dots showing. What are your expected winnings per play?

21. A multiple-choice test with five choices for each question is scored by giving 1 point for a correct answer, $\frac{-1}{2}$ points for a wrong answer, and 0 points if no answer is given. If a student guesses at each question on a 100-question test, how many points should he expect to get on the test?

22. You pay \$1 to enter a lottery in which there are three prizes, valued at \$1000, \$500, and \$100. If 10,000 people enter, what is your expected winning?

23. When a car arrives at a ferry terminal, the driver either pays cash, uses a commuter ticket, or buys a new book of commuter tickets. At the toll booth, it takes 5 seconds to use a commuter ticket, 35 seconds to pay cash, and 45 seconds to purchase a book of tickets. If 20% of the drivers pay cash, 4% purchase commuter ticket books, and 76% use commuter tickets, what is the average time spent by a car at a toll booth?

24. A data-processing program handles four types of transactions. Transactions of type A take 135 milliseconds to process; type B, 210 milliseconds; and type C, 300 milliseconds. If 85% of the transactions are of type A and 10% are of type B, what is the expected time for processing a transaction?

25. In a thirteen-card bridge hand, what is the expected number of aces?

26. Suppose that B is an event in a sample space S with $p(B) > 0$. Show that

    a.   If $A \subset B$, then $p(A \mid B) = \dfrac{p(A)}{p(B)}$

    b.   If $B \subset A$, then $p(A \mid B) = 1$

Computer Exercises for Chapter 6

1. Write a computer program to print all six permutations of the digits 1, 2, 3.

2. Write a computer program to compute the value of $C(n, r)$ where n and r are input by the user.

3. Write a computer program to print the first n rows of Pascal's Triangle, where n is input by the user.

• • • • **CHAPTER REVIEW EXERCISES**

1. Evaluate
   a. $P(9, 4)$
   b. $C(9, 4)$

2. The outcome of a horse race depends on which horses finish first, second, and third. If ten horses race, how many outcomes are there?

3. How many ways can you arrange the letters in the word *computer?* In *onomatopoeia?*

4. If six couples get together for bridge, how many different two-person teams are possible?

5. Answer Exercise 4 if women must team with women and men with men.

6. Answer Exercise 4 if no person may team with her spouse.

7. A box contains twelve white, eight black, and ten blue balls.

   a. How many ways can you choose a sample of three white, two black, and six blue balls?

   b. What is the probability that a sample of three balls contains one of each color?

   c. What is the probability that a sample of three balls contains one of each color, given that not all three are the same color?

8. a. How many ways can you answer a ten question true–false test?

   b. If you need at least seven correct answers to pass, what is the probability that you pass if you guess at each answer?

9. How many five-card poker hands contain all spades?

10. How many five-card poker hands contain all black cards?

11. How many 8-bit bytes contain an even number of 1's?

12. How many 8-bit bytes contain exactly five consecutive 0's?

13. How many 8-bit bytes have 1's and 0's alternating?

14. How many different five-letter arrangements are possible from the letters in the alphabet? How many of these begin with the letter q?

15. In how many orderings of the letters in the word *newelpost* does the w precede the p?

16. How many seven-card hands contain seven consecutive cards of the same suit if the ace counts as 1 or 14?

17. Use the Binomial Theorem or Pascal's Triangle to expand the following.

   a. $(x + 4)^7$
   b. $(3 - 2a)^5$

18. Find the coefficient of x^{20} in the expansion of $(x + 1)^{64}$.

19. Find the ninth term in the expansion of $(x - 5)^{17}$.

20. Show that $C(n + 1, r) = C(n, r - 1) + C(n, r)$.

21. Show that $C(2n, 2) = 2C(n, 2) + n^2$.

22. Show that $rC(n, r) = nC(n - 1, r - 1)$.

23. Show that
$$C(2n + 2, n + 1) = C(2n, n + 1) + 2C(2n, n) + C(2n, n - 1)$$

24. Show that the product of k consecutive integers is divisible by $k!$. (*Hint:* Look at $C(n + k, k)$.)

25. A marketing firm ran taste-tests to determine whether consumers prefer the new Coke over the old Coke or Pepsi. Five people participating in the tests were asked to taste seven colas and rank the best and the worst.

   a. How many ways can a particular taster choose the best cola?

   b. How many ways can all five tasters choose the best cola?

   c. If all the tasters are choosing randomly, what is the probability that they all choose the same cola as the best?

26. There are five judges for a writing contest in which thirteen papers were submitted.

   a. How many ways can one judge choose two papers for the first and second prizes?

   b. How many ways can all five judges choose two papers for the first and second prizes?

   c. If the judges choose the winners randomly, what is the probability that all five judges choose the same two papers?

27. What is the probability that a five-card poker hand contains four aces, given that it contains three aces?

28. What is the probability of rolling a pair with two dice, given that the total rolled is at least 9?

29. If one die is rolled, what is the expected value of the square of the number showing?

30. Four cards are dealt from a 52-card deck. What is the expected value of the square root of the number of spades?

7

GRAPHS

The origins of graph theory can be traced to the eighteenth century and a problem that was originally posed for amusement. The city of Königsberg in East Prussia (now Kaliningrad in the Soviet Union) is situated on the Pregel River. In the river there are two islands, which are connected to each other and to the mainland by seven bridges (Figure 7.1). The problem was to plan a walk through the city in such a way that each bridge was crossed exactly once. The problem came to the attention of the Swiss mathematician Leonhard Euler (1707–1783), who posed and solved a more general problem. In this chapter, we will study the general problem and its solution, as well as related topics.

Another classical problem is the problem of coloring a map of a state in such a way that no two adjacent counties have the same color. Figure 7.2 is a map of Vermont. It is easy to see that four colors are needed for this map. Lamoille, Orleans, and Caledonia counties must have different colors. Let us call the colors A, B, and C. If only these three colors are used, Washington county must have color B. Similarly, Chittenden county must then be colored C, but now there is no color available for Franklin county.

It was conjectured as early as 1852 that four colors suffice to color any map, no matter how complex. Unlike the Königsberg bridge problem, which was solved by Euler as soon as he had properly formulated it, the four-color problem remained unsolved until 1976, when the conjecture was proved by K. Appel and W. Haken of the University of Illinois. The solution involved extensive computations and required over 1200 hours of time on a high-speed computer.

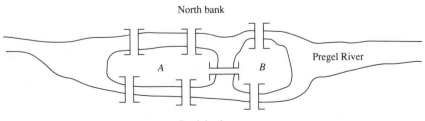

FIGURE 7.1   The bridges of Königsberg

FIGURE 7.2   Map of Vermont

Both problems involve land masses—joined by bridges in Königsberg, and by boundaries in Vermont. If we use one dot to represent each land mass and use a line connecting two dots to represent a bridge or a boundary, we obtain the diagrams shown in Figures 7.3 and 7.4. Diagrams of this type are called **graphs.** Graph theory deals with the techniques for solving problems that can be stated using graphs.

Although graph theory was once regarded as pure mathematics, the introduction of computers has brought it into the realm of the applied. This is because many

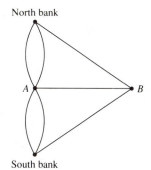

FIGURE 7.3   Graph for the Königsberg bridges

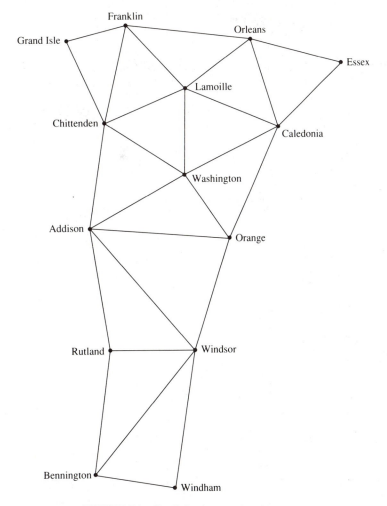

FIGURE 7.4   Graph for the counties of Vermont

methods of data organization can be described using graphs—especially the class of graphs called *trees.* In this chapter, we will introduce some general graph-theoretic concepts; then, in Chapter 8, we will look closely at the properties of trees.

7.1 • CONCEPTS AND TERMINOLOGY

In this section, we will introduce the basic terminology of graph theory and derive some fundamental facts about graphs.

Definition 7.1

A **graph** $G = (V, E, R)$ consists of a finite set V of **vertices** or **nodes,** a finite set E of **edges,** and a relation R on $E \times V$ called the **incidence relation** of G, such that for each edge $e \in E$, there are either one or two vertices $v \in V$ with $(e, v) \in R$. If $(e, v) \in R$, we say that e is **incident** to v. If v_1 and v_2 are vertices, and if there is an edge e incident to v_1 and v_2, we say that v_1 and v_2 are **adjacent.**

EXAMPLE 7.1   The Königsberg bridges can be represented by a graph in which $V = \{v_1, v_2, v_3, v_4\}$, $E = \{e_1, e_2, e_3, e_4, e_5, e_6, e_7\}$, and the incidence relation is given by

	v_1	v_2	v_3	v_4
e_1	1	1	0	0
e_2	1	1	0	0
e_3	1	0	1	0
e_4	0	1	0	1
e_5	0	1	0	1
e_6	0	0	1	1
e_7	0	1	1	0

where the 1 in the table position (e_i, v_j) indicates that edge e_i is incident to vertex v_j. This representation of the incidence relation is called an **incidence matrix.** The graph is illustrated in Figure 7.5. This is the same as Figure 7.3, but with the vertices and edges labeled.

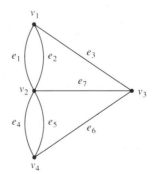

FIGURE 7.5   Graph for the Königsberg bridges

○

EXAMPLE 7.2   In the graph of Figure 7.5, e_1 and e_2 are incident to v_1 and v_2 but not to v_3. Vertex v_1 is adjacent to vertices v_2 and v_3. Vertex v_2 is adjacent to v_1, v_3, and v_4. However, v_1 is not adjacent to v_4.   ○

EXAMPLE 7.3   In the graph of Figure 7.4, Lamoille county is adjacent to Orleans, Caledonia, Washington, Chittenden, and Franklin counties but not to Windham county.   ○

Note that this usage of the word *graph* is different from its usage in the phrase *graph of a function.*

Any graph can be described by drawing a picture in which the vertices are points in the plane, and each edge is drawn as a line connecting the vertices to which it is incident. Clearly, all the information about the incidence relation is contained in the picture. In the remainder of this chapter, we will define graphs exclusively by drawing pictures.

In Definition 7.1, it is permissible for an edge to have only one vertex. Such an edge is called a **loop.** It is also possible for several edges to have the same two vertices. Such edges are called **multiple edges.** It is even possible for a vertex to have no edges incident to it. Such a vertex is called an **isolated vertex.** In the Königsberg bridge graph of Figure 7.5, there are multiple edges connecting v_1 with v_2, and also connecting v_2 with v_4. The graph in Figure 7.6 has multiple edges (both e_2 and e_3

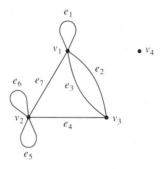

FIGURE 7.6   A graph with multiple edges and loops

connect v_1 with v_3), loops (e_1, e_5, and e_6), and an isolated vertex (v_4). In some classes of problems, such as map-coloring problems, loops do not arise.

Definition 7.2   **A simple graph** is a graph with no multiple edges and no loops.

EXAMPLE 7.4   The graph in Figure 7.4 is a simple graph.   ○

The terminology used in the literature of graph theory is not entirely consistent. Some authors use the term *graph* to mean what we have called a simple graph—that is, a graph without multiple edges or loops. These authors use terms such as *generalized graph* or *multigraph* to refer to what we have called a graph.

The Degree of a Vertex

Definition 7.3

The **degree** of a vertex in a graph is the number of edges incident to the vertex, with each loop at the vertex being counted twice.

You can think of the degree of a vertex as the number of different directions you can take if you leave the vertex along one of the edges.

EXAMPLE 7.5   In Figure 7.6, vertex v_1 has degree 5, vertex v_2 has degree 6, vertex v_3 has degree 3, and vertex v_4 (the isolated vertex) has degree zero.   ○

Theorem 7.1   In any graph, the sum of the degrees of the vertices equals twice the number of edges.

Proof   If you think of an edge as having two ends (which are the same if the edge is a loop), then each end contributes one count to the sum of the degrees of the vertices. So the sum of the degrees is the total number of ends, which is twice the number of edges.   ●

EXAMPLE 7.6   In Figure 7.6, there are seven edges. The sum of the degrees of the vertices is $5 + 6 + 3 + 0 = 14$, which is equal to 2×7.   ○

Theorem 7.2   In any graph, the number of vertices of odd degree is even.

Proof   Let x be the sum of the degrees of the even vertices, and let y be the sum of the degrees of the odd vertices. If E is the number of edges, then

$$x + y = 2E$$

by Theorem 7.1. Now x is even, since it is a sum of even numbers; and $2E$ is even. Therefore, y is even. But y is a sum of odd numbers; such a sum can be even only if it is the sum of an even number of odd numbers. Thus, the number of odd vertices is even.   ●

EXAMPLE 7.7   In Figure 7.6, there are two vertices of odd degree: v_1 and v_3.   ○

Some Special Graphs and Graph Isomorphism

Certain types of graphs occur with sufficient frequency that they have names. These are the complete graphs and the bipartite graphs.

Definition 7.4

A graph is said to be a **complete** graph if it is a simple graph and every vertex is adjacent to every other vertex. The number of vertices is called the **order** of the complete graph.

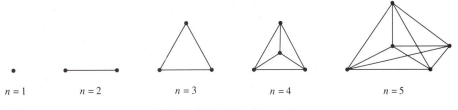

FIGURE 7.7 Complete graphs

In a complete graph, there is exactly one edge connecting each pair of vertices, and there are no loops. If n is the number of vertices, then the number of edges is equal to the number of pairs of vertices, which is $C(n, 2)$. Each vertex has one edge going from it to each other vertex, so the degree of each vertex is $n - 1$. Figure 7.7 illustrates complete graphs with one through five vertices.

It is apparent that any two complete graphs G_1 and G_2 having the same number of vertices are "alike" in the following sense: There are one-to-one correspondences between the sets of vertices of G_1 and G_2, and also between the sets of edges of G_1 and G_2, such that corresponding edges connect corresponding vertices. We can extend this idea to graphs in general.

Definition 7.5

Let $G_1 = (V_1, E_1, R_1)$ and $G_2 = (V_2, E_2, R_2)$ be graphs. The graph G_1 is **isomorphic** to the graph G_2 if there exist one-to-one, onto functions $f: V_1 \rightarrow V_2$ and $g: E_1 \rightarrow E_2$ such that $(e_1, v_1) \in R_1$ if and only if $(g(e_1), f(v_1)) \in R_2$. The pair of functions (f, g) is then called a **graph isomorphism.**

The definition says that two graphs are isomorphic if and only if their vertices and edges can be made to correspond in such a way that the incidence relation is preserved. In fact, two graphs are isomorphic if and only if the incidence matrix of one can be obtained from the incidence matrix of the other by a permutation of the rows and columns.

If two graphs are isomorphic, they have the same number of vertices and the same number of edges, but the converse is not true. Also, a graph isomorphism preserves degrees of vertices. That is, if (f, g) is a graph isomorphism, the degree of v is equal to the degree of $f(v)$. Figure 7.8 shows some pairs of isomorphic graphs, whereas the graphs in Figure 7.9 are not isomorphic. In Figure 7.8, the isomorphism consists of the functions f (which assigns v_i' to v_i) and g (which assigns e_i' to e_i). The graphs in Figure 7.9(a) cannot be isomorphic, because all of the vertices in the left graph have degree 2, whereas one of the vertices in the right graph has degree 3.

It is apparent that any two complete graphs of order n are isomorphic. It is convenient to speak of *the* complete graph of order n and give it a name; it is called K_n.

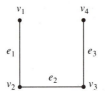
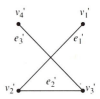

is isomorphic to

(a)

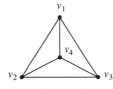

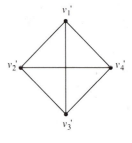

is isomorphic to

(b)

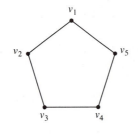

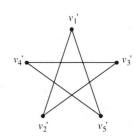

is isomorphic to

(c)

FIGURE 7.8   Some isomorphic graphs

is not isomorphic to

(a)

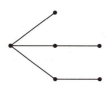
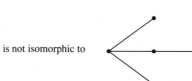

is not isomorphic to

(b)

FIGURE 7.9   Some nonisomorphic graphs

Definition 7.6

A graph G with set of vertices V is called **bipartite** if it is possible to write V as a union of two nonempty disjoint subsets V_1 and V_2, such that each edge of G connects a vertex of V_1 with a vertex of V_2.

Figure 7.10 shows several bipartite graphs. Note that a bipartite graph can contain no loops, since the two ends of every edge must be in different sets of vertices. The only complete graph that is also a bipartite graph is K_2, shown in Figure 7.7.

Note that the decomposition of V into V_1 and V_2 may not be unique. For instance, in Figure 7.10(b), either of the isolated vertices could have been placed in V_1 or V_2.

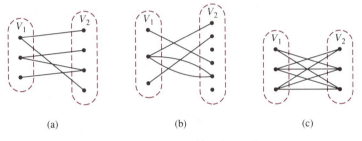

   (a)   (b)   (c)

FIGURE 7.10　Some bipartite graphs

Definition 7.7

The **complete bipartite graph** $K_{m,n}$ is the graph with vertices $V = V_1 \cup V_2$, where V_1 has m elements, V_2 has n elements, $V_1 \cap V_2 = \varnothing$, and each vertex of V_1 is connected to each vertex of V_2 by exactly one edge. The pair (m, n) is called the **order** of the complete bipartite graph.

Any two complete bipartite graphs of the same order (m, n) are isomorphic, so it is reasonable to use a name such as $K_{m,n}$. Also, $K_{m,n}$ is isomorphic to $K_{n,m}$. Figure 7.11 shows several complete bipartite graphs.

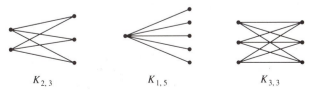

   $K_{2,3}$ $K_{1,5}$ $K_{3,3}$

FIGURE 7.11　Some complete bipartite graphs

Note that *complete* and *complete bipartite* are two different concepts: A complete bipartite graph is *not* a bipartite graph that is also a complete graph. The only graph that is both complete bipartite and complete is $K_2 = K_{1,1}$. None of the bipartite graphs in Figure 7.10 is complete bipartite.

Computer Representations of Graphs

There are several ways of representing a graph in a computer. The simplest is the **edge table.** The edge table for a graph lists each edge and identifies the vertices incident to it. Thus, an edge table is just a condensed form of the incidence matrix. The incidence matrix for the Königsberg bridge graph was given in Example 7.1. The corresponding edge table is as follows.

Edge	First Vertex	Second Vertex	
1	1	2	
2	1	2	
3	1	3	
4	2	4	— Indicates that e_4 connects v_2 to v_4
5	2	4	
6	3	4	
7	2	3	

In this table, one end of each edge is called the first vertex, and the other is called the second vertex, but it doesn't matter which is which. Although this method is simple, it is not always the most convenient. In many applications, it is important to be able to determine what vertices are adjacent to a given vertex. If the graph is given by an edge table, this can be determined only by searching through all edges.

Another way to store a graph is by using an **adjacency matrix.** In an adjacency matrix, both rows and columns correspond to vertices. An entry for a pair of vertices shows how many edges connect the vertices. The adjacency matrix for the Königsberg bridge graph is as follows.

	v_1	v_2	v_3	v_4	
v_1	0	2	1	0	
v_2	—	0	1	2	— Indicates that there are two edges connecting v_2 to v_4
v_3	—	—	0	1	
v_4	—	—	—	0	

Note that entries below the diagonal are not used. They are not needed because the number of edges connecting v_i to v_j is the same as the number connecting v_j to v_i. A nonzero number on the diagonal would indicate the presence of loops. If the graph is simple, all numbers on the diagonal of its adjacency matrix are zero, and all numbers above the diagonal are either zero or one.

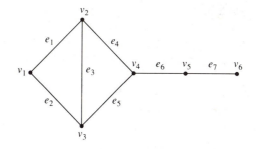

FIGURE 7.12   A simple graph

Although an adjacency matrix is sometimes easier to use than an incidence matrix, it is still necessary to search down an entire column to find all the vertices adjacent to a given vertex. Moreover, the adjacency matrix is wasteful of computer memory space if the number of edges is large. If the graph is a simple graph, the **adjacency list** method can be used. In this method, for each vertex we store a list of the vertices adjacent to it. Consider the simple graph in Figure 7.12. In this graph,

v_1 is adjacent to v_2 and v_3

v_2 is adjacent to v_1, v_3, and v_4

v_3 is adjacent to v_1, v_2, and v_4

v_4 is adjacent to v_2, v_3, and v_5

v_5 is adjacent to v_4 and v_6

v_6 is adjacent to v_5

The graph can be stored in a computer as shown in Figure 7.13, in which there are two structures: a vertex table and an adjacency list table. Each item in the vertex table contains a pointer to an item in the adjacency list table. Each item in the

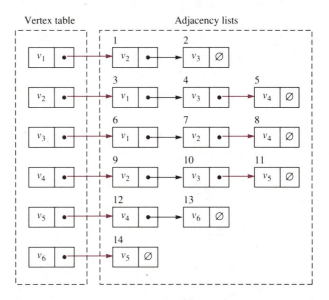

FIGURE 7.13   Adjacency list for the graph of Figure 7.12

adjacency list table identifies a vertex and also contains either a pointer to the next item or a null pointer (∅ or 0) if it is the last item in its list. (An arrangement of a list of items in which each item contains a pointer to the next item is called a **linked list**.) The entire structure can be written as two tables, or arrays:

Vertex table		Adjacency list table		
Vertex Number	Item Pointer	Item Number	Vertex Number	Item Pointer
1	1	1	2	2
2	3	2	3	0
3	6	3	1	4
4	9	4	3	5
		5	4	0
5	12	6	1	7
6	14	7	2	8
		8	4	0
		9	2	10
		10	3	11
		11	5	0
		12	4	13
		13	6	0
		14	5	0

Indicates that item 9 contains first vertex adjacent to vertex 4

Indicates that vertex 2 is adjacent to vertex 4 and that item 10 contains the next vertex adjacent to vertex 4

Note that the information in an adjacency list representation is redundant. For instance, the fact that v_2 and v_3 are adjacent is indicated in both the v_2 list (item 4) and in the v_3 list (item 7). Also note that the identifying numbers in column 1 are entirely arbitrary. If the list items were numbered in another way, a different but equally valid representation of the graph would be obtained.

EXERCISES 7.1

1. Draw the graphs given by each of the following incidence relations.

a.

	v_1	v_2	v_3	v_4
e_1	1	0	0	1
e_2	1	0	1	0
e_3	1	1	0	0
e_4	1	0	0	0
e_5	0	1	1	0
e_6	0	0	1	1

b.

	v_1	v_2	v_3	v_4
e_1	0	1	1	0
e_2	1	1	0	0
e_3	0	0	1	1
e_4	1	0	0	1
e_5	1	0	1	0
e_6	0	1	0	1

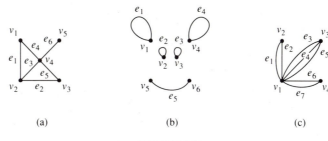

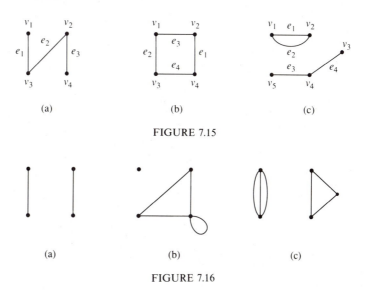

(a)   (b)   (c)

FIGURE 7.14

2. Write the incidence relations of the graphs in Figure 7.14.

3. In which of the graphs in Figure 7.15 is e_1 incident to v_1? To v_2? In which are v_1 and v_2 adjacent?

(a)   (b)   (c)

FIGURE 7.15

(a)   (b)   (c)

FIGURE 7.16

4. Which of the graphs in Figure 7.16 are simple?

5. Determine the degree of each of the vertices in Vermont (Figure 7.4).

6. Determine the degree of each of the vertices of the bipartite graphs of Figure 7.10.

7. Which of the pairs of graph in Figure 7.17 are isomorphic?

8. Draw $K_{4,3}$.

9. Classify all simple graphs with three vertices. That is, draw a set of simple graphs with three vertices, no two of them isomorphic to one another, such that every simple graph with three vertices is isomorphic to one of the graphs in the set.

10. Classify all simple graphs with four vertices.

11. If v is a vertex of the bipartite graph $K_{m,n}$, what is the degree of v?

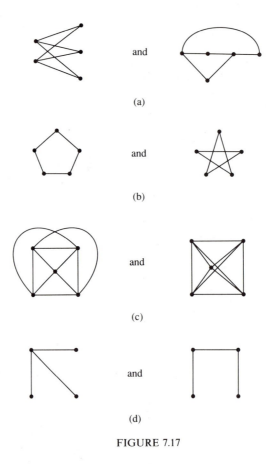

FIGURE 7.17

12. How many vertices does the bipartite graph $K_{m,n}$ have? How many edges?

13. Refer to a map of New York City. Draw a graph representing the bridges connecting Manhattan, the Bronx, Long Island, Staten Island, and Ward's Island.

14. Draw graphs representing the counties of **a.** Delaware, **b.** Rhode Island, **c.** New Hampshire.

15. Find the simplest map (of an imaginary state) that requires four colors.

16. Describe the graphs of Figure 7.15 by adjacency matrices.

17. Draw the graphs described by the following adjacency matrices:

a.

	v_1	v_2	v_3	v_4
v_1	1	0	2	1
v_2	—	0	1	1
v_3	—	—	2	0
v_4	—	—	—	0

b.

	v_1	v_2	v_3	v_4	v_5
v_1	0	2	1	0	0
v_2	—	0	1	2	1
v_3	—	—	0	1	0
v_4	—	—	—	0	0
v_5	—	—	—	—	0

c.

	v_1	v_2	v_3	v_4	v_5
v_1	0	1	1	1	1
v_2	—	0	1	0	0
v_3	—	—	0	1	0
v_4	—	—	—	0	1
v_5	—	—	—	—	0

18. Describe the graphs of Figure 7.15(a) and (b) by adjacency lists.

19. Describe the graph of Figure 7.14(a) by an adjacency list.

20. Draw the graph defined by the following adjacency list:

Vertex Number	Item Pointer	Item Number	Vertex Number	Item Pointer
1	12	1	1	3
2	11	2	6	0
3	1	3	4	10
4	8	4	2	0
5	6	5	4	4
6	5	6	3	0
		7	3	9
		8	2	7
		9	6	0
		10	5	0
		11	4	2
		12	3	0

21. Draw the graph defined by the following adjacency list:

Vertex Number	Item Pointer	Item Number	Vertex Number	Item Pointer
1	9	1	3	0
2	7	2	3	0
3	5	3	2	0
4	4	4	2	0
5	3	5	1	6
6	2	6	6	12
7	0	7	4	8
8	1	8	5	11
		9	2	10
		10	3	0
		11	1	0
		12	8	0

22. Describe the adjacency matrix of the graph K_n.

23. Describe the adjacency matrix of the graph $K_{m,n}$.

24. Let G be the graph whose vertices are the countries of the world and whose edges are common borders. What is the degree of Iceland? Of the United States?

7.2 • CONNECTIVITY

The graph in Figure 7.16(a) can be considered as consisting of two parts, with no edge connecting one part to the other. On the other hand, no such decomposition is possible for the graph in Figure 7.4 (Vermont). In the graph of Figure 7.4, it is possible to move from any vertex to any other by traveling along edges. The graph of Figure 7.4 is said to be *connected*, whereas that of Figure 7.16(a) is not. In this section, we will define this idea precisely. The concepts we develop will lead to a solution of the Königsberg bridge problem.

Walks and Paths

Definition 7.8

> Let G be a graph. A **walk** from a vertex u to a vertex v of G is an alternating sequence of vertices and edges
>
> $$v_0, e_1, v_1, e_2, v_2, \ldots, v_{n-1}, e_n, v_n$$
>
> such that $v_0 = u$, $v_n = v$, and each edge e_i is incident to the vertices v_{i-1} and v_i. If $u = v$, the walk is **closed;** otherwise it is **open.** The **length** of the walk is the number n of edges.

Note that a walk can consist of a single vertex v. This is the trivial walk from v to v, which has length zero.

EXAMPLE 7.8   In Figure 7.18(a), the sequence

$$v, a, w, b, x, b, w, e, y, d, z$$

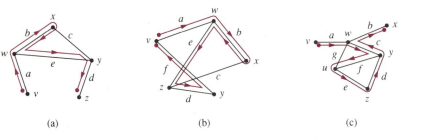
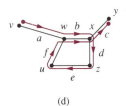

(a)   (b)   (c)   (d)

FIGURE 7.18   Walks in graphs

is a walk from v to z. In Figure 7.18(b), the sequence

$$v, a, w, b, x, b, w, e, z, d, y, f, v$$

is a walk from v to v. In part (c) of the figure, a walk from v to x is shown, and (d) shows a walk from v to y. The walk in (b) is closed; the others are open.   ○

A walk from u to v indicates a way of traveling from u to v along edges of the graph, but a walk is not necessarily an efficient way to go. For instance, the walk in Figure 7.18(a) backtracks, and the walks in (c) and (d) each traverse an edge twice. If we eliminate such inefficiencies from a walk, we obtain a *path*.

Definition 7.9   **A path** in a graph G is a walk in which no vertex or edge is repeated (except for the first and last vertices in the case of a closed walk). A closed path is called a **cycle.**

EXAMPLE 7.9   The walks in Figure 7.19 are paths.

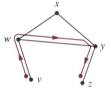
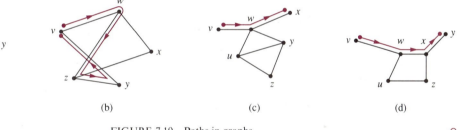

(a)   (b)   (c)   (d)

FIGURE 7.19   Paths in graphs   ○

In fact, each of the paths in Figure 7.19 was obtained from the corresponding walk in Figure 7.18 by deleting repeated vertices and edges. The next theorem shows that this can always be done.

Theorem 7.3   Every open walk from u to v in a graph G contains a path from u to v.

Proof   Let W be an open walk from u to v. Consider the set X of *all* open walks from u to v. Then $X \neq \varnothing$, since $W \in X$. Choose a walk W_1 from u to v having the shortest possible length. There is such a walk, since every nonempty set of positive integers has a smallest element (see Exercise 14, Section 2.6). Then W_1 must be a path. For if W_1 were not a path, then some vertex of W_1 would be repeated; that is, if

$$W_1 = v_0, e_1, v_1, e_2, v_2, \ldots, v_{n-1}, e_n, v_n$$

then $v_i = v_j$ for some $i < j$. But then if you delete the portion of the walk from v_i to v_j, you have a new walk from u to v that is shorter than W_1, contradicting the assumption that W_1 was the shortest walk.   ●

Again, the terminology of graph theory is not entirely standardized. Some authors use the term *path* to denote what we have called a walk. These authors use the term *simple path* to mean what we call a path.

Definition 7.10

A graph G is said to be **connected** if for every pair of vertices u, v in G, there is a path from u to v in G.

Clearly, we could have used *walk* in the definition in place of *path*.

EXAMPLE 7.10   The graphs in Figure 7.20 are connected. The graphs in Figure 7.21 are not connected.

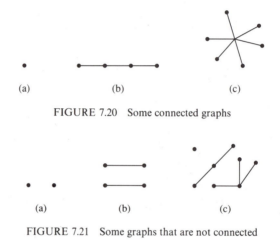

(a)   (b)   (c)

FIGURE 7.20   Some connected graphs

(a)   (b)   (c)

FIGURE 7.21   Some graphs that are not connected   o

Subgraphs

A glance at Figure 7.21 shows that each of the graphs in the figure is made up of two or more "parts" that are connected. These parts are called **components.** To show that every graph is made up of one or more connected components, we need the notion of a subgraph.

Definition 7.11

Let G be a graph whose set of vertices is V and whose set of edges is E. Let H be a graph whose set of vertices is W and whose set of edges is F. Then H is a **subgraph** of G if $W \subset V, F \subset E$, and an edge $e \in F$ is incident to a vertex $v \in W$ in the graph H if and only if e is incident to v in G.

Every subgraph of a graph G can be obtained from G by discarding zero or more vertices of G, then discarding all the edges incident to the discarded vertices, and finally discarding zero or more additional edges.

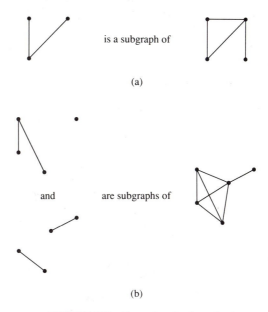

is a subgraph of

(a)

and   are subgraphs of

(b)

FIGURE 7.22   Examples of subgraphs

EXAMPLE 7.11   Figure 7.22 shows some examples of subgraphs.   ○

Definition 7.12   Let G be a graph whose set of vertices is V and whose set of edges is E. Let $W \subset V$. Then the subgraph of G **induced** by W is the subgraph whose vertices are the elements of W and whose edges are all edges of G that are incident only to vertices in W.

EXAMPLE 7.12   Figure 7.23 shows some graphs, subsets of the sets of vertices, and the induced subgraphs.   ○

To characterize the components of a graph, we need only to say what it means for two vertices to be in the same component. This is accomplished by the following theorem.

Theorem 7.4   Let G be a graph with set of vertices V. Define a binary relation R on V by the rule, $v\,R\,w$ if there is a path in G from v to w. Then R is an equivalence relation. The proof is left as an exercise.   ●

Since R is an equivalence relation, it partitions the set of vertices V into equivalence classes. Let us call the equivalence classes $V_1, \ldots, V_n$. Then if e is an edge of G, the vertices incident to e must be in the same class. The graph G is composed of the subgraphs induced by the equivalence classes $V_1, \ldots, V_n$. These are the components of the graph. This shows that every graph can be written as the union of connected components.

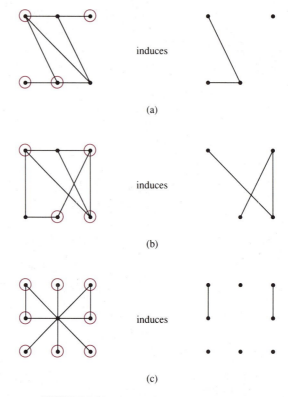

FIGURE 7.23   Examples of induced subgraphs

Here is a basic fact about connected graphs.

Theorem 7.5   Any connected graph with n vertices has at least $n - 1$ edges.

Proof   The proof is by induction on n. If $n = 1$, the theorem states that any connected graph with 1 vertex has at least zero edges, and this is certainly true. Now suppose that the statement is true for all graphs with fewer than n vertices. Let G be a graph having $n > 1$ vertices, and let v be any one of its vertices. Let H be the graph obtained by removing v from G. (This means that all edges incident to v must be removed also.) H may or may not be connected, but in any event it can be written as the union of one or more connected components $G_1, \ldots, G_m$. Since G is connected, the vertex v must have had at least one edge going to each of the components $G_1, \ldots, G_m$; so to make H, we had to remove at least m edges. Let the number of vertices of G_i be n_i. Then

$$\sum_{i=1}^{m} n_i = n - 1$$

because the components G_i account for all of the vertices of G except v. By the induction hypothesis, each G_i has at least $n_i - 1$ edges. It follows that the graph G

has at least

$$(n_1 - 1) + \cdots + (n_m - 1) + m = (n_1 + \cdots + n_m) - m + m$$
$$= (n - 1) - m + m$$
$$= n - 1$$

edges. ●

The idea behind the proof is illustrated in Figure 7.24. The graph shown has ten vertices. When the vertex v is removed from the graph, three edges are also removed, and the remainder consists of three components of 2, 3, and 4 vertices. The components have 1, 2, and 3 edges. The graph as a whole has

$$(1 + 2 + 3) + 3 = 9$$

edges.

Euler Circuits

We can use the concepts developed in this section to solve the Königsberg bridge problem. The problem is to find a walk through a graph G in which each edge is traversed exactly once. There are two cases to be considered, depending whether we require the walk to begin and end at the same vertex or allow it to begin and end at different vertices. (In the original statement of the Königsberg bridge problem, it is unclear which was intended.) Suppose G is a graph and W is a walk in G from a vertex v to a different vertex w. Then we can add one edge to the graph, an edge from w to v, and obtain a walk that starts and ends at v. The new edge increases the degrees of the vertices v and w by one each. So it is sufficient to examine the case in which the walk begins and ends at the same vertex.

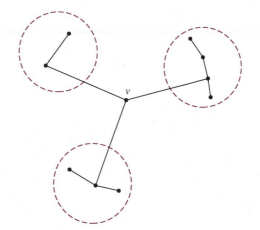

FIGURE 7.24   The proof of Theorem 7.5

Definition 7.13   An **Euler circuit** in a graph G is a closed walk in G in which each edge of G occurs exactly once.

The solution to the Königsberg bridge problem is contained in the following theorem.

Theorem 7.6   Let G be a connected graph. Then G has an Euler circuit if and only if every vertex has even degree.

Proof   If W is an Euler circuit in G, then whenever W enters a vertex along an edge, it must leave the vertex along a different edge. Each such event contributes 2 to the degree of the vertex. Since every edge is accounted for by W, the degree is even. This proves the "only if" part.

Now let G be a connected graph in which every vertex has even degree. We have to show that an Euler circuit exists, and we will do so by giving an algorithm for constructing one. Start at any vertex v. Build a walk starting at v by leaving along any edge and proceeding to a vertex w. If there are any unused edges at w, choose any one and add it to the walk. Continue in this way until you arrive at a vertex (call it u) where there are no unused edges. Then $u = v$, since if $u \neq v$, there would be an odd number of edges at u. In this way, we have constructed a closed walk, but it may not include all edges. If it does not, choose any vertex x on the walk where there is an unused edge, and repeat the process. The result will be a walk from x to x. Insert this walk into the first walk. The result will be a longer walk from v to v. Continue to expand the walk in this way until all edges have been used.   ●

The idea of the construction is shown in Figures 7.25 and 7.26. To find an Euler circuit in Figure 7.25, you can start at v_1. (Of course, you could start anywhere else, as well.) Choosing edges e_1, e_2, e_3, and e_4 leads back to v_1. This walk is not an Euler circuit. It includes only the edges shown as solid lines in Figure 7.26(a). To expand the walk, start at v_2 (you could also start at v_3) and choose the edges $e_5, e_6, e_7, e_8, e_9, e_{10}, e_{11}$. Insert these in the original walk. The result is

$$e_1, e_5, e_6, e_7, e_8, e_9, e_{10}, e_{11}, e_2, e_3, e_4$$

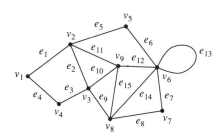

FIGURE 7.25   Find an Euler circuit

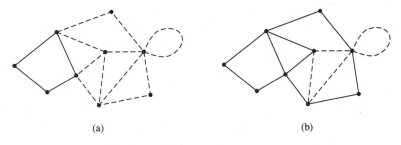

FIGURE 7.26   Two steps in finding an Euler circuit

This is still not an Euler circuit, as shown in Figure 7.26(b). Starting at v_9, you can choose edges $e_{12}, e_{13}, e_{14}, e_{15}$ and insert them at v_9. The final result is the circuit

$$e_1, e_5, e_6, e_7, e_8, e_9, e_{10}, e_{12}, e_{13}, e_{14}, e_{15}, e_{11}, e_2, e_3, e_4$$

A glance at Figure 7.3 now shows why there is no way to walk over each of the bridges of Königsberg once. In the graph, each of the four vertices has odd degree.

EXERCISES 7.2

1. In each of the graphs in Figure 7.27, a walk is indicated. Find a path contained in each walk.

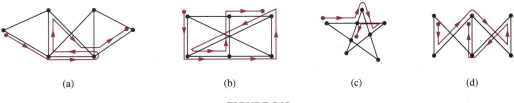

(a)　　　　　　　　(b)　　　　　　　　(c)　　　　　　　　(d)

FIGURE 7.27

2. Which of the graphs in Figure 7.28 are connected? How many components does each graph have?

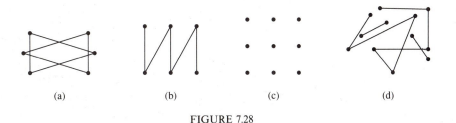

(a)　　　　　　　　(b)　　　　　　　　(c)　　　　　　　　(d)

FIGURE 7.28

3. In each of the graphs of Figure 7.29, draw the subgraph induced by the set of circled vertices.

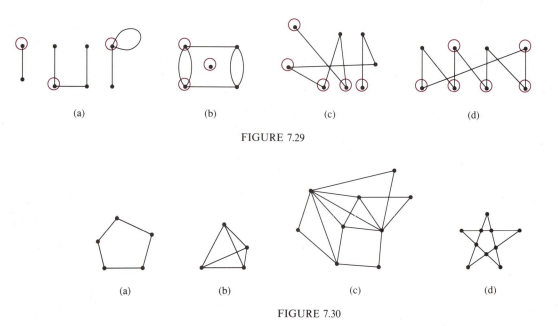

FIGURE 7.29

FIGURE 7.30

4. Which of the graphs in Figure 7.30 have Euler circuits? For each graph that has an Euler circuit, find one.

5. Prove Theorem 7.4.

6. Let G be a simple graph with n vertices. Show that if G has more than $C(n-1, 2)$ edges, then it is connected. (*Hint:* Let H be a nonconnected simple graph with n vertices and the largest possible number of edges. Then H has exactly two components. (Why?) Each of the two components must be a complete graph. (Why?) How many edges does H have?

7. Is it true that a graph G has an Euler circuit if and only if it is connected and every vertex has even degree? Explain.

8. For which integers n does the complete graph K_n have an Euler circuit?

9. For which pairs of integers m, n does the complete bipartite graph $K_{m,n}$ have an Euler circuit?

10. Can you find an Euler circuit for the bridges of New York City? (Refer to Exercise 13, Section 7.1.)

11. Show that a closed walk of odd length contains a cycle.

7.3 ● SOME CLASSICAL PROBLEMS

The Königsberg bridge problem is easy to state—and also easy to solve. But graph theory is a branch of mathematics in which there are many problems that are simple to state, but difficult to solve. The four-color problem is in this category. In this section, we will describe some important problems. Although complete solutions are beyond the scope of this book, we will obtain some interesting partial results.

Planarity

The graphs that arise in map-coloring problems all have the property that they can be drawn in the plane without edges crossing.

Definition 7.14

A graph G is called **planar** if it can be drawn in the plane in such a way that no edges cross. The graph is then said to be **embedded** in the plane.

When a graph is drawn on the plane without edges crossing, it divides the plane into regions bounded by the edges of the graph. The regions are called **faces.** The region outside the graph counts as a face. Another way to define planarity is to say that a graph is planar if it can be embedded in the surface of a sphere. The two definitions are equivalent, because if one point is removed from a spherical surface, the remaining surface can be flattened onto the plane. Thinking of a planar graph as drawn on a sphere is sometimes convenient, since on the sphere, no one face is distinguished as the outside.

If we draw a graph in the plane with edges crossing, this does not mean that the graph is not planar. For instance, the graph K_4 can be drawn with edges crossing, as shown in Figure 7.31(a), or without edges crossing, as in Figure 7.31(b). The graph is planar.

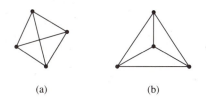

(a)   (b)

FIGURE 7.31   Two ways to draw K_4

Theorem 7.7   **(Euler's Theorem)**   Let G be a connected planar graph with p vertices and q edges, and suppose that it is embedded in the plane in such a way that there are r faces. Then

$$p - q + r = 2$$

Proof   The proof is by induction on q, the number of edges. If there is only one edge, the graph either has two vertices and one face (as in Figure 7.32(a)), or it has

(a)   (b)

FIGURE 7.32   Graphs with one edge

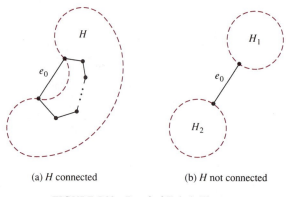

(a) H connected   (b) H not connected

FIGURE 7.33   Proof of Euler's Theorem

one vertex and two faces (as in Figure 7.32(b)). In either case, Euler's formula holds. Now suppose the formula is true for all graphs with fewer than q edges. Let G be a connected graph with p vertices, q edges, and r faces, and let e_0 be any edge. Let H be the graph obtained from G by deleting the edge e_0. Then H is either connected or has exactly two components. We need to treat these two cases separately.

(i) If H is connected, there is a path in H from one end of e_0 to the other (which may be of length zero if e_0 is a loop). See Figure 7.33(a), in which the dashed line encloses the graph H. The edge e_0 and the shortest such path enclose a face of G, which merges with the faces on the other side of e_0 when e_0 is removed. If H has p' vertices, q' edges, and r' faces, then

$$p' = p$$
$$q' = q - 1$$
$$r' = r - 1$$

By the inductive hypothesis,

$$p' - q' + r' = 2$$

and so

$$p - q + r = p' - (q' + 1) + (r' + 1)$$
$$= p' - q' + r'$$
$$= 2$$

(ii) If H is not connected, then it consists of two connected components H_1 and H_2, as shown in Figure 7.33(b). Euler's formula then applies to both H_1 and H_2 by the inductive hypothesis. Details of this case are left as an exercise. ●

Theorem 7.8   Let G be a planar simple graph with p vertices and q edges, embedded in the plane with r faces, and assume that $q > 1$. Then $3r \leq 2q$.

Proof   Each edge of G has two sides, and each side borders a face. (The two sides may border the same face.) If you add up the number of sides bordering all the faces, the sum s is equal to $2q$, since each of the q edges has two sides. Since G is simple and has at least two edges, each face must be bounded by at least 3 sides, so $s \geq 3r$. ●

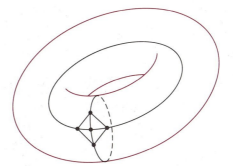

FIGURE 7.34   The graph K_5 embedded in a torus

Using Euler's Theorem and Theorem 7.8, it is easy to see that the complete graph K_5 is not planar. K_5 has 5 vertices and $C(5, 2) = 10$ edges. Euler's Theorem says that if K_5 were embedded in the plane with r faces,

$$5 - 10 + r = 2$$

so that there would be $r = 7$ faces. But Theorem 7.8 says that $3 \times 7 \leq 2 \times 10$, which is not true.

Although the graph K_5 cannot be embedded in the plane, it can be embedded in a torus, which is the surface of a doughnut. Figure 7.34 shows how this can be done.

Similar reasoning can be used to show that the graph $K_{3,3}$ is not planar. The details are left as exercises. You can make more nonplanar graphs from K_5 and $K_{3,3}$ by two methods: subdividing an edge by adding vertices, and adding more vertices and edges. Some examples are shown in Figure 7.35. The principal result in the theory of planar graphs, known as Kuratowski's Theorem, states that every non-planar graph can be obtained from K_5 or $K_{3,3}$ by these methods. That is, a graph is planar if and only if it does not contain a subgraph isomorphic to a graph that can be obtained from K_5 or $K_{3,3}$ by subdividing edges.

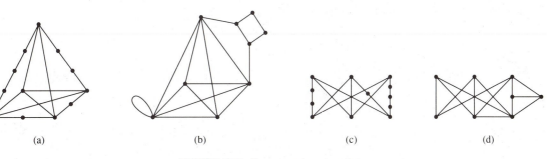

| | | | |
| (a) | (b) | (c) | (d) |

FIGURE 7.35   Some nonplanar graphs

Map Coloring

The problem of coloring a map so that no adjoining countries have the same color can be restated for arbitrary simple graphs in the following way. If G is a graph with set of vertices V, partition V into subsets $V_1, \ldots, V_n$ in such a way that the two ends of each edge lie in different subsets. If you have such a partition of V, you can assign a different color to each subset, and a map whose graph is G can then be colored using n colors.

Definition 7.15

A **coloring** of a simple graph G is a partitioning of the set V of vertices of G into subsets $V_1, \ldots, V_n$ such that for each edge e of G, the two vertices incident to e are in different subsets. The **chromatic number** of G is the smallest integer n such that there is a coloring of G consisting of n sets of vertices.

EXAMPLE 7.13   The chromatic number of the complete graph K_n is n; no two vertices can be in the same subset, because there is an edge joining them.   ○

EXAMPLE 7.14   The chromatic number of a bipartite graph $K_{m,n}$ is 2, for the set of vertices of a bipartite graph can be written as $V_1 \cup V_2$, where each edge of the graph joins a vertex of V_1 with a vertex of V_2.   ○

EXAMPLE 7.15   The **cyclic graph** C_n, consisting of n vertices and n edges in a single cycle (Figure 7.36) has chromatic number 2 if n is even and 3 if n is odd. If n is even, two colors can be alternated around the cycle. If n is odd, two colors can alternate for $n - 1$ vertices, with the remaining vertex getting the third color.

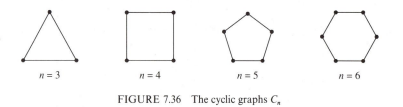

$n = 3$ $n = 4$ $n = 5$ $n = 6$

FIGURE 7.36   The cyclic graphs C_n ○

Clearly, any graph containing an odd cycle requires at least three colors. In fact, the presence of odd cycles characterizes these graphs.

Theorem 7.9   A simple graph has chromatic number 2 if and only if it does not contain an odd cycle.

Proof   We saw in Example 7.15 that if G contains an odd cycle, then G requires at least three colors. The proof of the converse is by induction on the number of vertices. A simple graph with two vertices can always be colored with two colors. Now suppose that the statement is true for all graphs having less than n vertices, and let G be a graph with n vertices containing no odd cycles. Choose any vertex v_0 of G, and let H be the graph obtained from G by deleting v_0 and the edges incident to it. Then H consists of one or more components $H_1, \ldots, H_n$ (see Figure 7.37). Of course, none of the graphs H_i contains an odd cycle either, so by the inductive hypothesis, each can be colored using only two colors. Pick a component H_i. Then all the vertices of H_i adjacent to v_0 in G must have the same color. For suppose that two such vertices w_1 and w_2 had different colors. A path in H_i from w_1 to w_2 would then

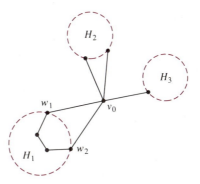

FIGURE 7.37   The proof of Theorem 7.9

need an odd number of edges, since the colors alternate. This path plus the edges connecting H_i to v_0 form an odd cycle in G.

Thus, by reversing the colorings of some of the components H_i if necessary, we can arrange that all of the vertices adjacent to v_0 are of the same color. We complete the coloring of G by giving v_0 the opposite color. ●

The four-color theorem, mentioned at the beginning of this chapter, can be stated as follows: The chromatic number of a planar simple graph is at most four. The hypothesis of planarity is essential. We have seen that the complete graph K_5 has chromatic number 5 and is not planar. Figure 7.38 is a map of a toroidal planet with five countries. The graph corresponding to the map is K_5. Each of the countries borders on every other country, so five colors are required to color the map.

The chromatic number of some categories of graphs is known, but the general problem of finding the chromatic number of an arbitrary graph is a difficult one. A brute-force approach would list all partitions of the set of vertices, find the partitions that determine colorings, and then select the smallest of the suitable partitions. However, this approach involves an enormous amount of computation, even for small numbers of vertices. No practical method for finding the chromatic number of an arbitrary graph has ever been found.

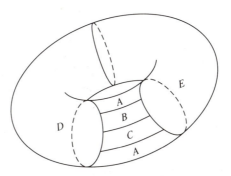

FIGURE 7.38   Five countries on a toroidal planet

The Traveling-Salesman Problem

A traveling salesman is based in city A in a region containing several cities connected by roads. He wishes to plan a trip, starting at city A and visiting each city exactly once, finally returning to city A. Can he do this? The problem can be simply stated in the language of graph theory. Construct a graph G in which the vertices are the cities of the region and the edges correspond to roads between cities. Then what the salesman wants is a cycle that contains each vertex.

Definition 7.16

A Hamiltonian cycle in a graph is a cycle that includes each vertex of the graph exactly once.

Graphs having Hamiltonian cycles are easily constructed: Start with a cycle, and add edges. In particular, the complete graphs K_n with $n > 2$ have Hamiltonian cycles. Some examples are shown in Figure 7.39: (a) is C_5, (c) is K_5, and (b) is intermediate between these extremes. However, not every graph has a Hamiltonian cycle.

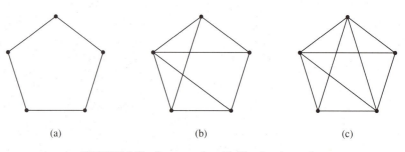

| (a) | (b) | (c) |

FIGURE 7.39   Some graphs with Hamiltonian cycles

EXAMPLE 7.16   The graph $K_{3,2}$ does not have a Hamiltonian cycle. To see this, refer to Figure 7.40. Starting at vertex v, there are two choices for the first edge and two choices for the second edge; the third edge is then determined. One of the four possibilities is shown in Figure 7.40, and the other three are similar. If, from w, you go back to v, you complete a cycle without passing through vertex x. On the other hand, if you go to x from w, you have to pass through y a second time to get back to v.

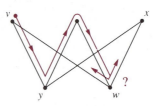

FIGURE 7.40   The graph $K_{3,2}$

○

Another category of graphs that do not have Hamiltonian cycles are graphs with bridges.

Definition 7.17   A **bridge** in a connected graph G is an edge having the property that, if it is removed, the graph is no longer connected.

EXAMPLE 7.17   In the graph in Figure 7.41, edges e, f, and g are bridges. Edges h, i, and j are not bridges.

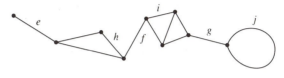

FIGURE 7.41   A graph with bridges   o

The general problem of determining whether a graph has a Hamiltonian cycle is very difficult. The brute-force approach—listing all permutations of the set of vertices and determining whether any are cycles—is computationally intractable. No practical method of determining the existence of a Hamiltonian cycle is known.

Returning to our salesman, we can pose an even more difficult problem. Each road in the region the salesman covers has a length. The problem is to find a route that has the smallest possible length, to minimize the salesman's travel costs. This is the most general form of the traveling-salesman problem.

Definition 7.18   A **weighted graph** is a graph G together with a real-valued function W, called the **weight function,** defined on its set of edges.

If G is a weighted graph with weight function W, and if H is a subgraph of G, then we will use the notation $W(H)$ to mean the sum of the weights of all the edges in the subgraph H.

EXAMPLE 7.18   Figure 7.42 shows several weighted graphs. The number shown by each edge is the value of a weight function on that edge.

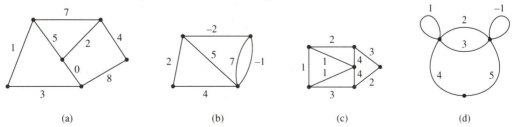

(a)   (b)   (c)   (d)

FIGURE 7.42   Some weighted graphs   o

EXAMPLE 7.19   Figure 7.43(a) is a weighted graph based on the complete graph K_4. There are three Hamiltonian cycles in the graph, shown in (b), (c), and (d). The sums of the weights for the three cycles are 12, 11, and 13. So the solution to the traveling-salesman problem for four cities is cycle (c).

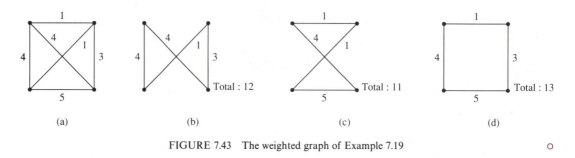

FIGURE 7.43   The weighted graph of Example 7.19   o

As you may expect, no practical general solution to the traveling-salesman problem has been found.

EXERCISES 7.3

1. For each of the graphs in Figure 7.44, determine the number of vertices, edges, and faces, and verify Euler's formula.

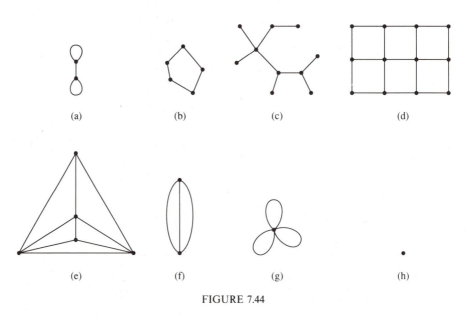

FIGURE 7.44

2. **a.** Draw a graph having at least 7 vertices in which the equation $3r = 2q$ holds, where r is the number of faces and q is the number of edges.

   b. Draw a graph having at least 7 vertices in which $3r < 2q$.

3. Complete the proof of Theorem 7.7. (*Hint:* If p_1 is the number of vertices in H_1 and p_2 is the number of vertices in H_2, how is p related to p_1 and p_2? Similarly for q and r.)

4. Where does the proof of Euler's Theorem break down if the graph is drawn on the torus instead of the plane?

5. Show that if G is a simple planar graph that does not contain a subgraph isomorphic to K_3, then $4r \leq 2q$, where q is the number of edges and r is the number of faces in a planar embedding of G.

6. Show that $K_{3,3}$ is not planar. (*Hint:* Use Euler's Theorem and the result of the previous exercise.)

7. Draw an embedding of $K_{3,3}$ in a torus.

8. Show that if $m > 1$, then $K_{m,n}$ has a Hamiltonian cycle if and only if $m = n$.

9. For each of the graphs of Figure 7.42, find the Hamiltonian cycle of least total weight.

10. For each of the following graphs, find a coloring of G using the smallest possible number of colors:

    a. Figure 7.42(a)   **b.** Figure 7.42(c)

    c. Figure 7.44(b)   **d.** Figure 7.44(c)

    e. Figure 7.44(d)   **f.** Figure 7.44(e)

11. Show that if a map is constructed by drawing circles in the plane, then the map can be colored using two colors.

12. Give an example of a graph G and an Euler circuit in G that is not a Hamiltonian cycle.

13. Give an example of a graph G and a Hamiltonian cycle in G that is not an Euler circuit.

7.4 • DIRECTED GRAPHS

In the graphs considered so far, an edge simply connects one vertex to another (or the same) vertex. Although we have talked about traveling along edges, no direction is associated with an edge itself. In some applications, however (particularly in data organization), an edge in a graph indicates some relation between objects. If the relation is not symmetric, the direction of the edge becomes significant.

Definition 7.19

A **directed graph** $G = (V, E, v_s, v_e)$ consists of a set V of vertices, a set E of edges, and two functions $v_s: E \to V$ and $v_e: E \to V$. The vertex $v_s(e)$ is the **start vertex** of e, and $v_e(e)$ is the **end vertex** of e. A directed graph is also called a **digraph**.

EXAMPLE 7.20   Let $V = \{v, w, x\}$, let $E = \{f, g, h, j, k\}$, and define v_s and v_e by

e	$v_s(e)$	$v_e(e)$
f	w	v
g	v	x
h	x	w
j	x	x
k	w	x

Then (V, E, v_s, v_e) is a directed graph. You can draw the directed graph by drawing a graph with these edges and vertices and marking each edge with an arrow pointing from the start vertex to the end vertex. The result is shown in Figure 7.45.

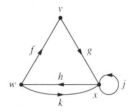

FIGURE 7.45   The digraph of Example 7.20   o

Let us compare the definition of directed graph with the definition of graph. If (V, E, v_s, v_e) is a directed graph, we can define a relation R on $E \times V$ by the rule

$$e \, R \, v \qquad \text{if and only if} \qquad v = v_s(e) \qquad \text{or} \qquad v = v_e(e)$$

Then (V, E, R) is a graph, which we call the **underlying graph** of the directed graph (V, E, v_s, v_e). If G is a directed graph, the underlying graph of G is obtained by "forgetting" the directions of the edges. Conversely, a graph can be made into a directed graph by specifying a direction for each edge.

Note that this definition of a directed graph allows loops and multiple edges, as does the definition of a graph. However, we do not count two edges with the same endpoints as "multiple" unless they have the same direction. For instance, the edges h and k in Figure 7.45 are not considered multiple. As with undirected graphs, we can define a simple directed graph to be a directed graph without loops or multiple edges.

Definition 7.20   A directed graph (V, E, v_s, v_e) is **simple** if

1. $v_s(e) \neq v_e(e)$ for all edges e, and
2. The function $v_s \times v_e \colon E \to V \times V$ defined by

$$(v_s \times v_e)(e) = (v_s(e), v_e(e))$$

is one-to-one.

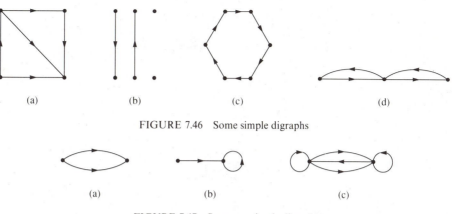

FIGURE 7.46 Some simple digraphs

FIGURE 7.47 Some nonsimple digraphs

The definition says that a directed graph is simple if it has no loops and no two distinct edges have both the same start point and the same end point. Figure 7.46 shows some simple directed graphs. Figure 7.47 shows some nonsimple directed graphs.

Indegree and Outdegree

The concept of degree of a vertex of a graph is easily generalized to directed graphs.

Definition 7.21

The **outdegree** of a vertex v in a directed graph G is the number of edges of G having v as start point.
  The **indegree** of a vertex v is the number of edges of G having v as end point.

You can think of the outdegree as the number of edges pointing out of the vertex, and the indegree as the number pointing in.

EXAMPLE 7.21 In Figure 7.48, vertex v has outdegree zero and indegree 2. Vertex w has outdegree 2 and indegree 1. Vertex z has outdegree 3 and indegree 1.

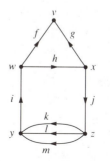

FIGURE 7.48 The graph of Examples 7.21 and 7.22

Theorem 7.10   In any directed graph G, the sum of the indegrees of the vertices and the sum of the outdegrees of the vertices are both equal to the number of edges of G.

The proof is left as an exercise.   ●

Connectedness in Digraphs

The concepts of walk, path, and cycle are easily applied to directed graphs.

Definition 7.22   Let G be a graph. A **directed walk** from a vertex u to a vertex v of G is an alternating sequence of vertices and edges

$$v_0, e_1, v_1, e_2, v_2, \ldots, v_{n-1}, e_n, v_n$$

such that the start vertex of e_i is v_{i-1} and the end vertex of e_i is v_i. A **directed path** is a directed walk that does not repeat any vertex or edge, except possibly for the first and last vertices. If the first and last vertices in a directed path are equal, the path is a **directed cycle.**

That is, a directed walk from u to v defines a route from u to v following a sequence of edges in the directions of the edges. When speaking of directed walks, paths, and cycles in a directed graph, we frequently refer to them simply as walks, paths, or cycles. The fact that they are directed is then inferred from the fact that the graph is a directed graph.

EXAMPLE 7.22   In Figure 7.48, the sequence

$$y, i, w, f, v$$

is a path from y to v. The sequence

$$z, l, y, i, w, h, x, j, z$$

is a cycle. The sequence

$$z, l, y, i, w, h, x, j, z, k, y$$

is a walk that is not a path, because the vertices z and y are repeated.   ○

Note that a cycle in a directed graph can have length 1; it then consists of a single loop. A cycle of length greater than 1 can contain no loops.

Definition 7.23   A directed graph G is **connected** if its underlying graph is connected. A directed graph G is **strongly connected** if for every pair of vertices v, w of G, there is a (directed) path from v to w.

It is obvious that every strongly connected digraph is connected, but the converse is not true.

EXAMPLE 7.23   The directed graph in Figure 7.49(a) is connected but not strongly connected. There is no directed path from v to u. However, the directed graph in Figure 7.49(b) is strongly connected.

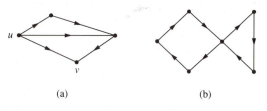

(a)   (b)

FIGURE 7.49   The graphs of Example 7.23   o

Digraphs and Relations

As seen in Chapter 3, directed graphs provide a means of representing a relation. Suppose R is a relation on a set X. Then $R \subset X \times X$. We can build a directed graph G from X and R as follows. The set of vertices of G is the set X. The set of edges of G is the set R of ordered pairs in the relation R. If $(a, b) \in R$, define $v_s(a, b) = a$ and $v_e(a, b) = b$. Then $G = (X, R, v_s, v_e)$ is a directed graph. The graph G contains an edge from a to b if and only if $a \, R \, b$. A directed graph constructed from a relation in this way may contain loops, but it will not contain multiple edges. Conversely, if G is a directed graph with set of vertices V, we can define a relation R on the set V by the rule: $u \, R \, v$ if and only if there is an edge in G from u to v.

EXAMPLE 7.24   Consider the relation on the set $\{a, b, c, d, e\}$ given by

	a	b	c	d	e
a	1	0	1	0	1
b	0	1	0	1	0
c	1	0	0	0	1
d	0	1	0	0	0
e	0	0	0	0	0

where a 1 in row x and column y indicates that the ordered pair (x, y) is in the relation, and a 0 indicates that the ordered pair is not in the relation. The directed graph for this relation has five vertices and eight edges. It is shown in Figure 7.50.

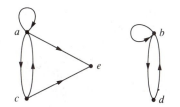

FIGURE 7.50   The graph of Example 7.24   o

EXAMPLE 7.25   The directed graph in Figure 7.51 represents the divisibility relation on the divisors of 30.

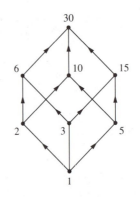

FIGURE 7.51   The divisors of 30   o

A directed graph can provide a simple way of illustrating a partial order relation. If G is a directed graph, define a relation P on the set V of vertices of G by the rule: $u\,P\,v$ if $u = v$ or there is a directed walk in G from u to v. The relation P is certainly reflexive. It is also transitive, since if there are walks from u to v and from v to w, they can be combined to form a walk from u to w. If P is a partial order relation (that is, if it is also antisymmetric), we say that the graph G **represents** the partial order P. The following theorem characterizes graphs that represent partial order relations.

Theorem 7.11   A directed graph G represents a partial order relation if and only if G contains no cycles of length greater than 1.

Proof   Let P be the relation constructed from G as described above. First, suppose G has a cycle of length greater than 1. Then the cycle must contain at least two distinct vertices; call them u and v. Then one part of the cycle is a path from u to v, and the other part is a path from v to u. This means that $u\,P\,v$ and $v\,P\,u$, so P is not antisymmetric.

Conversely, suppose that P is not antisymmetric. Then there exist two distinct vertices u and v with $u\,P\,v$ and $v\,P\,u$. By the construction of P, there are directed walks from u to v and from v to u. These can be combined to form a walk from u to u, which must contain a cycle of length at least 2.   ●

Computer Representations of Directed Graphs

The simplest way to represent a directed graph in a computer is to store a table defining the functions v_s and v_e. This is the same as the edge table for an undirected graph, but with the understanding that the first vertex column specifies the start vertex and the second vertex column specifies the end vertex. Other methods may be obtained by adapting the notions of adjacency matrix and adjacency list. When an adjacency matrix is used to represent an undirected graph, only the upper portion (on and above the diagonal) is used. By using the whole matrix, we can include directional information: The entry in row i and column j is the number of directed edges from v_i to v_j. A simple directed graph can be described by two adjacency lists:

one for the edges coming in to each vertex, and one for the edges going out of each vertex.

EXAMPLE 7.26   The graph of Figure 7.48 can be described by the following adjacency matrix:

	v	w	x	y	z
v	0	0	0	0	0
w	1	0	1	0	0
x	1	0	0	0	1
y	0	1	0	0	0
z	0	0	0	3	0

Indicates that there are three edges from z to y

Note that the sum of the entries in the matrix is the number of edges. Also, the sum of the numbers in a row is the indegree of the vertex corresponding to that row, and the sum of the numbers in a column is the outdegree of the vertex corresponding to that column.   o

EXAMPLE 7.27   The directed graph in Figure 7.52 can be described by the adjacency list in Figure 7.53.

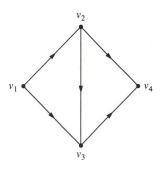

FIGURE 7.52   The graph of Example 7.27

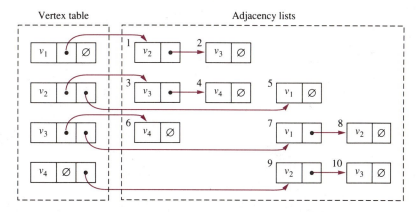

FIGURE 7.53   The adjacency list of Example 7.27

The list can be stored in the computer as the following tables:

Vertex Number	Out Pointer	In Pointer	Item Number	Vertex Number	Item Pointer
1	1	0	1	2	2
2	3	5	2	3	0
3	6	7	3	3	4
4	0	9	4	4	0
			5	1	0
			6	4	0
			7	1	8
			8	2	0
			9	2	10
			10	3	0

EXERCISES 7.4

1. Draw the following directed graphs:

   a. $V = \{s, t, u, w, x\}$, $E = \{a, b, c, d\}$, and the functions v_s and v_e are defined by

e	$v_s(e)$	$v_e(e)$
a	s	t
b	u	w
c	w	s
d	x	t

   b. $V = \{s, t, u, w, x, y, z\}$, $E = \{a, b, c, d, f, g\}$, and the functions v_s and v_e are defined by

e	$v_s(e)$	$v_e(e)$
a	s	t
b	u	z
c	w	s
d	w	w
f	u	z
g	x	t

2. For the directed graphs of Exercise 1, specify the underlying (undirected) graph (vertices, edges, incidence relation).

3. For each of the directed graphs in Figure 7.54, write the start and end functions in tabular form.

4. Which of the directed graphs in Figure 7.54 are simple?

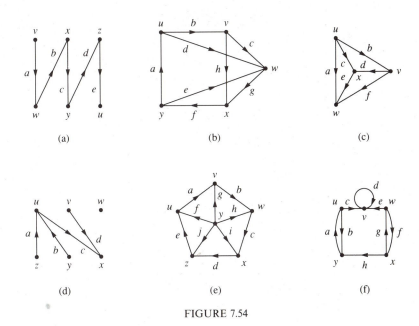

FIGURE 7.54

5. Find the indegree and the outdegree of each of the vertices in the directed graphs of Figure 7.54.

6. Which of the directed graphs in Figure 7.54 are strongly connected?

7. Draw the graph corresponding to the relation $<$ on the set $\{1, 2, 3, 4, 5\}$.

8. Draw the graph corresponding to the relation "divides" on the set $\{1, 2, 3, 4, 5\}$.

9. Draw the graph corresponding to the relation $\geq$ on the set $\{2, 3, 4, 5\}$.

10. Which of the directed graphs in Figure 7.54 represent partial order relations?

11. Prove Theorem 7.10.

12. When is the directed graph of an equivalence relation a simple directed graph?

13. Give an example of a directed graph having five vertices in which every directed path has length 1.

14. A graph is said to be **orientable** if it is possible to assign a direction to each edge in such a way that the resulting directed graph is strongly connected. Determine which of the graphs in Figure 7.55 are orientable. For each graph

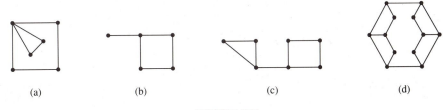

FIGURE 7.55

that is orientable, assign a direction to each edge so that it is strongly connected.

15. Show that a connected graph is orientable if and only if it contains no bridges. (*Hint:* Show first that if a connected graph has no bridges, then every edge is contained in a cycle.)

16. A **directed Euler circuit** in a directed graph is an Euler circuit in which each edge is traversed in its assigned direction; that is, it is a closed directed walk in the graph in which each edge occurs exactly once. Show that a connected directed graph has an Euler circuit if and only if, for every vertex v, indegree(v) = outdegree(v). (*Hint:* Modify the proof of Theorem 7.6.)

17. In each of the directed graphs of Figure 7.56, determine whether a directed Euler circuit exists. If so, find one.

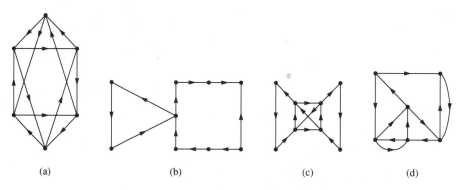

(a)   (b)   (c)   (d)

FIGURE 7.56

18. Prove or give a counterexample: The directed graph of a partially ordered set is connected.

19. Prove or give a counterexample: The directed graph of a totally ordered set is strongly connected.

20. Represent each of the directed graphs in Figure 7.54 using an adjacency matrix.

21. Represent the simple directed graphs in Figure 7.54(a) and (b) using the adjacency list method.

22. Draw the directed graphs represented by each of the following adjacency lists:

a.

Vertex Number	Out Pointer	In Pointer	Item Number	Vertex Number	Item Pointer
1	1	2	1	2	0
2	0	3	2	3	0
3	4	0	3	1	0
4	0	7	4	1	5
5	0	8	5	4	6
			6	5	0
			7	3	0
			8	5	0

b.

Vertex Number	Out Pointer	In Pointer	Item Number	Vertex Number	Item Pointer
1	1	11	1	2	6
2	2	12	2	3	7
3	3	13	3	4	8
4	4	14	4	5	9
5	5	15	5	1	10
			6	3	0
			7	4	0
			8	5	0
			9	1	0
			10	2	0
			11	4	16
			12	5	17
			13	1	18
			14	2	19
			15	3	20
			16	5	0
			17	1	0
			18	2	0
			19	3	0
			20	4	0

23. In the adjacency matrix of a directed graph,

   a. What is the meaning of the sum of the entries in the row corresponding to vertex v?

   b. What is the meaning of the sum of the entries in the column corresponding to vertex v?

7.5 • THE SHORTEST-PATH ALGORITHM

Suppose G is a weighted simple graph in which all weights are positive. We may think of the weight of each edge as being its length. Then the length of a path in the graph is the sum of the weights of the edges making up the path. Many optimization problems can be stated as finding the shortest path between two vertices in such a graph. For example, if the vertices of the graphs correspond to cities, the edges to major highways, and the weights to distances, the shortest path between two vertices is the shortest highway route between the cities. Or the vertices may be nodes in a communication network, the edges communication circuits, and the weights may be the costs of sending a message over a circuit. Then the shortest path between two vertices is the least-cost route for a message.

We will describe an algorithm due to E. Dijkstra that provides a means of finding the path of least length between two vertices. More precisely, if a is a vertex of G, the algorithm computes the lengths of the shortest paths from a to each of the other vertices. Let V be the set of vertices of G. The algorithm operates by constructing a sequence $S_1, \ldots, S_n$ of subsets of V such that $S_1 = \{a\}, S_n = V, S_{k+1}$ is obtained from

S_k by adding a carefully selected vertex, and at the kth step, the problem has been solved for the subgraph induced by the set of vertices S_k.

Although Dijkstra's algorithm can be stated very concisely, it is hard to understand in that form. Therefore, we will first illustrate the technique with an example.

EXAMPLE 7.28   Let G be the graph in Figure 7.57. We will find the lengths of the shortest paths from the vertex a to the other vertices. As we do so, we will put numerical labels (tentative path lengths) on each of the vertices and update them as we proceed.

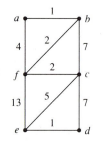

FIGURE 7.57   The graph of Example 7.28

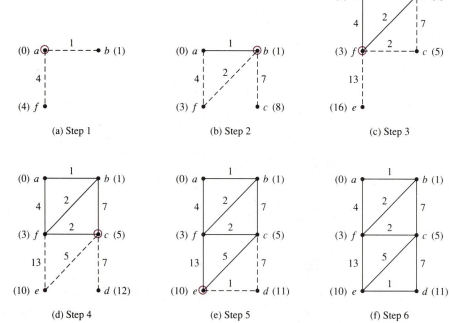

(a) Step 1

(b) Step 2

(c) Step 3

(d) Step 4

(e) Step 5

(f) Step 6

FIGURE 7.58   Dijkstra's algorithm for the graph of Figure 7.57

Set $S_1 = \{a\}$. The length of the shortest path from a to a is 0. There are two vertices adjacent to a—namely, b and f. These will be the candidates for the vertex to be added. Label a with 0 and label b and f with the lengths of the edges to a; the result is shown in Figure 7.58(a). Choose the candidate with the smallest label; this is b, so $S_2 = \{a, b\}$. Now c and f are adjacent to S_2; this is shown in Figure 7.58(b).

From Figure 7.58(b), we can derive an initial shortest-path estimate for vertex c: It is 8, obtained by adding the label of b (1) to the length of the edge from b to c (7). Moreover, we can improve on our estimate for f: The length of the path abf is the label of b (1) plus the length of the edge from b to f (2), yielding the estimate 3.

The candidates for the third vertex are c and f; f is chosen because it has the smallest label. Now $S_3 = \{a, b, f\}$. The result is shown in Figure 7.58(c). As in step 2, we improve the estimate for c ($5 = 3 + 2$), and produce an estimate for e ($16 = 3 + 13$). So $S_4 = \{a, b, f, c\}$. In subsequent steps, vertices c, e, and d are added in order. The final labels are the shortest path lengths from vertex a.

The steps can also be summarized in the following table, in which the selected vertex is indicated in boldface and the candidates for the next step are indicated by underlining. The entry ∞ (infinity) indicates that no finite shortest-path estimate has yet been made. In the notes to the right, $L_k(x)$ refers to the label value for vertex x at step k, and $w(x, y)$ is the weight of the edge from x to y.

Step	Selected Vertex	Label Values						Notes
		a	b	c	d	e	f	
1	a	**0**	$\underline{1}$	∞	∞	∞	$\underline{4}$	
2	b	0	**1**	$\underline{8}$	∞	∞	$\underline{3}$	$8 = L_1(b) + w(b, c)$ $3 = L_1(b) + w(b, f)$
3	f	0	1	$\underline{5}$	∞	$\underline{16}$	**3**	$5 = L_2(f) + w(f, c)$ $16 = L_2(f) + w(f, e)$
4	c	0	1	**5**	$\underline{12}$	$\underline{10}$	3	$12 = L_3(c) + w(c, d)$ $10 = L_3(c) + w(c, e)$
5	e	0	1	5	$\underline{11}$	**10**	3	$11 = L_4(e) + w(e, d)$
6	d	0	1	5	**11**	10	3	

Note that at each step, we did one of two things to each label: Either we left it alone (this was always the case once a vertex had been selected), or we replaced it with the sum of the weight of the selected vertex and the length of the appropriate edge, if the sum was smaller than the previous weight.

We will now state Dijkstra's algorithm formally. Let G be a weighted graph with vertices v_1 through v_n, and let $w(i, j)$ be the weight of the edge from v_i to v_j (or ∞ if there is no edge from v_i to v_j). Let $a = v_m$ be one of the vertices. The following steps show how to compute label values $L_k(i)$ so that $L_n(i)$ is the length of the shortest path from a to v_i.

Dijkstra's Shortest-Path Algorithm

Set $S_1 = \{a\}$.
Set $L_1(m) = 0$.
For $i \neq m$, set $L_1(i) = w(m, i)$.
For $k = 2$ through n, do the following:
    Let v_p be the vertex not in S_{k-1} having the least label.
    Set $S_k = S_{k-1} \cup \{v_p\}$
    For each m with $v_m \in S_k$, set
        $L_k(m) = L_{k-1}(m)$
    For each m with $v_m \notin S_k$, set
        $L_k(m) = \text{minimum } (L_{k-1}(m), L_{k-1}(v_p) + w(p, m))$
End "for" loop.

The fact that Dijkstra's algorithm produces the shortest path lengths is shown by the following theorem.

Theorem 7.12

At each step of Dijkstra's algorithm, the following are true:
(a) If $v_i \in S_k$, then $L_k(i)$ is the length of the shortest path in G from a to v_i.
(b) If $v_i \notin S_k$, then $L_k(i)$ is the length of the shortest path in G from a to v_i that does not contain any vertex of $V - S_k$ other than v_i itself.

Proof   The proof is by induction.

Step 1. $k = 1$. In this case, $S_1 = \{v_k\}$, and $L_1(k) = 0$, so (a) holds. Suppose $v_i \notin S_1$. Then there is a path from v_k to v_i that does not contain any other vertices of $V - S_1$, only if v_i is adjacent to v_k. So part (b) holds for $k = 1$.

Step 2. Assume that statements (a) and (b) are true at step k of the algorithm; we must show that they are also true at step $k + 1$.

First, we show that (a) is true at step $k + 1$. Let v be the vertex added in the $(k + 1)$st step. We must show that the label we assigned to v is the length of the shortest path from a to v. Suppose that there were a shorter path (call it P). Let w be the first vertex of P outside S_k. Then w was a candidate for inclusion in S_{k+1}, and moreover, there is a path from a to w whose length is less than the label assigned to v. This contradicts the choice of v.

Next, we show that (b) is true at step $k + 1$. Let w be a vertex adjacent to S_{k+1} (we can ignore all other vertices, since they had label ∞ at step k and still do at step $k + 1$). Let P be the shortest path from a to w, containing only vertices of S_{k+1} and w itself. Either P goes through v or it does not. If P does not contain v, then (b) is true by the inductive hypothesis. If P does contain v, then its length is the length of the path from a to v, plus the length of the edge from v to w; that is, $L_{k+1}(v) + w(v, w)$.   ●

With some additional record-keeping, Dijkstra's algorithm can provide not only the lengths of shortest paths from a vertex a, but also the shortest path itself. The additional work is the following: Whenever the label of a vertex is changed in any step, record for that vertex the identity of the vertex being added to the set S_k for that step. When the algorithm terminates, for each vertex v, we have the first vertex to go to on the path *back* to a; this is called the *predecessor* of v. By following the chain of predecessors, we can traverse the shortest path from a to v in the reverse direction.

EXAMPLE 7.29   Find the shortest path from a to d in Figure 7.57. Rework Example 7.28, recording the selected vertices whenever a label is changed; the result is shown in the following table.

Step	Selected Vertex	Label Values					
		a	b	c	d	e	f
1	a	**0**	$\underline{1}(a)$	∞	∞	∞	$4(a)$
2	b	0	**1**	$8(b)$	∞	∞	$\underline{3}(b)$
3	f	0	1	$\underline{5}(f)$	∞	$16(f)$	**3**
4	c	0	1	**5**	$12(c)$	$\underline{10}(c)$	3
5	e	0	1	5	$\underline{11}(e)$	**10**	3
6	d	0	1	5	**11**	10	3
Predecessor		—	a	f	e	c	b

The predecessor of d is e, the predecessor of e is c, the predecessor of c is f, the predecessor of f is b, and the predecessor of b is a. So the shortest path from a to d is $abfced$.   ○

EXERCISES 7.5

1. Use Dijkstra's algorithm to find the lengths of the shortest paths from vertex c of Figure 7.57 to all other vertices.

2. Use Dijkstra's algorithm to find the lengths of the shortest paths from vertex d of Figure 7.57 to all other vertices.

3. In Figure 7.59, find the lengths of the shortest paths from vertex a to all other vertices.

4. In Figure 7.59, find the lengths of the shortest paths from vertex z to all other vertices.

5. Find the shortest path from a to g in Figure 7.59.

6. Find the shortest path from z to g in Figure 7.59.

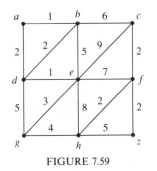

FIGURE 7.59

7. What happens when you apply Dijkstra's algorithm to a graph that is not connected?

8. Suppose that it is only required to find the distance from a vertex a of a weighted graph with positive weights to one other vertex z. Can Dijkstra's algorithm be shortened in this case? If so, how?

9. Suppose that G is a weighted graph with positive weights, in which no two edges have the same weight. Is the shortest path between two vertices necessarily unique?

10. Suppose that G is a *directed* weighted simple graph. Extend Dijkstra's algorithm to find the shortest directed paths from a vertex a to the other vertices.

11. Can Dijkstra's algorithm be applied to a weighted graph that is not simple?

12. Where in the proof of Theorem 7.12 did we use the fact that the weights are positive?

Computer Exercises for Chapter 7

1. Write a program to read an incidence matrix representation of a graph and print an adjacency matrix.

2. Write a program to read an adjacency matrix representation of a graph and print an incidence matrix. (*Hint:* The first step is to determine the number of edges.)

3. Write a program to read an adjacency matrix representation of a graph and determine whether the graph is a simple graph.

4. Write a program to read an adjacency matrix representation of a graph and determine whether the graph is connected.

5. Write a program to read an adjacency matrix representation of a graph and print an adjacency list representation.

6. Write a program to read an adjacency list representation of a graph and print an adjacency matrix representation.

7. Write a program to implement Dijkstra's shortest-path algorithm.

• • • • CHAPTER REVIEW EXERCISES

1. Draw the graphs given by each of the following incidence relations:

a.

	v_1	v_2	v_3	v_4
e_1	0	0	0	1
e_2	0	0	1	1
e_3	0	1	1	0
e_4	0	1	0	1
e_5	0	1	1	0
e_6	0	1	0	1

b.

	v_1	v_2	v_3	v_4
e_1	1	0	0	1
e_2	0	1	0	1
e_3	0	0	1	1
e_4	1	0	0	1
e_5	0	0	0	1
e_6	0	0	0	1

2. Write the incidence relations for each of the following graphs:

   a.   The graph of Figure 7.25   **b.**   The graph of Figure 7.45

3. Draw graphs representing the counties of **a.** Connecticut, **b.** Massachusetts.

4. Determine the degrees of each of the vertices in the graphs of the preceding exercise.

5. Describe the graph of Exercise 3a by an adjacency matrix.

6. Describe the graph of Exercise 3a by an adjacency list.

7. If v is a vertex of the cyclic graph C_n, what is the degree of v?

8. Draw the graphs described by the following adjacency matrices:

a.

	v_1	v_2	v_3	v_4
v_1	0	1	1	1
v_2	—	0	1	1
v_3	—	—	0	0
v_4	—	—	—	0

b.

	v_1	v_2	v_3	v_4	v_5
v_1	0	1	1	0	0
v_2	—	0	1	1	0
v_3	—	—	0	1	1
v_4	—	—	—	0	1
v_5	—	—	—	—	1

9. Draw the graph defined by the following adjacency list:

Vertex Number	Item Pointer	Item Number	Vertex Number	Item Pointer
1	12	1	1	0
2	11	2	5	1
3	8	3	4	4
4	7	4	6	0
5	3	5	5	0
6	2	6	1	5
		7	3	6
		8	2	9
		9	4	0
		10	3	0
		11	1	10
		12	2	13
		13	4	14
		14	6	0

10. Draw the graph defined by the following adjacency list:

Vertex Number	Item Pointer	Item Number	Vertex Number	Item Pointer
1	7	1	2	0
2	6	2	3	0
3	5	3	4	0
4	4	4	3	8
5	0	5	4	9
6	3	6	1	10
7	2	7	2	0
8	1	8	6	0
		9	7	0
		10	8	0

11. Which of the following graphs have Euler circuits? For each graph that has an Euler circuit, find one.
   a. The graph of Exercise 1a
   b. The graph of Exercise 1b
   c. The graph of Exercise 3a
   d. The graph of Exercise 3b
   e. The graph of Exercise 9
   f. The graph of Exercise 10

12. Draw the following directed graphs:
   a. $V = \{s, t, u, w, x\}$, $E = \{a, b, c, d\}$, and the functions v_s and v_e are defined by

e	$v_s(e)$	$v_e(e)$
a	x	u
b	w	t
c	u	s
d	t	x

   b. $V = \{s, t, u, w, x, y, z\}$, $E = \{a, b, c, d, f, g\}$, and the functions v_s and v_e are defined by

e	$v_s(e)$	$v_e(e)$
a	z	x
b	z	w
c	z	u
d	x	t
f	x	s
g	t	y

13. For the directed graphs of Exercise 12, write the incidence relation for the underlying (undirected) graphs.

14. Which of the directed graphs in Exercise 12 are simple?

15. Find the indegree and the outdegree of each of the vertices in the directed graphs of Exercise 12.

16. Which of the directed graphs in Exercise 12 are strongly connected?

17. Which of the directed graphs in Exercise 12 represent partial order relations?

18. Draw the directed graphs represented by each of the following adjacency lists:

a.

Vertex Number	Out Pointer	In Pointer		Item Number	Vertex Number	Item Pointer
1	1	6		1	2	0
2	2	7		2	3	0
3	3	8		3	4	0
4	4	9		4	5	0
5	5	10		5	1	0
				6	5	0
				7	1	0
				8	2	0
				9	3	0
				10	4	0

b.

Vertex Number	Out Pointer	In Pointer		Item Number	Vertex Number	Item Pointer
1	0	1		1	3	0
2	0	2		2	3	0
3	6	3		3	4	0
4	4	0		4	3	7
5	0	5		5	4	0
				6	1	8
				7	5	0
				8	2	0

19. Which of the pairs of graphs in Figure 7.60 are isomorphic?

20. Draw $K_{2,5}$.

21. Which of the graphs in Figure 7.16 are connected? How many components does each graph have?

22. In each of the graphs of Figure 7.61, draw the subgraph induced by the set of circled vertices.

23. For each of the graphs in Figure 7.61, determine the number of vertices, edges, and faces, and verify Euler's formula.

24. a. Draw a graph having at least 7 vertices in which the equation $3r = 2q$ holds, where r is the number of faces and q is the number of edges.

b. Draw a graph having at least 7 vertices in which $3r < 2q$.

25. For each of the graphs of Figure 7.62, find the Hamiltonian cycle of least total weight.

26. For each of the following graphs, find a coloring of G using the smallest possible number of colors:

a. Figure 7.30(a)   **b.** Figure 7.30(b)   **c.** Figure 7.30(c)   **d.** Figure 7.30(d)

27. Use Dijkstra's algorithm to find the lengths of the shortest paths in Figure 7.62(e) from the upper left vertex to all other vertices.

28. Use Dijkstra's algorithm to find the lengths of the shortest paths in Figure 7.62(e) from the upper right vertex to all other vertices.

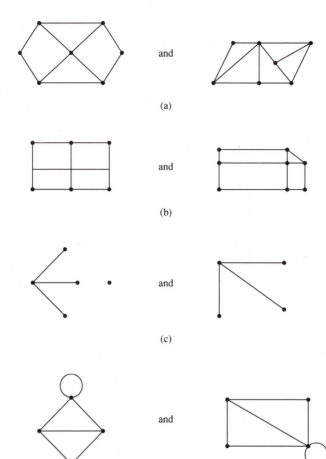

(a)

(b)

(c)

(d)

FIGURE 7.60

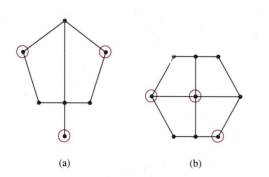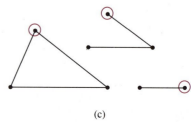

(a)   (b)   (c)

FIGURE 7.61

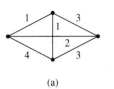
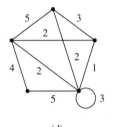

(a)

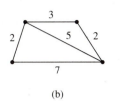
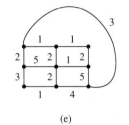

(b)

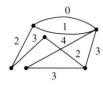
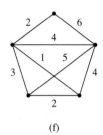

(c)

(d)

(e)

(f)

FIGURE 7.62

8

TREES

In the previous chapter, we studied a class of objects called graphs. We had originally encountered graphs informally in Chapter 3, where we used directed graphs to represent relations. We also encountered graphs in Chapter 6, where we used graphs as a tool in solving some counting problems. All of the graphs in Chapter 6 are simple connected graphs without cycles. Such graphs are called trees. These graphs are particularly important in organizing data for computer programming.

8.1 ● CHARACTERIZATIONS OF TREES

There are several ways to characterize the class of graphs called trees. We will take one characterization—the absence of cycles—as a definition and prove its equivalence to several other characterizations. Of course, any of these characterizations could have been chosen as the definition.

Definition 8.1   A **tree** is a connected graph containing no cycles.

Every tree is a simple graph, because loops and multiple edges in a graph form cycles.

EXAMPLE 8.1   The graphs in Figure 8.1 are trees.   ○

Theorem 8.1   In a tree with more than one vertex, there are at least two vertices with degree 1.

Proof   Let G be a tree. Since G is connected, there is a path in G from each vertex to every other vertex. Choose a longest path in G, and call its endpoints u and v. If v has degree more than 1, then there is an edge incident to it other than the last edge in the path. The vertex at the other end of this edge cannot be part of the path from u to v, for if it were, we would have a cycle. Consequently, we can extend the path by 1 by adding this edge. This contradicts the assumption that the path was the longest in the graph. Therefore, v has degree 1. Similarly for u.   ●

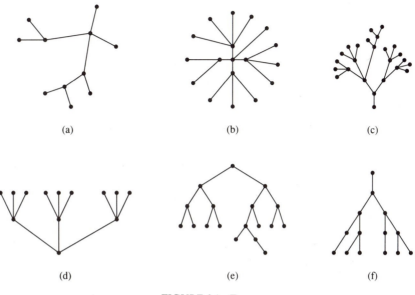

(a)   (b)   (c)

(d)   (e)   (f)

FIGURE 8.1   Trees

A vertex of degree 1 is often called a **leaf.** Figure 8.2 shows a tree with n vertices having only two vertices of degree 1. This tree is the **linear graph** on n vertices.

FIGURE 8.2   The linear graph on n vertices

We saw in Theorem 7.5 that a connected graph with n vertices must have at least $n - 1$ edges. The following theorems show that the property of having exactly $n - 1$ edges characterizes trees.

Theorem 8.2   A connected graph with n vertices is a tree if and only if it has exactly $n - 1$ edges.

Proof   Let G be a connected graph with n vertices and $n - 1$ edges. Let us show that it is a tree. If G has a cycle, we can remove one edge from the cycle. The resulting graph is still connected and contains only $n - 2$ edges, contradicting Theorem 7.5. So G has no cycles.

The proof of the converse is by induction on n, the number of vertices. A tree with one vertex can have no edges at all, since any edge in such a graph would be a loop and hence a cycle. Now suppose that the statement is true for all trees having fewer than n vertices, and let G be a tree with n vertices. Choose a vertex v of degree 1. If v and the edge incident to it are removed from G, the result H has $n - 1$ vertices. By the induction hypothesis, H has $(n - 1) - 1 = n - 2$ edges. Therefore, G has $(n - 2) + 1 = n - 1$ edges.   ●

Here are two more characterizations of trees. They can be understood by inspecting the examples in Figure 8.1.

Theorem 8.3   A simple graph is a tree if and only if there is exactly one path between each pair of vertices.

Proof   First suppose that there are vertices u and v with two distinct paths P and Q from u to v. Then there must be a first vertex w where the paths diverge. Let x be the first vertex of P, after w, which is also a vertex of Q. Then the portions of the paths between w and x can be combined to form a cycle (see Figure 8.3), so the graph is not a tree.

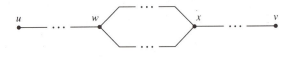

FIGURE 8.3   Forming a cycle from two paths

Conversely, suppose that G is not a tree. Then either G is not connected, or G contains a cycle. If G is not connected, there are vertices u and v with no path connecting them. If G contains a cycle, there are two paths connecting any two distinct vertices on the cycle. ●

Theorem 8.4   A graph G is a tree if and only if G is connected and every edge is a bridge.

Proof   Suppose G is a tree. Then G has $n - 1$ edges. Let e be any edge. If e is removed, the resulting graph H has fewer than $n - 1$ edges, so by Theorem 7.5, H cannot be connected. Therefore, e is a bridge.

Conversely, suppose that G is connected and that every edge of G is a bridge. If G were to contain a cycle, any edge in the cycle could be removed and the result would still be connected; that is, the edges in the cycle are not bridges. Therefore, G contains no cycles and is thus a tree. ●

Rooted Trees

A glance at Figure 8.1(c) and (d) shows the motivation for the name *tree*. Both of these trees have a "root" at the bottom. In the mathematical and computer science literature, however, trees are more frequently drawn with the root at the top, as in Figure 8.1(e) and (f).

Of course, what part of the tree is the "root" depends on how the graph is drawn and how it is interpreted. We can make the notion of root precise by using the concept of directed graph.

Definition 8.2

A **rooted tree** is a directed graph such that

1.   The underlying graph is a tree, and
2.   The indegree of each vertex is either zero or one.

In any tree, the number of vertices equals the number of edges plus one, and the sum of the indegrees of the vertices equals the number of edges. It follows that in a rooted tree, exactly one vertex has indegree zero, and all the other vertices have indegree one. The vertex with indegree zero is the **root** of the rooted tree.

EXAMPLE 8.2   The directed graphs in Figure 8.4 are rooted trees.

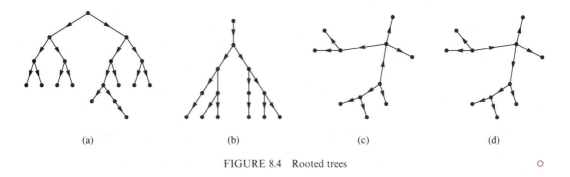

(a)   (b)   (c)   (d)

FIGURE 8.4   Rooted trees   ○

Note that any tree can be made into a rooted tree in many ways. The graphs of Figure 8.4(c) and (d) are two different rooted trees, both having the tree in Figure 8.1(a) as underlying graph. In fact, to make a rooted tree from a tree, you can choose any vertex v as the root and direct each edge outward from v.

Theorem 8.5   In any rooted tree, there is a unique directed path from the root to any other vertex.

Proof   Uniqueness of the path follows from Theorem 8.3. The proof of existence is by induction on the number of vertices. The statement is trivial for two vertices, since a rooted tree of two vertices is just a linear graph. So suppose that the statement is true for all rooted trees having fewer than n vertices, and assume that G is a rooted tree with n vertices. Since there are at least two vertices of degree 1 (Theorem 8.1), we can choose a vertex v of degree 1 that is not the root r. The vertex v has one edge coming into it; call this edge e. Let H be the directed graph obtained from G by deleting v and e. Then H is a rooted tree, and by the inductive hypothesis, there are unique directed paths from r to all the other vertices of H. So we only need to find a directed path from r to v. Let w be the start vertex of e. If $w = r$, we are done. Otherwise, there is a directed path from r to w, and we need only add the edge e to it to get a path from r to v.   ●

Theorem 8.5 shows that the following definition is meaningful.

Definition 8.3   The **depth** of a vertex v in a rooted tree is the length of the path from the root to v. The depth of a rooted tree is the maximum of the depths of its vertices.

In a rooted tree, the term **leaf** is used to refer to any vertex of degree 1, except that the root is not considered a leaf.

Much commonly used terminology applicable to rooted trees is derived from family trees. If there is an edge from a vertex u to a vertex v, u is said to be the **parent** of v, and v is said to be a **child** of u. If v and w have the same parent, they are said to be **siblings**. If there is a directed path from a vertex u to a vertex v, u is said to be an **ancestor** of v, and v is said to be a **descendent** of u.

A leaf of a rooted tree is a vertex with no children. Such a vertex is sometimes called a **terminal vertex**. An **internal vertex** of a rooted tree is any vertex other than the root and the terminal vertices.

EXAMPLE 8.3   In Figure 8.5, a is the parent of b and c, and b is a sibling of c. The vertex b is an ancestor of i, and i is a descendent of f, b, and a.

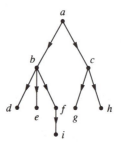

FIGURE 8.5   The graph of Example 8.3   ○

Computer Representations of Trees

In representing a tree, any of the methods discussed previously can be used. The adjacency matrix method is particularly inefficient for trees, since it uses a matrix containing n^2 entries to identify only $n - 1$ edges. The adjacency list method is usually used. For a rooted tree, the adjacency list takes on a particularly simple form. Since each vertex has indegree 1 (except for the root), the list of "in" vertices for each vertex contains only one item, which can be stored in the vertex table. For the same reason, each vertex is a member of at most one list of "out" vertices; so the pointer to the next item in the list can also be stored in the vertex table.

EXAMPLE 8.4   Figure 8.6 shows a rooted tree. The children of vertices v_1, v_2, and v_3 are arranged in linked lists by the colored arrows. The heavy lines point to the first, or oldest child. The tree is completely described by providing three data items for each vertex: the parent, the first child, and the next sibling. The data can be arranged as in the

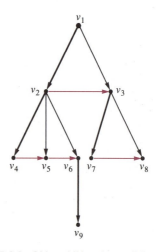

FIGURE 8.6   Oldest child and next sibling pointers

following table:

Vertex	Parent	First Child	Next Sibling
1	0	2	0
2	1	4	3
3	1	7	0
4	2	0	5
5	2	0	6
6	2	9	0
7	3	0	8
8	3	0	0
9	6	0	0

○

In fact, the information in such a table is redundant; the tree structure is completely defined by the Parent column alone. Moreover, the Vertex and Parent columns (less the entry for the root vertex) form the edge table for the tree! The remaining two columns can be reconstructed from the Vertex and Parent columns. The representation is not unique, since any child of a vertex can be taken to be the first child, and its siblings can be arranged in any order.

EXAMPLE 8.5   Construct oldest child and next sibling columns for the tree defined by the following table.

Vertex	Parent	First Child	Next Sibling
1	0		
2	1		
3	7		
4	2		
5	7		
6	1		
7	2		
8	7		

First, find the children of vertex v_1 by searching the Parent column for occurrences of 1. The children are v_2 and v_6. So if you take v_2 to be the first child of v_1, then v_6 is the next sibling of v_2. The result so far is as follows.

Vertex	Parent	First Child	Next Sibling
1	0	2	
2	1		6
3	7		
4	2		
5	7		
6	1		
7	2		
8	7		

The process is repeated for each vertex. The final result is

Vertex	Parent	First Child	Next Sibling
1	0	2	0
2	1	4	6
3	7	0	5
4	2	0	7
5	7	0	8
6	1	0	0
7	2	3	0
8	7	0	0

○

EXERCISES 8.1

1. Which of the graphs in Figure 8.7 are trees?

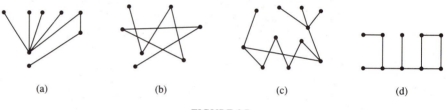

(a)   (b)   (c)   (d)

FIGURE 8.7

2. In each of the trees of Figure 8.8, find the path between the indicated vertices.

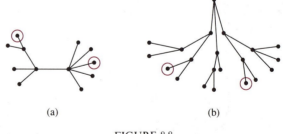

(a)   (b)

FIGURE 8.8

3. For which integers n is K_n a tree?

4. For which integers m, n is $K_{m,n}$ a tree?

5. Show that every tree is a planar graph. What is the number of faces in an embedding of a tree?

6. For which trees does an Euler walk exist? (An **Euler walk** is an open walk that contains each edge exactly once.)

7. Let G be a tree with n vertices. How many paths does G contain?

8. What is the chromatic number of a tree? Why?

9. For each of the trees in Figure 8.9, assign a direction to each edge in such a way that the resulting directed graph is a rooted tree with the indicated vertex as root.

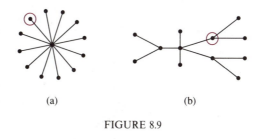

(a)   (b)

FIGURE 8.9

10. Which of the directed graphs in Figure 8.10 are rooted trees? For each of these graphs that is a rooted tree, find the root.

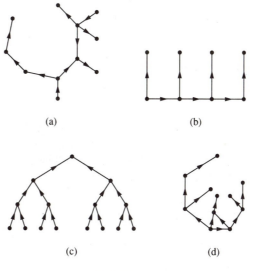

(a)   (b)

(c)   (d)

FIGURE 8.10

11. Draw a rooted tree with 10 vertices, having the largest possible number of leaves.

12. Draw a rooted tree with 10 vertices, having the largest possible number of internal vertices.

13. Give an example of a directed graph G that is not a rooted tree but has a vertex v with the property that for every other vertex w, there is a unique directed path from v to w.

14. For each of the rooted trees in Figure 8.11, list the parent, children, and siblings of each vertex.

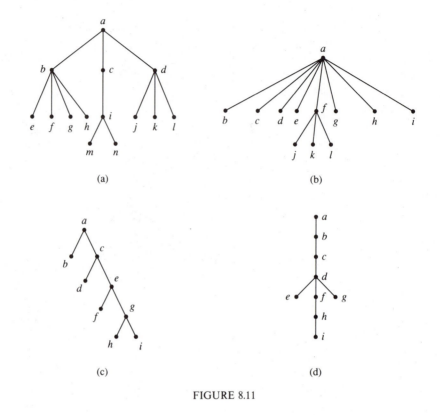

FIGURE 8.11

15. For each of the rooted trees in Figure 8.11, define the tree structure by a parent/oldest child/next sibling table.

16. For each of the rooted trees in Figure 8.11, how many oldest child/next sibling representations are possible?

17. Draw the trees defined by the following tables.

a.

Vertex	Parent	First Child	Next Sibling
1	0	2	0
2	1	4	3
3	1	7	0
4	2	0	5
5	2	0	6
6	2	0	0
7	3	0	8
8	3	0	9
9	3	0	0

b.

Vertex	Parent	First Child	Next Sibling
1	11	0	12
2	12	0	13
3	13	0	4
4	13	0	5
5	13	0	10
6	12	0	7
7	12	0	0
8	11	0	9
9	11	0	0
10	13	0	0
11	0	1	0
12	11	2	8
13	12	3	6

18. Reconstruct the oldest child and next sibling columns for the trees defined below, and draw the trees.

a.

Vertex	Parent	First Child	Next Sibling
1	2		
2	3		
3	4		
4	5		
5	0		
6	5		
7	5		
8	5		

b.

Vertex	Parent	First Child	Next Sibling
1	3		
2	3		
3	6		
4	3		
5	6		
6	0		
7	3		
8	6		
9	6		
10	6		

8.2 • SPANNING TREES

The concept of a spanning tree of a graph originated with optimization problems in communications networks. A communications network can be represented by a graph in which the vertices are the stations and the edges are the communication lines between stations. A subnetwork that connects all the stations without any redundancy will be a tree.

Definition 8.4

A tree T is a **spanning tree** of a graph G if T is a subgraph of G and every vertex of G is a vertex of T.

EXAMPLE 8.6   Figure 8.12 shows a graph and three different spanning trees for the graph. The spanning trees are made up of the edges shown as solid lines.

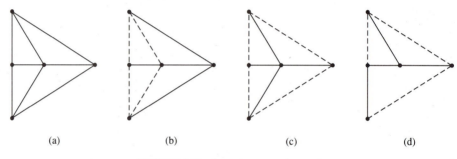

(a)   (b)   (c)   (d)

FIGURE 8.12   Spanning trees of a graph   o

The following fact about spanning trees will be useful.

Theorem 8.6   Let T be a spanning tree in a graph G, and let e be an edge of G not in the tree T. Let T' be the subgraph of G obtained by adding the edge e to T. Then T' contains a unique cycle C. If f is any edge of C, the graph obtained by deleting f from T' is a spanning tree of G.

The proof is left as an exercise.   ●

There are two algorithms for constructing a spanning tree for a graph: the **breadth-first** method and the **depth-first** method. In each method, we start at some vertex and build a rooted tree from it. Before describing the two methods precisely, it will help to look at a simple but extreme case. Consider the wheel in Figure 8.13(a). If you start at the center, you can build a spanning tree by including all the spokes (Figure 8.13(b)). This is the breadth-first method. Or you can start with one spoke, then travel along the outside (Figure 8.13(c)). This is the depth-first method. The breadth-first method yields a tree with small depth and large outdegrees for the vertices; the depth-first method yields a tree with a large depth and small outdegrees.

In the breadth-first method of finding a spanning tree, each vertex will be labeled with its depth in the resulting tree.

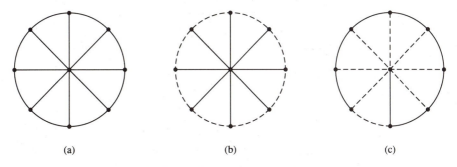

FIGURE 8.13 Two spanning trees for a wheel

Breadth-First Algorithm

1. Set the label of each vertex to "unlabeled."
2. Choose any vertex and label it 0. (This will be the root. At this point, the tree being built has zero edges.)
3. Set k equal to zero.
4. Repeat the following steps until the tree reaches $n - 1$ edges, where n is the number of vertices:
   a. For each vertex v labeled k, find all unlabeled vertices adjacent to v, label them $k + 1$, and add to the tree the edges from v to these vertices.
   b. Set k equal to $k + 1$.

EXAMPLE 8.7   In the graph of Figure 8.14(a), start at the center, vertex v_1. It is labeled 0. The vertices adjacent to v_1 are v_2, v_3, v_4, and v_5. These are labeled 1, and the edges from v_1 to v_2, v_3, v_4, and v_5 are added to the tree. At the next step, all the remaining vertices are labeled 2. The result, with labels, is shown in Figure 8.14(b).

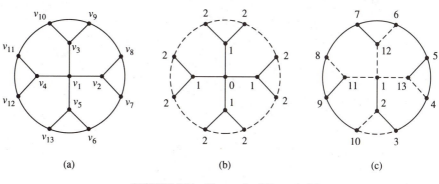

FIGURE 8.14 The graph of Example 8.7

The tree resulting from use of the breadth-first search method depends, of course, on the choice of root; it may also depend on the order in which the vertices are processed in step 4(a).

EXAMPLE 8.8 Figure 8.15 shows a graph and two different spanning trees, both obtained by the breadth-first method starting at the same root v_1. In Figure 8.15(b), vertex v_2 was processed before vertex v_3. In Figure 8.15(c), vertex v_3 was processed before vertex v_2.

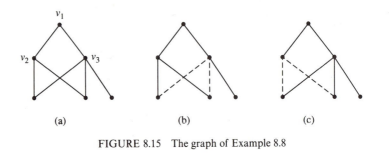

FIGURE 8.15   The graph of Example 8.8

In the depth-first method, each vertex is labeled with a sequence number. The sequence numbers indicate the order in which the vertices are added to the tree.

Depth-First Algorithm

1. Set the label of each vertex to "unlabeled."
2. Choose any vertex and label it 1. (This will be the root. At this point, the tree being built has zero edges.)
3. Set k equal to 1.
4. While k is less than the number of vertices in the graph, do the following steps:
   a. Let v be the vertex with the largest label that is adjacent to an unlabeled vertex. Let w be any one of the unlabeled vertices to which v is adjacent. Label w with $k + 1$ and add the edge from v to w to the tree.
   b. Set k equal to $k + 1$.

The tree that results from application of the depth-first method depends on the choice of the root and also on the choices of w in step 4(a).

EXAMPLE 8.9 Consider the same graph (Figure 8.14(a)) referred to in Example 8.7. Again start at v_1, which is labeled 1 (Figure 8.14(c)). From among the vertices adjacent to v_1, choose v_5 and label it 2. From among the vertices adjacent to v_5, choose v_6 and label it 3. The next choices are $v_7, v_8, v_9, v_{10}, v_{11}, v_{12}$, and v_{13}, labeled 4 through 10. At this point, the vertex with the largest label adjacent to an unlabeled vertex is v_{12}, whose label is 9. So v_4 is added and labeled 11. Finally, v_3 and v_2 are added and labeled 12 and 13.

Minimal Spanning Trees

If G is a graph of n vertices, any spanning tree for G will have $n - 1$ edges. If the length of each edge of the graph is 1, then the total length of any spanning tree is $n - 1$. By the criterion of length, all spanning trees are equivalent. If, however, the graph is **weighted** (that is, a number is assigned to each edge), the situation becomes more interesting. Weighted graphs arise naturally in many optimization problems.

EXAMPLE 8.10   An oil company has storage facilities at four locations. The company wants to interconnect the facilities by pipelines in such a way that oil can be transferred from any facility to any other. For environmental reasons, any pipeline built should follow a direct path, but due to variations in terrain, the cost of building lines varies widely. Figure 8.16 shows the four facilities as a graph. The weight on each edge of the graph is the cost of building a pipeline between the corresponding facilities, in millions of dollars. Choosing a set of pipelines to build is equivalent to choosing a spanning tree for the graph. The spanning tree with the smallest cost is the tree connecting vertices *A, B, C,* and *D* in that order.

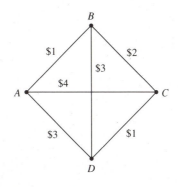

FIGURE 8.16   The graph of Example 8.10   o

Definition 8.5

Let *G* be a connected weighted graph with weight function *W*. **A minimal spanning tree** for *G* is a spanning tree *T* for *G* such that $W(T) \leq W(T')$ for every spanning tree *T'* of *G*.

In Example 8.10, it was obvious what the minimal spanning tree of the graph was, but only because there were two edges of weight 1 and it was possible to include them in a spanning tree. In general, the minimal spanning tree is not obvious. To find a minimal spanning tree, it is impractical to list all spanning trees and choose the one with the least weight. Two systematic methods for finding a minimal spanning tree are known: Prim's algorithm and Kruskal's algorithm. We will present Prim's algorithm and show that it does indeed construct a minimal spanning tree. Kruskal's algorithm will be presented in the exercises.

Prim's Algorithm for a Minimal Spanning Tree

1. Choose any vertex v, and let T_1 be the tree consisting of the vertex v alone.
2. Set $k = 1$.

3. While k is less than the number of vertices in the graph G, do the following steps:

    a. From among all edges of G that have one vertex in T_k and the other vertex not in T_k, choose an edge e having smallest weight. Build T_{k+1} by adding this edge to T_k.

    b. Set k equal to $k + 1$.

EXAMPLE 8.11   Consider the weighted graph in Figure 8.17(a). We will build a minimal spanning tree using Prim's algorithm and starting at the vertex on the left. Of the two edges leaving this vertex, choose the one of weight 2. This edge is labeled (1) in Figure 8.17(b). At the next step, we have three edges to choose from, of weights 4, 7, and 3; choose the edge of weight 3. This edge is labeled (2) in Figure 8.17(b). At the next step, we have four edges to choose from, of weights 4, 7, 2, and 1; choose the edge of weight 1. Continue in this way until all vertices are accounted for. The final result is shown in Figure 8.17(b). The weight of this spanning tree is $2 + 3 + 1 + 2 + 1 + 3 + 2 + 3 + 4 = 21$.

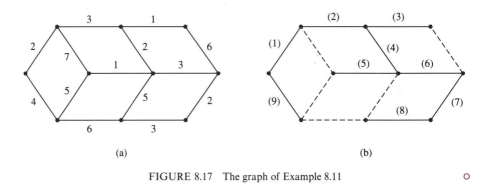

FIGURE 8.17   The graph of Example 8.11

Theorem 8.7   If G is a connected weighted graph with weight function W, then Prim's algorithm yields a minimal spanning tree of G.

Proof   Suppose G has n vertices, and let $T_1, \ldots, T_n$ be the sequence of trees constructed by Prim's algorithm. Each tree is obtained from the previous tree by adding an edge of minimal weight. We will show that there exists a sequence $M_1, \ldots, M_n$ of minimal spanning trees in G such that T_k is a subgraph of M_k for $k = 1, 2, \ldots, n$. Since T_n must equal M_n, this will prove the theorem.

The proof is by induction on k. In the case $k = 1$, the tree T_1 consists of a single vertex. If M is any minimal spanning tree, the vertex of T_1 must be in M. So we can choose $M_1 = M$, and T_1 is a subgraph of M_1. Now suppose that the tree T_k is contained in the minimal spanning tree M_k and that T_{k+1} is obtained from T_k by adding the edge e. There are two cases: e is an edge of M_k, or it is not. If e is an edge of M_k, choose $M_{k+1} = M_k$. If, however, e is not an edge of M_k, adding e to M_k creates a cycle C. The cycle C leaves T_k along e and must reenter T_k along some other edge— call it f. This state of affairs is shown in Figure 8.18.

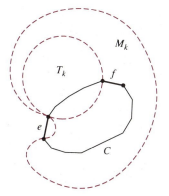

FIGURE 8.18   The proof of Theorem 8.7

Now choose M_{k+1} to be $M_k + e - f$. By Theorem 8.6, M_{k+1} is a spanning tree for G. Moreover, $W(e) \leq W(f)$, because e has minimal weight among all edges leaving T_k. (This was the criterion for the selection of e in Prim's algorithm.) So $W(M_{k+1}) \leq W(M_k)$. But M_k was a minimal spanning tree, so $W(M_{k+1}) = W(M_k)$ (in fact, $W(e) = W(f)$), and M_{k+1} is also a minimal spanning tree for G.   ●

In the proof of Theorem 8.7, the minimal spanning tree M_1 evolved into the minimal spanning tree M_n in a sequence of steps, each consisting of the replacement of at most one edge.

EXERCISES 8.2

1. Find spanning trees for each of the graphs in Figure 8.19
    a. By the breadth-first method
    b. By the depth-first method

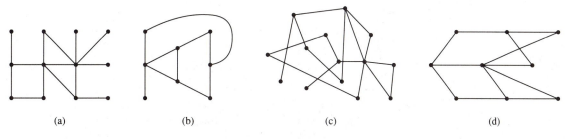

FIGURE 8.19

2. Describe the spanning tree obtained for K_n by the breadth-first method.
3. Describe the spanning tree obtained for K_n by the depth-first method.
4. Describe the spanning tree obtained for $K_{m,n}$ by the breadth-first method.

5. Describe the spanning tree obtained for $K_{m,n}$ by the depth-first method.

6. Prove Theorem 8.6.

7. Use Prim's algorithm to find minimal spanning trees for the weighted graphs of Figure 8.20.

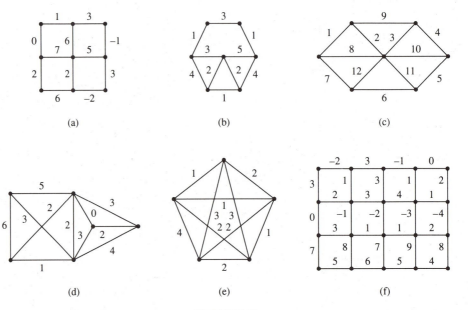

FIGURE 8.20

8. In **Kruskal's algorithm** for building a minimal spanning tree of a weighted graph G, a sequence of subgraphs $H_1, \ldots, H_{n-1}$ is constructed in the following way. H_1 consists of a single edge, an edge of least weight in G. H_k is formed from H_{k-1} by adding an edge of least weight among the remaining edges, provided that adding the edge does not form a cycle. (The subgraphs H_k need not be trees.) Use Kruskal's algorithm to construct minimal spanning trees for the graphs of Figure 8.20.

8.3 • BINARY TREES

One of the most common problems in data processing is searching a file for a record having a specified value in some field. For instance, a file of customer orders could be searched by name for all records pertaining to a particular customer, or searched by account number, or searched for all records pertaining to one product, or for orders taken by a particular sales representative.

One way to provide for efficient searching of a file on a field F is to sort the file on the field F. For instance, if F is customer name, the file could be sorted alphabetically on F; or if F is account number, the file could be sorted numerically on F. When the file is sorted on the field of interest, the technique of **binary search** can be used. In a binary search, you first look at the middle of the file and find out whether the desired record is in the top half or the bottom half. Then you repeat this process until the desired record is found (or shown not to exist).

EXAMPLE 8.12   Locate account number 037 in the following file of account numbers and customer names:

    002   Tell, William
    009   Boccanegra, Simon
    017   Miller, Luisa
    019   Budd, Billy
    023   Troyens, Les
    037   Lecouvreur, Adriana
    066   Chenier, Andrea
    076   Giovanni, Don
    098   Schicchi, Johnny
    101   Herring, Albert
    121   Lescaut, Manon
    133   Onegin, Eugene
    177   Grimes, Peter

The file is already sorted on account number, and there are 13 records. The middle of the file is record number 7,

    066   Chenier, Andrea

Since 066 is larger than 037, the required record is in the top half—that is, in the group

    002   Tell, William
    009   Boccanegra, Simon
    017   Miller, Luisa
    019   Budd, Billy
    023   Troyens, Les
    037   Lecouvreur, Adriana

The middle of this half is

    017   Miller, Luisa
    019   Budd, Billy

since there are an even number of records. You can choose either to compare to. If you choose the first (017), you find that the required record is in the bottom half,

    019   Budd, Billy
    023   Troyens, Les
    037   Lecouvreur, Adriana

It takes just two more comparisons to locate record 037.   ○

The advantage of binary search is its efficiency: If the size of the file is doubled, the number of comparisons is increased by only one. The number of comparisons required to find a record in a file of N records is approximately $\log_2(N)$. (For any positive number x, $\log_2(x) = y$ means that $2^y = x$.) For instance, a file of 1000 records can be searched in only 10 comparisons, since $2^{10} = 1024$. A file of 2000 records can be searched in 11 comparisons.

The disadvantage of binary search is the requirement that the file be sorted. If the data in the file rarely changes, and if there is only one field that is subject to search, this is not a problem. This is the case, for instance, with a telephone directory. But for the file to remain sorted when a record is added to the file, the new record must be placed in its proper position, and all records below that position must be moved down to make room. If a record is deleted, all records below the deletion must be moved up to fill the gap. Worst of all, if a search is to be done on some field other than the sort key, the file must be completely resorted according to the desired key.

An alternative to keeping a file in sorted order is to organize the records of the file as the vertices of a rooted tree. Pointers defining the tree structure (parent, children) can then be added to each record in the file. With the proper structure, a binary search can be conducted by starting at the root and following the edges of the tree. Several tree structures can exist simultaneously to permit searches on any of several key fields. It is necessary only to store an additional set of pointers for each additional key.

Definition 8.6

A **binary tree** is a rooted tree in which

1. Each vertex has zero, one, or two children, and
2. Each child of a vertex is identified as either a left child or a right child.

If there are two children of a vertex, one must be the left child and the other the right child.

EXAMPLE 8.13   Figure 8.21 shows some binary trees. In these figures, as in the remainder of this section, the left child is drawn to the left of the parent and the right child to the right.

○

It is important to note that a binary tree is more than just a rooted tree in which each vertex has at most two children. The identification of each child as a left child or a right child is significant. For instance, the trees in Figure 8.21(e) and (f) are identical as rooted trees, but different as binary trees.

Definition 8.7

Let v be the vertex of a binary tree T. If w is the left child of v, then w together with its descendants and their interconnecting edges form the **left subtree** of T at v. If w is the right child of v, then w together with its descendents and their interconnecting edges form the **right subtree** of T at v.

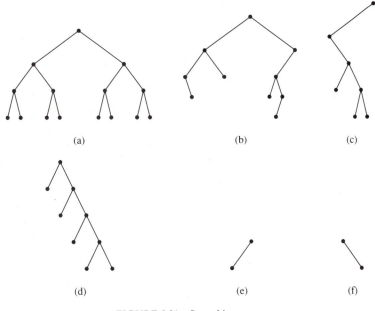

FIGURE 8.21   Some binary trees

EXAMPLE 8.14   In Figure 8.22, B is the left subtree of vertex w and A is the left subtree of vertex v. The tree C consisting of the single vertex y is the right subtree of vertex x. Both subtrees of vertex y are empty.

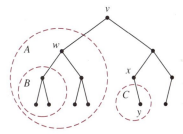

FIGURE 8.22   The tree of Example 8.14   o

Search Trees

To use a binary tree to guide a search, we must arrange the records of a file into the vertices of a binary tree in such a way that the tree structure is related to the order of the entries in the key field.

Definition 8.8   A **binary search tree** consists of a binary tree T, a totally ordered set X, and a one-to-one function $f: V \to X$ (where V is the set of vertices of T), such that if v is any vertex, the values of f on the left subtree of v are all less than $f(v)$ and the values of f on the right subtree of v are all greater than $f(v)$.

In typical applications, X is either a set of numbers or a set of character strings in lexicographic order.

EXAMPLE 8.15   Figure 8.23 shows the records of Example 8.12 arranged in a binary search tree by account number. The depth of the tree is 3. Any account record in the tree can be located in at most four comparisons, starting at the root and going left or right, depending whether the account number of the record being sought is less than or greater than the account number at the current vertex.

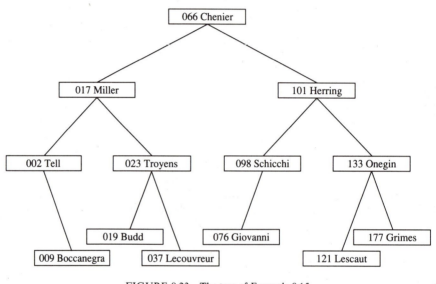

FIGURE 8.23   The tree of Example 8.15   o

The method for searching a binary tree for a record can be stated in the following way.

Binary Tree Search Algorithm

To search a tree for a key value x:
Step 1: Set v equal to the root.
Step 2: Repeat the following step until either success or failure occurs:
    If $x = f(v)$, then quit (success—v is the required record)
    else
        if $x < f(v)$, then
            if left child $(v) = \varnothing$, then quit (failure)
            else set $v =$ left child (v)
        else (this is the case $x > f(v)$)
            if right child $(v) = \varnothing$, then quit (failure)
            else set $v =$ right child (v).

EXAMPLE 8.16   In the tree of Figure 8.23, find the record (if any) with account number 50. To start, set v to "066 Chenier." Compare 50 to $f(v)$ (that is, to 66). 50 is less, and there is a left subtree, so set v to "017 Miller." Compare 50 to 17; 50 is greater, so go right by setting v to "023 Troyens." Compare 50 to 23; 50 is greater, so go right. Compare 50 to 37; 50 is greater, but there is no right subtree, so quit with failure. There is no record with account number 50.   ○

Building a Search Tree

In order to use the tree search method, it is first necessary to organize the records of a relation into a binary search tree. This can be done by a simple variant of the binary tree search algorithm. Take the first record and make it the root of the tree. For each additional record, use the search algorithm to find the leaf of the tree closest in value to the new record. Add the new record to that leaf as either a right child or left child, depending whether the value of the new record is greater than or less than the value of the leaf.

EXAMPLE 8.17   Arrange the numbers 6, 3, 5, 9, 8, 1, 12 into a binary search tree. The steps in the process are shown in Figure 8.24. The root is 6, then 3 is added as left child of 6. The next number, 5, is less than 6 but greater than 3, so it is added as a right child of 3. The next number, 9, is greater than 6, so it is added as a right child of 6, and so on. The final tree is shown in Figure 8.24(g).

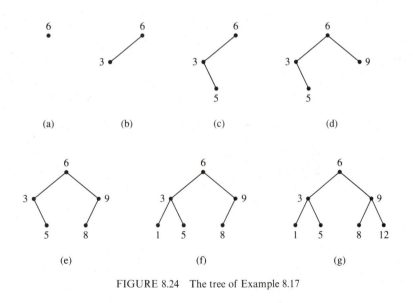

FIGURE 8.24   The tree of Example 8.17   ○

EXAMPLE 8.18   Arrange the records of Example 8.12 into a binary search tree alphabetically by customer name. The result is shown in Figure 8.25. The numbers in parentheses after each vertex indicate the order in which the vertices were added to the tree—that is, the order in which they appear in Example 8.12.

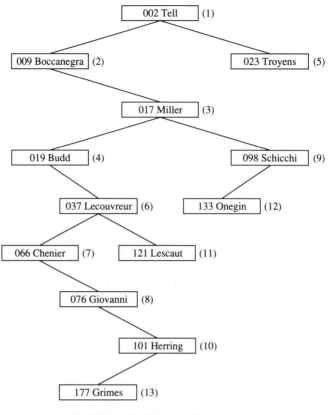

FIGURE 8.25   The tree of Example 8.18

Note that the tree in Figure 8.25 is rather unsatisfactory as a search tree. It has depth eight, while an efficient search tree for this set should have a depth of only four. The tree is said to be **unbalanced;** the tree in Figure 8.24(g) is **balanced.** Efficient use of tree search requires building a balanced tree and keeping it balanced as vertices are added and deleted. Algorithms for maintenance of binary search trees may be found in most advanced texts on data structures.

Computer Representations of Binary Trees

Representing a binary tree in a computer is simpler than representing a rooted tree. Since a vertex can have at most one sibling, it is not necessary to arrange the children of a vertex into a linked list, as we did in Example 8.4. For each vertex, three pointers are stored: parent, left child, and right child.

EXAMPLE 8.19   If we number the records of Example 8.12 from 1 to 13 in the order given, the binary tree of Figure 8.23 can be defined by the following table:

Vertex	Parent	Left Child	Right Child
1 (Tell)	3	0	2
2 (Boccanegra)	1	0	0
3 (Miller)	7	1	5
4 (Budd)	5	0	0
5 (Troyens)	3	4	6
6 (Lecouvreur)	5	0	0
7 (Chenier)	0	3	10
8 (Giovanni)	9	0	0
9 (Schicchi)	10	8	0
10 (Herring)	7	9	12
11 (Lescaut)	12	0	0
12 (Onegin)	10	11	13
13 (Grimes)	12	0	0

It is interesting to note that the parent–oldest-child–next-sibling method of representing a rooted tree T actually defines a binary tree B. The left child of a vertex in B represents the oldest child of the vertex in T, and the right child of a vertex in B represents the next sibling of the vertex in T. Thus, an arbitrary rooted tree can be represented by a binary tree.

EXERCISES 8.3

1. Use the binary search method to locate the number 41 in the list 1, 3, 5, 7, 11, 13, 17, 19, 23, 29, 31, 37, 41, 43, 45. List each comparison made.

2. Using binary search, what is the largest number of comparisons required to locate an entry in the index of this book?

3. Let T be a binary tree of depth N in which each path from the root to a leaf has the same length. How many vertices does T have? How many edges?

4. In the tree of Figure 8.23, search for an entry with account number 120. List each comparison step.

5. In the tree of Figure 8.25, search for entries with the names **a.** Jenufa, **b.** Siegfried, **c.** Padilla, **d.** Wozzeck. List each comparison step in each search.

6. Using the method described in the text, arrange the following lists of numbers into binary search trees:

   a. 36, 856, 561, 879, 945, 680, 966, 263, 870, 848, 67

   b. 434, 551, 447, 418, 961, 501, 194, 914, 589, 258, 975

7. Modify the tree in Figure 8.25 by adding four new records as follows:

   218 Ivnor, Monte
   018 Stuarta, Maria
   057 Bolena, Anna
   077 Wozzeck, Johann

8. Modify the tree of Figure 8.23 by adding the four records listed in the previous exercise.

9. If you use the method described in the text to construct a binary search tree from a file, and the file is already sorted on the key field, what kind of tree results?

10. Construct a parent/left child/right child table (see Example 8.19) for the following binary trees:

    a.   The tree of Figure 8.25   **b.**   The tree of Exercise 6a

    c.   The tree of Exercise 6b

8.4 ● TRAVERSING ROOTED TREES

When data is organized as a tree, it is often necessary to perform some operation on the data associated with each vertex of the tree. Since a computer must perform the operations sequentially, processing the whole tree is equivalent to defining an order on the set of vertices. The program is said to **traverse** the tree and to **visit** the vertices in the specified order. We will describe several ways to do this. The traversal algorithms apply to binary trees and also to a related class of trees called *ordered rooted trees*.

Definition 8.9   An **ordered rooted tree** is a rooted tree in which, for every vertex, a total order relation is defined on its children.

Of course, if a vertex is a leaf, it has no children, and so the empty relation is a total order relation on its children. So in specifying the order relations on children, we can ignore leaves. Whenever we draw a tree with the root at the top, we make it an ordered rooted tree by the following rule: The children of each vertex are ordered left to right. When a tree is represented in a computer using oldest-child and next-sibling pointers, it is made into an ordered rooted tree, as shown in Example 8.4.

EXAMPLE 8.20   Figure 8.26 shows two different ordered rooted trees. The underlying rooted trees in the figure are the same, but the order relations on the children of the vertices are different.

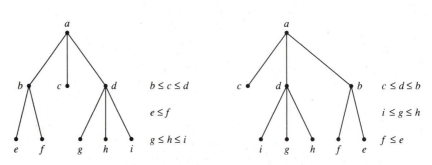

FIGURE 8.26   Two ordered rooted trees

Traversal Algorithms

The three methods of traversing an ordered rooted tree are called **preorder, inorder,** and **postorder.** They are defined as follows:

Preorder: To traverse a tree in preorder,

1. Visit the root, then
2. Traverse the first subtree in preorder, then the second subtree in preorder, etc.

Inorder: To traverse a tree in inorder,

1. Traverse the left subtree in inorder, then
2. Visit the root, then
3. Traverse the right subtree (and any remaining subtrees) in inorder

Postorder: To traverse a tree in postorder,

1. Traverse, in order, each of the subtrees in postorder, then
2. Visit the root.

Note that in a binary tree, the left subtree of a vertex may be empty. In an ordered rooted tree, the first subtree of a vertex is always considered the left subtree.

EXAMPLE 8.21   Figure 8.27 shows the development of preorder traversal for a tree. The tree is shown in Figure 8.27(a). There are three subtrees, whose roots are *b*, *c*, and *d* in that order.

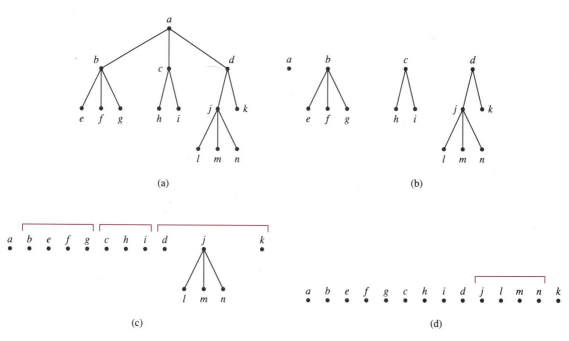

FIGURE 8.27   Preorder traversal

To traverse the tree in preorder, visit the root *a*, then traverse each of the three subtrees. The root *a* and the three subtrees are shown in Figure 8.27(b). Preorder traversal of the subtree rooted at *b* starts at *b* and continues with *e*, *f*, and *g*. The subtree at *c* is traversed in the order *c*, *h*, *i*. Finally, the subtree at *d* is traversed in the order *d*, the subtree at *j*, then *k*. This step is shown in Figure 8.27(c). The final result is shown in Figure 8.27(d). Figures 8.28 and 8.29 show inorder and postorder traversals of the same tree.

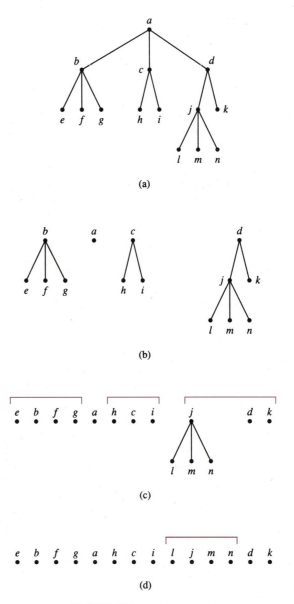

FIGURE 8.28   Inorder traversal

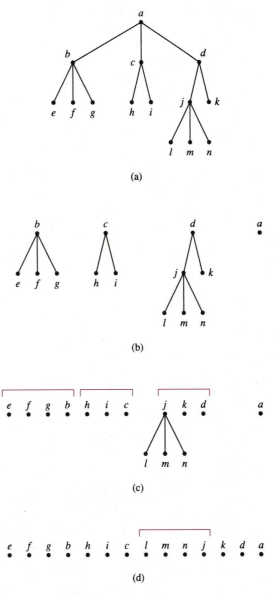

(a)

(b)

(c)

(d)

FIGURE 8.29   Postorder traversal   o

Figure 8.30 shows a simple geometrical interpretation of tree traversal. Draw a path around the tree, starting at the root and proceeding counterclockwise. You will round each leaf once and pass each internal vertex two or more times. For each of the traversal methods, you visit each leaf when you round it. For preorder traversal, visit each internal vertex when you pass it the first time. For inorder traversal, visit each internal vertex the second time you pass it. For postorder traversal, visit each internal vertex the last time you pass it.

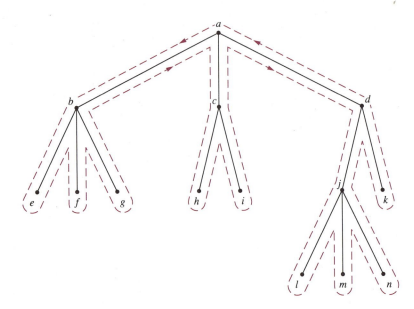

FIGURE 8.30   A geometrical interpretation

Inorder traversal of a binary search tree produces a list of the vertices in sorted order. But if the search tree is regarded as an ordered rooted tree instead of a binary tree, inorder traversal does not necessarily list the vertices in the proper order. To see this, look at Figure 8.23: Application of the definition of inorder traversal for ordered rooted trees will put 009 (Boccanegra) ahead of 002 (Tell). The problem is that, since there is no left child of vertex 002, the method treats the right child (009) as though it were the left child. To avoid confusion, it is sometimes useful to modify a binary search tree as follows: Whenever a vertex has only one child, add a "null" vertex (a vertex without a data item attached) for the other child. In Figure 8.23, the only null vertices needed are a left child for vertex 002 and a right child for vertex 098. With these vertices added, inorder traversal of the resulting ordered rooted tree lists the vertices in numerical order.

Application to Writing Arithmetic Expressions

Consider the arithmetic expression $(A + B * C) + D * (E + F)$, where, following computer programming practice, we have indicated multiplication explicitly by an asterisk. The expression indicates that certain operations are to be performed on A, B, C, D, E, F and intermediate results. The order of the operations is only partly specified by this expression and by the rules that multiplication is performed before addition and that addition is performed from left to right. For instance, B and C can be multiplied, the product added to A, then E and F added, the sum multiplied by D, and the two results added. Or E and F can be added, then B and C multiplied, then the sum of E and F multiplied by D, etc. The actual requirements for doing the indicated arithmetic can be expressed by a tree diagram, as shown in Figure 8.31. In this tree, the leaves correspond to variables, and internal vertices correspond to intermediate results. Any arithmetic expression can be represented by a tree in this way.

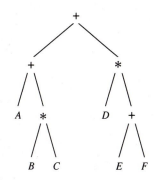

FIGURE 8.31   Tree for $(A + B * C) + D * (E + F)$

Now consider the reverse problem: Given such a tree, write the expression. To write the expression means to arrange the variables and operation symbols—that is, the vertices of the tree—in some order. This is the same as traversing the tree. If we traverse the tree of Figure 8.31 in inorder, the result is

$$A + B * C + D * E + F$$

which is our original expression, minus the parentheses. Unfortunately, this expression is not equivalent to our original if the usual operator hierarchy rules are followed.

Figure 8.32 shows the tree corresponding to the expression $(A + B * C) + (D * E + F)$. When the tree of Figure 8.32 is traversed in inorder, the result is the same as for Figure 8.31.

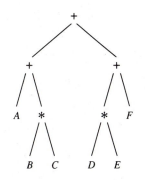

FIGURE 8.32   Tree for $(A + B * C) + (D * E + F)$

Now traverse the tree of Figure 8.31 in preorder. The result is

$$+ \quad + A * B C * D + E F$$

This expression uniquely determines the arithmetic operations to be performed. The operands for each operation are shown in the following diagram.

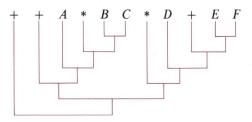

Writing an arithmetic expression with the operators first is called **prefix notation;** the conventional form is called **infix notation.** Prefix notation has the advantage that no parentheses are required. Prefix notation is also called **Polish notation.** It was first introduced by the Polish logician Jan Łukasiewicz (1878–1956) for use in propositional logic.

Now traverse the tree of Figure 8.31 in postorder. The result is

$$A\ B\ C * + D\ E\ F + * +$$

This form of the expression is called **postfix notation,** or **reverse Polish notation.** Like prefix notation, no parentheses are required for a unique interpretation.

Reverse Polish notation has a remarkable feature: When read from left to right, it provides directions for evaluating the expression. The computation uses a **stack.** A stack is a structure in which items are considered to be arranged vertically, and on which there are two operations: putting an item on top of the stack (**push**ing the item) and removing an item from the top of the stack (**pop**ping the item).

The algorithm for evaluating a reverse Polish expression is as follows. Start with an empty stack and scan the expression from left to right. Then

1. When you encounter a variable, push the value of the variable onto the stack.
2. When you encounter a unary operator (such as $-$), pop one value from the stack, perform the operation, and push the result.
3. When you encounter a binary operation, pop two values, perform the operation, and push the result.

At the end, the answer will be on top of the stack.

EXAMPLE 8.22   Evaluate $A\ B\ C * + D\ E - F + * +$ when $A = 2$, $B = 3$, $C = 4$, $D = -2$, $E = -3$, $F = 1$. The first three steps push 2, 3, and 4 onto the stack; then 4 and 3 are multiplied, leaving 12 and 2 on the stack. The complete process is shown below.

	A	B	C	$*$	$+$	D	E	$-$	F	$+$	$*$	$+$
Top of → stack	2	3	4 →	12 →	14	−2	−3 →	3	1 →	4 →	−8 →	6
		2	3	2		14	−2	−2	3	−2	14	
			2				14	14	−2	14		
									14			

Implementation of this method of calculation is very efficient. It is the basis of the design of many Hewlett-Packard calculators, the large-scale computers made by Burroughs (now UNISYS), and the FORTH programming language.

EXERCISES 8.4

1. For each of the trees in Figure 8.33, list the vertices in the order resulting from traversing the tree **a.** in preorder, **b.** in inorder, **c.** in postorder.

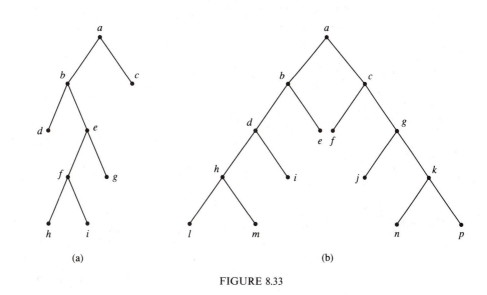

FIGURE 8.33

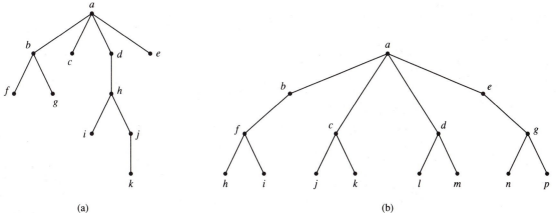

FIGURE 8.34

2. Do the same for the trees in Figure 8.34.
3. Draw trees representing the following arithmetic expressions:
   a. $(A + B/C) * (D/E - F)$ **b.** $A * B * C - D * (E + F)$
4. Draw trees representing the following arithmetic expressions:
   a. $A + (A/F + A * C)/D$ **b.** $((F + A) * B - C)/D$

5. Write each of the arithmetic expressions of Exercise 3: (1) in prefix notation; (2) in postfix notation.

6. Do the same for each of the arithmetic expressions of Exercise 4.

7. Using the method of Example 8.22, evaluate each of the expressions of Exercise 3 for the following values of the variables. Show the contents of the stack at each step.

   a. $A = 2, B = 3, C = -1, D = 5, E = -3, F = 11$

   b. $A = 7, B = -12, C = 5, D = 5, E = 13, F = -21$

8. Do the same for each of the arithmetic expressions of Exercise 4.

9. Write the expressions $A * (-B)$ and $(A - B) * C$ in postfix notation. Compare the sequences of operations that occur when using the method of Example 8.22 to evaluate them. Explain. How do most hand calculators address this problem?

10. Does prefix (Polish) notation lead to an evaluation method similar to that of Example 8.22? If so, describe it.

11. What vertices must be added to the tree of Figure 8.25 so that inorder traversal of the resulting ordered rooted tree produces alphabetical order? Write out the inorder traversal, including the null vertices.

12. When a binary tree is traversed in preorder, the result is *abdecfhjg*. When the tree is traversed in inorder, the result is *dbeahfjcg*. Find the tree.

Computer Exercises for Chapter 8

1. Write a program to read a representation of a graph and determine whether the graph is a tree.

2. Write a program to use Prim's algorithm to find a minimal spanning tree of a graph.

3. Write a program to construct a binary search tree for a file.

4. Write a program to search a binary tree for a record with a specified key value.

5. Write a program that accepts as input an arithmetic expression in reverse Polish notation and a set of values for the variables, then computes the value of the expression. You may limit variable names to a single letter.

• • • • CHAPTER REVIEW EXERCISES

1. Draw a rooted tree with 6 vertices, having the largest possible number of leaves.

2. Draw a rooted tree with 6 vertices, having the largest possible number of internal vertices.

3. For the rooted tree in Figure 8.5, list the parent, children, and siblings of each vertex.

4. Refer to the Chapter Review Exercises of Chapter 7. Which of the following graphs are trees?

   a. The graph of Exercise 1a

    b.   The graph of Exercise 1b

    c.   The graph of Exercise 12a

    d.   The graph of Exercise 12b

    e.   The graph of Exercise 18a

    f.   The graph of Exercise 18b

5. In each of the trees of Exercise 4, find the longest path.

6. Draw the trees defined by the following tables.

a.

Vertex	Parent	First Child	Next Sibling
1	0	6	0
2	1	0	7
3	4	9	0
4	6	5	0
5	4	0	3
6	1	8	2
7	1	0	0
8	6	0	4
9	3	0	0

b.

Vertex	Parent	First Child	Next Sibling
1	7	9	0
2	5	7	0
3	0	5	0
4	9	0	6
5	3	2	0
6	9	0	8
7	2	1	0
8	9	0	0
9	1	4	0

7. Reconstruct the oldest-child and next-sibling columns for the trees defined below, and draw the trees.

a.

Vertex	Parent	First Child	Next Sibling
1	5		
2	6		
3	5		
4	6		
5	6		
6	0		
7	6		
8	5		

b.

Vertex	Parent	First Child	Next Sibling
1	8		
2	8		
3	7		
4	7		
5	6		
6	9		
7	9		
8	9		
9	0		
10	3		

8. Find spanning trees for the graphs of Figure 7.30.

9. Refer to the Chapter Review Exercises of Chapter 7. Assign weights arbitrarily to the edges in the following graphs. Then use Prim's algorithm to find minimal spanning trees for the weighted graph.

    a.   The graph of Exercise 1a

    b.   The graph of Exercise 1b

    c.   The graph of Exercise 12a

    d.   The graph of Exercise 12b

   e.   The graph of Exercise 18a

   f.   The graph of Exercise 18b

10.   Construct a binary search tree containing the following items: 864, 565, 574, 697, 853, 113, 557, 262, 204, 896, 704, 482, 226, 708.

11.   List the vertices of Figure 8.23 in **a.** preorder, **b.** postorder.

12.   List the vertices of Figure 8.25 in **a.** preorder, **b.** postorder.

13.   For binary trees (a), (b), and (c) in Figure 8.24, identify the null vertices that need to be added so that inorder traversal of the resulting ordered rooted tree produces sorted order. In each case, list the vertices in sorted order, showing the positions of the null vertices.

14.   Using the method of Example 8.22, evaluate

$$A \mid B * (C \mid D * (E \mid B * (E/A \quad C)))$$

for the values $A = 5$, $B = -1$, $C = 2$, $D = -2$, $E = 10$. Show the state of the stack at each step.

15.   Dijkstra's shortest-path algorithm (see Section 7.5) constructs a spanning tree for a weighted graph—the predecessor of a vertex computed by the algorithm is the parent of the vertex in the tree. Is the tree so constructed necessarily a minimal spanning tree?

16.   When a binary tree is traversed in preorder, the result is *abdecfg*. When the tree is traversed in inorder, the result is *dbeagfc*. Find the tree.

RECURRENCE RELATIONS AND DYNAMICAL SYSTEMS

A **recurrence relation** is a formula that defines a number in a sequence of numbers in terms of previous numbers in the sequence. For instance, the formula $b_n = 2b_{n-1}$, with the starting point $b_0 = 1$, defines the sequence 1, 2, 4, 8, 16, ... Recurrence relations arise naturally in descriptions of many physical systems that evolve over time according to some law; such systems are called **dynamical systems.** Recurrence relations also arise frequently in determining the number of steps required to carry out certain computations. In this chapter, we will present some typical problems that give rise to recurrence relations, and we will describe some methods that can be used to analyze the most common kinds of relations.

9.1 • RECURRENCE RELATIONS

In this section, we will discuss a variety of problems that lead to recurrence relations.

EXAMPLE 9.1   A breeder has one mating pair of adult rabbits. Rabbits of this variety require one month to mature, after which each pair produces two young rabbits every month. Assuming that males and females are produced in equal numbers and that no rabbits die, how many pairs of rabbits does the breeder have after n months?

At each month, the number of pairs of young produced is equal to the number of mature pairs the previous month, and the number of mature pairs is equal to the number of mature pairs the previous month plus the number of young born the previous month. The evolution of this dynamical system may be described in the following table:

Month	Mature Pairs	Young Pairs
0	1	0
1	1	1
2	2	1
3	3	2
4	5	3
5	8	5
6	13	8
7	21	13

If we denote the sequence of numbers in the right-hand column by $F_0, F_1, F_2, \ldots,$ the sequence can be characterized by the equations

$$F_0 = 0$$
$$F_1 = 1$$
$$F_n = F_{n-1} + F_{n-2} \qquad \text{for } n = 2, 3, \text{ etc.}$$

The first two equations are **initial conditions,** and the third is a recurrence relation. This problem was first considered by the thirteenth-century Italian mathematician Leonardo Fibonacci, and the sequence 0, 1, 1, 2, 3, 5, 8, . . . is called the *Fibonacci Sequence,* seen in Example 3.5.   ○

It is by no means obvious how to write a formula, involving n alone, for the nth term in the Fibonacci Sequence, although we will derive one later in this chapter. In the next two examples, a formula for the nth term will be readily apparent.

EXAMPLE 9.2   A cell divides into two cells at the rate of one division per minute. If a culture is started with one cell and no cells die, how many cells are present after n minutes?

At each one-minute interval, the number of cells doubles. The number of cells present for $n = 0, \ldots, 4$ minutes is given in the following table:

Time (Minutes)	Number of Cells
0	1
1	2
2	4
3	8
4	16

If we denote the number of cells at time n by b_n, the sequence is characterized by the equations

$$b_0 = 1$$
$$b_n = 2b_{n-1}$$

It is apparent that b_n may be computed from the formula $b_n = 2^n$. This may be verified by mathematical induction. The verification is left as an exercise.   ○

EXAMPLE 9.3   The classical puzzle called the Towers of Hanoi (see Figure 9.1) consists of a board with three posts and a set of rings of increasing diameter that fit over a post. The problem is to move the rings from post A to post C in a sequence of moves. Each move consists of transferring one ring from one post to another, but no ring may be placed on top of a smaller ring. How many moves does it take to move a stack of n rings?

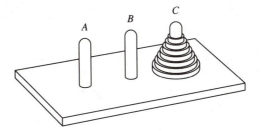

FIGURE 9.1   The Towers of Hanoi puzzle

If $n = 1$, only one move is required. If $n = 2$, the two rings may be moved in three steps:

$A \rightarrow B$ (small ring)

$A \rightarrow C$ (large ring)

$B \rightarrow C$ (small ring)

In general, if you have a method for moving a stack of $n - 1$ rings from one post to another, you can move n rings from post A to post C by moving the first $n - 1$ rings from A to B, then moving the nth (largest) ring from A to C, then moving the first $n - 1$ rings from B to C. If c_n is the number of moves needed to move n rings, the sequence c_n is characterized by

$$c_1 = 1$$
$$c_n = c_{n-1} + 1 + c_{n-1} = 2c_{n-1} + 1 \qquad \text{for} \qquad n = 2, 3, \text{etc.}$$

The values of c_n can be computed from these equations; they are as follows:

n	c_n
1	1
2	$2 \times 1 + 1 = 3$
3	$2 \times 3 + 1 = 7$
4	$2 \times 7 + 1 = 15$
5	$2 \times 15 + 1 = 31$

It is apparent that $c_n = 2^n - 1$. This may also be verified by mathematical induction; it is left as an exercise.   ○

EXAMPLE 9.4   A bank pays 4.2% interest on deposits, compounded quarterly. A woman deposits $100 on the last day of each quarter. How much money does she have in the bank at the end of n quarters? At the end of 0 quarters, she has only her initial $100 deposit. At the end of the nth quarter, she has the money that was on deposit at the end of the previous quarter, plus the interest earned on this money, plus her new $100 deposit. If d_n is the amount on deposit at the end of the nth quarter, then

$$d_0 = 100$$

$$d_n = d_{n-1} + \frac{0.042}{4}d_{n-1} + 100$$

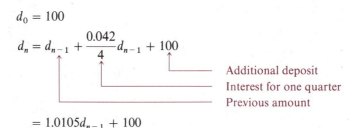

Additional deposit
Interest for one quarter
Previous amount

$$= 1.0105d_{n-1} + 100$$

Her account balances over the first eighteen months are as follows.

At End of Quarter	Account Balance
0	100.00
1	201.05
2	303.16
3	406.34
4	510.61
5	615.97
6	722.44

○

EXAMPLE 9.5   A population of cells in a culture is growing in the following way: A new cell, resulting from cell division, requires two minutes to mature. At one-minute intervals, 20% of the mature cells die, 20% divide into two immature cells, and the remaining 60% do neither. How does the population evolve?

If we let a_n be the number of mature cells present after n minutes and b_n be the number of immature cells (cells one minute old) at the same time, then

$$a_n = 0.6a_{n-1} + b_{n-1} \qquad (1)$$

Number of cells that have just matured
Number of surviving older cells

$$b_n = 2 \times 0.2a_{n-1}$$

Number of older cells that divided

$$= 0.4a_{n-1} \qquad (2)$$

Equation (2) can be written
$$b_{n-1} = 0.4a_{n-2} \qquad (3)$$

If you substitute Equation (3) in Equation (1), you get

$$a_n = 0.6a_{n-1} + 0.4a_{n-2} \qquad (4)$$

This is a recurrence relation for a_n. Suppose that at time $n = 1$, there are 1100 mature cells and 400 immature cells in the culture. That is, $a_1 = 1100$ and $b_1 = 400$.

From Equation (2), we can compute

$$400 = b_1 = 0.4a_0$$

so

$$a_0 = \frac{400}{0.4} = 1000$$

From the initial conditions $a_0 = 1000$ and $a_1 = 1100$ and Equation (4), the number of mature cells present after n minutes is

n	a_n
0	1000
1	1100
2	1040
3	1076
4	1054.4
5	1067.36
6	1059.584
7	1064.249
8	1061.450
9	1063.129
10	1062.166
11	1062.726
12	1062.363

The numbers of cells are graphed in Figure 9.2. Of course, the fractions are not significant, since a_n is a number of cells and is therefore an integer. But then, the

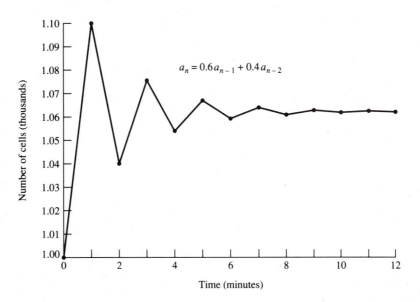

FIGURE 9.2 The sequence of Example 9.5

formula is only an approximation to the actual behavior of the culture. Within ten minutes, the culture has reached a stable state in which birth of new cells is matched by death of old ones.   ○

EXAMPLE 9.6   Suppose that in Example 9.5, only 15% of the mature cells divide, instead of 20%. Then 65% of the mature cells survive without dividing. The recurrence relation would be

$$a_n = 0.65a_{n-1} + 0.3a_{n-2}$$

With $a_0 = 1000$ and $a_1 = 1100$, the first few values of a_n are

n	a_n
0	1000
1	1100
2	980
3	1009
4	939.7
5	937.8
6	892.1
7	877.2
8	843.0
9	823.1
10	794.9

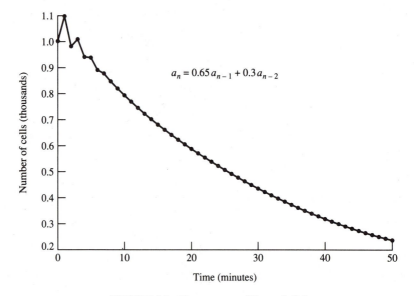

FIGURE 9.3   The sequence of Example 9.6

The values of a_n for larger values of n are graphed in Figure 9.3. The population dies out.   ○

Recurrence relations can be used to define functions of more than one integer variable. For instance, the binomial coefficient $C(n, k)$ is defined by the initial conditions

$$C(n, 0) = 1 \quad \text{and} \quad C(n, n) = 1$$

and the recurrence relation

$$C(n, k) = C(n - 1, k) + C(n - 1, k - 1)$$

This was proved in Theorem 6.4.

EXAMPLE 9.7   For $n \geq 0$ and $k \geq 0$ with $k \leq n$, let $D(n, k)$ be the number of k-element subsets of $\{1, 2, \ldots, n\}$ that do not contain any pair of consecutive numbers. Then $D(n, 0) = 1$; and $D(n, n) = 0$, except that $D(0, 0) = 1$ and $D(1, 1) = 1$. To derive a recurrence relation for $D(n, k)$, suppose that S is a k-element subset of $\{1, 2, \ldots, n\}$ that does not contain any pair of consecutive numbers. Then S either contains n or does not. In the first case, S cannot contain $n - 1$, so S consists of n and a $(k - 1)$-element subset of $\{1, 2, \ldots, n - 2\}$; there are $D(n - 2, k - 1)$ possibilities in this case. In the second case, S is a k-element subset of $\{1, 2, \ldots, n - 1\}$; there are $D(n - 1, k)$ possibilities for this. Thus, $D(n, k)$ is characterized by the initial conditions stated above and the recurrence relation

$$D(n, k) = D(n - 1, k) + D(n - 2, k - 1)$$

To compute the values of $D(n, k)$ for some small values of n and k, first write down the initial conditions in an array:

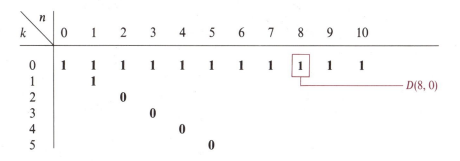

k \ n	0	1	2	3	4	5	6	7	8	9	10
0	1	1	1	1	1	1	1	1	1	1	1
1		1									
2			0								
3				0							
4					0						
5						0					

$D(8, 0)$

Then fill in the second row by computing

$$D(2, 1) = D(2 - 1, 1) + D(2 - 2, 1 - 1)$$
$$= D(1, 1) + D(0, 0) = 1 + 1 = 2$$
$$D(3, 1) = D(3 - 1, 1) + D(3 - 2, 1 - 1)$$
$$= D(2, 1) + D(1, 0) = 2 + 1 = 3$$

and so on. After filling in each row, we can compute the numbers in the next row. The result is

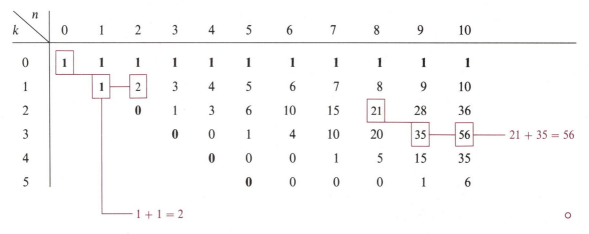

k \ n	0	1	2	3	4	5	6	7	8	9	10	
0	1	1	1	1	1	1	1	1	1	1	1	
1		1	2	3	4	5	6	7	8	9	10	
2			0	1	3	6	10	15	21	28	36	
3				0	0	1	4	10	20	35	56	21 + 35 = 56
4					0	0	0	1	5	15	35	
5						0	0	0	0	1	6	

$1 + 1 = 2$

○

EXERCISES 9.1

1. Find the values s_2 through s_7 in each of the following cases:

   a. $s_n = 2s_{n-1} + 5$ and $s_1 = 1$
   b. $s_n = 3s_{n-1} - 2$ and $s_1 = 3$

   c. $s_n = \dfrac{s_{n-1}^2}{2}$ and $s_1 = 2$
   d. $s_n = 2s_{n-1} + n$ and $s_1 = 2$

2. Find the values s_2 through s_7 in each of the following cases:

   a. $s_n = 2s_{n-1} + 3s_{n-2}$ and $s_0 = 1$, $s_1 = 1$
   b. $s_n = 3s_{n-1} - s_{n-2}$ and $s_0 = 2$, $s_1 = -1$

   c. $s_n = \dfrac{s_{n-1}}{2} + ns_{n-2}$ and $s_0 = 0$, $s_1 = 1$

3. How many ways s_n can the number n be written as a sum of 1's and 2's? Determine s_n for $n = 1, 2, 3$, and 4, and find a recurrence relation for s_n. (Note that order is significant.)

4. How many n-digit sequences of 1's and 0's do *not* contain the sequence 00? Find a recurrence relation for the answer. (*Hint:* If a sequence does not contain 00, it is either of the form $1x$ or $01x$, where x is a sequence of 0's and 1's that does not contain 00.)

5. How many n-digit sequences of 1's and 0's do not contain the sequence 000?

6. A bank is paying interest at 6% compounded quarterly. A man deposits $100, then makes an additional deposit each quarter, increasing his deposit by $10 each time. Find a recurrence relation for his account balance at the end of n quarters. How much does he have in the account after three years?

7. You owe $20,000 on your home mortgage. The annual interest rate is 12%, and you make monthly payments of $220. Write a recurrence relation for the amount still owed after n months. How much do you owe after six months?

8. One line divides the plane into two regions, two lines (if they are not parallel) divide the plane into 4 regions, and three lines (if no two are parallel and they do not all intersect in a point) divide the plane into seven regions (see Figure 9.4). Into how many regions do n lines divide the plane if no two are parallel and no three intersect in a point? Find a recurrence relation for the answer. (*Hint:* When the nth line is added, how many regions are split in two?)

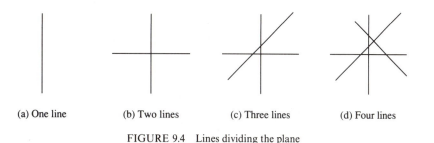

(a) One line   (b) Two lines   (c) Three lines   (d) Four lines

FIGURE 9.4   Lines dividing the plane

9. A retirement annuity of $100,000 per annum is earning 6% interest and is issuing monthly checks of $1000. Find a recurrence relation for the amount remaining after n months.

10. In Example 9.5, suppose that only 15% of the mature cells die at each one-minute interval. Determine the population of mature cells at one-minute intervals up to 10 minutes. How is the population behaving?

11. A **derangement** of a finite set X is a permutation f of X such that $f(x) \neq x$ for all $x \in X$. Find initial conditions and a recurrence relation for the number D_n of derangements of a set of n elements. (*Hint:* Assume $X = \{1, 2, \ldots, n\}$. There are $n - 1$ possibilities for $f(n)$, of which $f(n) = 1$ is typical. If $f(n) = 1$, consider two cases: $f(1) = n$ and $f(1) \neq n$.)

12. Show that $D_n = n!\left(1 - \dfrac{1}{1!} + \dfrac{1}{2!} - \dfrac{1}{3!} + \cdots + \dfrac{(-1)^n}{n!}\right)$. (See Exercise 11.) (*Hint:* Use mathematical induction.)

13. Use mathematical induction to verify the formula for b_n in Example 9.2.

14. Use mathematical induction to verify the formula for c_n in Example 9.3.

15. In Example 9.5, suppose that the fraction of mature cells that die is α and the fraction of mature cells that divide is β, where $0 \leq \alpha \leq 1$ and $0 \leq \beta \leq 1$. Find recurrence relations for

   a. a_n, the number of mature cells at time n

   b. b_n, the number of immature cells at time n

   c. $s_n = a_n + b_n$, the total number of cells at time n

16. One method for sorting a list of n numbers is to find the smallest of the n numbers and exchange it with the first number, then find the smallest of the remaining $n - 1$ numbers and exchange it with the second number, and so on until the last number is reached. This method is called **selection sort.** If c_n is the

number of comparisons needed to sort a list of n numbers using this method, find a recurrence relation for c_n.

17.   You fill in n graduate school applications and address the n corresponding envelopes. Show that if you place the applications randomly in the envelopes, the probability that each application will be placed in the wrong envelope is

$$\sum_{k=2}^{n} \frac{(-1)^k}{k!}$$

9.2 • FIRST-ORDER LINEAR RELATIONS

By solving a recurrence relation for given initial conditions, we mean finding a formula for the nth term of the sequence that involves only n and the initial conditions, and not the previous term. There is no general method for finding the solution of a recurrence relation; however, there are methods for solving certain classes of simple relations. To identify these classes, we need some terminology.

Definition 9.1

A recurrence relation of the form

$$s_n = f(s_{n-1}, \ldots, s_{n-k})$$

is said to be of **order** k. If f is a linear function of the arguments $s_{n-1}, \ldots, s_{n-k}$—that is, if

$$f(s_{n-1}, \ldots, s_{n-k}) = g_1(n)s_{n-1} + \cdots + g_k(n)s_{n-k} + h(n)$$

where g_i and h are functions of n—then the recurrence relation is said to be **linear.** If the relation is linear and $h(n) = 0$ for all n, the relation is said to be **homogeneous.** If the coefficient functions g_i are constant, the recurrence relation is said to have **constant coefficients.**

EXAMPLE 9.8   The recurrence relation

$$s_n = s_{n-1} + ns_{n-2} + n^2$$

is linear and of order 2, but it is not homogeneous and does not have constant coefficients. In this case,

$$g_1(n) = 1$$
$$g_2(n) = n$$
$$h(n) = n^2$$

○

EXAMPLE 9.9   The recurrence relation

$$s_n = 5s_{n-1} - 7s_{n-2} + 3s_{n-3} + n^2 - n$$

is linear of order 3 with constant coefficients, but it is not homogeneous.

○

EXAMPLE 9.10   The recurrence relation

$$s_n = -7s_{n-2} + 5s_{n-3}$$

is a third-order linear homogeneous relation with constant coefficients.   ○

EXAMPLE 9.11   The recurrence relation

$$s_n = s_{n-1}s_{n-2} + s_{n-3}$$

is of order 3 and is not linear.   ○

The Case of Constant Coefficients

Now consider the first-order linear recurrence relations with constant coefficients. These have the form

$$s_n = as_{n-1} + h(n)$$

where a is a constant. Consider the case in which $h(n)$ is also a constant b. Then the recurrence relation has the form

$$s_n = as_{n-1} + b$$

Starting with s_0, we can write out the first few terms of the sequence:

$$
\begin{aligned}
s_0 &= s_0 \\
s_1 &= as_0 + b \\
s_2 &= as_1 + b \\
    &= a(as_0 + b) + b \\
    &= a^2 s_0 + ab + b \\
    &= a^2 s_0 + b(a + 1) \\
s_3 &= a^3 s_0 + a^2 b + ab + b \\
    &= a^3 s_0 + b(a^2 + a + 1) \\
s_4 &= a^4 s_0 + a^3 b + a^2 b + ab + b \\
    &= a^4 s_0 + b(a^3 + a^2 + a + 1)
\end{aligned}
$$

A simple pattern is apparent:

$$s_n = a^n s_0 + b \sum_{i=0}^{n-1} a^i$$

Of course, this formula can be verified by mathematical induction, and this is left as an exercise.

It was shown in Example 2.20 that

$$\sum_{i=0}^{n-1} a^i = \frac{a^n - 1}{a - 1}, \qquad \text{for } a \neq 1$$

By using this result in the formula for s_n, we obtain a complete solution of the recurrence relation $s_n = as_{n-1} + b$.

Theorem 9.1   The general solution of the recurrence relation

$$s_n = as_{n-1} + b$$

is as follows.

Case 1: $a = 1$.

$$s_n = s_0 + nb$$

Case 2: $a \neq 1$.

$$s_n = a^n \left[s_0 + \frac{b}{a-1} \right] - \frac{b}{a-1}$$

Proof   We saw earlier that

$$s_n = a^n s_0 + b \sum_{i=0}^{n-1} a^i$$

If $a = 1$, this is just $s_0 + nb$. If $a \neq 1$, we can use the formula for the sum of a finite geometric series and obtain

$$s_n = a^n s_0 + b \frac{a^n - 1}{a - 1}$$

The result follows from combining the two terms containing a^n.   ●

EXAMPLE 9.12   Solve the recurrence relation

$$s_n = s_{n-1} + 3$$

with initial condition $s_0 = 5$. In this case, $a = 1$ and $b = 3$. The solution is

$$s_n = 5 + 3n$$

To verify that $5 + 3n$ is indeed the solution to the recurrence relation $s_n = s_{n-1} + 3$, observe that

$$\begin{aligned} s_n &= s_{n-1} + 3 \\ &= 5 + 3(n - 1) + 3 \\ &= 5 + 3n \end{aligned}$$ ○

EXAMPLE 9.13   Solve the recurrence relation of Example 9.4,

$$d_n = 1.0105d_{n-1} + 100$$

with initial condition $d_0 = 100$.

In this case, $a = 1.0105$ and $b = 100$. The solution is

$$s_n = (1.0105)^n \left[100 + \frac{100}{0.0105} \right] - \frac{100}{0.0105}$$

$$= 9623.81(1.0105)^n - 9523.81.$$ ○

When using Theorem 9.1, it is good practice to check the accuracy of your work by comparing the values of the formula with the first few terms of the sequence as computed directly from the recurrence relation. In Example 9.13, the comparison is

n	d_n from Example 9.4	$9623.81(1.0105)^n - 9523.81$
0	(100)	(100)
1	201.05	201.05
2	303.16	303.16
3	406.34	406.34

The 100 in parentheses is the value of s_0 and is not computed. The remaining numbers in the second column are computed using the recurrence relation, and those in the third column are computed from the formula.

EXAMPLE 9.14   Solve the recurrence relation

$$s_n = 2s_{n-1} - 1$$

with initial condition $s_0 = 1$.

In this case, $a = 2$ and $b = -1$. The solution is

$$s_n = 2^n \left[1 + \frac{-1}{2-1} \right] - \frac{-1}{2-1} = 1$$

○

The solution in Example 9.14 is the constant sequence $1, 1, 1, \ldots$. We also say that 1 is a *fixed point* of the recurrence relation $s_n = 2s_{n-1} - 1$. In general, the number x is a **fixed point** of a recurrence relation if the sequence $x, x, x, \ldots$ is a solution of the relation. If a recurrence relation describes a dynamical system, a fixed point of the relation corresponds to a stable state of the system (that is, a state that does not change over time).

Some interesting observations can be made about the recurrence relation of Theorem 9.1 in the case when $a \neq 1$. First, the number $\frac{-b}{a-1}$ is a fixed point of the recurrence relation, because

$$a\left[\frac{-b}{a-1} \right] + b = \frac{-b}{a-1}$$

Second, every solution of the recurrence relation is of the form

$$Aa^n + \frac{-b}{a-1}$$

where $A = s_0 + \frac{b}{a-1}$. Moreover, terms of the form Aa^n are the solutions of the homogeneous recurrence relation $s_n = as_{n-1}$, obtained from $s_n = as_{n-1} + b$ by dropping the term b.

These circumstances suggest that in solving linear homogeneous recurrence relations of higher order, we should look for solutions that are exponentials. They also suggest that for nonhomogeneous relations, we should look for solutions that are a sum of two parts, one of which is a solution of a homogeneous relation. We will follow up on these ideas in the sections that follow.

EXERCISES 9.2

1. Determine the order of each of the following recurrence relations. Classify each relation as linear or nonlinear. For each linear relation, classify it as homogeneous or nonhomogeneous, and as having constant coefficients or not having constant coefficients.

   a. $s_n = \dfrac{s_{n-1}}{s_{n-2}}$

   b. $s_n = 3s_{n-2} - 4s_{n-3}$

   c. $s_n = -1.6s_{n-1} + \dfrac{4.2}{s_{n-2}}$

2. Do the same for the recurrence relations

   a. $s_n = 6s_{n-3} + \dfrac{2}{n} - 5s_{n-1}$

   b. $s_n = 2ns_{n-1} + 3n^2 s_{n-2} - \dfrac{n^2 - n}{2}$

   c. $s_n = 8s_{n-1} + 5s_{n-2} - 7s_{n-3} - 18s_{n-4}$

3. Solve each of the following recurrence relations for the given initial conditions.

   a. $s_n = 3s_{n-1} - 2;\ s_0 = -2$

   b. $s_n = s_{n-1} + 2;\ s_0 = 5$

   c. $s_n = 0.5s_{n-1} + 2;\ s_0 = 1$

   d. $s_n = -0.1s_{n-1};\ s_0 = 10$

4. Solve each of the following recurrence relations for the given initial conditions.

   a. $s_n = 0.2s_{n-1} + 0.4;\ s_0 = -0.25$

   b. $s_n = -s_{n-1} + 1;\ s_0 = 0$

   c. $s_n = -s_{n-1} + 3;\ s_0 = 1$

5. Solve the recurrence relation of Exercise 6 of Section 9.1.

6. Solve the recurrence relation of Exercise 8 of Section 9.1.

7. If $a = 1$, when does the recurrence relation $s_n = as_{n-1} + b$ have a fixed point? What is it?

8. Does the recurrence relation $s_n = 3s_{n-1} + 5s_{n-2} - 13s_{n-3} + 2$ have a fixed point? If so, what is it?

9. Does the recurrence relation $s_n = 3 - \dfrac{1}{s_{n-1}}$ have any fixed points? If so, what are they?

10. Verify by mathematical induction that the solution of the recurrence relation $s_n = as_{n-1} + b$ is

$$s_n = a^n s_0 + b \sum_{i=0}^{n-1} a^i$$

11. Solve the recurrence relation $s_n = 2s_{n-2} - \tfrac{3}{4}$ for the initial conditions $s_0 = 2$, $s_1 = \tfrac{1}{2}$.

9.3 • LINEAR HOMOGENEOUS RELATIONS

The general form of a linear homogeneous recurrence relation with constant coefficients is

$$s_n = a_1 s_{n-1} + a_2 s_{n-2} + \cdots + a_k s_{n-k}$$

$$= \sum_{i=1}^{k} a_i s_{n-i}$$

If $a_k \neq 0$, this relation is of order k. From our experience with the first-order case, we may guess that the relation has solutions of the form $s_n = Ar^n$, for some suitable choices of A and r. A few examples will illustrate the kinds of situations that can occur. We will not concern ourselves with initial conditions at first, but we will return to them later.

EXAMPLE 9.15   Consider the second-order relation

$$s_n = 5s_{n-1} - 6s_{n-2}$$

Suppose that there is a solution of the form

$$s_n = Ar^n$$

Then by substituting this expression in the recurrence relation, we obtain

$$Ar^n = 5Ar^{n-1} - 6Ar^{n-2}$$

Dividing each term of this equation by Ar^{n-2} leaves

$$r^2 = 5r - 6$$

or

$$r^2 - 5r + 6 = 0$$

The solutions of this quadratic equation are $r = 2$ and $r = 3$. We can conclude that $s_n = Ar^n$ is a solution of the recurrence relation $s_n = 5s_{n-1} - 6s_{n-2}$ whenever $r = 2$ or $r = 3$. Note that A can have any value whatsoever! The solutions we have found include all sequences of the form $A(2)^n$ and $A(3)^n$, but these are not all of the solutions. It is easy to check that any sequence of the form

$$s_n = A(2)^n + B(3)^n$$

is also a solution and this is left as an exercise.   ○

EXAMPLE 9.16   Solve the recurrence relation of Example 9.15 if the initial conditions are $s_0 = 3$ and $s_1 = 5$.

If one of the solutions found in Example 9.15 satisfies these conditions, then the initial conditions require that

$$s_0 = A(2)^0 + B(3)^0 = 3$$
$$s_1 = A(2)^1 + B(3)^1 = 5$$

That is,

$$A + B = 3$$

and

$$2A + 3B = 5$$

The solution of this system of two equations in two unknowns is $A = 4, B = -1$. So

$$s_n = 4(2)^n + (-1)3^n = 4(2)^n - 3^n$$

is a solution of $s_n = 5s_{n-1} - 6s_{n-2}$ with initial conditions $s_0 = 3$ and $s_1 = 5$.   ○

In Example 9.15, we saw that from the two solutions $s_n = 2^n$ and $s_n = 3^n$, we can build new solutions by multiplying these solutions by constants and adding the results. This is true of linear homogeneous equations generally, even when the coefficients are not constant.

Theorem 9.2   If the sequences u_n and v_n are solutions of the linear homogeneous recurrence relation

$$s_n = g_1(n)s_{n-1} + \cdots + g_k(n)s_{n-k}$$

then the sequence

$$t_n = Au_n + Bv_n$$

is also a solution for any numbers A and B.

Proof   To say that u_n and v_n are solutions of the recurrence relation means that

$$u_n = g_1(n)u_{n-1} + \cdots + g_k(n)u_{n-k}$$

and

$$v_n = g_1(n)v_{n-1} + \cdots + g_k(n)v_{n-k}$$

Multiply these equations by A and B, respectively, and add the results. The sum is

$$Au_n + Bv_n = g_1(n)(Au_{n-1} + Bv_{n-1}) + \cdots + g_k(n)(Au_{n-k} + Bv_{n-k})$$

which means that

$$t_n = g_1(n)t_{n-1} + \cdots + g_k(n)t_{n-k}$$ ●

The sequence t_n is said to be a **linear combination** of the sequences u_n and v_n.

Characteristic Polynomials

Now let us return to the case of constant coefficients.

Definition 9.2

If

$$s_n = a_1 s_{n-1} + a_2 s_{n-2} + \cdots + a_k s_{n-k}$$

is a linear homogeneous recurrence relation with constant coefficients, then the polynomial

$$x^k - a_1 x^{k-1} - a_2 x^{k-2} - \cdots - a_k$$

is called the **characteristic polynomial** of the relation.

EXAMPLE 9.17   Here are some examples of linear homogeneous recurrence relations with constant coefficients, and their characteristic polynomials.

Recurrence Relation	Characteristic Polynomial
$s_n = 5s_{n-1}$	$x - 5$
$s_n = 3s_{n-1} - 7s_{n-2}$	$x^2 - 3x + 7$
$s_n = -4s_{n-1} + 3s_{n-2} + 2s_{n-3}$	$x^3 + 4x^2 - 3x - 2$

Example 9.15 suggests the following theorem.

Theorem 9.3   If r is a root of the characteristic polynomial of a linear homogeneous recurrence relation with constant coefficients, then $s_n = r^n$ is a solution of the relation.

Proof   Suppose that the recurrence relation is

$$s_n = a_1 s_{n-1} + a_2 s_{n-2} + \cdots + a_k s_{n-k}$$

If r is a root of the characteristic polynomial, then

$$r^k - a_1 r^{k-1} - a_2 r^{k-2} - \cdots - a_k = 0$$

This can be rewritten as

$$r^k = a_1 r^{k-1} + a_2 r^{k-2} + \cdots + a_k$$

Multiply both sides of this equation by r^{n-k}. The result is

$$r^n = a_1 r^{n-1} + a_2 r^{n-2} + \cdots + a_k r^{n-k}$$

which shows that $s_n = r^n$ satisfies the relation.   ●

Theorems 9.2 and 9.3 together give us a method for solving linear homogeneous recurrence relations with constant coefficients. The next three examples illustrate this method.

EXAMPLE 9.18   Find a solution of the recurrence relation

$$s_n = \frac{3}{2}s_{n-1} - \frac{1}{2}s_{n-2}$$

satisfying the initial conditions $s_0 = 3$, $s_1 = 2$.

The characteristic polynomial is

$$x^2 - \frac{3}{2}x + \frac{1}{2}$$

which has roots $x = 1$ and $x = \frac{1}{2}$. So by Theorem 9.3, the sequences $s_n = 1^n$ and $s_n = (\frac{1}{2})^n$ are solutions. By Theorem 9.2, there are more solutions of the form $s_n = A(1)^n + B(\frac{1}{2})^n$. From the initial conditions, we get

$$s_0 = A(1)^0 + B\left(\frac{1}{2}\right)^0 = 3$$

$$s_1 = A(1)^1 + B\left(\frac{1}{2}\right)^1 = 2$$

That is,

$$A + B = 3$$

$$A + \frac{1}{2}B = 2$$

The solution of this system is $A = 1$, $B = 2$, so the solution of the recurrence relation satisfying $s_0 = 3$ and $s_1 = 2$ is

$$s_n = 1(1)^n + 2\left(\frac{1}{2}\right)^n$$

$$= 1 + \left(\frac{1}{2}\right)^{n-1}$$

○

EXAMPLE 9.19   Find a formula for the nth Fibonacci number F_n defined in Example 9.1.

We have to solve the recurrence relation $F_n = F_{n-1} + F_{n-2}$ with initial conditions $F_0 = 0$ and $F_1 = 1$. The characteristic polynomial is $x^2 - x - 1$, which has roots

$$\frac{1 + \sqrt{5}}{2} \quad \text{and} \quad \frac{1 - \sqrt{5}}{2}$$

The solution is of the form

$$F_n = A\left(\frac{1 + \sqrt{5}}{2}\right)^n + B\left(\frac{1 - \sqrt{5}}{2}\right)^n$$

for suitable constants A and B. From the initial conditions, we get the equations

$$F_0 = A + B = 0$$

$$F_1 = A\frac{1 + \sqrt{5}}{2} + B\frac{1 - \sqrt{5}}{2} = 1$$

The solution of this system is $A = \dfrac{1}{\sqrt{5}}$, $B = \dfrac{-1}{\sqrt{5}}$. The final result is

$$F_n = \frac{1}{\sqrt{5}}\left[\left(\frac{1 + \sqrt{5}}{2}\right)^n - \left(\frac{1 - \sqrt{5}}{2}\right)^n\right]$$

○

It is worth noting that although the irrational number $\sqrt{5}$ appears in this expression, it always cancels out when the expression is evaluated. The same thing happens when the roots of the characteristic polynomial are complex numbers. Although $i = \sqrt{-1}$ may appear in the formula for the nth term of a sequence, it always cancels out when the formula is evaluated.

EXAMPLE 9.20   Find some solutions of the recurrence relation

$$s_n = -s_{n-2}$$

The characteristic polynomial of this relation is

$$x^2 + 1$$

which has roots i and $-i$. All sequences of the form

$$s_n = Ai^n + B(-i)^n$$

are solutions. If the initial conditions are $s_0 = a$ and $s_1 = b$, where a and b are real, then using the technique of Example 9.18, you can find that $A = \frac{a-bi}{2}$ and $B = \frac{a+bi}{2}$. Therefore,

$$s_n = \frac{a - bi}{2}i^n + \frac{a + bi}{2}(-i)^n$$

To see that i always cancels, you can evaluate this expression for $n = 0, 1, 2,$ and 3. Details are left as an exercise.   ○

We now have a strategy for finding solutions of linear homogeneous recurrence relations with constant coefficients: Find the roots of the characteristic polynomial, write down a linear combination of the sequences corresponding to each of the roots, write down an equation for each initial condition, and solve the resulting equations. However, as the following example shows, this strategy does not always work.

EXAMPLE 9.21   Find a solution of the recurrence relation

$$s_n = 4s_{n-1} - 4s_{n-2}$$

satisfying the initial conditions $s_0 = 0$, $s_1 = 2$.

The characteristic polynomial is

$$x^2 - 4x + 4$$

which has only one root, namely $x = 2$. So sequences of the form $s_n = A(2)^n$ are solutions. If one of these solutions satisfies the initial conditions $s_0 = 0$, $s_1 = 2$, we would have

$$s_0 = A(2)^0 = 0$$
$$s_1 = A(2)^1 = 2$$

The first equation requires that $A = 0$, and the second requires that $A = 1$. Both cannot be true at the same time, so the solution of the problem, with the given initial conditions, is not of the form $s_n = A(2)^n$.   o

Although every sequence of the form $s_n = A(2)^n$ is a solution of the recurrence relation $s_n = 4s_{n-1} - 4s_{n-2}$, no sequence of this form satisfies the initial conditions $s_0 = 0$, $s_1 = 2$. This leads us to look for solutions of other forms.

Other Forms of Solutions

In Example 9.21, we can write down the first few terms of the sequence by computing them directly from the recurrence relation, and compare them to the sequence 2^n:

n	$s_n = 4s_{n-1} - 4s_{n-2}$	2^n	$\dfrac{s_n}{2^n}$
0	(0)	1	0
1	(2)	2	1
2	8	4	2
3	24	8	3
4	64	16	4
5	160	32	5

The numbers in parentheses are the initial conditions. The ratio $\dfrac{s_n}{2^n}$ is not a constant A; rather, it is equal to n. The solution of $s_n = 4s_{n-1} - 4s_{n-2}$ satisfying $s_0 = 0$ and $s_1 = 2$ is $s_n = n(2)^n$. In fact, from Theorem 9.2, it follows that any sequence of the form

$$s_n = A(2)^n + Bn(2)^n$$

is a solution of the recurrence relation.

EXAMPLE 9.22   Find a solution of $s_n = 4s_{n-1} - 4s_{n-2}$ satisfying the initial conditions $s_0 = 2$, $s_1 = -2$.

We just saw that any sequence of the form $s_n = A(2)^n + Bn(2)^n$ is a solution. From the initial conditions, we get the equations

$$s_0 = A(2)^0 + B(0)(2)^0 = 2$$
$$s_1 = A(2)^1 + B(1)(2)^1 = -2$$

That is,

$$A \qquad = 2$$
$$2A + 2B = -2$$

The solution of this system of equations is $A = 2$, $B = -3$, so the solution of the recurrence relation is

$$s_n = 2(2)^n - 3n(2)^n = (2 - 3n)2^n \qquad \qquad \circ$$

We have seen that the problem of finding solutions to linear homogeneous recurrence relations with constant coefficients is reduced to the problem of finding roots of polynomials. The basic result in the theory of polynomial equations is the Fundamental Theorem of Algebra, stated in Chapter 1. The theorem states that every polynomial

$$f(x) = x^n - a_1 x^{n-1} - a_2 x^{n-2} - \cdots - a_n$$

can be factored into linear factors

$$f(x) = \prod_{j=1}^{n} (x - r_j)$$

where $r_1, \ldots, r_n$ are the roots of the polynomial $f(x)$. The roots of $f(x)$ need not be distinct. If we collect factors containing the same root, $f(x)$ can be rewritten in the form

$$f(x) = (x - r_1)^{m_1}(x - r_2)^{m_2} \ldots (x - r_k)^{m_k}$$

where $r_1, \ldots, r_k$ are the distinct roots of $f(x)$. The number m_i is called the **multiplicity** of the root r_i. For example, the polynomial

$$g(x) = x^5 - 12x^4 - 57x^3 - 134x^2 + 156x - 72$$

has only two roots, 2 and 3. It can be written as a product of linear factors in the form

$$g(x) = (x - 3)^2(x - 2)^3$$

The root 2 has multiplicity 3, and the root 3 has multiplicity 2.

We can now state the general solution to the problem of linear homogeneous recurrence relations with constant coefficients. However, the proof is beyond the scope of this book.

Theorem 9.4 Let

$$s_n = a_1 s_{n-1} + a_2 s_{n-2} + \cdots + a_k s_{n-k}$$

be a linear homogeneous recurrence relation with constant coefficients, and let $r_1, \ldots, r_j$ be the distinct roots of its characteristic polynomial, having multiplicities

$m_1, \ldots, m_j$, respectively. Then for each $i = 1, \ldots, j$, each of the sequences

$$r_i^n, nr_i^n, n^2r_i^n, \ldots, n^{m_i-1}r_i^n$$

is a solution of the recurrence relation. Every solution of the recurrence relation is a linear combination of these sequences.   ●

EXAMPLE 9.23   Find all solutions of the recurrence relation

$$s_n = 12s_{n-1} + 57s_{n-2} + 134s_{n-3} - 156s_{n-4} + 72s_{n-5}$$

The characteristic polynomial is

$$x^5 - 12x^4 - 57x^3 - 134x^2 + 156x - 72$$

which, as we have seen, has roots 3 and 2, of multiplicities 2 and 3, respectively. The general solution is

$$s_n = A(3)^n + Bn(3)^n + C(2)^n + Dn(2)^n + En^2(2)^n$$ ○

EXERCISES 9.3

1. Find the characteristic polynomials for each of the following recurrence relations.

   a. $s_n = -15s_{n-1}$ **b.** $s_n = -6s_{n-1} + 5s_{n-2}$

   c. $s_n = s_{n-1} + 7s_{n-2} - 13s_{n-3}$

2. Find the fourth number in the Fibonacci Sequence using the formula of Example 9.19.

3. Show that the sequence $s_n = A(2)^n + B(3)^n$ is a solution of the recurrence relation of Example 9.15.

4. Solve the recurrence relation $s_n = \frac{1}{4}s_{n-1} + \frac{1}{8}s_{n-2}$ for the initial conditions $s_0 = 2, s_1 = 1.5$.

5. Solve the recurrence relation $s_n = \frac{s_{n-2}}{4}$ for the initial conditions $s_0 = 2, s_1 = 3$.

6. Solve the recurrence relation of Example 9.5 for the initial conditions given in the example.

7. Solve the recurrence relation of Example 9.6 for the initial conditions given in the example.

8. Solve the recurrence relation $s_n = -s_{n-1} - \frac{s_{n-2}}{4}$ for the initial conditions $s_0 = 4, s_1 = 4$.

9. Solve the recurrence relation $s_n = 2s_{n-1} - 2s_{n-2}$ for the initial conditions $s_o = 2, s_1 = 1$.

10. Solve the recurrence relation $s_n = 6s_{n-1} - 11s_{n-2} + 6s_{n-3}$ for the initial conditions $s_0 = 1, s_1 = 1, s_2 = 0$. (*Hint:* 1 is a root.)

11. Solve the recurrence relation $s_n = 3s_{n-1} - 4s_{n-3}$ for the initial conditions $s_0 = 0, s_1 = 1, s_2 = 2$. (*Hint:* 2 and -1 are roots.)

12. In Example 9.20, verify the formulas given for A and B, and use the expression for s_n derived in the example to compute the first six terms of the sequence.

13. Solve the recurrence relation $s_n = -(s_{n-1} + s_{n-2})$ for three different sets of initial conditions of your choice,

   a. Directly using the recurrence relation, and

   b. Using the method of this section.

   Explain the relationship between the results.

9.4 • NONHOMOGENEOUS SYSTEMS

A nonhomogeneous linear recurrence relation is a recurrence relation of the form

$$s_n = g_1(n)s_{n-1} + \cdots + g_k(n)s_{n-k} + h(n)$$

in which the term $h(n)$ is not the constant zero. In Section 9.2, we solved the problem for the case of a first-order relation with constant coefficients and constant h term (that is, for a relation of the form $s_n = as_{n-1} + b$). The general solution, for $a \neq 1$, was written as follows:

$$s_n = a^n\left[s_0 + \frac{b}{a-1}\right] - \frac{b}{a-1}$$

We noticed that the first term is a solution of the homogeneous relation $t_n = at_{n-1}$. In fact, if we subtract any two solutions of $s_n = as_{n-1} + b$, the difference is a solution of $t_n = at_{n-1}$, because the constant term cancels. This situation holds in general, even if the coefficients are not constant, and provides the clue to solving nonhomogeneous systems.

Theorem 9.5 Suppose that the sequences u_n and v_n are solutions of the linear relation

$$s_n = g_1(n)s_{n-1} + \cdots + g_k(n)s_{n-k} + h(n)$$

Then the sequence $u_n - v_n$ is a solution of the associated homogeneous relation

$$t_n = g_1(n)t_{n-1} + \cdots + g_k(n)t_{n-k}$$

Proof Since u_n and v_n are solutions of

$$s_n = g_1(n)s_{n-1} + \cdots + g_k(n)s_{n-k} + h(n)$$

we have

$$u_n = g_1(n)u_{n-1} + \cdots + g_k(n)u_{n-k} + h(n)$$

and

$$v_n = g_1(n)v_{n-1} + \cdots + g_k(n)v_{n-k} + h(n)$$

By subtracting these two equations, we get

$$(u_n - v_n) = g_1(n)(u_{n-1} - v_{n-1}) + \cdots + g_k(n)(u_{n-k} - v_{n-k})$$

which shows that $u_n - v_n$ is a solution of the homogeneous relation

$$t_n = g_1(n)t_{n-1} + \cdots + g_k(n)t_{n-k} \qquad \bullet$$

Theorem 9.5 provides a method of finding a solution of a nonhomogeneous system: Find *any one* solution of the nonhomogeneous relation and add to it solutions of the associated homogeneous relation. The solution of the homogeneous relation can be chosen to make the sum fit any given initial conditions.

EXAMPLE 9.24   Find a solution of the recurrence relation

$$s_n = 2s_{n-1} + n$$

satisfying $s_0 = 0$.

In this case, $h(n) = n$. Since $h(n)$ is a polynomial of degree 1, let's look for a solution that is a polynomial of degree 1 (that is, $s_n = an + b$). If $s_n = an + b$ is a solution of $s_n = 2s_{n-1} + n$, we must have

$$
\begin{aligned}
an + b &= 2(a(n-1) + b) + n \\
&= 2an - 2a + 2b + n
\end{aligned}
$$

for all integers n. By collecting and combining terms, we can rewrite this equation as

$$2a - b = n(a + 1)$$

In order for this equation to hold for all integers n, it must hold for $n = 0$ and $n = 1$, so

$$2a - b = 0$$

and

$$2a - b = a + 1$$

from which we can see that $a = -1$ and $b = -2$. So

$$s_n = -n - 2 = -(n + 2)$$

is one solution of the recurrence relation $s_n = 2s_{n-1} + n$.
   It follows from Theorem 9.5 that all solutions of the relation are of the form

$$s_n = A(2)^n - (n + 2)$$

because $A(2)^n$ is a solution of the associated homogeneous relation $t_n = 2t_{n-1}$.
   To find A, use the initial condition $s_0 = 0$. Then

$$s_0 = A(2)^0 - (0 + 2) = 0$$

It follows that $A = 2$, so the solution satisfying $s_0 = 0$ is

$$s_n = 2(2)^n - (n + 2) \qquad \circ$$

EXAMPLE 9.25   Find a solution of the recurrence relation

$$s_n = 5s_{n-1} - 6s_{n-2} + 4$$

for the initial condition $s_1 = 2$ and $s_2 = -4$.

Since $h(n)$ is the constant 4, we guess that there is a solution that is a constant c. If there is, c must satisfy

$$c = 5c - 6c + 4$$

so $2c = 4$ and $c = 2$. The associated homogeneous relation

$$t_n = 5t_{n-1} - 6t_{n-2}$$

has characteristic polynomial $x^2 - 5x + 6$, which has the roots $x = 2$ and $x = 3$. So all solutions of the relation $s_n = 5s_{n-1} - 6s_{n-2} + 4$ are of the form

$$s_n = A(2)^n + B(3)^n + 2$$

From the initial conditions, we get two equations in A and B that can be solved simultaneously. The solutions are $A = 3$ and $B = -2$. The final result is the sequence

$$s_n = 3(2)^n - 2(2)^n + 2 \qquad \circ$$

In general, if the recurrence relation has constant coefficients and the term $h(n)$ is a polynomial in n of degree m, there is likely to be a solution that is also a polynomial in n of degree m. This strategy works most of the time, but not always. When it fails, it is necessary to try polynomials of higher degree.

EXAMPLE 9.26 Find a solution to the recurrence relation $s_n = s_{n-1} + n$ for the initial condition $s_1 = 1$.

Note that the solution is just a formula for the sum of the first n integers, since

$$s_1 = 1$$
$$s_2 = 1 + 2$$
$$s_3 = 1 + 2 + 3$$

etc.

Suppose that there is a solution of the form $s_n = an + b$. Then for every integer n, we would have

$$an + b = [a(n - 1) + b] + n$$

which simplifies to

$$0 = -a + n$$

This equation cannot hold for all n, no matter what value is chosen for the constant a, so there is no solution of the form $s_n = an + b$. The next form to try is a quadratic polynomial:

$$s_n = an^2 + bn + c$$

If there is a solution of this form, then

$$an^2 + bn + c = a(n - 1)^2 + b(n - 1) + c + n$$

This can be simplified to

$$(2a - 1)n = a - b$$

and this will hold for all n if

$$2a - 1 = 0$$

and

$$a - b = 0$$

The solution of this system of two equations is $a = \frac{1}{2}$, $b = \frac{1}{2}$. The polynomial $s_n = \frac{1}{2}n^2 + \frac{1}{2}n$ satisfies the recurrence relation and, in fact, satisfies the given initial condition. This formula for $1 + 2 + \cdots + n$ was proved in Example 2.16 by mathematical induction.   ○

EXERCISES 9.4

1. Solve the recurrence relation $s_n = \dfrac{s_{n-1} + n}{2}$ for the initial condition $s_0 = 1$.

2. Solve the recurrence relation $s_n = 1.01s_{n-1} - 0.05n$ for the initial condition $s_0 = 100$.

3. Solve the recurrence relation $s_n = s_{n-1} + n$ for the initial conditions
    a. $s_0 = 5$ b. $s_0 = -3$

4. Solve the recurrence relation $s_n = -s_{n-1} - \dfrac{s_{n-2}}{4} + 1$ for the initial conditions $s_0 = 4$, $s_1 = 4$.

5. Solve the recurrence relation $s_n = 2s_{n-1} - 2s_{n-2} - 2n$ for the initial conditions $s_0 = 2$, $s_1 = 1$.

6. Solve the recurrence relation $s_n = 3s_{n-1} - 4s_{n-3} + 2$ for the initial conditions $s_0 = 0$, $s_1 = 1$, $s_2 = 2$.

7. Under what circumstances does $s_n = as_{n-1} + bn$ fail to have a solution that is a polynomial in n of degree 1?

8. Under what circumstances does $s_n = as_{n-1} + bs_{n-2} + c$ fail to have a constant solution?

9. Under what circumstances does $s_n = as_{n-1} + bs_{n-2} + c$ fail to have either a constant solution or a solution that is a linear polynomial in n?

10. Use the methods of this section to find a formula for $s_n = \sum\limits_{i=1}^{n} i^2$. (*Hint:* $s_n = s_{n-1} + n^2$.)

9.5 • NUMERICAL EXPERIMENTATION

In the previous sections, we saw how to find explicit solutions for several simple types of recurrence relations. But not every recurrence relation can be solved explicitly. If the relation is not linear, or if its coefficients are not constant, or if the nonhomogeneous term $h(n)$ is not a polynomial, it is generally not possible to find an explicit solution. In such cases, it is still possible to gain some information about the behavior of the solution by computing a large number of terms and graphing the

	A	B	C	D	E	F
1						
2						
3						
4						
5						
6						
7						
8						
9						
10						
11						

FIGURE 9.5   A spreadsheet

values. If a pattern can be detected, it may be possible to guess a solution and then verify it by mathematical induction. Even when an explicit solution is available, a graph can provide insight into the behavior of the solution.

The class of computer programs called *spreadsheets* are particularly well adapted to experimenting with recurrence relations. A spreadsheet program manipulates an array of entries, called *cells,* which may contain numbers or rules for performing computations. The columns of a spreadsheet are called A, B, C, . . . , and the rows are numbered 1, 2, 3, ... Each cell in the spreadsheet is identified by its column and row—for instance A2 or B1. A typical spreadsheet layout is shown in Figure 9.5. Operations that can be performed on spreadsheets include putting a number in a cell, putting a formula in a cell, and copying a cell or a rectangular range of cells to another part of the spreadsheet.

In the following examples, we will see how spreadsheets can be used with several relations. (Of course, the same computations can be done by writing custom programs.) The spreadsheet commands are for Lotus 1-2-3, a product of Lotus Development Corporation. Commands for other products may differ from those given.

EXAMPLE 9.27   Use a spreadsheet to make a table of Fibonacci numbers.

We will put the numbers 0, 1, 2, . . . in column A and the corresponding Fibonacci numbers in column B. The steps are as follows:

Step 1: Put 0 in cell A1.

Step 2: Put the formula $+A1+1$ in cell A2. (The number 1 appears in A2, because this is the sum of 1 and the contents of A1.)

Step 3: Copy cell A2 to the range A3 through A11. (The numbers 2 through 10 appear in column A. When the formula $+A1+1$ is copied to cell A3, it is automatically changed to $+A2+1$, and so on. It is this automatic adjustment feature of spreadsheets that makes them useful for working with recurrence relations.)

Step 4: Put 0 and 1 in cells B1 and B2, respectively.

Step 5: Put $+B2+B1$ in cell B3. (The number 1 appears in cell B3.)

Step 6: Copy cell B3 to the range B4 through B11. (The Fibonacci numbers appear in column B.)

The final appearance of the spreadsheet is shown in Figure 9.6. Of course, any larger number could be used in place of 11 in order to carry the computation beyond 11 rows.

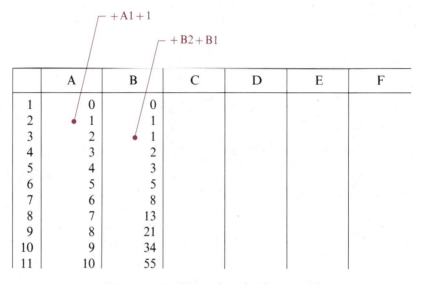

FIGURE 9.6   The Fibonacci numbers in a spreadsheet

EXAMPLE 9.28   Set up a spreadsheet to compute terms of the recurrence relation

$$s_n = as_{n-1} + bs_{n-2} + c$$

As in Example 9.27, we start by putting 0 in cell A1 and $+A1+1$ in cell A2, then copying cell A2 down the column. This fills in the A column with $0, 1, 2, \ldots$, which will be the value of n. Put s_0 and s_1 in cells B1 and B2. Let's store $a, b,$ and c in the top row beside s_0, so put $a, b,$ and c in cells C1, D1, and E1. Put the formula

$$+C\$1*B2+D\$1*B1+E\$1$$

in cell B3 and copy it down the B column. (The $ in C$1, D$1, and E$1 suppresses automatic adjustment of the cell address when the formula is copied.) The values of s_n appear in column B. Whenever any of the parameters $s_0, s_1, a, b,$ or c is changed, all the values of s_n are automatically recomputed. Carry out this procedure for the recurrence relation of Example 9.5,

$$s_n = 0.6s_{n-1} + 0.4s_{n-2}$$

in which $s_0 = 1000, s_1 = 1100, a = 0.6, b = 0.4, c = 0$. The result is shown in Figure 9.7.

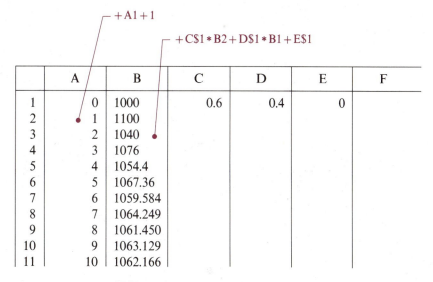

FIGURE 9.7   A spreadsheet for Example 9.28

EXAMPLE 9.29   Describe the behavior of $s_n = 1.75s_{n-1} + bs_{n-2}$ for values of b near -0.75, for the initial conditions $s_0 = 8$ and $s_1 = 6.5$.

We compute the values of s_n for $b = -0.740, -0.745, -0.75, -0.755$, and -0.760, and for n in the range 0 to 50. A spreadsheet to do the computations is shown in Figure 9.8. The graphs of the sequences are shown in Figure 9.9.

Note that the sequence grows rapidly for $b = -0.740$ and $b = -0.745$, approaches the constant 2 for $b = -0.750$, and drops rapidly to zero for $b = -0.755$ and $b = -0.760$. To what can we attribute this behavior? The characteristic polynomial of the recurrence relation is $x^2 - 1.75x + b$, and its roots are

b	r_1	r_2
-0.740	0.7149	1.0351
-0.745	0.7314	1.0186
-0.750	0.75	1.
-0.755	0.7719	0.9781
-0.760	0.8000	0.9500

The general solution of the relation is of the form $s_n = Ar_1^n + Br_2^n$. The powers of the smaller root r_1 drop rapidly to zero, so for large values of n, the behavior of the solution is determined by the larger root r_2. For $b > -0.75$, r_2 is greater than 1, and its powers grow large as n increases, whereas for $b < -0.75$, r_2 is less than 1, and its powers go to zero. This example illustrates that when one of the roots of the characteristic polynomial is close to 1, the behavior of the solution for large n can change radically in response to small changes in the coefficients.

+ B$1 * B4 + B$2 * B3

+ C$1 * C4 + C$2 * C3

	A	B	C	D	E	F
1	A:	1.75	1.75	1.75	1.75	1.75
2	B:	−0.74	−0.745	−0.75	−0.755	−0.76
3	0	8	8	8	8	8
4	1	6.5	6.5	6.5	6.5	6.5
5	2	5.455	5.415	5.375	5.335	5.295
6	3	4.73625	4.63375	4.53125	4.42875	4.32625
7	4	4.251737	4.074887	3.898437	3.722387	3.546737
8	5	3.935715	3.678909	3.423828	3.170471	2.918840
9	6	3.741216	3.402300	3.067871	2.737923	2.412450
10	7	3.634699	3.213237	2.800903	2.397659	2.003469
11	8	3.592223	3.088452	2.600677	2.128771	1.672609
12	9	3.596714	3.010929	2.450508	1.915117	1.404429
13	10	3.636003	2.968230	2.337881	1.744233	1.186568
...	...	...	...	...	...	...
43	40	9.682905	4.724413	2.000060	0.649011	0.086649
44	41	10.02256	4.812347	2.000045	0.634741	0.082170
45	42	10.37413	4.901920	2.000033	0.620793	0.077945
46	43	10.73803	4.993162	2.000025	0.607159	0.073954
47	44	11.11470	5.086102	2.000019	0.593829	0.070181
48	45	11.50458	5.180774	2.000014	0.580796	0.066612
49	46	11.90814	5.277208	2.000010	0.568052	0.063234
50	47	12.32585	5.375437	2.000008	0.555590	0.060034
51	48	12.75822	5.475495	2.000006	0.543403	0.057001
52	49	13.20576	5.577416	2.000004	0.531485	0.054127
53	50	13.66899	5.681234	2.000003	0.519830	0.051401

FIGURE 9.8   A spreadsheet for Example 9.29

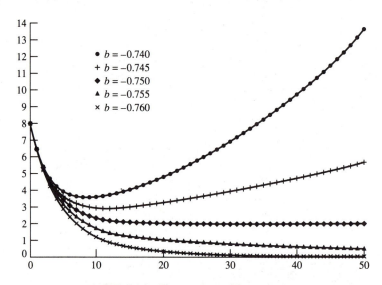

FIGURE 9.9   The sequences of Example 9.29

EXAMPLE 9.30   Describe the effect on the solution of $s_n = as_{n-1} - s_{n-2}$ of varying the initial conditions and of varying the coefficient a.

If we choose $a = 1.95$ and $s_0 = 1$ and let s_1 vary from 0.6 to 1.4, we get the sequences shown in Figure 9.10. The sequences lie on curves that are smooth waves, called *sine waves*. All of the waves have the same period (number of points from one peak to the next), which is about 28. Changing the initial conditions affects only the amplitude (height) of the waves and their phase (where the peaks occur). On the other hand, if we hold the initial conditions constant and vary the coefficient a between 1.94 and 1.96, we get the sequences shown in Figure 9.11. The period of

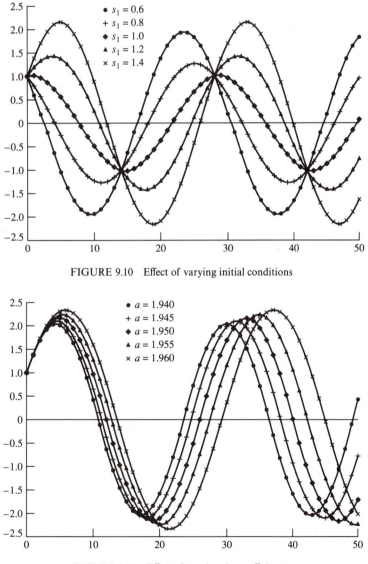

FIGURE 9.10   Effect of varying initial conditions

FIGURE 9.11   Effect of varying the coefficient a

the wave changes in response to changes in a. The amplitudes in Figure 9.11 also vary, but this variation could be eliminated by selecting different initial conditions.

To what can we attribute this behavior? The characteristic polynomial is $x^2 - ax + 1$, and its roots are

$$r_1 = \frac{a + \sqrt{a^2 - 4}}{2} \quad \text{and} \quad r_2 = \frac{a - \sqrt{a^2 - 4}}{2}$$

If $|a| < 2$, these roots are complex numbers of absolute value 1. The argument of the root r_1 and the period of the wave for the various values of the coefficient a are given in the following table:

a	$\text{Arg}(r_1)$ (Degrees)	Period
1.940	14.07	25.59
1.945	13.47	26.73
1.950	12.84	28.04
1.955	12.18	29.56
1.960	11.48	31.36

The argument of the other root, r_2, is the negative of the argument of r_1. The period is equal to 360 degrees divided by the argument of r_1.   ○

In general, when the characteristic polynomial of a relation has complex roots (which always come in pairs), the solution will exhibit oscillations whose periods are determined by the arguments of the complex roots. If the absolute value of a root is greater than 1, the magnitude of the corresponding oscillation grows, whereas if the absolute value is less than 1, the oscillation dies out. This phenomenon is illustrated in the following example.

EXAMPLE 9.31   Describe the behavior of $s_n = 1.5s_{n-1} - 0.85s_{n-2}$ for the initial conditions $s_0 = 1$ and $s_1 = 0.95$.

The result of computing and graphing 50 values is shown in Figure 9.12. The roots of the characteristic polynomial are $0.75 \pm 0.536i$, which have absolute value 0.922 and argument ± 35.6 degrees. The period is 10.12. The solution oscillates with this period while decaying to zero.   ○

Finally, let's look at a nonhomogeneous case.

EXAMPLE 9.32   Describe the behavior of $s_n = 1.0s_{n-1} - 0.85s_{n-2} + 0.5n^2$ for the initial conditions $s_0 = 1400$ and $s_1 = 1000$.

The result for 50 values of s_n is shown in Figure 9.13. The roots of the characteristic polynomial of the associated homogeneous relation $t_n = 1.0t_{n-1} - 0.85t_{n-2}$ are $0.5 \pm 0.7746i$, which have absolute value 0.9219 and argument ± 57.16 degrees. The period is 6.30. Any solution of the given recurrence relation is the sum of a

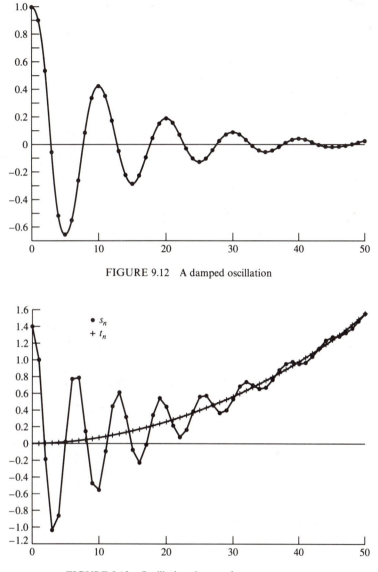

FIGURE 9.12   A damped oscillation

FIGURE 9.13   Oscillations in a nonhomogeneous case

polynomial in n and a solution of the homogeneous relation. The polynomial solution may be found by the methods of Section 9.4; it is

$$u_n = 0.588n^2 + 0.969n + 0.901$$

This is also graphed in Figure 9.13. The sequence s_n approaches t_n as n grows large. This behavior is typical of solutions of nonhomogeneous systems with constant coefficients, whenever all the roots of the characteristic polynomial have absolute value less than 1.   ○

EXERCISES 9.5

See the computer exercises at the end of this chapter.

9.6 • ADDITIONAL TOPICS

Another approach to recurrence relations involves the use of generating functions of sequences. A generating function is a sum of infinitely many terms. To see that such a sum can be meaningful in some situations, we consider the infinite geometric series.

Infinite Geometric Series

Recall that if $a \neq 1$, the sum of the finite geometric series

$$1 + a + a^2 + \cdots + a^{n-1}$$

is given by the formula

$$\sum_{i=0}^{n-1} a^i = \frac{a^n - 1}{a - 1}$$

If $|a| < 1$, the term a^n gets closer and closer to 0 as n grows large; consequently, the sum of a geometric series with n terms approaches $\frac{-1}{a-1} = \frac{1}{1-a}$ as n grows. In this case, it is meaningful to speak of the sum of an infinite geometric series, and we write

$$\sum_{i=0}^{\infty} a^i = \frac{1}{1-a}, \qquad |a| < 1$$

Return of the Fibonacci Sequence

Let us look at the Fibonacci Sequence 0, 1, 1, 2, 3, 5, ... from a different point of view. We begin by writing down the infinite sum

$$G(z) = F_0 + F_1 z + F_2 z^2 + F_3 z^3 + F_4 z^4 + F_5 z^5 + F_6 z^6 + \cdots$$
$$= 0 + z + z^2 + 2z^3 + 3z^4 + 5z^5 + 8z^6 + \cdots$$
$$= \sum_{i=0}^{\infty} F_i z^i$$

This is not a geometric series, so we have no reason to believe that this expression has meaning for any number z. But suppose that it does mean something. How can we find out what the meaning is? Let's look at the functions $zG(z)$ and $z^2G(z)$. We can see that

$$zG(z) = F_0 z + F_1 z^2 + F_2 z^3 + F_3 z^4 + F_4 z^5 + F_5 z^6 + F_6 z^7 + \cdots$$
$$z^2 G(z) = \qquad F_0 z^2 + F_1 z^3 + F_2 z^4 + F_3 z^6 + F_4 z^6 + F_5 z^7 + \cdots$$

Then

$$G(z) - zG(z) - z^2G(z) = F_0 + F_1z - F_0z$$
$$+ (F_2 - F_1 - F_0)z^2$$
$$+ (F_3 - F_2 - F_1)z^3$$
$$+ (F_4 - F_3 - F_2)z^4$$
$$+ \cdots$$

Now $F_0 = 0$ and $F_1 = 1$, and all the expressions in parentheses are zero, because the recurrence relation for the Fibonacci Sequence is

$$F_n = F_{n-1} + F_{n-2}$$

So we get

$$G(z) - zG(z) - z^2G(z) = F_0 + F_1z - F_0z = z$$

That is,

$$G(z) = \frac{z}{1 - z - z^2}$$

Recall that the characteristic polynomial of the recurrence relation for the Fibonacci Sequence is $x^2 - x - 1$. The denominator $1 - z - z^2$ of the formula for $G(z)$ is *not* the characteristic polynomial, but it is related to the characteristic polynomial in the following way. Substitute $z = \frac{1}{x}$; then

$$1 - z - z^2 = 1 - \frac{1}{x} - \frac{1}{x^2}$$
$$= \frac{x^2 - x - 1}{x^2}$$

Let

$$r_1 = \frac{1 + \sqrt{5}}{2} \qquad \text{and} \qquad r_2 = \frac{1 - \sqrt{5}}{2}$$

be the roots of the characteristic polynomial. Then the characteristic polynomial can be factored as

$$x^2 - x - 1 = (x - r_1)(x - r_2)$$

so

$$1 - z - z^2 = \frac{(x - r_1)(x - r_2)}{x^2}$$
$$= \left(1 - \frac{r_1}{x}\right)\left(1 - \frac{r_2}{x}\right)$$
$$= (1 - r_1z)(1 - r_2z)$$

and

$$G(z) = \frac{z}{(1 - r_1 z)(1 - r_2 z)}$$

When two fractions $\frac{a}{b}$ and $\frac{c}{d}$ are added, the sum is $\frac{ad+bc}{bd}$, whose denominator bd is the product of the denominators of the factors. This suggests that our formula for $G(z)$ can be written as a sum

$$G(z) = \frac{z}{(1 - r_1 z)(1 - r_2 z)} = \frac{A}{(1 - r_1 z)} + \frac{B}{(1 - r_2 z)}$$

which can be simplified to

$$z = A(1 - r_2 z) + B(1 - r_1 z)$$

Since this must be true for all values of z (in particular for $z = 0$ and $z = 1$), we get the equations

$$0 = A + B$$
$$1 = -r_2 A - r_1 B$$

These can be solved for A and B, remembering that

$$r_1 = \frac{1 + \sqrt{5}}{2} \qquad \text{and} \qquad r_2 = \frac{1 - \sqrt{5}}{2}$$

to get $A = \dfrac{1}{\sqrt{5}}$ and $B = \dfrac{-1}{\sqrt{5}}$. We arrive at

$$G(z) = \frac{1}{\sqrt{5}} \left[\frac{1}{(1 - r_1 z)} - \frac{1}{(1 - r_2 z)} \right]$$

Finally, we observe that $\frac{1}{1-r_1 z}$ is the sum of the infinite geometric series $1 + r_1 z + r_1^2 z^2 + \cdots$, and similarly for $\frac{1}{1-r_2 z}$. So

$$G(z) = F_0 + F_1 z + F_2 z^2 + F_3 z^3 + F_4 z^4 + F_5 z^5 + F_6 z^6 + \cdots$$
$$= \frac{1}{\sqrt{5}} (1 + r_1 z + r_1^2 z^2 + \cdots - 1 - r_2 z - r_2^2 z^2 - \cdots)$$

Comparing coefficients of like powers of z, we see that

$$F_n = \left(\frac{1}{\sqrt{5}} \right)(r_1^n - r_2^n)$$

which is, of course, the same formula we derived in Example 9.19.

Thus, we have arrived at the solution of the Fibonacci recurrence relation by assuming the existence of a function defined as an infinite sum and then computing with it. Such a function is called a *generating function*.

Generating Functions

Let $a_0, a_1, a_2, \ldots$ be any finite or infinite sequence of numbers. Then the expression

$$G(z) = \sum_{i=0}^{\infty} a_i z^i$$
$$= a_0 + a_1 z + a_2 z^2 + \cdots$$

is called the **generating function** for the sequence $a_0, a_1, a_2, \ldots$. An expression of the form

$$\sum_{i=0}^{\infty} a_i z^i$$

is called a **formal power series** in the variable z. There is no guarantee that a formal power series defines any function at all, so the term *generating function* is not entirely appropriate. In some cases, of course, the expression does define a function. If the series is finite (that is, if $a_i = 0$ for all but a finite number of subscripts i), then $G(z)$ is a function whose domain is $\mathbf{C}$; in fact, it is a polynomial. If all of the a_i are equal to 1 and if $|z| < 1$, the series is a geometric series and has the sum $G(z) = \frac{1}{1-z}$.

Although the formula $\frac{1}{1-z}$ is defined for all complex numbers except 1, the infinite sum $1 + z + z^2 + \cdots$ is equal to $\frac{1}{1-z}$ only when $|z| < 1$. The study of formal power series and their relation to functions is a central topic in the theory of complex numbers and requires calculus. For purposes of discrete mathematics, it suffices to compute with formal power series as though they were functions. This is what we did to rediscover the formula for the nth term of the Fibonacci Sequence.

In the next example, we use the method of generating functions to rediscover the formula for the nth term in a sequence generated by a linear first-order recurrence relation with constant coefficients and constant nonhomogeneous term.

EXAMPLE 9.33   Solve the recurrence relation $s_n = as_{n-1} + b$. Write down the generating function

$$G(z) = s_0 + s_1 z + s_2 z^2 + \cdots$$

Then

$$aG(z) = as_0 + as_1 z + as_2 z^2 + \cdots$$

Recall that $1 + z + z^2 + \cdots = \frac{1}{1-z}$, so

$$\frac{b}{1-z} = b + bz + bz^2 + \cdots$$

By adding, we get

$$aG(z) + \frac{b}{1-z} = (as_0 + b) + (as_1 + b)z + (as_2 + b)z^2 + \cdots$$

$$= s_1 + s_2 z + s_3 z^2 + \cdots \qquad \text{Using the recurrence relation } s_n = s_{n-1} + b$$

$$= \frac{G(z) - s_0}{z}$$

This equation can be solved for $G(z)$, and the result is

$$G(z) = \frac{(b - s_0)z + s_0}{(1 - z)(1 - az)}$$

As we saw in the case of the Fibonacci numbers, such a fraction can be written as a sum

$$\frac{(b - s_0)z + s_0}{(1 - z)(1 - az)} = \frac{A}{1 - z} + \frac{B}{1 - az}$$

which can be written in the form

$$(b - s_0)z + s_0 = A(1 - az) + B(1 - z)$$

Since this is true for all z, we let $z = 0$ and $z = 1$ to get

$$A + B = s_0$$
$$b = A(1 - a)$$

These equations can be solved for A and B, yielding

$$A = \frac{-b}{(a - 1)}$$

$$B = s_0 + \frac{b}{a - 1}$$

When we combine these facts, we see that

$$G(z) = s_0 + s_1 z + s_2 z^2 + \cdots$$
$$= \frac{-b}{a - 1}(1 + z + z^2 + \cdots) + \left[s_0 + \frac{b}{a - 1} \right](1 + az + a^2 z^2 + \cdots)$$

By comparing coefficients of like powers of z, it follows that

$$s_n = a^n \left[s_0 + \frac{b}{a - 1} \right] - \frac{b}{a - 1}$$

which is, of course, the same result obtained in Theorem 9.1.   ○

Describing the Growth of Functions

We have seen several examples of functions $f(n)$ whose values grow large as n grows large. In Example 9.3, we saw that the number of moves $f(n)$ required to solve the Towers of Hanoi puzzle with n rings is given by $f(n) = 2^n - 1$. If n is large, this is very close (on a percentage basis) to 2^n. We also know that the sum of the finite geometric series $g(n) = 1 + a + a^2 + \cdots + a^n$ is

$$\frac{a^{n+1} - 1}{a - 1}, \qquad \text{for } a \neq 1$$

If $a > 1$, $g(n)$ grows large as n grows large. For large values of n, it is the term a^{n+1} alone that determines how $g(n)$ behaves. It is instructive to compare $g(n)$

to a^n:

$$\frac{g(n)}{a^n} = \frac{a^{n+1} - 1}{a^n(a-1)}$$

$$= \frac{a - \dfrac{1}{a^n}}{a - 1}$$

$$= \frac{1 - \dfrac{1}{a^{n+1}}}{1 - \dfrac{1}{a}}$$

If n is large, $\dfrac{1}{a^{n+1}}$ is close to 0, so $\dfrac{g(n)}{a^n}$ is very close to $\dfrac{1}{1 - \frac{1}{a}}$ and $g(n)$ is very close

to $\dfrac{a^n}{1 - \frac{1}{a}}$

Thus, $g(n)$ can be thought of as behaving like a^n, if the factor $\dfrac{1}{1 - \frac{1}{a}}$ is ignored.

If the value of a function f is thought of as an estimate of the work required to accomplish a task, as in the case of the Towers of Hanoi puzzle, the unit in which work is measured may be of no interest. The "shape" of the curve alone gives us interesting information as to how the complexity of the task grows with increasing values of n. We now make this idea precise with the following definition.

Definition 9.3   A function f is said to be **asymptotically bounded** by a function g if there are positive numbers n_0 and M such that $|f(n)| \le Mg(n)$ for all $n \ge n_0$. The set of all functions asymptotically bounded by g is denoted by $\mathbf{O}(g)$, read "big-oh (g)." If f is asymptotically bounded by g, we write $f \in \mathbf{O}(g)$.

The number M is called a **scale factor** for f. To say that f is asymptotically bounded by g means that the growth of f is no faster than the growth of g, given the scale factor M. The purpose of n_0 in the definition is to exclude exceptions for small values of n. The relation between f and g is illustrated in Figure 9.14.

EXAMPLE 9.34   Let $f(n) = 5n^2 + 2n + 1$ and $g(n) = n^2$. Then $f \in \mathbf{O}(g)$, because for any real number $n > 1$,

$$5n^2 + 2n + 1 \le 5n^2 + 2n^2 + n^2$$
$$= 8n^2$$

$M = 8$ is the scale factor, and $n_0 = 1$.   ○

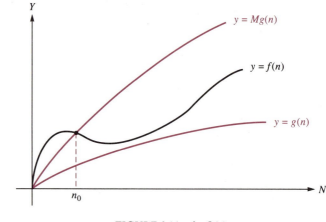

FIGURE 9.14 $f \in \mathbf{O}(g)$

EXAMPLE 9.35   Let $f(x) = 3x^2 + 2x$, and let $g(x) = x^2$. Then f is asymptotically bounded by g, because if $x > 2$,

$$2x < x^2 \qquad \text{and} \qquad 3x^2 + 2x < 3x^2 + x^2 = 4x^2$$

Here, $M = 4$ and $n_0 = 2$.   ○

EXAMPLE 9.36   For the number of moves $f(n)$ required to solve the Towers of Hanoi puzzle, we can write $f(n) \in \mathbf{O}(2^n)$. In this case, the factor M in Definition 9.3 is 1 and $n_0 = 0$. This is because

$$2^n - 1 \leq 1(2^n)$$

for all $n \geq 0$.   ○

EXAMPLE 9.37   If $a > 1$ and if $f(n)$ is the sum of the finite geometric series $1 + a + a^2 + \cdots + a^n$; that is,

$$f(n) = \sum_{i=0}^{n} a^i$$

then $f(n) \in \mathbf{O}(a^n)$. In this case, the scale factor is $\frac{1}{a-1}$, and again $n_0 = 0$.   ○

EXAMPLE 9.38   Let $f(n)$ be a polynomial of degree k in n—that is,

$$f(n) = a_k n^k + a_{k-1} n^{k-1} + \cdots + a_1 n + a_0$$

with $a_k \neq 0$. Show that $f(n) \in \mathbf{O}(n^k)$. Look at the ratio

$$\frac{|f(n)|}{n^k}$$

If n is positive, the ratio is equal to

$$\left| \frac{a_k n^k + a_{k-1} n^{k-1} + \cdots + a_1 n + a_0}{n^k} \right| = \left| a_k + \frac{a_{k-1}}{n} + \cdots + \frac{a_1}{n^{k-1}} + \frac{a_0}{n^k} \right|$$

$$\le |a_k| + \left| \frac{a_{k-1}}{n} \right| + \cdots + \left| \frac{a_1}{n^{k-1}} \right| + \left| \frac{a_0}{n^k} \right| \qquad \text{By the triangle inequality}$$

$$\le |a_k| + |a_{k-1}| + \cdots + |a_1| + |a_0|$$

$$= \sum_{i=0}^{k} |a_i| \qquad \text{Because } \frac{1}{n^i} \le 1 \text{ for any positive integer } i.$$

Set

$$M = \sum_{i=1}^{k} |a_i|$$

Then

$$\frac{|f(n)|}{n^k} \le M \qquad \text{and} \qquad |f(n)| \le M n^k$$

This shows that $f(n) \in \mathbf{O}(n^k)$. Here, n_0 is any positive number. ○

EXAMPLE 9.39   Show that the amount of work required to search a sorted array of n items using the binary search technique is $\mathbf{O}(\log_2(n))$. For each n, there is a unique integer m such that $2^{m-1} < n \le 2^m$. The search can be accomplished in at most m comparisons. Now suppose that the amount of work required to accomplish a comparison (measured in whatever units are appropriate) is K. Then the total amount of work is Km. Since $2^m < 2n$, we have

$$m < \log_2(2n) = 1 + \log_2(n)$$
$$< 2 \log_2(n) \qquad \text{Since } 1 < \log_2 n \text{ for } n > 2$$

and therefore,

$$Km < 2K \log_2(n)$$

In this case, n_0 can be taken to be any number greater than 2. ○

Note that in Example 9.39, the value of K is quite irrelevant; therefore, the unit in which computational work is measured (such as milliseconds) is also irrelevant. As long as the work required is proportional to the number of comparisons, the conclusion applies. The result provides qualitative information on how the workload grows with the size of the array to be searched; it is independent of the speed of the computer used to do the search.

EXAMPLE 9.40   A function f is said to be **bounded** if there exists a number M such that $|f(n)| \le M$ for all n. So every bounded function is asymptotically bounded by the constant function 1; in fact, $\mathbf{O}(1)$ is the set of all bounded functions. ○

Suppose that a sequence of numbers s_n is defined by a linear recurrence relation of degree k with constant coefficients, a nonhomogeneous term that is a polynomial in n, and some initial conditions. Then we can count the number of additions and multiplications required to compute each term of the sequence from the preceding n terms. If this number is K, then $K(n - k)$ operations are required to find the term s_n, starting from the initial condition. We can say that the amount of work required to compute s_n directly is $\mathbf{O}(n)$. If we succeed in solving the relation (that is, in finding a formula for s_n), we can compute s_n in a number of arithmetic operations that does not depend on n. Solving the relation reduces the work from $\mathbf{O}(n)$ to $\mathbf{O}(1)$.

EXERCISES 9.6

1. Find the sum of the following infinite series:

    a. $\sum_{i=0}^{\infty} (\frac{1}{2})^i$

    b. $\sum_{j=0}^{\infty} (\frac{1}{6})^j$

2. Find the sum of the following infinite series:

    a. $\sum_{k=0}^{\infty} \dfrac{2.5}{(-0.6)^k}$

    b. $\sum_{i=5}^{\infty} (0.25)^i$

3. Use the method of generating functions to find solutions of the following recurrence relations:

    a. $s_n = 3s_{n-1} - 2s_{n-2}, s_0 = 0, s_1 = 1$

    b. $s_n = \dfrac{s_{n-2}}{4}, s_0 = 1, s_1 = 1$

    c. $s_n = 1.5s_{n-1} + s_{n-2}, s_0 = 1, s_1 = 1$

4. Use the method of generating functions to find solutions of the following recurrence relations:

    a. $s_n = 2s_{n-1} - n, s_0 = 2$
    b. $s_n = 5s_{n-1} - 6s_{n-2} + 2n, s_0 = 0, s_1 = 1$

5. Let $h(n) = n - 15$, and let $f(n) = 13n + 7$. Show that $f \in \mathbf{O}(h)$.

6. If $f(n) = n + 20$ and $g(n) = n^2 - 4$, show that $f \in \mathbf{O}(g)$.

7. Let $f(n) = 3n^2 + 2$ and $g(n) = n^2 - 10$. Show that $f \in \mathbf{O}(g)$.

8. Show that if f is any polynomial function, then $f\left(\dfrac{1}{n}\right) \in \mathbf{O}(1)$.

9. Show that $\log_2(n) \in \mathbf{O}(n)$.

10. Let $f(n) = an + b$ and $g(n) = cn + d$, and assume that $a \neq 0$ and $c \neq 0$. Show that $f \in \mathbf{O}(g)$ and $g \in \mathbf{O}(f)$.

11. Show that $n^2 \notin \mathbf{O}(n)$.

12. A simple strategy for sorting an array of numbers, called **insertion sort,** is to arrange the first two entries in the proper order; then for each succeeding number, find its proper place, move everything below this place down one

position to make room for the new number; and insert the number. Show that the amount of work required to sort n items using this method is $\mathbf{O}(n^2)$.

13. Another strategy for sorting an array of numbers, called **merge sort**, is to divide the list into two equal parts, sort each part (using the merge sort strategy), then merge the two parts. (Try this with paper and pencil and about 20 numbers.) Show that the amount of work required to sort n items using merge sort is $\mathbf{O}(n \log_2(n))$.

14. Let $f(x)$ be a polynomial of degree n. Compare the number of multiplications required to evaluate f at a point

   a. Using substitution

   b. Using Horner's method (described in Exercise 19 of Section 3.1)

   c. Comment on the efficiency (in terms of $\mathbf{O}$) of each method of evaluating a polynomial.

Computer Exercises for Chapter 9

1. Compute and graph 50 terms in the sequence defined by the recurrence relation $s_n = 0.77s_{n-1} + 3$, with initial condition $s_0 = 1$.

2. Compute and graph 50 terms in the sequence defined by the recurrence relation $s_n = -0.9s_{n-1} + 3$, with initial condition $s_0 = 5$.

3. Compute and graph 50 terms in the sequence defined by the recurrence relation $s_n = 1.02s_{n-1} + 0.05$, with initial condition $s_0 = 1$.

4. Compute and graph 100 terms in the sequence defined by the recurrence relation $s_n = 1.8s_{n-1} - 0.8s_{n-2}$, with initial conditions $s_0 = 1$, $s_1 = 1.5$. Explain the features of the graph in relation to the roots of the characteristic polynomial.

5. Compute and graph 100 terms in the sequence defined by the recurrence relation $s_n = -0.3s_{n-1} + 0.7s_{n-2}$, with initial conditions $s_0 = 1$, $s_1 = 1.5$. Explain the features of the graph in relation to the roots of the characteristic polynomial.

6. Compute and graph 100 terms in the sequence defined by the recurrence relation $s_n = 2.8s_{n-1} - 2.8s_{n-2} + s_{n-3}$, with initial conditions $s_0 = 5$, $s_1 = 5$, $s_2 = 4.8$. Explain the features of the graph in relation to the roots of the characteristic polynomial.

7. Compute and graph 100 terms in the sequence defined by the recurrence relation $s_n = 0.8s_{n-1} - 0.8s_{n-2} - s_{n-3}$, with initial conditions $s_0 = -5$, $s_1 = 5$, $s_2 = -3$. Explain the features of the graph in relation to the roots of the characteristic polynomial.

8. Compute and graph 100 terms in the sequence defined by the recurrence relation

$$s_n = 3.664s_{n-1} - 5.346244s_{n-2} + 3.664s_{n-3} - s_{n-4}$$

with initial conditions $s_0 = 3.5$, $s_1 = 3.5$, $s_2 = 2.8$, $s_3 = 2.1$. Explain the features of the graph in relation to the roots of the characteristic polynomial.

9. Describe how the character of the solutions of the recurrence relation of Exercise 4 changes in response to small changes in the coefficients of the relation.

10. Describe how the character of the solutions of the recurrence relation of Exercise 5 changes in response to small changes in the coefficients of the relation.

11. Describe how the character of the solutions of the recurrence relation of Exercise 6 changes in response to small changes in the coefficients of the relation.

12. Describe how the character of the solutions of the recurrence relation of Exercise 7 changes in response to small changes in the coefficients of the relation.

• • • • CHAPTER REVIEW EXERCISES

1. Find the values s_2 through s_5 in each of the following cases:

    a. $s_n = 1.5s_{n-1} - 4$ and $s_1 = 13$

    b. $s_n = -0.5s_{n-1} + 1.5$ and $s_1 = 0$

    c. $s_n = \dfrac{1}{s_{n-1}} + 1$ and $s_1 = 2$

    d. $s_n = -2s_{n-1} + n$ and $s_0 = 0$

2. Find the values s_2 through s_5 in each of the following cases:

    a. $s_n = -s_{n-1} + s_{n-2}$ and $s_0 = 0, s_1 = 1$

    b. $s_n = 4s_{n-1} + s_{n-2}$ and $s_0 = -2, s_1 = 1$

    c. $s_n = 2s_{n-1} + n^2 s_{n-2}$ and $s_0 = 1, s_1 = 1$

3. How many ways s_n can the number n be written as a sum of 1's, 2's, and 3's, if order is significant? Determine s_n for $n = 1, 2, 3,$ and 4, and find a recurrence relation for s_n.

4. How many n-digit sequences of 1's and 0's do not contain the sequence 111? Find a recurrence relation for the answer.

5. How many n-digit sequences of 1's and 0's do not contain the sequence 001?

6. A bank is paying interest at 6%, compounded quarterly. A man deposits $100, then makes an additional deposit of $100 each month. Find a recurrence relation for his account balance at the end of n quarters. How much does he have in the account after three years?

7. You owe $80,000 on your home mortgage. The annual interest rate is 12%, and you make monthly payments of $1100. Write a recurrence relation for the amount still owed after n months. How much do you owe after 8 months?

8. A retirement annuity of $150,000 is earning 7% interest and is issuing monthly checks of $1200. Find a recurrence relation for the amount remaining after n months.

9. Determine the order of each of the following recurrence relations. Classify each relation as linear or nonlinear. For each linear relation, classify it as homogeneous or nonhomogeneous, and as having constant coefficients or not having constant coefficients.

a. $s_n = \dfrac{2}{s_{n-2}} + 1$

b. $s_n = 3s_{n-2} - 4s_{n-3} + s_{n-5}$

c. $s_n = 6s_{n-1}^2 + 4s_{n-2}^3$

d. $s_n = 6s_{n-2} + n^3 - 4n^2 + 4 - 5s_{n-1}$

10. Solve each of the following recurrence relations for the given initial conditions.

   a. $s_n = 1.2s_{n-1} + 7; s_0 = 1$

   b. $s_n = -3s_{n-1} - 1; s_0 = 0$

   c. $s_n = s_{n-1} - 2; s_0 = 10$

   d. $s_n = -0.1s_{n-1} + 0.1; s_0 = 5$

11. Does the recurrence relation $s_n = s_{n-1}^2 - 3s_{n-1} + 2$ have any fixed points? If so, what are they?

12. Find the characteristic polynomials for each of the following recurrence relations.

   a. $s_n = 13s_{n-1}$

   b. $s_n = 5s_{n-1} - 4s_{n-2}$

   c. $s_n = -s_{n-1} - 6s_{n-2} - 21s_{n-3}$

13. Show that the nth Fibonacci number is the nearest integer to

$$\frac{1}{\sqrt{5}}\left(\frac{1 + \sqrt{5}}{2}\right)^n$$

14. Solve the recurrence relation $s_n = 3s_{n-1} + \frac{1}{2}$ for the initial conditions $s_0 = 1$.

15. Solve the recurrence relation $s_n = 0.4s_{n-1} + 1.2s_{n-2}$ for the initial conditions $s_0 = 1, s_1 = 1$.

16. Solve the recurrence relation $s_n = \frac{3}{2}s_{n-1} - \frac{3}{4}s_{n-2} + \frac{1}{8}s_{n-3}$ for the initial conditions $s_0 = 2, s_1 = 3$, $s_2 = 4$.

17. Solve the recurrence relation of Example 9.5, assuming that at time 0 there are 1500 mature cells and 200 immature cells.

18. Solve the recurrence relation of Example 9.6, assuming that at time 0 there are 1500 mature cells and 200 immature cells.

19. Solve the recurrence relation $s_n = 3s_{n-1} - 2.25s_{n-2}$ for the initial conditions $s_0 = 0, s_1 = -1$.

20. Solve the recurrence relation $s_n = s_{n-1} - \dfrac{s_{n-2}}{2}$ for the initial conditions $s_0 = 0, s_1 = 2$.

21. Solve the recurrence relation $s_n = \frac{3}{2}s_{n-1} - \frac{1}{2}s_{n-2}$ for the initial conditions $s_0 = 2, s_1 = 0$.

22. Solve the recurrence relation $s_n = 3s_{n-1} - 4s_{n-3}$ for the initial conditions $s_0 = 1, s_1 = 0, s_2 = -1$. (*Hint:* 2 and -1 are roots.)

23. Solve the recurrence relation $s_n = \dfrac{s_{n-1} + 2n}{2}$ for the initial condition $s_0 = 0$.

24. Solve the recurrence relation $s_n = 2s_{n-1} - 0.05n$ for the initial condition $s_0 = -10$.

25. Solve the recurrence relation $s_n = s_{n-1} - n$ for the initial conditions

   a. $s_0 = 0$

   b. $s_0 = 10$

26. Solve the recurrence relation $s_n = 3s_{n-1} - 2.25s_{n-2} + 1$ for the initial conditions $s_0 = 4, s_1 = 0$.

27. Solve the recurrence relation $s_n = 2s_{n-1} - 2s_{n-2} + n$ for the initial conditions $s_0 = 0, s_1 = 0$.

28. Solve the recurrence relation $s_n = 5s_{n-1} - 8s_{n-2} + 4s_{n-3}$ for the initial conditions $s_0 = 0, s_1 = 0$, $s_2 = 1$.

29. Find the sum of the following infinite series:

   a. $\displaystyle\sum_{i=0}^{\infty}(0.625)^i$

   b. $\displaystyle\sum_{j=2}^{\infty}\frac{1}{6^j}$

   c. $\displaystyle\sum_{k=0}^{\infty}\frac{-2}{7^k}$

   d. $\displaystyle\sum_{i=4}^{\infty}\left(\frac{2}{3}\right)^i$

30. Use the method of generating functions to find solutions of the following recurrence relations:

    a. $s_n = 2s_{n-1} - 4s_{n-2}$, $s_0 = 1$, $s_1 = 1$ **b.** $s_n = 9s_{n-2}$, $s_0 = 1$, $s_1 = 1$

    c. $s_n = s_{n-1} + s_{n-2} + 1$, $s_0 = 0$, $s_1 = 1$ **d.** $s_n = -3s_{n-1} + 2n$, $s_0 = 0$

31. Let $g(n) = 2n^2 - 10$, and let $f(n) = 3n^2 + 2n + 1$. Show that $f \in \mathbf{O}(g)$.

32. If $f(n) = n^2 + 2$ and $g(n) = n^3 - 4$,

    a.   Show that $f \in \mathbf{O}(g)$ **b.**   Show that $g \notin \mathbf{O}(f)$

33. Show that if $f(n) = (-1)^n$, then $f \in \mathbf{O}(1)$.

34. Show that the amount of work required to sort a list of n numbers using the selection sort (described in Exercise 16 of Section 9.1) is $\mathbf{O}(n^2)$.

Part III

ADVANCED TOPICS

10

TOPICS FROM NUMBER THEORY

Classical number theory is the study of the properties of the integers. The subject is as old as Euclid and Pythagoras; it is the original discrete mathematics. In the earlier chapters, you have encountered some number-theoretic concepts, including greatest common divisor and least common multiple. In this chapter, we will examine these ideas in detail and relate them to some useful algorithms that are easily implemented on computers.

10.1 • DIVISIBILITY

In Chapter 3, we introduced the concept of relation and gave many examples. One of these was the relation of divisibility.

Definition 10.1

The integer a **divides** the integer b, written $a \mid b$, if there exists an integer x such that $b = ax$.

Note that x may be positive or negative or even zero. It is not required that x be unique.

EXAMPLE 10.1 The following are true:

$$2 \mid 4 \quad -3 \mid 6 \quad 2 \mid -8 \quad -5 \mid -10$$

Also, $a \mid 0$ for every integer a, since $0 = 0a$. In particular, $0 \mid 0$. ○

Note that the fact that 0 divides 0 does not mean that zero can be divided by zero. To say that b can be divided by a means that there is a unique x (namely, the quotient $\frac{b}{a}$) such that $b = ax$, but if $a = b = 0$, x is not unique. So "a divides b" and "b can be divided by a" do not mean the same thing.

Here are some basic facts about the divisibility relation:

Theorem 10.1 If a, b, and c are integers, then

(a) If $a \mid b$, then $a \mid bc$
(b) If $a \mid b$ and $a \mid c$, then $a \mid (b + c)$ and $a \mid (b - c)$

(c)   If $a\,|\,b$ and $b\,|\,c$, then $a\,|\,c$

(d)   If $c \neq 0$, then $a\,|\,b$ if and only if $ac\,|\,bc$

Proof   We will prove part (a) only; the proofs of parts (b), (c), and (d) are left as exercises. Suppose $a\,|\,b$. Then there is an integer x such that $b = ax$. It follows that $bc = (ax)c = a(xc)$. This shows that $a\,|\,bc$.   ●

Note that the divisibility relation is reflexive (property (a) with $b = a$ and $c = 1$) and transitive (property (c)), but not symmetric.

It follows from part (b) of Theorem 10.1 that if an integer x divides all of the terms of an equation but one, then x divides the remaining term also. For example, if $a = b + c$, $x\,|\,a$ and $x\,|\,b$, then $x\,|\,c$. We will frequently use part (b) in this form.

Common Divisors

Let a and b be two integers that are not both zero, and let X be the set of common divisors of a and b. We know that X is not empty, since $1 \in X$; and X is certainly a finite set. (If $a \neq 0$, then a has only finitely many divisors, and the common divisors are among these.) Therefore, X has a greatest element, which is called the **greatest common divisor**, or **g.c.d.**, of a and b. The g.c.d. of a and b is written g.c.d. (a, b) or just (a, b).

Note that if $a \neq 0$, then $(a, 0) = |a|$.

EXAMPLE 10.2   Find the greatest common divisor of 12 and 15.

The divisors of 12 are

$$-12, -6, -4, -3, -2, -1, 1, 2, 3, 4, 6, 12$$

and the divisors of 15 are

$$-15, -5, -3, -1, 1, 3, 5, 15$$

The common divisors of 12 and 15 are

$$-3, -1, 1, 3$$

and the greatest of these is 3.   ○

Clearly, you can always find the g.c.d. of two integers by listing all the common divisors and choosing the largest (as in Example 10.2), but this is hardly efficient. There is a much more efficient way, due to Euclid (Book VII, Proposition 2). It is based on the following theorem.

Theorem 10.2   Let a and b be integers, with $b > 0$. Divide a by b, giving quotient q and remainder r, so that

$$a = bq + r \qquad \text{and} \qquad 0 \leq r < b$$

Then $(a, b) = (b, r)$.

Proof   Let X be the set of common divisors of a and b, and let Y be the set of common divisors of b and r. It suffices to show that $X = Y$ (that is, a and b have exactly the same common divisors as b and r). If c is a divisor of a and b, it follows from the equation $a = bq + r$ and Theorem 10.1 that c is a divisor of r also. By the same reasoning, every common divisor of b and r is a divisor of a.   ●

Theorem 10.2 can be used to reduce the problem of finding the g.c.d. of a and b to the simpler problem of finding the g.c.d. of b and r. The problem is simpler because the numbers are smaller, and the theorem tells us that the simpler problem has the same answer as the original one. The process of finding the g.c.d. of two integers by repeated application of Theorem 10.2 is called the **Euclidean Algorithm.**

EXAMPLE 10.3   Find the g.c.d. of 1281 and 243. By dividing 1281 by 243, we get quotient 5 and remainder 66, that is,

$$1281 = 243(5) + 66$$

Therefore, $(1281, 243) = (243, 66)$. Repeat the process until the remainder is 0.

$$243 = 66(3) + 45$$
$$66 = 45(1) + 21$$
$$45 = 21(2) + 3$$
$$21 = 3(7) + 0$$

So $(1281, 243) = (243, 66) = (66, 45) = (45, 21) = (21, 3) = 3$.   ○

The Euclidean Algorithm always works, because the sequence of remainders is a decreasing sequence of nonnegative integers—the remainder from a division is always less than the divisor. In general, the g.c.d. of a and b is the last nonzero remainder resulting from repeated division. Equivalently, it is the number used as divisor when the remainder of 0 is reached.

The Euclidean Algorithm provides a way to prove the following important theorem about divisors.

Theorem 10.3   If a and b are integers and not both zero, then every common divisor of a and b divides (a, b).

Proof   Carry out the Euclidean Algorithm for a and b; the result will look like

$$a = bq_1 \quad + r_1$$
$$b = r_1 q_2 \quad + r_2$$
$$r_1 = r_2 q_3 \quad + r_3$$
$$\cdots$$
$$r_{n-2} = r_{n-1} q_n + r_n$$
$$r_{n-1} = r_n q_{n+1} + 0$$

so $r_n = (a, b)$. Now suppose that $d \mid a$ and $d \mid b$. Then repeated application of Theorem 10.1 shows that $d \mid r_1, d \mid r_2, \ldots, d \mid r_n$.   ●

In fact, we saw in Example 10.2 that the common divisors of 12 and 15 are -3, -1, 1, and 3. These are exactly the divisors of 3, which is the g.c.d. of 12 and 15.

Theorem 10.4   If a and b are not both zero, and if $m \neq 0$, then

$$(ma, mb) = m(a, b)$$

Proof   Write out the Euclidean Algorithm for a and b as in the proof of Theorem 10.3. Then multiply both sides of every equation by m. The result is the Euclidean Algorithm applied to ma and mb, and the last nonzero remainder is $mr_n = m(a, b)$.   ●

From Theorem 10.4, it is easy to see that

$$\left(\frac{a}{m}, \frac{b}{m} \right) = \frac{(a, b)}{m}$$

In particular, if $d = (a, b)$, then $\left(\dfrac{a}{d}, \dfrac{b}{d} \right) = \dfrac{(a, b)}{d} = 1$.

EXAMPLE 10.4   The Euclidean Algorithm to find $(42, 15)$ is

$$42 = 15(2) + 12$$
$$15 = 12(1) + 3$$
$$12 = 3(4) + 0$$

so $(42, 15) = 3$. Multiply all equations by 2 and obtain

$$84 = 30(2) + 24$$
$$30 = 24(1) + 6$$
$$24 = 6(4) + 0$$

which shows that $(84, 30) = 6$. Also note that

$$\left(\frac{84}{2}, \frac{30}{2} \right) = \frac{(84, 30)}{2} = \frac{6}{2} = 3$$

and that

$$\frac{(84, 30)}{6} = \left(\frac{84}{6}, \frac{30}{6} \right) = (14, 5) = 1$$ ○

Definition 10.2   The numbers a and b are **relatively prime** if $(a, b) = 1$.

In earlier chapters, we had occasion to speak of fractions in lowest terms. A fraction is in lowest terms if its numerator and denominator are relatively prime.

Theorem 10.5   If $a \,|\, bc$ and $(a, b) = 1$, then $a \,|\, c$.

Proof   Since $(a, b) = 1$, it follows from Theorem 10.4 that

$$(ac, bc) = (a, b)c = c$$

Now $a \mid ac$ and $a \mid bc$, so by Theorem 10.3, $a \mid c$.   ●

The Least Common Multiple

A concept closely related to greatest common divisor is the **least common multiple,** or **l.c.m.** If a and b are nonzero integers, let X be the set of positive common multiples of a and b. Then X is nonempty (because $|ab| \in X$), so X contains a least element, which is called the *least common multiple* of a and b. If either a or b is zero, then the only common multiple of a and b is zero, so l.c.m. (a, b) is defined to be zero. The l.c.m. of a and b is denoted by l.c.m.(a, b) or by $[a, b]$.

Theorem 10.6   Suppose that a and b are not both zero, and that $m = $ l.c.m.(a, b). If x is a common multiple of a and b, then $m \mid x$. That is, every common multiple of a and b is a multiple of the least common multiple.

Proof   If either a or b is zero, then all common multiples of a and b are zero, so the statement is trivial. So assume that neither a nor b is zero. Divide x by m, obtaining quotient q and remainder r. Then

$$x = mq + r, \qquad \text{where } 0 \le r < m$$

Now $a \mid x$ and $b \mid x$, and also $a \mid m$ and $b \mid m$; so by Theorem 10.1, $a \mid r$ and $b \mid r$. That is, r is a common multiple of a and b. But m is the least positive common multiple of a and b, so $r = 0$. Therefore $x = mq$, which means that $m \mid x$.   ●

Now we can derive the relationship between the g.c.d. and the l.c.m. of a and b.

Theorem 10.7   Suppose a and b are positive integers. Then

$$\text{l.c.m.}(a, b) = \frac{ab}{\text{g.c.d.}(a, b)}$$

Proof   Let $d = (a, b)$. We show that $\frac{ab}{d} = [a, b]$ in two steps. First, we show that $\frac{ab}{d} \ge [a, b]$; then we show that $\frac{ab}{d} \le [a, b]$. For the first part, note that

$$\frac{ab}{d} = \left(\frac{a}{d}\right)b = \left(\frac{b}{d}\right)a$$

and that both $\frac{a}{d}$ and $\frac{b}{d}$ are integers; therefore, $\frac{ab}{d}$ is a common multiple of a and b and must be at least as large as the least common multiple. For the second part, note that ab is a common multiple of a and b; so by Theorem 10.6, $\frac{ab}{[a, b]}$ is an integer. Then the number

$$\frac{a}{\dfrac{ab}{[a, b]}} = \frac{[a, b]}{b}$$

is an integer, because $b \mid [a, b]$. This shows that $\frac{ab}{[a,b]}$ is a divisor of a. Similar reasoning shows that $\frac{ab}{[a,b]}$ is a divisor of b. Since every common divisor of a and b is less than or equal to the greatest common divisor, we have

$$\frac{ab}{[a, b]} \leq (a, b) = d$$

which is equivalent to $\frac{ab}{d} \leq [a, b]$. This completes the second part.   ●

EXERCISES 10.1

1. Which of the following are true?
    a. $5 \mid 257$ b. $-15 \mid -195$ c. $17 \mid 0$
    d. $0 \mid 17$ e. $-1 \mid 1$

2. Use the Euclidean Algorithm to find the greatest common divisor of each of the following pairs of numbers:
    a. 25 and 10   b. 127 and 37   c. 143 and 11

3. Use the Euclidean Algorithm to find the greatest common divisor of each of the following pairs of numbers:
    a. 245 and 47   b. 33,647 and 3672   c. 792 and 5604

4. Prove parts (b), (c), and (d) of Theorem 10.1.

5. Is the converse of part (a) of Theorem 10.1 true? Prove or give a counterexample.

6. Is the converse of part (b) of Theorem 10.1 true? Prove or give a counterexample.

7. Show that if $d \mid a$, $d \mid b$, and $d \mid c$, then $d \mid (ax + by + cz)$ for any integers x, y, and z.

8. Show that if $a \mid b$, $b \mid c$, and $c \mid d$, then $a \mid d$.

9. Suppose that a and b are relatively prime. Show that the g.c.d. of $a + b$ and $a - b$ is either 1 or 2.

10. Show that $(a, a + 2)$ is 2 if a is even and 1 if a is odd.

11. Suppose that x, y, u, and v are integers such that $xv - yu = 1$, and let
$$m = xa + yb$$
$$n = ua + vb$$
Show that $(m, n) = (a, b)$. (*Hint:* Solve the system of equations for a and b.)

12. Suppose that $\frac{a}{b}$ and $\frac{c}{d}$ are fractions in lowest terms. Show that $\frac{a}{b} + \frac{c}{d}$ cannot be an integer unless $b = d$.

13. Show that any two consecutive terms in the Fibonacci Sequence are relatively prime.

14. What happens when you apply the Euclidean Algorithm to a pair of consecutive numbers from the Fibonacci Sequence?

15. Suppose that a and b are any two integers, not both zero. How are the g.c.d. and l.c.m. of a and b related?

16. Show that if a, b, and k are integers with $k \geq 2$ and $a \equiv b \pmod{k}$, then $(a, k) = (b, k)$.

17. Refer to Exercise 28 of Section 3.2. Since the function f is one-to-one and onto, it has an inverse g. Show how the Euclidean Algorithm can be used to compute $g(\frac{a}{b})$, where $\frac{a}{b}$ is a fraction in lowest terms. Use the method to compute $g(\frac{13}{5})$.

18. Let a and b be positive integers with $a \geq b$.

    a. Write $a = bq + r$ with $0 \leq r < b$. Show that $r < \frac{a}{2}$.

    b. Show that in any line (other than the first line) of the Euclidean Algorithm for a and b, the product of the dividend and divisor is less than half the same product for the line above. That is, $r_k r_{k+1} < \frac{1}{2} r_{k-1} r_k$ for each k.

    c. Show that the number of steps in the Euclidean Algorithm for a and b is $\mathbf{O}(\log_2(a))$.

10.2 • PRIMES AND FACTORIZATION

In this section, we will review the concept of prime and derive some additional facts about prime numbers.

Definition 10.3

An integer p is prime if $|p| > 1$ and the only positive divisors of p are 1 and $|p|$.

Note that 0, 1, and -1 are not primes, but that both 5 and -5 are primes. Also, a prime number is relatively prime to every prime number except itself and its negative.

The following theorem and proof appear in Euclid's *Elements*.

Theorem 10.8 There are infinitely many primes.

Proof See Example 4.26. ●

Theorem 10.9 If p is a prime and $p \mid ab$, then $p \mid a$ or $p \mid b$.

Proof If p does not divide a, then $(p, a) = 1$; so by Theorem 10.5, $p \mid b$. ●

More generally, if a prime p divides a product $a_1 \ldots a_n$, then it divides at least one of the factors a_i. This may be shown by mathematical induction and is left as an exercise.

We saw in Chapter 2 that every integer greater than 1 can be written as a product of primes. We can now show that except for signs and the order of the factors, there is only one way to do this. This is known as the **Unique Factorization Theorem.**

Theorem 10.10 Let a be an integer greater than 1. Then there is exactly one way to write a in the form

$$a = p_1^{k_1} p_2^{k_2} \cdots p_n^{k_n}$$

where $p_1, \ldots, p_n$ are positive primes, $p_1 < p_2 < \cdots < p_n$, and $k_1, \ldots, k_n$ are positive integers.

Proof Suppose that

$$a = p_1^{k_1} p_2^{k_2} \cdots p_n^{k_n}$$

and

$$a = q_1^{h_1} q_2^{h_2} \cdots q_m^{h_m}$$

are two factorizations of a into primes. If p_i is any of the primes $p_1, \ldots, p_n$, then

$$p_i \mid q_1^{h_1} q_2^{h_2} \cdots q_m^{h_m}$$

It follows from Theorem 10.9 that p_i must divide one of the factors $q_j^{h_j}$, so $p_i = q_j$. So every prime in the list $p_1, \ldots, p_n$ must occur in the list $q_1, \ldots, q_m$, and similar reasoning applied to q_j shows that the converse must be true too. So $m = n$, and the two lists of primes are the same. Thus, the two factorizations may be written

$$a = p_1^{k_1} p_2^{k_2} \cdots p_n^{k_n} = p_1^{h_1} p_2^{h_2} \cdots p_n^{h_n}$$

It remains to show that the exponents of each prime are the same. Apply Theorem 10.5. Since $p_i^{k_i} \mid p_1^{h_1} p_2^{h_2} \cdots p_n^{h_n}$ and $p_i^{k_i}$ is relatively prime to all of the other primes, it follows that $p_i^{k_i} \mid p_i^{h_i}$; so $k_i \le h_i$. By the same reasoning, $h_i \le k_i$, so $k_i = h_i$. ●

Relation to g.c.d. and l.c.m.

The prime factorizations of two numbers have a simple relation to their greatest common divisor and least common multiple. Suppose a and b are two integers greater than 1, and let $p_1, \ldots, p_n$ be all of the primes occurring in a or b. Then by the uniqueness of prime factorization, we can write

$$a = p_1^{k_1} p_2^{k_2} \cdots p_n^{k_n}$$
$$b = p_1^{h_1} p_2^{h_2} \cdots p_n^{h_n}$$

The exponents k_i and h_i may be positive or zero, since a given prime may occur in one number and not in the other. You can construct the prime factorization of the g.c.d. of a and b by using the smaller of the two exponents for each prime. Similarly, the l.c.m. is constructed by using the larger of the two exponents for each prime.

EXAMPLE 10.5 The prime factorizations of 4116 and 5600 are

$$4116 = 2^2 3^1 7^3 = 2^2 3^1 5^0 7^3$$
$$5600 = 2^5 5^2 7^1 = 2^5 3^0 5^2 7^1$$

Therefore, the g.c.d. of 4116 and 5600 is

$$2^2 3^0 5^0 7^1 = 28$$

and the l.c.m. is

$$2^5 3^1 5^2 7^3 = 823200$$

A Counting Problem

The prime factorization of a number can be used to answer some questions about the divisors of the number. First, let's count the divisors. If the prime factorization of a number a is

$$a = p_1^{k_1} p_2^{k_2} \ldots p_n^{k_n}$$

then each of the divisors of a can be constructed by writing the same primes, each with an exponent no larger than its exponent in the factorization of a. For instance, for p_1, we can choose any exponent in the range $0, 1, \ldots, k_1$, and there are $k_1 + 1$ possible choices. Similarly, there are $k_2 + 1$ choices for the exponent of p_2, and so forth. Thus, the total number of positive divisors of a is

$$(k_1 + 1)(k_2 + 1) \ldots (k_n + 1) = \prod_{i=1}^{n} (k_i + 1)$$

EXAMPLE 10.6   Find the number of positive divisors of 14,000. Since $14{,}000 = 2^4 5^3 7^1$, the number of positive divisors is

$$(4 + 1)(3 + 1)(1 + 1) = 40$$

Sum of Divisors

We can also find the sum of the positive divisors of a. This sum is usually denoted by $\sigma(a)$. If a is a power of a prime p (that is, $a = p^k$), finding $\sigma(a)$ is simple: The positive divisors of a are

$$1, p, p^2, \ldots, p^k$$

and their sum is

$$\sigma(p^k) = 1 + p + p^2 + \cdots + p^k = \frac{p^{k+1} - 1}{p - 1}$$

because the sum is a finite geometric series. For example, $\sigma(16) = 1 + 2 + 2^2 + 2^3 + 2^4 = 31$.

Now suppose that $a = bc$, where $(b, c) = 1$. Then every divisor d of a can be written in a unique way as $d = xy$, where x is a divisor of b and y is a divisor of c. The factor x contains only the primes that occur in b; the factor y contains only the primes that occur in c. Thus, if B is the set of positive divisors of b and C is the set of positive divisors of c, then we have a one-to-one correspondence between the set of positive divisors of a and the set $B \times C$. This reasoning shows that if $(b, c) = 1$, then

$$\sigma(bc) = \sigma(b)\sigma(c)$$

EXAMPLE 10.7   Consider the factorization of 60 into 4×15, which are relatively prime. The divisors of 4 are 1, 2, and 4, and the divisors of 15 are 1, 3, 5, and 15. Every divisor of 60 is the product of a divisor of 4 and a divisor of 15, as can be seen in the table.

Divisors of 15

		1	3	5	15
Divisors of 4	1	1	3	5	15
	2	2	6	10	30
	4	4	12	20	60

Using this result, it is easy to write down a general formula for the sum of the divisors of a number. Suppose

$$a = p_1^{k_1} p_2^{k_2} \cdots p_n^{k_n}$$

is the prime factorization of a. Since $p_i^{k_i}$ and $p_j^{k_j}$ are relatively prime if p_i and p_j are different primes,

$$\sigma(a) = \sigma(p_1^{k_1})\sigma(p_2^{k_2}) \cdots \sigma(p_n^{k_n})$$
$$= \prod_{i=1}^{n} \frac{p_i^{k_i+1} - 1}{p_i - 1}$$

EXAMPLE 10.8   Find the sum of the positive divisors of 4116.

From the prime factorization $4116 = 2^2 3^1 7^3$, we can compute that

$$\sigma(4116) = \frac{2^3 - 1}{2 - 1} \times \frac{3^2 - 1}{3 - 1} \times \frac{7^4 - 1}{7 - 1} = 7 \times 4 \times 400 = 11,200$$

The Folklore of Prime Numbers

The number 6 has the property that the sum of its proper divisors (that is, of the positive divisors strictly less than itself) is again 6: $1 + 2 + 3 = 6$. The number 28 has the same property: $1 + 2 + 4 + 7 + 14 = 28$. Numbers with this property are called **perfect** numbers. A number n is perfect if $\sigma(n) = 2n$. Thus, $\sigma(6) = 12 = 2 \times 6$, and $\sigma(28) = 56 = 2 \times 28$.

A rule for constructing even perfect numbers was given by Euclid: If p and $2^p - 1$ are prime, then $2^{p-1}(2^p - 1)$ is perfect. This is easy to verify (it is left as an exercise); in fact, it can be shown that all even perfect numbers are of this form. It is not known whether there are any odd perfect numbers.

One may ask, "For which primes p is the number $M_p = 2^p - 1$ prime?" Prime numbers of this form are known as *Mersenne primes*, after the French amateur mathematician Marin Mersenne (1588–1648). The numbers M_2, M_3, M_5, and M_7 are primes, but $M_{11} = 2047$ is not. The search for large prime numbers has concentrated on finding Mersenne primes. Edouard Lucas (1842–1891), a French mathematician, showed that

$$M_{127} = 170, 141, 183, 460, 469, 231, 731, 687, 303, 715, 884, 105, 727$$

is prime. Thereafter, no larger primes were found until computers became available. In 1953, Raphael M. Robinson used a computer to find several new primes, including M_{2281}. The largest known Mersenne prime is $M_{216,091}$, found by Chevron

Geosciences Co. in 1985, using a Cray X-MP supercomputer. The largest known prime is $(391581) \times 2^{216193} - 1$, found in 1989 at the Amdahl Benchmark Center.

There are many open questions relating to prime numbers. For instance, are there infinitely many twin primes? Twin primes are primes that differ by 2, such as 3 and 5, 11 and 13, 41 and 43, 1,000,000,009,649 and 1,000,000,009,651. Is it true that every even number greater than 2 is the sum of two primes? This statement is known as **Goldbach's conjecture.**

EXERCISES 10.2

1. Find the prime factorizations of
   a. 2310
   b. 4752
   c. 48,510
   d. 86,625
   e. 1,543,500
   f. 174,636,000

2. Use the prime factorizations from the exercise above to find the g.c.d. and l.c.m. of the following pairs of numbers:
   a. 2310 and 86,625
   b. 4752 and 1,543,500
   c. 48,510 and 174,636,000
   d. 4752 and 86,625

3. Prove that if a prime p divides a product of n factors, then it divides at least one of them. (*Hint:* Use induction on n.)

4. Find the number of positive divisors of each of the numbers in Exercise 1.

5. Find the sum of the positive divisors of each of the numbers in Exercise 1.

6. Find a formula for the total number of divisors (positive and negative) of a number a.

7. What is the sum of all the divisors (positive and negative) of a?

8. Let $d(a)$ be the number of positive divisors of a. Show that if $(a, b) = 1$, then $d(ab) = d(a)d(b)$.

9. Show that if p and $2^p - 1$ are prime, then $2^{p-1}(2^p - 1)$ is perfect.

10. Find the first five even perfect numbers.

11. The **Euler φ-function** $\varphi(n)$ is defined for positive integers n by the following rule: $\varphi(n)$ is the number of positive integers less than or equal to n and relatively prime to n.
    a. Find $\varphi(n)$ for $n = 1, 2, \ldots, 12$.
    b. If p is a prime, what is $\varphi(p)$?
    c. If p is prime, show that $\varphi(p^k) = p^k - p^{k-1}$.

12. Refer to Exercise 11.
    a. If $a = p_1^{k_1} p_2^{k_2} \ldots p_n^{k_n}$ is the prime factorization of a, show that

    $$\varphi(a) = a \prod_{i=1}^{n} \left(1 - \frac{1}{p_i}\right)$$

    (*Hint:* Of the numbers from 1 to a, $\dfrac{1}{p_i}$ of them are divisible by p_i, $\dfrac{1}{p_i p_j}$ of them are divisible by both p_i and p_j, etc. Use the inclusion–exclusion principle of Section 6.1.)

b. Show that if a and b are relatively prime, then $\varphi(ab) = \varphi(a)\varphi(b)$.

13. Show that if n is a positive integer and $\sigma(n) = n + 1$, then n is prime.

14. Find $\varphi(n)$ for each of the numbers in Exercise 1.

10.3 • LINEAR DIOPHANTINE EQUATIONS

A **Diophantine equation** is an equation for which a solution in integers is desired. A linear Diophantine equation is an equation of the form

$$a_1 x_1 + a_2 x_2 + \cdots + a_n x_n = c$$

where $a_1, a_2, \ldots, a_n$ and c are integers. We wish to find integers $x_1, \ldots, x_n$ for which the equation is true.

If $n = 1$, this is not an interesting question; if $a_1 \neq 0$, there is only one solution to the equation $a_1 x_1 = c$, and the solution either is an integer or it is not. For instance, the solution of the equation $2x = 6$ is $x = 3$, which is an integer, but the solution of $2x = 5$ is $x = \frac{5}{2}$, which is not.

Let's look next at equations with two unknowns. The equation

$$5x - 17y = 3$$

has the integer solution

$$x = 21, \quad y = 6$$

and also the solution

$$x = 38, \quad y = 11$$

The integer solutions correspond to those points on the line $5x - 17y = 3$ having integer coordinates. Some of these points are shown in Figure 10.1.

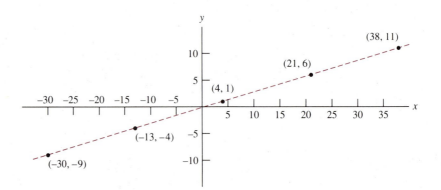

FIGURE 10.1   Some integer solutions of $5x - 17y = 3$

Not every linear Diophantine equation has a solution. For example, the equation

$$4x + 6y = 5$$

can have no solution in integers at all, for if it did, the left side of the equation would be an even integer, whereas 5 is odd. The line $4x + 6y = 5$ does not pass through any point with integer coordinates.

If the equation

$$ax + by = c$$

has a solution in integers, then $(a, b) \mid c$. This follows from Theorem 10.1, since $(a, b) \mid a$ and $(a, b) \mid b$.

A Criterion for the Existence of a Solution

For a method of finding solutions to linear Diophantine equations, we turn to the Euclidean Algorithm. In the Euclidean Algorithm, we start with a and b, and determine the g.c.d. of a and b. By working backwards through the steps of the algorithm, it is possible to write (a, b) in terms of a and b. The process is best illustrated by examples.

EXAMPLE 10.9   Find a solution of the equation

$$15x + 6y = 3 \tag{1}$$

The Euclidean Algorithm for the coefficients 15 and 6 is

$$15 = 6(2) + 3 \tag{2}$$
$$6 = 3(2) + 0 \tag{3}$$

This shows that $(15, 6) = 3$. Equation (2) may be written in the form

$$15(1) + 6(-2) = 3$$

so $x = 1$, $y = -2$ is a solution.   ○

EXAMPLE 10.10   Find a solution of the equation

$$35x - 16y = 1 \tag{1}$$

The Euclidean Algorithm applied to 35 and 16 is

$$35 = 16(2) + 3 \tag{2}$$
$$16 = 3(5) + 1 \tag{3}$$
$$3 = 1(3) + 0 \tag{4}$$

So $(35, 16) = 1$. To find an integer solution of (1), rewrite Equations (2) and (3) in the form

$$3 = 35(1) - 16(2) \tag{5}$$
$$1 = 16(1) - 3(5) \tag{6}$$

Substitute the right side of Equation (5) for 3 in Equation (6). The result is

$$1 = 16(1) - [35(1) - 16(2)](5)$$
$$= 16[1 + (2)(5)] - 35(1)(5)$$
$$= 16(11) - 35(5)$$

So $x = 11$, $y = 5$ is a solution of Equation (1).   ○

EXAMPLE 10.11   Find a solution of the equation

$$35x - 16y = 5$$

For this we can use the result of the previous example. We have seen that

$$1 = 16(11) - 35(5)$$

Multiply both sides of this equation by 5. The result is

$$5 = 16(11)(5) - 35(5)(5)$$
$$= 16(55) - 35(25)$$

So $x = 55$, $y = 25$ is a solution of $35x - 16y = 5$.   ○

Our observations can be summarized in the following theorem.

Theorem 10.11   The equation

$$ax + by = c \tag{1}$$

has a solution in integers if and only if $(a, b) \mid c$.

Proof We have already seen that the condition $(a, b) \mid c$ is necessary, because if a solution exists, every divisor of a and b is a divisor of c. So let $d = (a, b)$, and assume $d \mid c$. It is enough to show that there is a solution x_1, y_1 of the equation

$$ax_1 + by_1 = d \tag{2}$$

for then $x = \left(\dfrac{c}{d}\right)x_1$, $y = \left(\dfrac{c}{d}\right)y_1$ is a solution of Equation (1). The proof that (2) can be solved in integers can be done by induction on the number of steps in the Euclidean Algorithm for a and b.

Step 1: If there is only one step in the Euclidean Algorithm for a and b—that is, if the result of the algorithm is

$$a = bq + 0$$

then $b \mid a$, so $d = (a, b) = b$. Consequently,

$$a(1) + b(1 - q) = a + b - bq = d$$

and $x = 1$, $y = 1 - q$ is a solution.

Step 2: Suppose that a solution of $ax_1 + by_1 = d$ exists whenever $(a, b) \mid d$, and the number of steps in the Euclidean Algorithm for a and b has fewer than n steps.

(This is the induction hypothesis.) Suppose that the Euclidean Algorithm for a and b has n steps. Then its first step is of the form

$$a = bq + r$$

The Euclidean Algorithm for b and r has $n - 1$ steps, so by the induction hypothesis, there are integers x_2 and y_2 such that

$$bx_2 + ry_2 = d$$

Substituting $r = a - bq$ yields

$$d = bx_2 + (a - bq)y_2$$
$$= a(y_2) + b(x_2 - qy_2)$$

so that $x_1 = y_2$, $y_1 = x_2 - qy_2$ is the required solution.   ●

Theorem 10.11 provides an answer to a question raised in Chapter 1. Recall that the number system $\mathbf{Z}_n$ consists of the equivalence classes on the set $\mathbf{Z}$ relative to the equivalence relation of congruence modulo n. If $[a] \in \mathbf{Z}_n$ and $a \neq 0$, when does $[a]$ have a multiplicative inverse? If $[x]$ is a multiplicative inverse for $[a]$ in $\mathbf{Z}_n$, then $[a][x] = [1]$ in $\mathbf{Z}_n$, which means that

$$ax \equiv 1 \ (\mathrm{mod}\ n)$$

The congruence means that $n | (ax - 1)$. That is, there is an integer y such that

$$ax - 1 = ny$$

or

$$ax - ny = 1$$

By Theorem 10.11, this equation has an integer solution if and only if $(a, n) = 1$, that is, if a and n are relatively prime. Finding the multiplicative inverse of $[a]$ in $\mathbf{Z}_n$ is equivalent to finding a solution of the linear Diophantine equation $ax - ny = 1$. For instance, in $\mathbf{Z}_9$, $[3]$ does not have a multiplicative inverse, but $[4]$ does. To find the inverse of $[4]$, solve the equation $4x - 9y = 1$.

An Algorithm for Finding a Solution

Theorem 10.11 provides a way to find a solution to any linear Diophantine equation in two unknowns that has a solution, but the process of repeated substitution is tedious when done by hand and cannot easily be reduced to a computer program. In order to be able to write a program, we need to find a way to organize the computations neatly. That is, we need to recognize a pattern. To find the pattern, let's look carefully at a longer example.

EXAMPLE 10.12   Find an integer solution of the equation

$$365x + 67y = 1$$

The Euclidean Algorithm yields

$$365 = 67(5) + 30$$
$$67 = 30(2) + 7$$
$$30 = 7(4) + 2$$
$$7 = 2(3) + 1$$
$$2 = 1(2) + 0$$

The result of repeated substitution from bottom to top is as follows. The quotients from applying the Euclidean Algorithm are shown in boldface type.

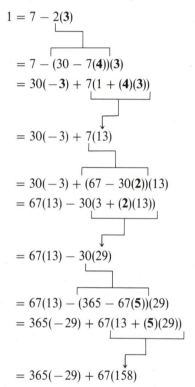

$$1 = 7 - 2(\mathbf{3})$$

$$= 7 - (30 - 7(\mathbf{4}))(\mathbf{3})$$
$$= 30(-3) + 7(1 + (\mathbf{4})(\mathbf{3}))$$

$$= 30(-3) + 7(13)$$

$$= 30(-3) + (67 - 30(\mathbf{2}))(13)$$
$$= 67(13) - 30(3 + (\mathbf{2})(13))$$

$$= 67(13) - 30(29)$$

$$= 67(13) - (365 - 67(\mathbf{5}))(29)$$
$$= 365(-29) + 67(13 + (\mathbf{5})(29))$$

$$= 365(-29) + 67(158)$$

So $x = -29$, $y = 158$ is a solution of $365x + 67y = 1$.   ○

In Example 10.12, the process of repeated substitution included only three essential computations:

$$1 + (4)(3) = 13$$
$$3 + (2)(13) = 29$$
$$13 + (5)(29) = 158$$

The rest was housekeeping, which served to tell us which of the numbers 29 and 158 is x, which is y, and what signs are required. If we ignore the housekeeping, the entire computation can be arranged alongside the Euclidean Algorithm in the following way. Again, the quotients are boldface.

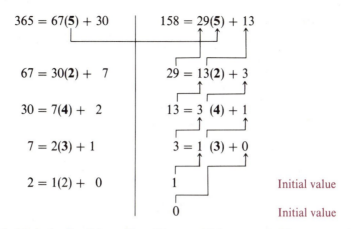

$$365 = 67(5) + 30 \qquad 158 = 29(5) + 13$$

$$67 = 30(2) + 7 \qquad 29 = 13(2) + 3$$

$$30 = 7(4) + 2 \qquad 13 = 3 \ (4) + 1$$

$$7 = 2(3) + 1 \qquad 3 = 1 \ (3) + 0$$

$$2 = 1(2) + 0 \qquad 1 \qquad \text{Initial value}$$

$$0 \qquad \text{Initial value}$$

The left side is the Euclidean Algorithm, and it is computed from top to bottom. The right side is computed from bottom to top, starting with the initial values 0 and 1 and using the quotients in reverse order. The final quotient, 2, is not used at all. The initial values 0 and 1, and the step $3 = 1(3) + 0$, are a convenient way of starting the computation. The last two results, 29 and 158, are the numbers that appear in the solution.

Of course, the penalty for discarding the housekeeping information at the beginning is that we must reconstruct it in the end. But since there are only eight possible combinations (x vs. y, x positive or negative, y positive or negative), this can be done easily by trial and error.

EXAMPLE 10.13   Find a solution of the equation

$$130x - 38y = 6$$

The result of using the Euclidean Algorithm and then working backwards is

$$130 = 38(3) + 16 \qquad 24 = 7(3) + 3$$

$$38 = 16(2) + 6 \qquad 7 = 3(2) + 1$$

$$16 = 6(2) + 4 \qquad 3 = 1(2) + 1$$

$$6 = 4(1) + 2 \qquad 1 = 1(1) + 0$$

$$4 = 2(2) + 0 \qquad 1 \qquad \text{Initial value}$$

$$0 \qquad \text{Initial value}$$

From the Euclidean Algorithm, we see that $(130, 38) = 2$. Since $2 \mid 6$, we can build a solution of the equation

$$130x_1 - 38y_1 = 2$$

using the numbers 24 and 7. By trial and error, we find that $x_1 = -7$ and $y_1 = -24$

works. To find a solution of

$$130x - 38y = 6$$

multiply x_1 and y_1 by $\frac{6}{2} = 3$. The final result is

$$x = -21, \qquad y = -72$$ ○

EXERCISES 10.3

1. Which of the following equations have integral solutions? If a solution exists, find one.

    a. $38x + 156y = 19$ **b.** $84x - 51y = 20$

    c. $-7182x + 24815y = 3$ **d.** $228x - 37y = 1$

    e. $247x + 19y = -57$

2. Which of the following equations have integral solutions? If a solution exists, find one.

    a. $539x + 193y = 0$ **b.** $-13x - 27y = 5$

    c. $74x + 14y = -52$ **d.** $52{,}351x + 5768y = -2$

3. Show that if p is a prime, every nonzero element of $\mathbf{Z}_p$ has a multiplicative inverse.

4. Show that if every nonzero element of $\mathbf{Z}_n$ has a multiplicative inverse, then n is prime.

5. Which elements of $\mathbf{Z}_{15}$ have multiplicative inverses?

6. Which elements of $\mathbf{Z}_{25}$ have multiplicative inverses?

7. Find multiplicative inverses for $[2]$ and $[3]$ in $\mathbf{Z}_7$.

8. Find a multiplicative inverse for $[17]$ in $\mathbf{Z}_{100}$.

10.4 ● MORE ABOUT LINEAR DIOPHANTINE EQUATIONS

In the last section, we saw how to determine whether a linear Diophantine equation $ax + by = c$ has a solution, and if so, how to find one. But we also saw that if a solution exists, it isn't unique; for instance, we noted that $5x - 17y = 3$ has solutions $x = 21, y = 6$ and also $x = 38, y = 11$. There are many others. In this section, we will see how to find all the solutions of such an equation.

Finding All the Solutions of $ax + by = c$

Suppose that x_1, y_1 and x_2, y_2 are both solutions of the equation $ax + by = c$. That is,

$$ax_1 + by_1 = c$$

and

$$ax_2 + by_2 = c$$

By taking the difference of these equations, we see that

$$a(x_1 - x_2) + b(y_1 - y_2) = 0$$

That is, the numbers $x_1 - x_2, y_1 - y_2$ are a solution of the associated homogeneous equation

$$au + bv = 0$$

Conversely, if x_1 and y_1 satisfy $ax + by = c$ and if u and v satisfy $au + bv = 0$, then

$$x = x_1 + u, \qquad y = y_1 + v$$

is a solution of $ax + by = c$.

So our problem is to find all solutions of $au + bv = 0$ (that is, of $au = -bv$). Let $d = (a, b)$. Then the equation $au = -bv$ may be rewritten

$$\left(\frac{a}{d}\right)u = -\left(\frac{b}{d}\right)v$$

Since $(\frac{a}{d}, \frac{b}{d}) = 1$, it follows that $(\frac{a}{d}) \mid v$, so $v = (\frac{a}{d})t$ for some integer t. Thus

$$\left(\frac{a}{d}\right)u = -\left(\frac{b}{d}\right)\left(\frac{a}{d}\right)t$$

and

$$u = -\left(\frac{b}{d}\right)t$$

Therefore, if x_1, y_1 is one solution of $ax + by = c$, then all the solutions of $ax + by = c$ are of the form

$$x = x_1 - \left(\frac{b}{d}\right)t$$

$$y = y_1 + \left(\frac{a}{d}\right)t$$

where t ranges over the integers.

EXAMPLE 10.14   Find all solutions of $3x - 2y = 1$.

One solution is $x = 1$, $y = 1$. To find all the solutions, we have to find the solutions to $3u - 2v = 0$ (that is, $3u = 2v$). Since $(3, 2) = 1$, these are given by

$$u = 2t$$
$$v = 3t$$

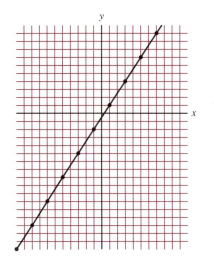

FIGURE 10.2   The solutions of $3x - 2y = 1$

so the solutions of $3x - 2y = 1$ are given by

$$x = 1 + 2t$$
$$y = 1 + 3t$$

Some of these, for $t = 0$, 1, and 2, are

t	x	y
0	1	1
1	3	4
2	5	7

The solutions are the points on the line $3x - 2y = 1$ having integer coordinates. They are shown in Figure 10.2.   ○

EXAMPLE 10.15   Find all solutions of the equation

$$130x - 38y = 6$$

In Example 10.13, we saw that one solution is

$$x = -21, \qquad y = -72$$

Since $(130, 38) = 2$, all solutions are of the form

$$x = -21 - \left(-\frac{38}{2}\right)t = -21 + 19t$$

$$y = -72 + \left(\frac{130}{2}\right)t = -72 + 65t$$

For $t = -1, 0$, and 1, the solutions are

t	x	y
-1	-40	-137
0	-21	-72
1	-2	-7

○

EXAMPLE 10.16   A student bought several textbooks at \$20 each and some videotapes at \$17 each. The total bill was \$94. How many of each did she buy?

If we let x equal the number of textbooks and y equal the number of videotapes, the problem is to find a solution of

$$20x + 17y = 94 \tag{1}$$

in which x and y are nonnegative integers. The technique of this section can be used to find the solution

$$x_1 = 6, \qquad y_1 = -7 \tag{2}$$

of the equation

$$20x_1 + 17y_1 = 1 \tag{3}$$

To find a solution of (1), multiply (2) by 94, obtaining $x = 94(6) = 564$, $y = 94(-7) = -658$:

$$20(564) + 17(-658) = 94$$

Since $(20, 17) = 1$, all solutions of (1) are of the form

$$x = 564 - 17t \tag{4}$$
$$y = -658 + 20t \tag{5}$$

We are looking for a solution with $x \geq 0$ and $y \geq 0$—that is,

$$564 - 17t \geq 0 \tag{6}$$
$$-658 + 20t \geq 0 \tag{7}$$

From Equations (6) and (7), it follows that $t \leq 33.2$ and $t \geq 32.5$, so the only possibility is $t = 33$. Substituting $t = 33$ in Equations (4) and (5) gives

$$x = 3, \qquad y = 2$$

The student bought three books and two videotapes.   ○

EXAMPLE 10.17   In how many ways can a 100-foot wall be built from 6-foot and 10-foot sections, ignoring the order of the sections?

The problem is equivalent to finding the number of nonnegative solutions of the equation

$$6x + 10y = 100 \tag{1}$$

where x is the number of 6-foot sections and y is the number of 10-foot sections. Now $(6, 10) = 2$, and one solution of the equation

$$6x_1 + 10y_1 = 2$$

is

$$x_1 = 2, \qquad y_1 = -1 \tag{2}$$

To find a solution to (1), multiply (2) by $\frac{100}{(6,\,10)} = 50$:

$$x = 100, \qquad y = -50 \tag{3}$$

and all solutions of (1) are of the form

$$x = 100 - \frac{10}{2}t = 100 - 5t$$

$$y = -50 + \frac{6}{2}t = -50 + 3t$$

Since we need $x \geq 0$ and $y \geq 0$, we need to find the integers t that satisfy

$$100 - 5t \geq 0 \tag{4}$$
$$-50 + 3t \geq 0 \tag{5}$$

Inequalities (4) and (5) require that $t \leq 20$ and $t \geq 16.7$, respectively, so the allowable values of t are 17, 18, 19, and 20. The corresponding ways of building the wall are

t	Number of 6-ft Sections	Number of 10-ft Sections
17	15	1
18	10	4
19	5	7
20	0	10

EXERCISES 10.4

1. Find all integer solutions of the following equations:
    a. $20x - 17y = 3$ b. $-35x + 65y = 15$ c. $36x + 51y = -6$
2. Find all nonnegative integer solutions of the following equations:
    a. $5x + 2y = 20$ b. $-35x + 65y = 15$ c. $36x + 51y = -6$
3. Show that $(a, b, c) = ((a, b), c)$.
4. Find the following g.c.d.'s:
    a. $(35, 91, 56)$ b. $(78, 130, 195)$
    c. $(77, 21, 33)$ d. $(128, 36, 73)$

5. A shopper bought some pads at \$1.98 each and some pens at \$2.59 each. The total bill was \$26.81. How many pads and pens did he buy?

6. A 229-foot cable is to be assembled from 9-foot and 17-foot sections. Can this be done? If so, how?

7. A walkway of length 12 feet 8 inches is to be built from concrete slabs, which come in 30-inch and 36-inch lengths. Can this be done? If so, how?

8. In how many ways can \$13.85 be paid out using only dimes and quarters?

9. The curator of reptiles fed the boa constrictor yesterday, and she is going to feed the python tomorrow. She feeds the boa once every 35 days and the python once every 29 days. When will she feed both snakes on the same day?

10. *The classical monkey problem.* Five men and a monkey are stranded on a jungle island. During the day, they collect coconuts. At the end of the day, they agree to divide them equally among themselves in the morning, and they retire for the night. At midnight, the first man awakens; not trusting the others, he divides the coconuts into five equal piles (there is one left over, which he throws to the monkey), hides one pile, combines the remaining four piles, and goes back to sleep. At 1, 2, 3, and 4 A.M., respectively, each of the other men does the same in turn. At dawn, they all awake and divide the remaining coconuts equally (there is one left over, which they throw to the monkey). What is the smallest number of coconuts that the men could have collected?

10.5 • ANOTHER METHOD FOR DIOPHANTINE EQUATIONS

In this section, we will return to the Euclidean Algorithm and the problem of finding a solution to a linear Diophantine equation in two variables. In Section 10.3, we started at the bottom (where the g.c.d. is found) and worked upward. Here, we will demonstrate another method, starting at the top and working down. The results will lead, in the next section, to some remarkable results on a totally different subject: approximating real numbers by rational numbers.

Return of the Euclidean Algorithm

If a and b are positive integers, the result of applying the Euclidean Algorithm to a and b has the structure

$$a = bq_1 \quad + r_1 \tag{1}$$
$$b = r_1 q_2 \quad + r_2 \tag{2}$$
$$r_1 = r_2 q_3 \quad + r_3$$
$$\cdots$$

$$r_{n-2} = r_{n-1} q_n + r_n$$
$$r_{n-1} = r_n q_{n+1} + 0$$

If we define $r_{-1} = a$ and $r_0 = b$, then each of the lines in the algorithm is of the form

$$r_{i-2} = r_{i-1} q_i + r_i \tag{3}$$

Equation (1) can be written in the form

$$r_1 = a - bq_1 \tag{4}$$

This shows how to write r_1 in terms of a and b. Similarly, Equation (2) can be written in the form

$$r_2 = b - r_1 q_2 \tag{5}$$

Now substitute the right side of Equation (4) for r_1 in Equation (5). The result is

$$
\begin{aligned}
r_2 &= b - r_1 q_2 \\
&= b - (a - bq_1)q_2 \\
&= a(-q_2) + b(1 + q_1 q_2)
\end{aligned}
$$

which shows how to write r_2 in terms of a and b. The substitution can be continued, eventually resulting in an equation expressing r_n in terms of a and b. The technique is best illustrated by an example.

EXAMPLE 10.18   Find a solution of the equation $26x + 10y = 2$.

The Euclidean Algorithm applied to 26 and 10 is

$$
\begin{aligned}
26 &= 10(2) + 6 \\
10 &= 6(1) + 4 \\
6 &= 4(1) + 2 \\
4 &= 2(2) + 0
\end{aligned}
$$

Working from the top down, we can compute

$$
\begin{aligned}
6 &= 26 - 10(2) \\
4 &= 10 - 6(1) \\
&= 10 - [26 - 10(2)](1) \\
&= 26(-1) + 10[1 + (2)(1)] \\
&= 26(-1) + 10(3) \\
2 &= 6 - 4(1) \\
&= [26 - 10(2)] - [26(-1) + 10(3)](1) \\
&= 26[1 - (-1)] + 10[(-2) - (3)(1)] \\
&= 26(2) + 10(-5)
\end{aligned}
$$

So one solution of $26x + 10y = 2$ is $x = 2$, $y = -5$.   ○

As with the computations in Example 10.12, the computations in Example 10.18 are confusing and error-prone. To use the method reliably, and to be able to write a computer program for it, we need a systematic way of arranging the work. In the general case, the Euclidean Algorithm is made up of equations of the form

$$r_{i-2} = r_{i-1}q_i + r_i \tag{1}$$

where q_i is the quotient and r_i is the remainder in the ith division step. We want to find sequences of integers x_i and y_i such that

$$r_i = ax_i + by_i \tag{2}$$

If we write Equation (1) in the form

$$r_i = r_{i-2} - r_{i-1}q_i$$

and substitute for r_{i-2} and r_{i-1} using Equation (2), we get

$$\begin{aligned} r_i &= r_{i-2} - r_{i-1}q_i \\ &= (ax_{i-2} + by_{i-2}) - (ax_{i-1} + by_{i-1})q_i \\ &= a(x_{i-2} - x_{i-1}q_i) + b(y_{i-2} - y_{i-1}q_i) \end{aligned}$$

So from Equation (2),

$$x_i = x_{i-2} - x_{i-1}q_i \tag{3}$$
$$y_i = y_{i-2} - y_{i-1}q_i \tag{4}$$

Equations (3) and (4) are similar to recurrence relations and can be used to compute x_i and y_i from a suitable starting point. To find the starting point, observe that

$$\begin{aligned} r_{-1} &= a = a(1) + b(0), &\text{so} && x_{-1} &= 1, & y_{-1} &= 0 \\ r_0 &= b = a(0) + b(1), &\text{so} && x_0 &= 0, & y_0 &= 1 \end{aligned}$$

The computation of Example 10.18 can now be arranged in the following scheme:

i	-1	0	1	2	3	4
q_i			2	1	1	2
y_i	0	1	-2	3	-5	13
x_i	1	0	1	-1	2	-5

You should work through this table carefully, using Equations (3) and (4), to see the pattern. For instance, $13 = 3 - (-5)(2)$ and $-5 = -1 - (2)(2)$. The numbers we need to solve the Diophantine equation are in the next-to-last column, since the g.c.d. is the next-to-last remainder in the Euclidean Algorithm.

Note that we have written the row for y_i above the line for x_i. This may seem odd, but there is a reason for it that will become apparent in the next section. Also, in both the x_i and y_i rows, the signs of the numbers alternate. This observation allows us to simplify the pattern further. Let us define

$$h_i = (-1)^{i+1}x_i$$
$$k_i = (-1)^i y_i$$

Then

$$\begin{aligned} h_{-1} &= (-1)^0 x_{-1} &= 1 \\ h_0 &= (-1)^1 x_0 &= 0 \\ k_{-1} &= (-1)^{-1} y_{-1} &= 0 \\ k_0 &= (-1)^0 y_0 &= 1 \end{aligned}$$

So the initial values for h and k are the same as the initial values for x and y. The recurrence relation for h_i can be found by computing

$$
\begin{aligned}
h_i &= (-1)^{i+1}x_i \\
&= (-1)^{i+1}(x_{i-2} - x_{i-1}q_i) \\
&= (-1)^{i+1}x_{i-2} + (-1)^i x_{i-1}q_i \\
&= h_{i-2} + h_{i-1}q_i
\end{aligned}
\tag{5}
$$

A similar computation shows that

$$
k_i = k_{i-2} + k_{i-1}q_i
\tag{6}
$$

If a and b are positive, all of the quotients q_i will be positive, and Equations (5) and (6) show that all of the numbers h_i and k_i will be positive for $i \geq 1$. This confirms our observation that the signs of the x_i and y_i alternate. Equations (5) and (6) are easier to use than Equations (3) and (4). The computations of Example 10.18, written using h_i and k_i, look like this, where we have underlined the numbers that must be negated to get x_i and y_i:

i	-1	0	1	2	3	4
q_i			2	1	1	2
k_i	<u>0</u>	1	<u>2</u>	3	<u>5</u>	13
h_i	1	<u>0</u>	1	<u>1</u>	2	<u>5</u>

Note that the h_i and k_i will always be nonnegative integers as long as all the quotients are positive. Again, you should study the table carefully to see the pattern. The k_i and h_i in the last column, when multiplied by the g.c.d. of the coefficients of the equation, give us the coefficients. In the case of Example 10.18, the coefficients are 26 and 10, and their g.c.d. is 2. Multiplying 13 and 5 by 2 gives 26 and 10. This feature is a useful check on accuracy when the computations are done manually.

EXAMPLE 10.19   Solve $62x + 23y = 1$.

The Euclidean Algorithm applied to 62 and 23 is

$$
\begin{aligned}
62 &= 23(2) + 16 \\
23 &= 16(1) + 7 \\
16 &= 7(2) + 2 \\
7 &= 2(3) + 1 \\
2 &= 1(2) + 0
\end{aligned}
$$

The computation to find h_i and k_i is

i	-1	0	1	2	3	4	5
q_i			2	1	2	3	2
k_i	<u>0</u>	1	<u>2</u>	3	<u>8</u>	27	<u>62</u>
h_i	1	<u>0</u>	1	<u>1</u>	3	<u>10</u>	23

So $x = h_4 = -10$ and $y = k_4 = 27$ is a solution of $62x + 23y = 1$.   ○

This method of solving a linear Diophantine equation in two variables requires twice as much arithmetic as the method of Section 10.3. But it has an advantage for computer programming, in that it is not necessary to store the quotients in an array or stack as they are computed, then retrieve them in the reverse order. Each quotient can be used (and discarded) as soon as it is computed.

EXAMPLE 10.20   Find a solution of

$$-13x + 263y = 5 \tag{1}$$

The quotients resulting from applying the Euclidean Algorithm to 263 and 13 are 20, 4, and 3, and the g.c.d. is 1. A solution of $263x' + 13y' = 1$ can be computed as follows:

i	-1	0	1	2	3
q_i			20	4	3
k_i	0	1	<u>20</u>	81	<u>263</u>
h_i	1	<u>0</u>	1	<u>4</u>	13

From the numbers in the next-to-last column, we see that

$$263(-4) + 13(81) = 1 \tag{2}$$

To find a solution of (1), multiply (2) by 5 and reverse the terms:

$$-13((-81)(5)) + 263((-4)(5)) = 5$$

so $x = -81(5) = -405$ and $y = (-4)(5) = -20$.   ○

EXERCISES 10.5

1. Use the method of this section to find integer solutions of the following equations:

    a. $53x + 17y = 1$ **b.** $128x + 52y = 8$ **c.** $108x - 47y = 3$

2. Use the method of this section to find integer solutions of the following equations:

    a. $539x - 2313y = -17$ **b.** $287x + 210y = 5$

10.6 ● CONTINUED FRACTIONS

If a and b are positive integers, the sequence of quotients obtained by applying the Euclidean Algorithm to a and b can be used to write the fraction $\frac{a}{b}$ in an interesting form, in which all numerators are equal to 1. For the numbers in

Example 10.19, this is:

$$\frac{62}{23} = 2 + \frac{16}{23}$$

$$= 2 + \frac{1}{\dfrac{23}{16}}$$

$$= 2 + \frac{1}{1 + \dfrac{7}{16}}$$

$$= 2 + \frac{1}{1 + \dfrac{1}{\dfrac{16}{7}}}$$

$$= 2 + \frac{1}{1 + \dfrac{1}{2 + \dfrac{2}{7}}}$$

$$= 2 + \frac{1}{1 + \dfrac{1}{2 + \dfrac{1}{\dfrac{7}{2}}}}$$

$$= 2 + \frac{1}{1 + \dfrac{1}{2 + \dfrac{1}{3 + \dfrac{1}{2}}}}$$

An expression of this form is called a *continued fraction*. Writing expressions of this type is awkward and takes up a lot of space. So it is useful to have a short notation.

Definition 10.4   Let $q_1, q_2, \ldots, q_n$ be real numbers, and assume that $q_2, \ldots, q_n$ are positive. Then the **continued fraction** defined by $q_1, q_2, \ldots, q_n$ is the number

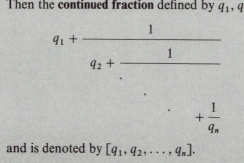

and is denoted by $[q_1, q_2, \ldots, q_n]$.

The condition that $q_2, \ldots, q_n$ are positive guarantees that the expression in the definition will not involve division by zero. In most cases of interest, the q_i's are integers. However, the extension of the definition to nonintegers will be useful in proving statements about continued fractions.

EXAMPLE 10.21   From the continued fraction for $\frac{62}{23}$ shown earlier, we can write shorter continued fractions by discarding some of the quotients:

$$[2] = 2$$

$$[2, 1] = 2 + \frac{1}{1} = 3$$

$$[2, 1, 2] = 2 + \cfrac{1}{1 + \cfrac{1}{2}} = \frac{8}{3}$$

$$[2, 1, 2, 3] = 2 + \cfrac{1}{1 + \cfrac{1}{2 + \cfrac{1}{3}}} = \frac{27}{10}$$

$$[2, 1, 2, 3, 2] = \frac{62}{23}$$

○

Convergents of Continued Fractions

Note that the fractions $2, 3, \frac{8}{3}, \frac{27}{10}$, and $\frac{62}{23}$ can be read from the table in Example 10.19. They are exactly the ratios

$$\frac{k_i}{h_i}$$

Figure 10.3 shows the positions of these fractions on the number line. They appear on alternating sides of $\frac{62}{23}$ and get rapidly closer to $\frac{62}{23}$. In fact, $\frac{27}{10}$ differs from $\frac{62}{23}$ by less than 0.0044, a remarkably good approximation for a fraction whose denominator is as small as 10. Another example of a good approximation is the well-known fraction $\frac{22}{7}$ for the number π. The error $|\frac{22}{7} - \pi|$ is less than 0.0013.

These observations motivate the following definition.

Definition 10.5   If $[q_1, q_2, \ldots, q_n]$ is a continued fraction, then the numbers

$$[q_1]$$
$$[q_1, q_2]$$
$$\ldots$$
$$[q_1, q_2, \ldots, q_{n-1}]$$
$$[q_1, q_2, \ldots, q_{n-1}, q_n]$$

are called the **convergents** of the continued fraction.

FIGURE 10.3   Convergents for [2, 1, 2, 3, 2]

The convergents are so called because, in a sense that we will make precise presently, they converge to the number represented by the continued fraction.

Continued fractions provide a way to find very good rational approximations to any real number, including, as we shall see, the fraction $\frac{22}{7}$ for π. First, however, we need to derive some basic facts about continued fractions.

Theorem 10.12   Let $[q_1, q_2, \ldots, q_n]$ be a continued fraction. Define

$$h_{-1} = 1 \qquad h_0 = 0 \qquad h_i = h_{i-2} + h_{i-1}q_i$$
$$k_{-1} = 0 \qquad k_0 = 1 \qquad k_i = k_{i-2} + k_{i-1}q_i$$

Then

$$[q_1, q_2, \ldots, q_n] = \frac{k_n}{h_n}$$

Proof   The proof is by induction on n.

Step 1: If $n = 1$, the statement to be proved is

$$q_1 = \frac{k_1}{h_1} = \frac{q_1}{1}$$

which is true.

Step 2: Assume that the statement is true for all continued fractions of length shorter than n (induction hypothesis). Define a new continued fraction

$$[Q_1, Q_2, \ldots, Q_{n-1}]$$

by

$$Q_i = q_i \qquad \text{for } i = 1, 2, \ldots, n - 2$$
$$Q_{n-1} = q_{n-1} + \frac{1}{q_n}$$

(Here we are using the fact that the numbers appearing in a continued fraction need not be integers.) Then

$$[q_1, q_2, \ldots, q_n] = [Q_1, Q_2, \ldots, Q_{n-1}]$$

Define the numbers H_i and K_i for the continued fraction $[Q_1, Q_2, \ldots, Q_{n-1}]$ in the same way that the h_i and k_i are defined for $[q_1, q_2, \ldots, q_n]$. Then $H_i = h_i$ and $K_i = k_i$

for $i = 1, 2, \ldots, n - 2$. (This is because $Q_i = q_i$ for these i.) It follows that

$$[q_1, q_2, \ldots, q_n] = [Q_1, Q_2, \ldots, Q_{n-1}]$$

$$= \frac{K_{n-1}}{H_{n-1}} \qquad \text{By the induction hypothesis}$$

$$= \frac{K_{n-3} + Q_{n-1}K_{n-2}}{H_{n-3} + Q_{n-1}H_{n-2}}$$

$$= \frac{k_{n-3} + \left(q_{n-1} + \dfrac{1}{q_n}\right)k_{n-2}}{h_{n-3} + \left(q_{n-1} + \dfrac{1}{q_n}\right)h_{n-2}}$$

$$= \frac{k_{n-3} + k_{n-2}q_{n-1} + \dfrac{k_{n-2}}{q_n}}{h_{n-3} + h_{n-2}q_{n-1} + \dfrac{h_{n-2}}{q_n}}$$

$$= \frac{k_{n-1} + \dfrac{k_{n-2}}{q_n}}{h_{n-1} + \dfrac{h_{n-2}}{q_n}}$$

$$= \frac{k_{n-1}q_n + k_{n-2}}{h_{n-1}q_n + h_{n-2}}$$

$$= \frac{k_n}{h_n}$$

which is exactly what is to be shown. ●

Theorem 10.13 Suppose that $[q_1, q_2, \ldots, q_n]$ is a continued fraction, define h_i and k_i as above, and set $r_i = \dfrac{k_i}{h_i}$. That is, r_i is the ith convergent of $[q_1, q_2, \ldots, q_n]$. Then

(a) $h_2 < h_3 < h_4 < \cdots$
(b) $h_{i-1}k_i - k_{i-1}h_i = (-1)^i$
(c) $r_i - r_{i-1} = \dfrac{(-1)^i}{h_i h_{i-1}}$
(d) $r_1 < r_3 < r_5 < \cdots > r_6 > r_4 > r_2$

Proof Part (a) is easily shown by induction and is left as an exercise. To show part (b), recall that

$$h_i = h_{i-2} + h_{i-1}q_i$$
$$k_i = k_{i-2} + k_{i-1}q_i$$

The number q_i can be eliminated from these equations by multiplying the first equation by k_{i-1}, multiplying the second equation by h_{i-1}, and subtracting. The result is

$$k_{i-1}h_i - h_{i-1}k_i = k_{i-1}h_{i-2} - h_{i-1}k_{i-2}$$

which can be written

$$-(h_{i-1}k_i - k_{i-1}h_i) = h_{i-2}k_{i-1} - k_{i-2}h_{i-1}$$

which shows that the numbers $h_{i-1}k_i - k_{i-1}h_i$ alternate in sign as i increases. Now (b) can easily be proved by mathematical induction.

To show part (c), we can compute

$$\begin{aligned}
r_i - r_{i-1} &= \frac{k_i}{h_i} - \frac{k_{i-1}}{h_{i-1}} \\
&= \frac{k_i h_{i-1} - k_{i-1}h_i}{h_i h_{i-1}} \\
&= \frac{(-1)^i}{h_i h_{i-1}}
\end{aligned}$$

Finally, for part (d) we can compute

$$\begin{aligned}
r_{i+1} - r_{i-1} &= (r_{i+1} - r_i) + (r_i - r_{i-1}) \\
&= \frac{(-1)^{i+1}}{h_{i+i}h_i} + \frac{(-1)^i}{h_i h_{i-1}} \\
&= (-1)^i \frac{\dfrac{-1}{h_{i+1}} + \dfrac{1}{h_{i-1}}}{h_i}
\end{aligned}$$

Since $h_{i-1} < h_{i+1}$, the numerator

$$\frac{-1}{h_{i+1}} + \frac{1}{h_{i-1}}$$

is positive. This means that $r_{i+1} - r_{i-1}$ is positive if i is even (so $r_1 < r_3 < \cdots$), and $r_{i+1} - r_{i-1}$ is negative if i is odd (so $r_2 > r_4 > \cdots$).   ●

Continued Fractions for Real Numbers

Theorem 10.13 gives a complete description of the behavior of the convergents of a continued fraction. The odd-numbered convergents increase; the even-numbered convergents decrease; all the even-numbered convergents are larger than all the odd-numbered convergents; and the difference between two successive convergents is 1 divided by the product of their denominators. The following examples show how these facts can be used to obtain good rational approximations to real numbers.

EXAMPLE 10.22   Find a continued fraction for $\sqrt{2} = 1.41421\ldots$.

We begin by writing (approximately)

$$\sqrt{2} = 1 + 0.41421$$

Now note that $0.41421 = \frac{1}{2.41423}$, so we can write (still approximately)

$$\sqrt{2} = 1 + \cfrac{1}{2 + 0.41423}$$

Again, $0.41423 = \frac{1}{2.41412}$, so

$$\sqrt{2} = 1 + \cfrac{1}{2 + \cfrac{1}{2.41412}}$$

At this point, we suspect that there is a pattern here. It appears that if we continue the process, we will get

$$\sqrt{2} = [1, 2, 2, 2, 2, \ldots]$$

Of course, we are also seeing effects of roundoff errors. To verify that the pattern is valid, we can compute

$$\sqrt{2} \doteq 1 + (\sqrt{2} - 1)$$
$$= 1 + \cfrac{1}{\cfrac{1}{\sqrt{2} - 1}}$$

which can be simplified to

$$\sqrt{2} = 1 + \cfrac{1}{1 + \sqrt{2}}$$

The right side of this equation can be substituted for $\sqrt{2}$ in the denominator, giving

$$\sqrt{2} = 1 + \cfrac{1}{1 + \left[1 + \cfrac{1}{1 + \sqrt{2}}\right]}$$

This substitution can be repeated as often as desired, confirming the pattern we suspected.

Now let's compute some convergents:

i	-1	0	1	2	3	4	5	6
q_i			1	2	2	2	2	2
k_i	0	1	1	3	7	17	41	99
h_i	1	0	1	2	5	12	29	70

The convergents are $1, \frac{3}{2}, \frac{7}{5}, \frac{17}{12}, \frac{41}{29}, \frac{99}{70}$. The differences between these numbers and $\sqrt{2}$ are

i	r_i	$r_i - \sqrt{2}$
1	1	-0.41421
2	$\frac{3}{2}$	0.08579
3	$\frac{7}{5}$	-0.01421
4	$\frac{17}{12}$	0.00245
5	$\frac{41}{29}$	-0.00042
6	$\frac{99}{70}$	0.00007

As predicted by Theorem 10.13, the signs of the errors alternate, with $\sqrt{2}$ sandwiched between the even and odd convergents:

$$1 < \frac{7}{5} < \frac{41}{29} < \sqrt{2} < \frac{99}{70} < \frac{17}{12} < \frac{3}{2}.$$

○

In general, if x is any real number, the terms q_i in the continued fraction for x can be obtained by the following recurrence relations:

$$x_1 = x$$
$$q_i = \lfloor x_i \rfloor \qquad \text{Greatest integer function}$$
$$x_{i+1} = \frac{1}{x_i - q_i}$$

If x is a rational number $\frac{a}{b}$, this process will terminate, because it is equivalent to applying the Euclidean Algorithm to a and b. If x is irrational, the process will continue indefinitely. Of course, the number of terms we can compute correctly is limited by how accurately we know x and how accurately we do the computations.

EXAMPLE 10.23   Find some terms in the continued fraction for π.

The successive values of x_i and q_i can be determined approximately using a pocket calculator. We can compute the convergents of the continued fraction as we go along. The results, using a calculator carrying 10 decimal digits of accuracy, are

i	x_i	q_i	k_i	h_i	$\dfrac{k_i}{h_i} - \pi$
			0	1	
			1	0	
1	3.141592654	3	3	1	-0.141592653
2	7.062513305	7	22	7	0.001264489
3	15.99659454	15	333	106·	-0.000083220
4	.1.003417095	1	355	113	0.0000002667
5	292.6462647	292	103993	33102	-0.0000000006

The approximation $\frac{355}{113}$ for π is truly remarkable, being accurate to better than one part in 11 million. With some additional terms, the continued fraction for π continues as follows:

$$\pi = [3, 7, 15, 1, 292, 1, 1, 1, 2, 1 \ldots]$$

No pattern in the terms of this fraction has ever been recognized. ○

EXERCISES 10.6

1. Write the following continued fractions in short notation:

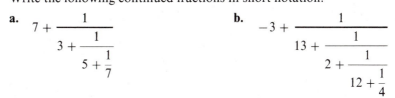

2. Write the following continued fractions in long form:
   a. $[1, 2, 3, 4, 5]$ b. $[-1, 3, 2, 16]$
   c. $[0, 721]$ d. $[3, 1.2, 2.3, \frac{13}{5}]$

3. Compute all of the convergents of each of the continued fractions in Exercise 1.

4. Compute all of the convergents of each of the continued fractions in Exercise 2.

5. Find the continued fraction for $\sqrt{5}$. Identify the pattern in the terms, verify it, and compute the first six convergents. Compute the differences between the convergents and $\sqrt{5}$.

6. Prove part (a) of Theorem 10.13.

7. Complete the proof of part (b) of Theorem 10.13.

8. Find the continued fraction for $\sqrt{3}$. Identify the pattern in the terms, verify it, and compute the first six convergents. Compute the differences between the convergents and $\sqrt{3}$.

9. Let x be the number represented by the infinite continued fraction $[3, 3, 3, 3, \ldots]$. Show that $x = 3 + \frac{1}{x}$. (*Hint:* Write out the continued fraction in long form.) Find x. Compute the first six convergents.

10. Let x be the number represented by the infinite continued fraction $[2, 3, 1, 3, 1, 3, 1, \ldots]$. Find an equation satisfied by x, and find x. Compute the first six convergents.

11. Find the first six convergents of the continued fraction for the number $e = 2.7182818285927 \ldots$ (the base of the natural logarithms).

Computer Exercises for Chapter 10

1. Write a program to find the g.c.d. of two integers using the Euclidean Algorithm. Your program should correctly handle the case in which one or both of the inputs is zero.

2. Set up a spreadsheet to find the g.c.d. of two integers.

3. A *recursive programming language* is one in which a procedure may call itself; Pascal is such a language. Write a program that uses a recursive procedure call to solve a linear Diophantine equation. (*Hint:* If $a = bq + r$, and if x' and y' is a solution of $bx' + ry' = c$, then $x = y'$ and $y = x' - qy'$ is a solution of $ax + by = c$.)

4. Write a program to find the g.c.d. of several integers.

5. Write a program to use the method of Section 10.3 to find a solution of the equation $ax + by = c$. (*Hint:* You may wish to identify classes of trivial cases, such as $c = 0$, and dispose of them separately.)

6. Is the method of Section 10.3 for solving linear Diophantine equations amenable to use of a spreadsheet program? Explain.

7. Write a program to use the method of Section 10.5 to solve a linear Diophantine equation in two variables. Do not store the quotients.

8. Set up a spreadsheet to use the method of Section 10.5 to solve a linear Diophantine equation in two variables.

9. Set up a spreadsheet to compute the convergents of the continued fraction for a rational number, for which you enter the numerator and denominator.

10. Write a program (or set up a spreadsheet) to find the terms and convergents of the continued fraction for a real number. Of course, when you enter the number as a decimal (such as 23.456), you are really entering a rational number, so the process should terminate. What really happens? Why?

• • • • CHAPTER REVIEW EXERCISES

1. Which of the following are true?
   a. $15 \mid -245$
   b. $12 \mid 4404$
   c. $-131 \mid 131$
   d. $5465 \mid 739{,}663$

2. Use the Euclidean Algorithm to find the greatest common divisor of each of the following pairs of numbers:
   a. 16 and 3
   b. 380 and 165
   c. 2026 and 1697
   d. 9751 and 6804
   e. 1393 and 225
   f. 3927 and 1189

3. Find the prime factorizations of
   a. 284,130
   b. 3,303,300
   c. 11,858
   d. 3456
   e. 56,250
   f. 4985

4. Use the prime factorizations from Exercise 3 to find the g.c.d. and l.c.m. of the following pairs of numbers:
   a. 56,250 and 4985
   b. 11,858 and 284,130

5. Find the number of positive divisors of each of the numbers in Exercise 3.

6. Find the sum of the positive divisors of each of the numbers in Exercise 3.

7. Which of the following equations have integral solutions? If any solution exists, find all solutions.

    a. $13x + 156y = 26$ **b.** $55x - 48y = 3$ **c.** $-3932x + 180y = 12$

8. Which of the following equations have integral solutions? If any solution exists, find all solutions.

    a. $181x - 585y = 1$ **b.** $85x - 17y = 39$ **c.** $300x + 148y = -52$

9. Which elements of $\mathbf{Z}_{14}$ have multiplicative inverses?

10. Which elements of $\mathbf{Z}_{42}$ have multiplicative inverses?

11. Find the multiplicative inverse of $[11]$ in $\mathbf{Z}_{14}$.

12. Find the multiplicative inverse of $[37]$ in $\mathbf{Z}_{42}$.

13. Find all nonnegative integer solutions of the equations of Exercise 7.

14. Find the following g.c.d.'s:

    a. $(135, 191, 256)$ **b.** $(28, 120, 216)$ **c.** $(87, 22, 39)$ **d.** $(322, 168, 629)$

15. A hobbyist bought some NAND gates at $0.27 each and some inverters at $0.25 each. She paid $3.68 total. How many NAND gates and how many inverters did she buy?

16. You do laundry every five days. Today is Monday, and you are doing laundry. When will you next do laundry on a Monday?

17. A laboratory computer is receiving signals from one test instrument at 10.735-second intervals and from another test instrument at 3.515-second intervals. Sometimes the computer receives signals from both instruments at exactly the same time. How frequently do these events occur?

18. Four days ago, you invested $10,000 in 90-day notes with automatic renewal after each 90-day period. Today, you invested $15,000 in 4-week notes, also with automatic renewal. When will both securities mature on the same day?

19. Write the following continued fractions in short notation:

    a. $-1 + \dfrac{1}{2 + \dfrac{1}{8 + \dfrac{1}{2}}}$ **b.** $5 + \dfrac{1}{7 + \dfrac{1}{9 + \dfrac{1}{11 + \dfrac{1}{13}}}}$

20. Write the following continued fractions in long form:

    a. $[2, 8, 2, 8]$ **b.** $[7, 9]$ **c.** $[2, 1, 1, 1, 4, 1]$ **d.** $[0, 1, 2, 3, 2, 1]$

21. Compute all of the convergents of each of the continued fractions in Exercise 19.

22. Compute all of the convergents of each of the continued fractions in Exercise 20.

23. Find the continued fraction for $\sqrt{6}$. Identify the pattern in the terms, verify it, and compute the first six convergents. Compute the differences between the convergents and $\sqrt{6}$.

24. Find the continued fraction for $\sqrt{11}$. Identify the pattern in the terms, verify it, and compute the first six convergents. Compute the differences between the convergents and $\sqrt{11}$.

25. Let x be the number represented by the infinite continued fraction $[1, 1, 2, 2, 2, 2, 2, \ldots]$. Find x. Compute the first six convergents.

26. Find the first six convergents of the continued fraction for $\sqrt[3]{2}$.

11

LANGUAGES AND GRAMMARS

Throughout the preceding chapters, we used symbols to represent mathematical and logical concepts: numbers, sets, functions, propositions, predicate expressions. In each case we combined primitive symbols such as letters and digits into meaningful sequences that express ideas. Up to this point, we relied on our intuition to determine which sequences of symbols were truly meaningful. In this chapter, we will formalize the notion of "meaningful string of symbols" by precisely defining the concept of language. Thus, we shall be able to define the language of numbers, of set theory, of propositional logic, and of predicate logic. Languages defined by precise definitions are called **formal languages,** to distinguish them from the natural languages used in human conversation. Although we will draw on our experiences with natural languages such as English, the problem of characterizing natural languages is immensely more complex than that of characterizing mathematical languages.

11.1 ● VOCABULARIES, SENTENCES, AND LANGUAGES

Consider a work written in a natural language. It is made up of **sentences,** each of which is a sequence of **words.** To know the language, you must know the words of the language, and which sequences of words constitute sentences. Of course, you must also understand the **meaning** of the words to understand the language. Nevertheless, it is possible to imagine a person, or a machine, that could formulate grammatically correct sentences without regard for meaning. Some computer programs have been written that do this, the most famous of which is ELIZA. ELIZA receives sentences from the user and transforms them according to formal rules, relying on the user to supply meaning. To the untrained user, the computer appears to be a psychoanalyst carrying on a conversation.

Formal Languages

In the formal languages of mathematics and computer programming, a formula or a program statement is a sequence of symbols. Thus, the following concept of language applies both to natural languages and to formal ones.

Definition 11.1

> Let V be a set of symbols, and let V^* be the set of all finite sequences of elements of V. (The set V may be finite or infinite.) A **language** L is a subset of V^*. V is called the **vocabulary** of the language. The elements of L are the **sentences** of the language.

When the language is used to express logical or mathematical statements, a sentence in the language is often called a **well-formed formula,** or **wff** (pronounced woof).

EXAMPLE 11.1   Let $V = \{a, b\}$ and $L = \{\varnothing, a, b, ab, ba\}$. Then L is a language with vocabulary V: It contains five sequences of symbols from the vocabulary. The empty set $\varnothing$ is the sequence of no symbols.   ○

This definition encompasses both natural languages such as English and formal languages such as those of mathematics and computing. We may consider English as a language in one of two ways. If we take the vocabulary of English to be the set of all (correctly spelled) words, then the sentences of the English language consist of all sequences of words that obey the rules of English grammar. Or we may take the vocabulary to be the set of letters of the alphabet, plus blanks and punctuation marks. Then the sentences of the English language are the sequences of letters and other symbols that make up correctly spelled words arranged in accordance with the rules of English grammar.

To consider English as a language according to Definition 11.1, we have to refer to rules of English grammar. Although these rules are discussed in schoolbooks and taught throughout elementary and high school, writing down a complete set of rules for English grammar is no easy task. In fact, characterizing the grammars of natural languages is a central problem of theoretical linguistics.

Fortunately, the grammars of the formal languages of mathematics and computing are much simpler than those of natural languages. They can be characterized by relatively short sets of rules. The following examples illustrate the kinds of rules involved.

EXAMPLE 11.2   Let the vocabulary V consist of the symbols 0, 1, 2, 3, 4, 5, 6, 7, 8, 9, and $-$. We will define a language consisting of sequences of these symbols that represent integers in simple, common forms. Thus, 0, 7, 38, 7395773987758937, and -5 will belong to this language. The sequences 00, -037, and $9-1$ will not belong, because although they represent integers, the representation is not the simplest, most common way of writing them. $69-$ will not be in the language, because it does not represent an integer at all. Although $+8$ represents an integer, it will not be in the language, because it contains a symbol $(+)$ that is not in the vocabulary. It is easy to formulate a set of rules for such a language. A sequence of symbols of the vocabulary is in the language if

   a.   It is the sequence 0, or
   b.   It is a finite sequence of digits that does not begin with 0, or
   c.   It consists of a $-$ followed by a finite sequence of digits that does not start with a 0.   ○

EXAMPLE 11.3   Let the vocabulary V consist of the digits 0 through 9 and the symbols $+, -, .,$ and E. Consider a language consisting of all sequences of these symbols that represent a rational number in scientific notation. (Of course, only those rational numbers that have exact representations as terminating decimals can be represented in a language of this type. The number $\frac{1}{3}$, for instance, cannot.) Such a language may include $0, -00, -5.387, 2.6E+3,$ and $-0.00567E-02,$ among others. Most computers are equipped with programs to recognize sentences in such languages and translate the sentences into the computer's internal number format. Formulating the grammar of such a language is left as an exercise.   ○

EXAMPLE 11.4   Consider a vocabulary consisting of articles, nouns, and verbs. The articles are "a" and "the." The nouns are "cheese," "mouse," "cat," "dog," and "dogcatcher." The verbs are "chased" and "ate." The sentences of the language are all sequences of the following form: article, noun, verb, article, noun. Then the following are sentences of the language (where we follow the customary rules of capitalization and punctuation):

> The mouse ate the cheese.
> A cat ate the mouse.
> The dog chased the cat.
> The dogcatcher chased a cheese.
> A mouse chased the dogcatcher.   ○

This example seems to suggest that English grammar could be characterized by simply classifying all words, then enumerating all legal sentence structures. However, it is not so simple; the rules for declension of nouns (cat, cats, cat's, cats') and conjugation of verbs (go, went, gone) are elaborate enough.

Grammars

Both mathematical and natural languages have grammatical rules that allow sentences of arbitrary complexity. Consider the grammar of Example 11.2 again. If S is a sentence of this language other than $\varnothing$, and if d is a digit, then Sd is a sentence. Thus, since -127 is a sentence of the language, so is -1272. The following example shows that such rules occur in natural languages as well.

EXAMPLE 11.5   Consider a vocabulary consisting of articles, prepositions, nouns, transitive verbs, and intransitive verbs. Let the sentences of the language be sequences of the following form: subject, intransitive verb, prepositional phrase. The subject must be either an article plus a noun or an article plus a noun plus an adjectival phrase. An adjectival phrase is a sequence consisting of a subject and a transitive verb. A prepositional phrase is a sequence consisting of a preposition, an article, and a noun. Then with a suitable vocabulary, our language includes the following sentences:

> The cheese was on the table.
> The cheese the mouse ate was on the table.
> The cheese the mouse the cat chased ate was on the table.
> The cheese the mouse the cat the dog the dogcatcher the supervisor fired caught chased chased ate was on the table.   ○

The key feature of this example is that the definition of "subject" is recursive: It refers to the term "adjectival phrase," whose definition in turn refers to "subject." Here are two well-known examples of recursive definitions in mathematics.

EXAMPLE 11.6   The definition of the factorial function given in Example 3.6 is a recursive definition. The definition is:

$$0! = 1$$
$$N! = N \times (N - 1)!$$

o

EXAMPLE 11.7   The definition of the Fibonacci numbers F_n, given in Example 3.5, is a recursive definition. The definition is:

$$F_0 = 0$$
$$F_1 = 1$$
$$F_n = F_{n-1} + F_{n-2}$$

o

Example 11.5 illustrates that as the number of levels of recursion in the structure of a sentence increases (five levels, in this case), comprehension becomes more difficult. Fortunately, our ability to understand highly nested sentences is much better for the simpler grammars of logical and mathematical statements than for natural languages. For instance, the algebraic expression

$$(x - (y + (z - (u + (v - w)))))$$

is fairly clear, and we can easily do algebraic manipulations with it, although it has the same level of nesting as the last sentence in Example 11.5.

Programming Languages

The preceding examples show that in order to characterize natural and formal languages, it is necessary to consider classes of words and parts of sentences. We can characterize the parts and then state rules for combining them. In order to talk about classes of words and parts of sentences, it is useful to have names for them. In studying English grammar, we use a variety of names, such as adjective, noun, preposition, subject, verb, verb phrase, and prepositional phrase. For the formal languages of logic and mathematics, there are similar names: digit, number, variable, connective, operator, expression. In the case of programming languages (in which a sentence is a program), we have declarations, assignment statements, control statements, comment statements, and the like.

The BASIC programming language is a simple language implemented in some form for almost all computers available today. Like many natural languages, BASIC is plagued by a proliferation of dialects; rarely are the BASICs of two different computers exactly alike. A sentence in the BASIC language (that is, a program) is a sequence of statements. Each statement consists of a line number (which is a positive integer) followed by a keyword and parameters. Line numbers, when required, must be in increasing order. If the keyword is omitted, the keyword LET is assumed. The allowable parameters depend on the keyword. Thus, to define a dialect of BASIC, it is necessary to provide a list of all the keywords and to define the parameters for each.

EXAMPLE 11.8   The following is a simple BASIC program, illustrating some of the most common statement types:

```
10 DIM A(10)
15 REM THIS PROGRAM ADDS A COLUMN OF NUMBERS
20 SUM = 0
30 FOR I = 1 TO 10
40      INPUT A(I)
50      SUM = SUM + A(I)
60 NEXT I
70 PRINT SUM
80 END
```

Statements 20 and 50 are assignment statements; the keyword LET is assumed.

○

EXAMPLE 11.9   The following is an equivalent program written in the Pascal programming language:

```
program sum(input, output);
{This program adds a column of numbers}
    var a: array [1 .. 10] of real;
        sum: real; i: integer;
begin
    sum := 0;
    for i := 1 to 10 do begin
        readln(a[i]);
        sum := sum + a[i]
    end;
    writeln(sum)
end.
```

Note that every variable used in the program is declared. Also, the semicolon is used to separate statements.

○

EXAMPLE 11.10   The following is an equivalent program in the C programming language:

```
extern double atof( );
main( )
{
    int i; real sum, a[9];
    sum = 0;
    char buffer[20];
    for (i = 0; i < 10; i ++) {
        gets (buffer, 20); /* read input line */
        a[i] = atof(buffer); /* convert to internal form */
        sum = sum + a[i];
    }
    printf("%8.4",sum);
}
```

In this language, the semicolon is used as a statement terminator rather than a statement separator as in Pascal. Another conspicuous difference is that the input of a number must be done in two steps. The first step, "gets" (for get string), obtains the character string typed by the user. The second step, "atof" (ASCII to floating), converts the character string to the computer's internal representation of a real number.   ○

The variety of language features in these examples suggests that several different methods of defining languages may be needed, depending on the characteristics of the language. In the following sections, we describe the three most common methods of specifying languages: production rules, syntax diagrams, and statement patterns.

EXERCISES 11.1

1. For the vocabulary V in Example 11.1, define a language L different from that in the example.

2. Given the vocabulary $V = \{a, b, +, \#\}$,
   a. Exhibit three elements of the set V^*
   b. Give three different examples of languages using this vocabulary.

3. a. Describe an English sentence structure that admits a recursive definition such as that in Example 11.5.
   b. Write four sentences illustrating the recursion to the third level.

4. Do the same as Exercise 3 for a foreign language.

5. Give informal descriptions (such as that in Example 11.4) of the grammars of classes of English sentences containing these sentences:
   a. The dog chased the cat.
   b. The cat was chased by the dog.
   c. The book was written by the author.

6. Give an informal description of the grammar of the language of Example 11.3.

7. Compare these features of the BASIC, Pascal and C languages:
   a. Separation of statements
   b. Rules for comment statements
   c. Facilities for building compound sentences

11.2 ● SPECIFYING LANGUAGES: PRODUCTIONS

We have seen that we can describe languages by giving names to parts of sentences (expression, number, subject, predicate) and then providing definitions for these names. One way of providing a definition of a sentence part is to write a list of all the ways the part can be formed. Such a list of rules is called a *grammar*. Both linguists and computer scientists have investigated many ways to define grammars. It has turned out that all of the formal languages of computing and mathematics may be defined by a single class of grammars, called *context-free phrase structure grammars*.

Definition 11.2

A **context-free phrase structure grammar** consists of the following items:

1. A symbol, called the **root symbol**
2. A set of symbols, called the **nonterminal symbols,** that includes the root symbol
3. Another set of symbols, called the **terminal symbols,** that will form the vocabulary of the language
4. A set of **productions,** which are rules of the form

$$S ::= E$$

where S is a nonterminal symbol and E is a finite sequence of symbols, both terminal and nonterminal

The rule means that the sequence E may be substituted for the symbol S. There must be at least one such rule for each nonterminal symbol. The language defined by the grammar is the set of all sequences of terminal symbols that can be obtained from the root symbol in a finite number of steps, where each step consists of replacing a nonterminal symbol S with a sequence E. In each step, the sequence E must be one for which $S ::= E$ is one of the productions of the grammar.

EXAMPLE 11.11   Consider a grammar that contains only one nonterminal symbol S, two terminal symbols x and y, and the two rules $S ::= xS$ and $S ::= y$. Then S must be the root symbol. To find the sentences of the language, note that:

> y is a sentence, since it can be obtained from S by one application of the second rule.
>
> xy is a sentence since it can be obtained by one application of the first rule and one of the second rule.

Similarly, xxy and $xxxy$ are sentences. The sentences of the language consist of zero or more x's followed by a single y.   o

When the grammar contains several rules for the same nonterminal symbol S, such as

$$S ::= E_1$$
$$S ::= E_2$$
$$S ::= E_3$$

we will abbreviate this to

$$S ::= E_1 \,|\, E_2 \,|\, E_3$$

The use of the symbols $::=$ and $|$ in this context has become customary because these symbols rarely occur in programming or mathematical applications.

EXAMPLE 11.12   Consider the grammar with root symbol S, one other nonterminal symbol T, terminal symbols x and y, and the rules

$$S ::= xS \mid Ty$$
$$T ::= x \mid Ty$$

To build a sentence, we must start with S, use the first S rule ($S ::= xS$) zero or more times, then use the second S rule ($S ::= Ty$); then use the first T rule zero or more times, then the second T rule. The sentences are strings consisting of one or more x's followed by one or more y's, such as xy, xxy, xyy, and so forth.

○

The definition of the ALGOL programming language is normally presented as a context-free phrase structure grammar. When programming language statements are described by a set of productions, the definition is said to be in **Backus–Naur form,** or BNF. The name refers to John Backus and Peter Naur of IBM, the developers of the FORTRAN programming language. Ironically, most descriptions of FORTRAN do not use BNF.

Classification of Grammars

The type of grammar of Definition 11.2 is called a phrase structure grammar because all of the rules are definitions of types of phrases; the nonterminal symbols represent phrase types. It is said to be context-free because if $S ::= E$ is a production, the replacement of S with E is allowed anywhere S occurs, without regard to the rest of the sequence. As the term *context-free* suggests, there are other types of phrase-structure grammars.

Definition 11.3   A **Type 0 grammar** is a phrase structure grammar in which the left side of a production (the string to be substituted for) is allowed to be any string of terminal and nonterminal symbols, not just one nonterminal symbol.

A **Type 1 grammar** is a type 0 grammar in which the length of the string on the right of each production is greater than or equal to the length of the string on the left side.

A **Type 2 grammar,** also called a **context-free grammar,** is a type 1 grammar in which the left side of each production is a single nonterminal symbol. (This is the case of Definition 11.2.)

A **Type 3 grammar,** also called a **regular grammar,** is a type 2 grammar in which the right side of each production has at most one nonterminal symbol, which must be at the extreme right.

Type 0 and type 1 grammars are very difficult to analyze and are important primarily in linguistics. The languages of logic, mathematics, and programming (with some exceptions) are readily described by context-free grammars. Regular grammars are of interest because the languages they define can be recognized by a class of simple computing machines, the finite state machines, which we consider in the next chapter.

Recursive Productions

The key to defining useful languages is to use productions in which nonterminal symbols occur on the right-hand side.

EXAMPLE 11.13   Consider the following grammar:

Root symbol: S
Nonterminal symbols: S
Terminal symbols: a, b
Productions:

$$(1) \qquad S ::= ab$$
$$(2) \qquad S ::= aSb$$

We would normally write this as

$$S ::= ab \,|\, aSb$$

Then the sentences of the language defined by this grammar are *ab, aabb, aaabbb, aaaabbbb,* and so on. The sentence *aabb* is derived from S by first applying rule (2), obtaining *aSb,* then applying rule (1). *aaabbb* is obtained by applying rule (2) twice, then rule (1).   ○

Note that it is the occurrence of the nonterminal symbol S on the right side of rule (2) that enables us to generate arbitrarily large sentences. Rule (2) is an example of a **recursive production.**

One of the strings of symbols that may appear on the right side of a production is the *empty string,* that is, the string that consists of no symbols at all. The empty string can be very useful in defining grammars. The following example demonstrates the use of the empty string.

EXAMPLE 11.14   Consider the representations of integers discussed in Example 11.2. We can define this language by the following grammar, where $\varnothing$ denotes the empty string:

$$N ::= SDX$$
$$S ::= - \,|\, \varnothing$$
$$X ::= \varnothing \,|\, 0X \,|\, DX$$
$$D ::= 1 \,|\, 2 \,|\, 3 \,|\, 4 \,|\, 5 \,|\, 6 \,|\, 7 \,|\, 8 \,|\, 9$$

Here N is the root symbol; S, D, and X are nonterminal symbols; and the terminal symbols are the minus sign $(-)$ and the digits 0 through 9. Using these rules, we can build the number -1307 from N in the following nine steps:

	Root Symbol	N
(1)	$N ::= SDX$	SDX
(2)	$S ::= -$	$-DX$
(3)	$D ::= 1$	$-1X$
(4)	$X ::= DX$	$-1DX$
(5)	$D ::= 3$	$-13X$
(6)	$X ::= 0X$	$-130X$
(7)	$X ::= DX$	$-130DX$
(8)	$D ::= 7$	$-1307X$
(9)	$X ::= \varnothing$	-1307

The grammars encountered in both mathematical and natural languages contain large numbers of nonterminal symbols corresponding to classes of words or symbols, or to types of phrases. These classes also have common English names: adjective, prepositional phrase, participle, term, factor, expression. In describing such grammars, it is helpful to use the common names as the nonterminal symbols of the language. Some convention is needed to indicate when the names are being used as nonterminal symbols. One common practice is to set the names in a different font, such as bold or italic. Another method, very frequently used in programming language manuals, is to enclose the names between the symbols $<$ and $>$; for example, $<$factor$>$, $<$expression$>$.

The following example defines a language whose sentences are the representations of decimal numbers and fractions in scientific notation, in the format recognized by such languages as FORTRAN and BASIC—for instance, $-127.648E-13$.

EXAMPLE 11.15   Floating-point numbers are defined by the following productions:

$<$number$>$	$::= <$sign$> <$integer$> <$fraction$> <$exponent$>$
$<$sign$>$	$::= + \| - \| <$empty$>$
$<$integer$>$	$::= <$empty$> \| <$digit$> <$integer$>$
$<$fraction$>$	$::= <$empty$> \| . \| . <$integer$>$
$<$exponent$>$	$::= <$empty$> \| E <$sign$> <$integer$>$
$<$digit$>$	$::= 0 \| 1 \| 2 \| 3 \| 4 \| 5 \| 6 \| 7 \| 8 \| 9$

This example shows another common name for the empty string: $<$empty$>$.

All of the languages of logic, mathematics, and computing that we have encountered contain the concept of an **expression:** a string of symbols that can be enclosed in parentheses or brackets, then used as a variable in constructing larger expressions. The definition of expression is the most fundamental recursive definition in these languages. The idea is adequately illustrated by considering expressions in elementary algebra, as in the following example.

EXAMPLE 11.16   An algebraic expression in the variables a through z is defined by this grammar:

$\langle$expression$\rangle$ ::= $\langle$term$\rangle$ |
                      $\langle$expression$\rangle$ + $\langle$term$\rangle$ |
                      $\langle$expression$\rangle$ − $\langle$term$\rangle$

$\langle$term$\rangle$ ::= $\langle$factor$\rangle$ |
                      $\langle$term$\rangle$ * $\langle$factor$\rangle$ |
                      $\langle$term$\rangle$ / $\langle$factor$\rangle$

$\langle$factor$\rangle$ ::= $\langle$variable$\rangle$ | ($\langle$expression$\rangle$)

$\langle$variable$\rangle$ ::= $a\,|\,b\,|\,c\,|\,d\,|\,e\,|\,f\,|\,g\,|\,h\,|\,i\,|\,j\,|\,k\,|\,l\,|\,m\,|\,n\,|\,o\,|\,p\,|\,q\,|\,r\,|\,s\,|\,t\,|\,u\,|\,v\,|\,w\,|\,x\,|\,y\,|\,z$

Thus, algebraic expressions include

x

$x + y * z$

$(x + y * z/w - u/x/v) * z/a/(b * c + (((d))))$ ○

Parsing Trees

Another way to show how productions are applied to form a sentence is by using a graph showing the effects of each substitution. Such a graph is called a **parsing tree** for the sentence. The parsing tree is built by starting with the root symbol, then drawing below each symbol what is substituted for it. To build the parsing tree for Example 11.14, we start with N and draw the first substitution, SDX, below it:

In the next three steps, we draw − below S, 1 below D, and DX below X, obtaining

Continuing in this manner, we can complete the parsing tree. It looks like this:

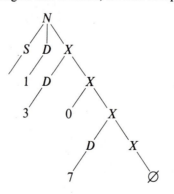

It is not always possible to draw a unique parsing tree leading to a sentence of a language, as the following example shows.

EXAMPLE 11.17   Consider a grammar with one nonterminal symbol S, one terminal symbol x, and the production

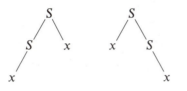

$$S ::= x \mid Sx \mid xS$$

The sentences of the language are the sequences of one or more x's. Every sentence except x can be derived in more than one way, as we can see for xx:

○

Ambiguities of this sort present a problem in any attempt to translate or interpret the language. If there are two ways to parse the sentence, the sentence may have two meanings. Therefore, for definitions of mathematical and programming languages, it is preferable to use only grammars in which such ambiguities do not occur. Such grammars are called *unambiguous*.

Definition 11.4   A context-free phrase structure grammar is called **unambiguous** if each sentence of the language has exactly one parsing tree. Otherwise it is called **ambiguous.**

Note that it is the grammar that is said to be ambiguous, not the language defined by the grammar. Indeed, it is possible to have two different grammars, one ambiguous and the other unambiguous, defining the same language. If either of the productions $S ::= xS$ or $S ::= Sx$ is deleted from the grammar of Example 11.17, the resulting grammar defines the same language and is unambiguous. For instance, if only the rule $S ::= xS$ is used, the sentence xxx can be constructed in one way only:

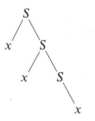

The problem of determining whether a context-free phrase structure grammar is unambiguous is one of a class of mathematical problems that have been shown to be unsolvable. That is, it is not possible to write a computer program that accepts any grammar as its input and determines whether the grammar is unambiguous.

EXERCISES 11.2

1. What are the sentences of the language defined by the grammar

    root symbol: S

    $S ::= xy \mid xS$

2. What are the sentences of the language defined by the grammar

    root symbol: S

    $S ::= aS \mid bS \mid c$

3. What are the sentences of the language defined by the grammar

    root symbol: S

    $S ::= aSa \mid bSb \mid <\text{empty}>$

4. What are the sentences of the language defined by the grammar

    root symbol: S

    $S ::= x \mid xT$

    $T ::= y \mid Sy$

5. For the grammar of Exercise 1, draw parsing trees for each of the following sentences:

    a. xxy **b.** $xxxy$

6. For the grammar of Exercise 3, draw parsing trees for each of the following sentences:

    a. aa **b.** $abba$ **c.** $babbbbab$

7. Refer to Example 11.16. Draw parsing trees for each of the following expressions:

    a. x **b.** $x + y * z$ **c.** $(x + y) * z$

8. Write a grammar to define the language whose sentences are all sequences of an even number of x's.

9. Write a grammar to define the language whose sentences are all sequences of an odd number of x's.

10. Write a grammar for a subset of English including statements of the form

    $<\text{noun phrase}> <\text{transitive verb}> <\text{noun phrase}>$

    including the articles "a" and "the," adjectives, and prepositional phrases modifying nouns. Write several sentences in the language and draw the parsing trees for each.

11. The grammar for floating point numbers in Example 11.15 allows the empty string, ., and .E as a floating point number. Write a different grammar that requires that (1) there must be at least one digit before the exponent, and (2) if E is present, there must be at least one digit after it.

12. Expand the grammar for algebraic expressions of Example 11.16 to include the exponentiation operation. You may represent the exponentiation operation by ^ (as in BASIC) or by ** (as in FORTRAN).

13. Consider the language consisting of all strings of one or more *a*'s, followed by one or more *b*'s, followed by one or more *c*'s, where either (1) the number of *a*'s equals the number of *b*'s, or (2) the number of *b*'s equals the number of *c*'s. For instance, *aaabbbcc* and *abbcc* belong to the language, but *abbccc* does not. Write a grammar for this language. Then draw two parsing trees for the sentence *aabbcc*. (This language is **inherently ambiguous**—that is, every grammar that defines the language is ambiguous.)

14. Write a grammar for algebraic expressions in Polish notation, including the unary operator − and the binary operators +, −, *, and /. Does the use of − as both a unary and a binary operator introduce any ambiguity? If so, how can it be resolved?

15. Some calculators, particularly those made by Hewlett-Packard, require that expressions to be evaluated be entered in reverse Polish notation. Write a grammar for the set of arithmetic expressions that can be evaluated by such a calculator. Consider integer arithmetic only. Treat ENTR and +/− as single terminal symbols; they correspond to single keys on the calculator. Ignore any limitation on number size or depth of nesting built into the calculator.

11.3 • OTHER WAYS OF SPECIFYING LANGUAGES

The productions of a context-free phrase structure grammar are a very powerful way of describing grammatical rules, but they can be hard to read and understand when written as sequences of symbols. In this section, we look at two widely used alternatives: syntax diagrams and statement patterns.

Syntax Diagrams

A **syntax diagram,** also called a **flow diagram,** is a graphic method of presenting a production or a group of related productions. For instance, the rule of Example 11.13 can be written as shown in Figure 11.1. In this diagram, the productions correspond to the routes through the diagram from start to finish. Nonterminal symbols are written in boxes, and terminal systems are written in ovals. Diagrams of this type were popularized in the definition of the programming language Pascal.

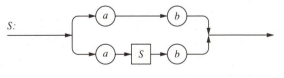

FIGURE 11.1   A simple syntax diagram

The following example illustrates how a grammar with several productions can be drawn.

EXAMPLE 11.18   The productions of Example 11.14 can be written as shown in Figure 11.2.

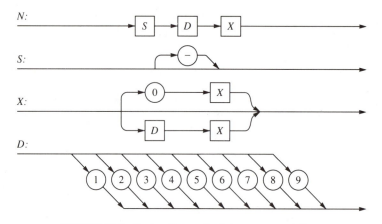

FIGURE 11.2   Syntax diagram for integer representations   o

In Figure 11.2, it is possible to travel through the *S* diagram from start to finish without hitting any symbol. Such a route corresponds to the empty string.

Any context-free phrase structure grammar can be described by a set of such diagrams. All that is necessary is to draw one diagram for each nonterminal symbol; each diagram will have one path for each production. However, we have more flexibility with syntax diagrams than we do with productions, since we can draw loops. Consider the simple diagram in Figure 11.3.

If *D* represents any digit, this diagram defines the set of all sequences of one or more digits. The number of digits depends on the number of times we choose to pass around the lower loop as we travel through the diagram. Similarly, the diagram in Figure 11.4 defines the set of sequences of zero or more digits. The difference between the two diagrams is that it is possible to pass through the second diagram without hitting any symbol.

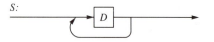

FIGURE 11.3   Diagram for a string of one or more digits

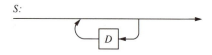

FIGURE 11.4   Diagram for a string of zero or more digits

The following example illustrates how several productions can be combined into one diagram.

EXAMPLE 11.19   The grammar of Example 11.15, defining floating-point numbers, can be defined by the diagrams in Figure 11.5.

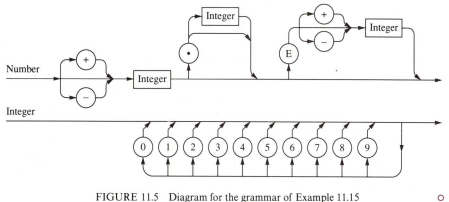

FIGURE 11.5   Diagram for the grammar of Example 11.15

It is not always advantageous to minimize the number of diagrams used to present a grammar, for if the diagram becomes too complex, it can be harder to understand than the corresponding list of productions. This is shown in the following example.

EXAMPLE 11.20   The definition of an expression in Example 11.16 can be drawn as three diagrams (one for each production), as shown in Figure 11.6. It can also be drawn as a single diagram, as in Figure 11.7.

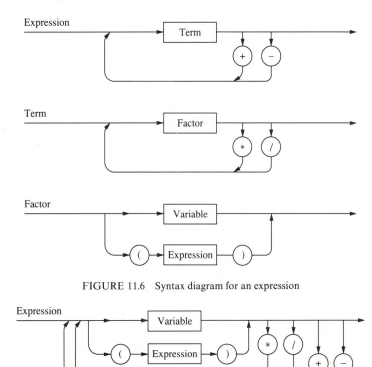

FIGURE 11.6   Syntax diagram for an expression

FIGURE 11.7   Another syntax diagram for an expression

Although the three diagrams in Figure 11.6 can be combined into one, this is done at the expense of two useful names: *term* and *factor*. In general, the best presentation of a grammar by diagrams is one in which there is one diagram for every useful phrase type. This principle is illustrated very well in the *Pascal Report*. The grammar of the entire Pascal programming language is presented in 17 diagrams, occupying only 3 pages.

Statement Patterns

Unlike natural languages and the programming languages ALGOL and Pascal, the COBOL, FORTRAN, and BASIC languages are not conveniently defined by productions or by syntax diagrams. The difference is in the use of recursion. ALGOL and Pascal were designed to support recursive algorithms, and the language definitions are themselves recursive. In both these languages, a statement can be a compound statement made up of a series of statements. For example, the diagram for ⟨statement⟩ in Pascal includes the diagram for a compound statement shown in Figure 11.8. In languages such as ALGOL and Pascal, procedures (subroutines) can contain other procedures. Moreover, a procedure can call itself.

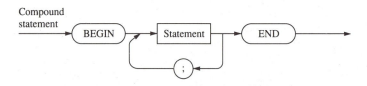

FIGURE 11.8   Syntax diagram for a compound statement

In the COBOL, FORTRAN, and BASIC languages, a program consists of a series of statements. Example 11.8 is a simple BASIC program consisting of eight statements.

Books and manuals on FORTRAN, COBOL, and BASIC use a simple device, the statement pattern, to define the programming language. The definition of the language is a list of statement types. Each statement type is identified by a keyword or other conspicuous feature—such as the keywords DIM, FOR, and NEXT, and the equal sign after SUM, in Example 11.8. Each statement, in turn, is defined by a statement pattern that shows what is allowed to follow the keyword. Choices are listed vertically in curly brackets: { and }. An optional item is enclosed in square brackets: [and]. When an item is permitted to be repeated any number of times, the square bracket is followed by an ellipsis, thus: [____] .... A statement definition using these conventions is called a **statement pattern.** These conventions lack the expressive power of productions and syntax diagrams, but when adequate for a language, they are very easy to understand.

EXAMPLE 11.21   The format of the INPUT statement in BASIC is defined by the pattern

INPUT ["prompt string";] variable [,variable] . . .

The interpretation of the pattern is that the word INPUT is required, followed by an optional prompt string enclosed in quotes and followed by a semicolon, then one or more variables separated by commas.   ○

The following example shows a more complex definition.

EXAMPLE 11.22   Suppose that "integer" denotes a string of zero or more decimal digits. Then using a statement pattern, the floating point numbers of Example 11.15 may be defined by

$$\left[\begin{Bmatrix} + \\ - \end{Bmatrix}\right] \text{ integer } [.[\text{integer}]] \left[E\begin{bmatrix}\begin{Bmatrix} + \\ - \end{Bmatrix}\end{bmatrix} [\text{integer}]\right]$$

The interpretation of the pattern is as follows. The number starts with an optional sign, which may be either $+$ or $-$. An integer is required to be next. Next is an optional decimal point; if it is present, an integer may follow. The optional final part, if present, must start with the letter E, which may be followed by a $+$ or $-$ sign, and then by an integer.   ○

EXAMPLE 11.23   Every COBOL program begins with a group of lines called an **identification division.** The statement pattern defining the identification division is

```
IDENTIFICATION DIVISION.
PROGRAM-ID. program-name.
[AUTHOR. [comment-entry] . . .]
[INSTALLATION. [comment-entry] . . .]
[DATE-WRITTEN. [comment-entry] . . .]
[DATE-COMPILED. [comment-entry] . . .]
[SECURITY. [comment-entry] . . .]
```
○

Note that this statement pattern actually defines a sequence of statements, which must appear in the specified order. However, all of the statements except the first two are optional. The following is an example of an identification division:

```
IDENTIFICATION DIVISION.
PROGRAM-ID. PRODUCTION-COST-ESTIMATION.
AUTHOR. PAULA HIRSCHFELDER.
DATE-WRITTEN. 12/17/78.
```

EXAMPLE 11.24   The COBOL statement pattern for an instruction to write a line to a printer is as follows:
WRITE record-name [FROM identifier = 1]

$$\left[\begin{Bmatrix} \text{BEFORE} \\ \text{AFTER} \end{Bmatrix} [\text{ADVANCING}] \begin{Bmatrix} \left[\begin{bmatrix} \text{identifier-2} \\ \text{integer} \end{bmatrix} \begin{Bmatrix} \text{LINE} \\ \text{LINES} \end{Bmatrix}\right] \\ \begin{Bmatrix} \text{mnemonic-name} \\ \text{PAGE} \end{Bmatrix} \end{Bmatrix}\right]$$

$$\left[[\text{AT}] \begin{Bmatrix} \text{END-OF-PAGE} \\ \text{EOP} \end{Bmatrix} \text{imperative-statement}\right]$$

The pattern indicates that everything except "WRITE record-name" is optional. A typical "WRITE" statement using several of the optional features is as follows:

> WRITE OUTPUT-RECORD AFTER ADVANCING 3 LINES
> AT END-OF-PAGE GOTO NEXT-PAGE-PROCESSING.

An equivalent statement, omitting optional words, is

> WRITE OUTPUT-RECORD AFTER 3 END-OF-PAGE
> GOTO NEXT-PAGE-PROCESSING.   ○

EXERCISES 11.3

1. Draw syntax diagrams to define the grammar of Exercise 1, Section 11.2.
2. Draw syntax diagrams to define the grammar of Exercise 2, Section 11.2.
3. Draw syntax diagrams to define the grammar of Exercise 3, Section 11.2.
4. Draw syntax diagrams to define the grammar of Exercise 4, Section 11.2.
5. Draw syntax diagrams to define the grammar of Exercise 8, Section 11.2.
6. Draw syntax diagrams to define the grammar of Exercise 9, Section 11.2.
7. Draw syntax diagrams to define the grammar of Exercise 11, Section 11.2.
8. Write a grammar for the language defined by the syntax diagram of Figure 11.9.

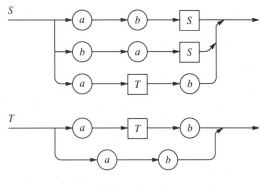

FIGURE 11.9

9. Write a grammar for the language defined by the syntax diagram of Figure 11.10.

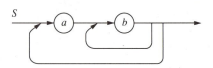

FIGURE 11.10

10. Write a grammar for the language defined by the syntax diagram of Figure 11.11.

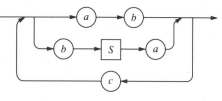

FIGURE 11.11

11. Define the grammar of Exercise 11 of Section 11.2 by statement patterns.

12. Write a grammar for the COBOL identification division of Example 11.23.

13. Define the COBOL identification division of Example 11.23 by a syntax diagram.

14. Write a grammar for the COBOL "WRITE" statement of Example 11.24.

15. Define the COBOL "WRITE" statement of Example 11.24 by a syntax diagram.

11.4 • THE LANGUAGES OF LOGIC AND MATHEMATICS

We are now in a position to state formal definitions of the underlying language of logic and mathematics. The language, although primitive, is nevertheless adequate to represent all the concepts and statements of ordinary mathematics. In principle, all ordinary mathematical statements could, by interpretation of all the definitions involved in the statement, be reduced to a sentence in this language. We will define the language in three stages—propositional logic, predicate logic, and set theory— with additional symbols and rules introduced at each stage.

Definition 11.5

The **language of propositional logic** is defined by the following grammar:

Terminal symbols: $\neg, \vee, (,)$, and an infinite set of propositional variables $p, q, r, \ldots$

Nonterminal symbol: S (the root symbol)
                      $<$propositional variable$>$

Productions:
   $<$propositional variable$> ::= p\,|\,q\,|\,r\,|\ldots$
   $S ::= \;<$propositional variable$>\,|\,(S \vee S)\,|\,\neg S$

EXAMPLE 11.25   The following are sentences of the language of propositional logic as defined in Definition 11.5:

   $\neg p$

   $(p \vee s)$

   $(\neg(\neg p \vee s) \vee \neg r)$ ○

At this point, the language can be extended by defining the two propositional constants, the truth values T and F:

T   means $(p \vee \neg p)$

F   means $\neg T$

Additional binary connectives may be defined by the customary formulas:

$p \wedge q$ means $\neg(\neg p \vee \neg q)$

$p \Rightarrow q$ means $(\neg p \vee q)$

$p \Leftrightarrow q$ means $(p \Rightarrow q) \wedge (q \Rightarrow p)$

Also, parentheses and the usual operator hierarchy and association rules may be introduced to improve readability.

EXAMPLE 11.26   The following are sentences of the language of propositional logic extended by the definitions just described.

$p \Rightarrow (q \Leftrightarrow p)$

$r \wedge q \wedge p \Rightarrow p \vee q \vee r$

$\neg p \Rightarrow (p \Rightarrow q)$ ○

So far, we have only propositional variables. To proceed to predicate logic, we need individual and predicate variables.

Definition 11.6   The **language of predicate logic** is defined by the following grammar:

Terminal symbols:
The terminal symbols of the language of propositional logic, plus:

The quantifiers ∀ and ∃;
The comma (,);
An infinite set of individual variables $x, y, z, \ldots$;
For each positive integer n, an infinite set of n-ary predicate variables $P, Q, R, \ldots$

Nonterminal symbols:
The nonterminal symbols of the language of proposition logic, plus:

<individual variable>
<1-ary predicate variable>
<2-ary predicate variable>
⋮

Productions:
The productions of the language of propositional logic, plus:
<individual variable> ::= $x \mid y \mid z \mid \ldots$
<1-ary predicate variable> ::= $P, Q, \ldots$

<2-ary predicate variable> ::= $R, T, \ldots$
$\vdots$

S ::= <1-ary predicate variable> (<individual variable>)
S ::= <2-ary predicate variable> (<individual variable>,
<individual variable>)
$\vdots$

S ::= $\forall$<individual variable> S
S ::= $\exists$<individual variable> S

EXAMPLE 11.27   The following formulas are sentences of the language of predicate logic:

$\forall x p$

$\exists x Q(x)$

$(\neg p \vee (\forall x Q(x) \vee R(x, z)))$

$\forall x (\neg P(x) \vee \exists y T(x, y))$ o

The definitions to extend the language of propositional logic apply, as does the introduction of operator hierarchy.

The language of mathematics can now be completed by adding only two symbols, the equals sign ($=$) and the $\in$ (is an element of) relation of set theory.

Definition 11.7   **The language of mathematics** is defined by the following grammar:

Terminal symbols:
The terminal symbols of the language of predicate logic, plus the symbols $\in$ and $=$
Nonterminal symbols:
The nonterminal symbols of the language of predicate logic
Productions:
The productions of the language of predicate logic, plus:
S ::= (<individual variable> $=$ <individual variable>)
S ::= (<individual variable> $\in$ <individual variable>)

EXAMPLE 11.28   The following formulas are sentences of the language of mathematics:

$(\neg (x \in y) \vee (x = z))$

$\forall y \forall x (\neg (x = y) \vee (y = x))$ o

The language of mathematics includes, of course, all the formulas of logic. For example, the second formula of Example 11.28 may be written

$\forall y \forall x ((x = y) \Rightarrow (y = x))$

which is just the symmetric law for equality.

In order to write mathematical sentences concisely, we need one more type of definition, the introduction of names. When it has been proven for some property P that there exists one and only one object x having that property, it is convenient to give that object a name. The simplest case is that of the empty set $\varnothing$. $\varnothing$ is characterized by the property

$$\forall y \neg (y \in x)$$

Therefore a statement $A(\varnothing)$ about the empty set may be restated, without the use of the name $\varnothing$, as

$$\forall x (\forall y \neg (y \in x) \Rightarrow A(x))$$

EXAMPLE 11.29   The statement

$$w \in \varnothing$$

may be translated as

$$\forall x (\forall y \neg (y \in x) \Rightarrow (w \in x)) \qquad\qquad\qquad ○$$

Thus, in principle, all mathematical statements can be reduced to sentences in the language of mathematics. The theorems of mathematics are those sentences in the language that can be deduced from the axioms of equality and of set theory using valid arguments. Of course, even in the most trivial of cases, it is not practical to do so, but the theoretical possibility raises the question: "Is it possible to write a computer program to examine a sentence in the language of mathematics and determine whether or not it is a theorem?"

Around 1900, it was still hoped in mathematical circles that the answer might be "Yes." (Although computers did not exist in 1900, the notion of an algorithm did.) It is now known that the answer is "No." The question plays a key role in the problem of the theoretical limits of computers, which is considered in the next chapter.

EXERCISES 11.4

1. Which of the following formulas are sentences of the language of propositional logic? Of predicate logic? Of mathematics?

   a. $(p \Rightarrow \forall x q)$ **b.** $\exists x P(x)$ **c.** $q \Rightarrow \forall p (p \Rightarrow q)$

2. Which of the following formulas are sentences of the language of propositional logic? Of predicate logic? Of mathematics?

   a. $\forall P \forall Q (P(x) \wedge Q(x) \Rightarrow Q(x) \wedge (P)x))$

   b. $(p \vee \neg q) \vee \neg s$ **c.** $\forall x (x \in y) \Rightarrow (x = z)$

3. The following formulas are common mathematical statements about sets written in the formal language of mathematics as defined in the section. Restate each formula in everyday mathematical notation or in plain English.

a. $(\neg(z \in x) \vee (z \in y))$

b. $(w \in x) \wedge \forall z(\neg(z \in x) \vee (z \in y)) \Rightarrow (w \in y)$

4. Restate in everyday mathematical notation or in plain English:

$$\forall x \forall y \forall z (\forall w((w \in x) \Rightarrow (w \in y)) \Rightarrow (\forall w((w \in y) \Rightarrow (w \in z))$$
$$\Rightarrow \forall w((w \in x) \Rightarrow (w \in z))))$$

Computer Exercises for Chapter 11

1. Examine a COBOL reference manual and compare the language's features for comment statements, compound statements, and conditional statements (IF . . . THEN) to the corresponding features of C, Pascal, and BASIC.

2. Examine the assembly language manual for your computer. What is the vocabulary? What method is used to describe the grammar of the language? Discuss relative merits of BNF, syntax diagrams, and statement patterns for defining the language.

3. Examine a book or reference manual on the Ada programming language. What method is used to define the grammar of the language? Compare the language to Pascal and C in the areas of comment statements, compound statements, use of the semicolon, and conditional statements.

4. Examine the operating system reference manual for your computer. The manual will define a language of operating system commands. Write the grammar of this language in BNF. Define the same grammar by a set of syntax diagrams.

• • • • CHAPTER REVIEW EXERCISES

1. Give informal descriptions (such as that in Example 11.4) of the grammars of classes of English sentences containing these sentences:

a. To err is human.   b. To be is to do.

c. Digging a hole is hard work.

2. What are the sentences of the language defined by the grammar

       root symbol: S

    $S ::= a \,|\, aT$

    $T ::= b \,|\, bS$

3. What are the sentences of the language defined by the grammar

       root symbol: S

    $S ::= b \,|\, Tb \,|\, bT$

    $T ::= x \,|\, Tx$

4. What are the sentences of the language defined by the grammar

    root symbol: S

    $S ::= aSb \,|\, Sb \,|\, <empty>$

5. Write a grammar to define the language whose sentences are sequences of x's and y's where the number of x's and the number of y's are both odd.

6. Write a grammar to define the language whose sentences are sequences of x's and y's where all of the y's are together (e.g., $xyyyxxxx$).

7. For the grammar of Exercise 4, draw parsing trees for each of the following sentences:

    a. *abb*   **b.** *aabb*   **c.** *bb*

8. For the grammar of Exercise 5, draw parsing trees for each of the following sentences:

    a. *xyxx*   **b.** *xyxyyx*

9. Draw syntax diagrams to define the language of Exercise 4.

10. Draw syntax diagrams to define the language of Exercise 5.

11. Write a context-free grammar for the language defined by the syntax diagram of Figure 11.12.

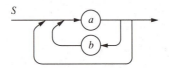

FIGURE 11.12

12. Write a context-free grammar for the language defined by the syntax diagram of Figure 11.13.

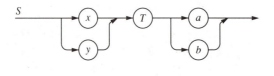

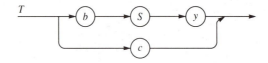

FIGURE 11.13

13. Define the language of Exercise 3 by statement patterns.

14. Define the language of Exercise 6 by statement patterns.

MACHINES AND COMPUTATION

The study of the theoretical limits of computing machines predates by many years the construction of practical computers. By the turn of the century, mathematicians were interested in whether there were "effective processes" (what we now call algorithms) for solving certain classes of problems. For example, is there an algorithm for determining whether a formula of propositional logic is a tautology? There is—it is truth-table analysis. Is there an algorithm for determining whether a formula of predicate logic is valid? It is now known that no such algorithm exists. Is there an algorithm for determining whether any mathematical statement is a theorem? It has been shown that the answer to this question is "No." In this chapter, we will provide a basis for the study of these kinds of questions by examining the relationship of algorithms to mathematical models of computing machines.

12.1 • MODELS OF COMPUTATION

To show that there is an algorithm for solving a particular problem, it is only necessary to produce one. But to show that no such algorithm exists, an analysis of what can and cannot be done by algorithms is required. For this, we need a precise definition of *algorithm*. The definition is provided by proposing some kind of abstract machine, then defining an algorithm to be any process that can be performed on the machine. It is then necessary to determine mathematically what the machine can and cannot do.

In this section, we give some examples of abstract machines and discuss their relationship to practical computers. Later we will determine some basic facts about their capabilities.

The modern computer, whether a personal computer or large mainframe, consists of a few basic subsystems: a processor, a central memory, and input/output (I/O) devices, connected as shown in Figure 12.1. The execution of a computer program can be thought of as a series of steps carried out by the central processor:

Read an item from an input device
Take some action that depends on the contents of the central memory
Change the contents of central memory
(Possibly) write an item to an output device

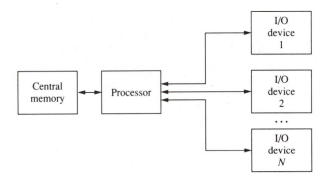

FIGURE 12.1   The components of a computer

This sequence of events is shown in Figure 12.2.

At each step, how the input is processed and what output is generated depends on the state of the central memory—that is, on its contents. If the memory contains M bits, at any time it can be in any of 2^M different states.

Although the various concepts of abstract machines predated the invention of memories organized as bits, the concept of the state of the machine at any time is found in the earliest models of computing machines. In general, an **abstract machine** consists of a processor which can be in any of N states, together with one or more input and output devices. The abstract machines differ from one another in the number and type of input-output devices, and in the operations that are allowed on these devices.

The input and output devices found on modern computers include a wide variety of equipment, including keyboards, displays, punched-card readers, printers, magnetic tapes, and magnetic disks. But in discussing mathematical models of

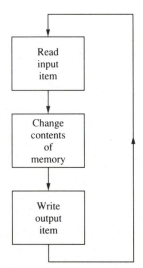

FIGURE 12.2   Events in computer program execution

computing machines, it would be preferable not to have to deal with such variety. A small number of simple, idealized devices would be easier to handle. To identify our ideal input and output devices, we can examine actual devices and pick out their common features.

Most types of computer I/O devices are **serial** in nature. That is, they transmit or receive information as a sequence of items. Devices such as keyboards, displays, and printers transmit and receive information as a sequence of characters. Tape drives store information as a sequence of characters. In fact, a conventional character-oriented device such as a keyboard or a printer could be replaced by a tape containing the sequence of characters to be typed or printed. These considerations lead to the choice of tape as the "ideal" device for abstract models of computing machines.

Figure 12.3 shows the general concept of an abstract computing machine consisting of a processor, a memory, and one or more tape drives. We must assume that each "reel" of tape is infinitely long; otherwise, the tape could hold only a fixed amount of information and could be considered to be just an extension of central memory. Abstract machines differ from one another in the number of tape drives provided, in whether the tapes are infinite in both directions or only one, in the set of characters that can be written on the tape, and in the set of operations allowed on the tape. Operations on tape drives include reading, writing, advancing, backspacing, and rewinding. On most modern tape drives, writing information at a point on the tape causes all data already written beyond that point to become inaccessible. However, "random access" tape drives, in which information in the middle of a tape could be overwritten without affecting other data, were made by some manufacturers before the cost of disk drives became low enough to be affordable by all users.

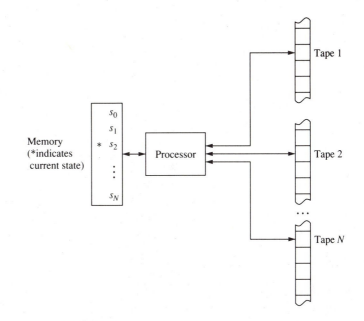

FIGURE 12.3   The general abstract computing machine

EXERCISES 12.1

1. If a machine has a 32-bit memory, how many states can it have?

2. Consider the keyboard of a computer or terminal as an input tape. What is the set of allowed characters?

3. Consider a computer printer as an output tape. What is the set of allowed characters?

4. Discuss the relationship of other computer I/O devices (disk drives, cathode-ray tube displays, plotters, etc.) to the concept of abstract machine considered in this section.

12.2 ● FINITE STATE MACHINES

A finite state machine is an abstract machine with one input tape that can be read in the forward direction only. The finite state machine is the simplest of the abstract computing machines. In fact, it is simple enough that we can define its capabilities precisely.

Definition 12.1

A finite state machine consists of

a. An **input alphabet** A, which is a finite set of symbols
b. A finite set S, called the set of **states** of the machine
c. An element $s_0 \in S$, called the **initial state**
d. A function $F: A \times S \to S$, called the **transition function**

EXAMPLE 12.1   Let $A = \{a, b, c\}$ be the input alphabet. Let $S = \{T, U, V\}$ be the set of states, with $s_0 = T$ as the initial state. Define $F: A \times S \to S$ by the following table:

Symbols

		a	b	c
	T	U	V	T
States	U	V	T	U
	V	T	V	V

The table shows that $F(a, T) = U$, $F(b, T) = V$, and so on. The alphabet A, the set of states S, the initial state T, and the transition function F form a finite state machine.

○

You can think of a finite state machine as operating in the following way. The machine is provided with a finite sequence of characters $a_0, a_1, a_2, \ldots, a_n$ on its input tape. The characters a_i are elements of the input alphabet A. The machine

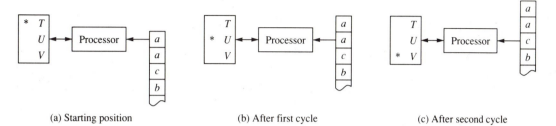

(a) Starting position   (b) After first cycle   (c) After second cycle

FIGURE 12.4   The machine of Example 12.1

starts operating in its initial state s_0. At the first step, the first input character a_0 is read, then the machine shifts to the state $s_1 = F(a_0, s_0)$, defined by the first input character and the initial state. At the second step, the machine reads the next character from the tape and shifts to the state $s_2 = F(a_1, s_1)$. The machine continues to operate in this way until all characters in the input sequence have been processed. The result of the computation is the state in which the machine finishes.

Figure 12.4 shows the operation of the machine of Example 12.1 when given an input tape with the characters a, a, c, b. The machine starts in state T with the tape positioned at the first character, a. Since $F(a, T) = U$, the machine changes to state U and advances the tape. At the second step, the machine is in state U, and the input is again a. Since $F(a, U) = V$, the machine shifts to state V and advances the tape. With this input, the machine remains in state V for two more steps and then halts.

It is sometimes convenient to think of a finite state machine as providing an answer to a yes-or-no question. We can do this by dividing the set of states into two classes: the accepting or "yes" states, and the rejecting or "no" states. If the machine stops in an accepting state, it is said to have accepted its input, or to have answered "yes." If it stops in a rejecting, or "no" state, it is said to have rejected its input, or to have answered "no." For instance, in Example 12.1, we may choose T and U to be the accepting states and V to be the rejecting state. Then the machine rejects the input a, a, c, b. With such a convention, we can design machines to make decisions about input sequences.

EXAMPLE 12.2   Let the input alphabet of a machine consist of the letters a and b only. Describe a machine that determines whether the input tape contains both a's and b's. We need four states: S (start), A (all a's), B (all b's), and M (mixed). We choose M as the accepting state (yes, there are both a's and b's). The transition function F is defined by the following table:

	Symbols	
	a	b
S	A	B
A	A	M
B	M	B
M	M	M

States

The machine starts in state S. If the first input character is an a, the machine enters state A, and it will stay there as long as the input characters continue to remain a's. Similarly if the first input character is a b. Once the machine has encountered both an a and a b, it enters state M and stays there.   o

We can depict the operation of a finite state machine for a particular input sequence by writing the input characters and below them the states through which the machine passes. For instance, given the input sequence $aaaaabaab$, the machine of Example 12.2 operates as follows:

		a	a	a	a	a	b	a	a	b
S	A	A	A	A	A	M	M	M	M

Start $\longrightarrow$ $\overset{}{\llcorner}$ $F(a, S)$ $\overset{}{\llcorner}$ $F(b, A)$

Notice that the initial state stands separately on the left.

In Examples 12.1 and 12.2, the transition function was presented by a table. Another way to present a transition function is by drawing a **state diagram,** or **transition diagram,** in which the states are shown as labeled circles and the transitions are shown by labeled arrows. The start state is indicated by an arrow from a solid dot. It is customary to indicate the accepting states by double circles. The state diagram for Example 12.2 is shown in Figure 12.5.

The accepting state M (indicating that both a's and b's have been encountered) is drawn as a double circle. The arrow from S to A is labeled a to show that when input a is encountered in state S, the machine moves to state A. In a transition diagram for a finite state machine, the labels on the arrows leaving each state must account for all characters in the input alphabet. If several input characters cause the same transition from a state, we can reduce clutter by drawing one arrow and writing all the corresponding characters beside it.

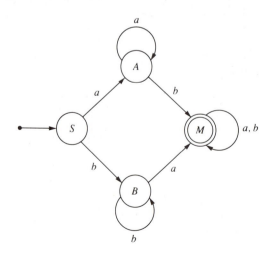

FIGURE 12.5   State diagram for Example 12.2

EXAMPLE 12.3   Consider a simple vending machine that dispenses one item priced at 20 cents. As a customer deposits coins, the machine must keep track of the total amount deposited. As soon as the customer has deposited at least 20 cents, the machine dispenses the item and returns any change due. This machine has four states: I (idle), $S5$ (5 cents deposited), $S10$ (10 cents deposited), and $S15$ (15 cents deposited). The transition function is described by the following table:

		Nickel	Dime	Quarter
	I	$S5$ (0)	$S10$ (0)	I (5)
	$S5$	$S10$ (0)	$S15$ (0)	I (10)
States	$S10$	$S15$ (0)	I (0)	I (15)
	$S15$	I (0)	I (5)	I (20)

The numbers in parentheses are the amounts of change returned to the customer. They are not a part of the definition of the transition function. The state diagram for this machine is shown in Figure 12.6.

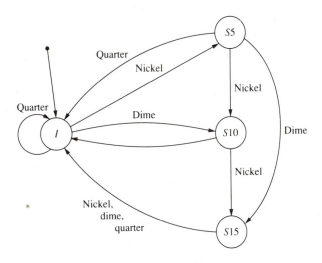

FIGURE 12.6   State diagram for Example 12.3   o

Machines with Outputs

Example 12.3 suggests that we consider a slightly more powerful abstract machine: a finite state machine with outputs. In the example, each transition was accompanied by another action: returning change to the customer. In 7 of the 12 entries, the amount returned is zero.

Definition 12.2   **A finite state machine with outputs** consists of

    a.   An **input alphabet** A, which is a finite set of symbols
    b.   An **output alphabet** B, which is also a finite set of symbols
    c.   A finite set **S**, called the set of **states** of the machine
    d.   An element $s_0 \in \mathbf{S}$, called the **initial state**
    e.   A function $F: A \times \mathbf{S} \to \mathbf{S}$, called the **transition function**
    f.   A function $G: A \times \mathbf{S} \to B$, called the **output function**

You can think of a finite state machine with outputs as operating in the following way. The machine is provided with a finite sequence of characters $a_0, a_1, a_2, \ldots, a_n$ on its input tape. The machine starts operating in its initial state s_0. At the first step, the first input character a_1 is read, then the machine shifts to the state $s_1 = F(a_0, s_0)$, defined by the first input character and the initial state. It also writes the output character $G(a_0, s_0)$ to the output tape. At the second step, the machine reads the next character from the input tape and shifts to the state $s_2 = F(a_1, s_1)$, writing the character $G(a_1, s_1)$ to the output tape. The machine continues to operate in this way until all characters in the input sequence have been processed. The result of the computation is the sequence of characters written to the output tape.

We can depict the operation of a finite state machine with outputs by listing the characters on the input tape and, below them, the new states and the output characters produced. For the machine of Example 12.3, if the input is nickel, nickel, quarter, nickel, dime, dime, dime, dime, then the operation is

Inputs		n	n	q	n	d	d	d	d
States	I	$S5$	$S10$	I	$S5$	$S15$	I	$S10$	I
Outputs		0	0	15*	0	0	5*	0	0*

The output alphabet in the example consists of the symbols

$$0, 5, 10, 15, 20, 0*, 5*, 10*, 15*, 20*$$

where the * indicates that an item is dispensed to the customer.

You can think of a machine with accepting and rejecting states as a special case of a machine with outputs where $G(a, s) = $ "YES" if s is an accepting state, and $G(a, s) = $ "NO" if s is a rejecting state. The machine accepts an input if the last symbol it writes on its output tape is "YES."

In designing a machine with outputs, it is sometimes desirable to have the machine produce no output at all for some combinations of input character and state. Such a machine may be included in the concept of Definition 12.2 by allowing the output alphabet B to contain a "null" character. A null character is a character that is to be ignored when the output is interpreted. The following example illustrates this idea and a set of accepting states.

EXAMPLE 12.4   This machine examines a sequence of keyboard characters (letters, digits, punctuation marks, the space, special characters) and determines whether it contains an integer (a sequence of digits, or a sequence of digits preceded by a minus sign, with all other characters blank). The input alphabet consists of all keyboard characters. The output alphabet consists of the digits 0 through 9, the minus sign, and NULL. The transition and output functions are defined by the following table. The output character is written in parentheses.

		Symbols		
		$-$	Digit	Blank	Other
	S	$M(-)$	P(digit)	S(NULL)	X(NULL)
	P	X(NULL)	P(digit)	P'(NULL)	X(NULL)
	P'	X(NULL)	X(NULL)	P'(NULL)	X(NULL)
States	M	X(NULL)	N(digit)	X(NULL)	X(NULL)
	N	X(NULL)	N(digit)	N'(NULL)	X(NULL)
	N'	X(NULL)	X(NULL)	N'(NULL)	X(NULL)
	X	X(NULL)	X(NULL)	X(NULL)	X(NULL)

The initial state is S. The state P indicates an apparent positive number, and P' indicates a completed positive number. The state M indicates that the last character was a minus sign and that a digit is therefore expected. The state N indicates an incomplete negative number, and the state N' indicates a completed negative number. Finally, the state X indicates that the input is not a number. The minus sign (if any) and the digits are copied to the output tape. The accepting states are P, P', N, and N'. If the machine halts in an accepting state, the input contains a number and the number (after ignoring NULLs) appears on the output. If the machine halts in a rejecting state, the input does not contain a number, and the output is meaningless.

○

Finite state machines are a very useful way of describing many physical devices, from vending machines to telephone keypads. They also provide a systematic way to organize a computer program that simulates a physical device. But there are many tasks that cannot be performed by such machines. The following example illustrates their limitations.

EXAMPLE 12.5   Suppose a machine is to be provided with an input string that consists of some number of a's followed by some number of b's. The task is to determine whether the number of a's is equal to the number of b's. We will show that no finite state machine can perform this task. Suppose such a machine exists, and let N be the number of states in the machine. Provide an input that consists of $(N + 1)$ a's followed by some number of b's. Denote by S_k the state of the machine after it has processed the kth a, and let S_0 be the start state. Then $S_0, S_1, \ldots, S_N$ is a list of $N + 1$

states of the machine. Since there are only N different states, by the Pigeonhole Principle, at least two of the states in the list must be the same. Suppose that these are S_j and S_k: $S_j = S_k$. By returning to state S_j when the kth a is input, the machine has "lost" $(k - j)$ a's, and there is no way that it can distinguish the actual input of $N + 1$ a's from an input of $(N + 1) - (k - j)$ a's. Therefore, there is no way to determine whether the number of b's equals the number of a's.   ○

The idea behind Example 12.5 is this: Since a finite state machine does not have access to any memory besides its own internal memory (set of states), it can be defeated by posing a problem that requires more memory than the machine contains.

EXERCISES 12.2

1. Refer to Example 12.1. If T is the accepting state, find the result of running the machine with the following inputs:

    a. a, a, c, b **b.** a, b, c, b **c.** b, b, b

2. Refer to Example 12.1. If T is the accepting state, find the result of running the machine with the following inputs:

    a. c, b, a, c, b, a, a **b.** c, c, b, b, c, b, a **c.** c, c, c, c, c, c, b

3. Refer to Example 12.3. What is the input alphabet in this example? What is the set of states? What is the initial state?

4. Refer to Example 12.3. List the sequence of states and outputs corresponding to the following sequences of inputs:

    a. Nickel, nickel, nickel, nickel, nickel

    b. Quarter, nickel, quarter, nickel, dime

5. Refer to Example 12.3. List the sequence of states and outputs corresponding to the following sequences of inputs:

    a. Dime, dime, dime, quarter, nickel

    b. Nickel, dime, nickel, dime, nickel, dime, nickel

6. Draw the state diagram for the machine of Example 12.1.

7. Draw the state diagram for the machine of Example 12.4.

8. Describe each of the following devices as a finite state machine. In each case, specify the input alphabet, the output alphabet (if any), the set of states, the initial state, the transition function, and the output function (if any).

    a. An electrical fuse

    b. An electrical circuit breaker (A circuit breaker has three positions: on, off, and tripped.)

    c. A microwave oven with a digital keypad control

9. Repeat Exercise 8 for the following:

    a. A burglar alarm with a keypad control. To arm the alarm, you enter the code 7, 9. To disarm the alarm or to reset it after a burglary, you enter the code 5, 4, 2, 7.

b. A tape recorder

10. Draw the state diagram for the machine of Exercise 8a.

11. Draw the state diagram for the machine of Exercise 8b.

12. Identify three additional everyday devices that operate as finite state machines. For each device, specify the input alphabet, the output alphabet (if any), the set of states, the initial state, the transition function, and the output function (if any). Draw the state diagrams.

12.3 • LANGUAGE RECOGNITION

One of the central problems of computer systems is recognition and interpretation of user inputs. This includes both computer programs written in programming languages, such as FORTRAN or Pascal, and commands typed into utility programs, such as spreadsheets or database programs. The acceptable statements or commands are a language defined by a grammar. Abstract computing machines provide a way of approaching the problem. For each class of abstract machines, there is a class of languages that the machines can recognize. We will illustrate this with some simple examples using finite state machines.

EXAMPLE 12.6   Consider the language over the alphabet $\{a, b\}$ consisting of the sentences ab and ba only. This language may be defined by a grammar with terminal symbols a and b, nonterminal symbol S, and production

$$S ::= ab \,|\, ba$$

Design a finite state machine that determines whether its input belongs to the language. Call the start state S. When the machine is in state S, either an a or a b can be the start of a sentence of the language, so we need two more states, A and B. In state A, only a b is acceptable, and in state B, only an a is acceptable. So we draw two more states, A' and B', which are accepting states. We have now arrived at Figure 12.7. This diagram does not yet define a finite state machine, since it does not

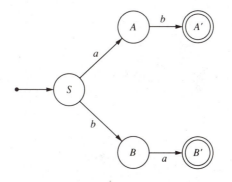

FIGURE 12.7   Example 12.6—first step

indicate what happens if the machine encounters an *a* while in state *A*, a *b* while in state *B*, or anything while in state *A'* or *B'*. In all of these cases, the input does not belong to the language. We complete the diagram by adding a state *T* and drawing lines to it for all the unacceptable inputs. The result is Figure 12.8. The state *T* is called a "trap" state or a "black hole." Whenever the machine enters this state, it can never leave.

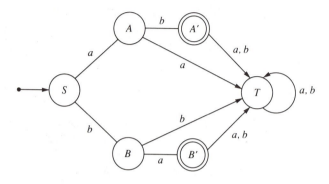

FIGURE 12.8   Example 12.6—complete design   o

The machine constructed in Example 12.6 accepts exactly those input strings that belong to the language. We say that the machine **recognizes** the language.

The machine of Figure 12.8 can be simplified by combining states *A'* and *B'* into a single "end" state *E*. The result of doing so is shown in Figure 12.9.

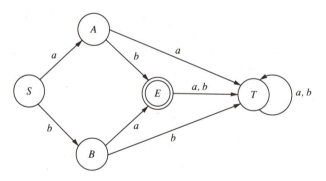

FIGURE 12.9   Example 12.6—simplified machine

EXAMPLE 12.7   Let the alphabet be {*a*, *b*, *c*}, and consider the language consisting of the sentences *abc*, *ab*, and *c*. The language may be defined by a grammar with the production

$$S ::= abc \mid ab \mid c$$

Again, we can build a state diagram by starting with a start state *S* and adding states for each of the ways to build a sentence. With the trap state *T* added at the end, we obtain the diagram of Figure 12.10. The machine having this state diagram recognizes the language {*abc*, *ab*, *c*}.   o

The machine of Figure 12.10 could be simplified by combining states C and D. However, there is no way to combine state B with state C or D without changing the language recognized by the machine.

The languages in Examples 12.6 and 12.7 were finite languages: They each contained a finite number of sentences. Here are two examples of infinite languages that can be recognized by finite state machines.

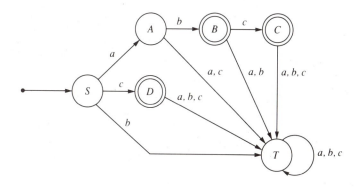

FIGURE 12.10   A machine for the language of Example 12.7

EXAMPLE 12.8   Consider the grammar with nonterminal symbol S, terminal symbols a, b, and c, and the production

$$S ::= abS \mid c$$

The language generated by this grammar consists of the sentences of the form $(ab)^n c$, where $n \geq 0$. That is, it consists of c, abc, $ababc$, $abababc$, and so on. The language is recognized by the machine in Figure 12.11. If the input is of the form $(ab)^n c$, the machine will loop between states S and A until the c is encountered, then go to state E. Otherwise, the machine will enter the trap state T.

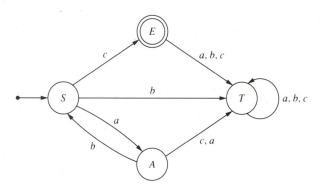

FIGURE 12.11   The machine of Example 12.8   o

EXAMPLE 12.9   Design a machine to examine a string of 0's and 1's and determine whether there is an even number of 1's.

The strategy is to keep track of the number of 1's encountered so far. We need two states: E to indicate that there has been an even number of 1's, and F to indicate that there has been an odd number. If the machine halts in state E, the total number of 1's in the string was even. So E will be the accepting state. The machine changes state only when it encounters a 1 on the input tape. The state diagram for this machine is shown in Figure 12.12.

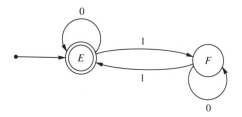

FIGURE 12.12   The machine of Example 12.9   o

We can write down a grammar for the language consisting of strings of 0's and 1's having an even number of 1's. We use two nonterminal symbols E and F; E will produce strings with an even number of 1's, and F will produce strings with an odd number of 1's. The initial symbol is E. The productions of the grammar are:

$$E ::= 0E \mid 1F \mid 0$$
$$F ::= 0F \mid 1E \mid 1$$

In Example 12.9, there is a one-to-one correspondence between the states of the machine (E, F) and the nonterminal symbols of the grammar (also called E, F). There is also a correspondence between transitions in the diagram and certain productions: The transition from E to E marked 0 corresponds to the production $E ::= 0E$, and the transition from E to F marked 1 corresponds to the production $E ::= 1F$.

You may have noticed that all of the grammars in Examples 12.6 through 12.9 are regular (or type 3) grammars. Recall from Chapter 11 that a regular grammar is one for which each production is of the form

$$A ::= x_1 \cdots x_n B$$

where A and B are nonterminal symbols and the x_i's are terminal symbols. It is rarely possible to achieve such a perfect correspondence as that exhibited in Example 12.9. However, by looking for such correspondences, we shall be able to see the relationship between finite state machines and regular grammars.

Finite and Infinite Languages

We have seen in the preceding examples that finite state machines can recognize both finite and infinite languages. The languages of Examples 12.6 and 12.7 are finite, and the languages of Examples 12.8 and 12.9 are infinite. We can see how the machines of Examples 12.8 and 12.9 can recognize infinite languages by examining their state diagrams, Figures 12.11 and 12.12. Both contain loops on paths from the start state to the accepting state. In Figure 12.11, the loop passes between states S and A. With suitable input (namely, alternating a's and b's), the machine can alternate between states S and A an arbitrarily large number of times before exiting the loop to state E. In Figure 12.12, the machine completes a loop from E to F and back for every pair of 1's that it encounters on input. The existence of such loops characterizes machines that accept infinite languages.

Theorem 12.1   Let M be a finite state machine with N states. Then the following three statements are equivalent:
  - (a) In the state diagram of M, there is a path from the start state to an accepting state that contains a loop.
  - (b) The machine accepts at least one input string with length greater than or equal to N.
  - (c) The language recognized by the machine is infinite.

Proof   We will show that (a) implies (c), (c) implies (b), and (b) implies (a).

First, we show that (a) implies (c). If the state diagram has a loop on a path from the start state to an accepting state, then there is a portion of the diagram that looks like Figure 12.13. Then the machine accepts the input $x_1 \ldots x_n z_1 \ldots z_k$. But it also accepts the inputs

$$x_1 \ldots x_n y_1 \ldots y_m z_1 \ldots z_k$$
$$x_1 \ldots x_n y_1 \ldots y_m y_1 \ldots y_m z_1 \ldots z_k$$
$$x_1 \ldots x_n y_1 \ldots y_m y_1 \ldots y_m y_1 \ldots y_m z_1 \ldots z_k$$

In fact, it accepts all sentences of the form

$$x_1 \ldots x_n (y_1 \ldots y_m)^j z_1 \ldots z_k$$

where j is any positive integer. So the language recognized by the machine is infinite.

The implication (c) $\Rightarrow$ (b) is trivial. If there are infinitely many sentences recognized by the machine, at least one of them must have more than N symbols.

To complete the proof, we need to show that (b) implies (a). Let $x_1 \ldots x_k$ be a sentence accepted by the machine, with $k \geq N$. Let S_0 be the start state of the

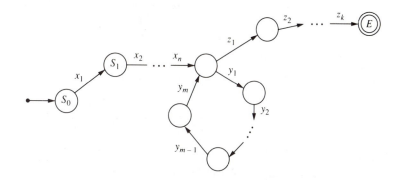

FIGURE 12.13   A loop in a state path

machine, and let $S_0, S_1, \ldots, S_k$ be the sequence of states through which the machine passes as it accepts $x_1 \ldots x_k$. That is, $S_i = F(x_i, S_{i-1})$, where F is the transition function of the machine. By the Pigeonhole Principle, two of the states $S_0, S_1, \ldots, S_k$ must be the same. Assume $S_m = S_n$, where $m < n$. Then the sequence of states from S_m to S_n is the path loop we are looking for.   ●

This theorem shows that if a machine M recognizes a finite language, the number of states of M must be at least one greater than the longest string in the language.

EXERCISES 12.3

1. List the sequence of states if the machine of Example 12.6 is given each of the following inputs:

   a. *a*   **b.** *aa*   **c.** *bb*   **d.** *aba*

2. List the sequence of states if the machine of Example 12.7 is given each of the following inputs:

   a. *aabc*   **b.** *cab*   **c.** *ac*   **d.** *bb*

3. List the sequence of states if the machine of Example 12.8 is given each of the following inputs:

   a. *a*   **b.** *ab*   **c.** *abac*

   d. *ababab*   **e.** *abababc*   **f.** *bababac*

4. List the sequence of states if the machine of Example 12.9 is given each of the following inputs:

   a. 00110101   **b.** 11010101

   c. 01111101   **d.** 100000000011

5. Refer to Figure 12.14. List the sequence of states if the machine of Figure 12.14 is given each of the following inputs. Which are accepted by the machine?

   a. *abcded*   **b.** *abcabcded*

   c. *abcde*   **d.** *deabda*

6. Describe the language recognized by the machine of Figure 12.14. Is the language finite or infinite? Explain.

7. Refer to Figure 12.15. List the sequence of states if the machine of Figure 12.15 is given each of the following inputs. Which are accepted by the machine?

   a. *ac*   **b.** *ca*   **c.** *acb*

   d. *cbc*   **e.** *cbcbcbc*   **f.** *abacbb*

8. Describe the language recognized by the machine of Figure 12.15. Is the language finite or infinite? Explain.

9. Draw state diagrams for machines to recognize the following languages:

   a. $\{a, b, c, d\}$ **b.** $\{abc, bcd, cda, dab\}$

   c. $\{a, ab, abc, abcd\}$ **d.** $\{aaab, abbb\}$

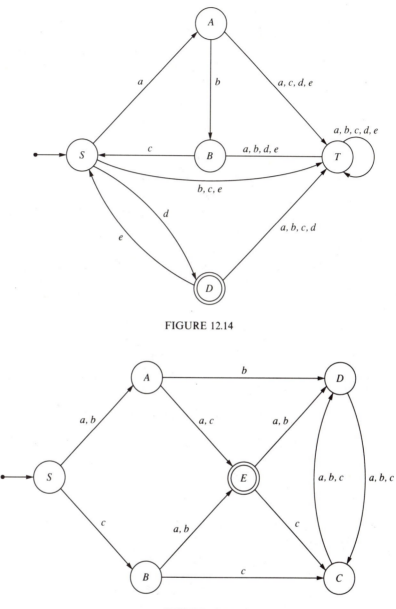

FIGURE 12.14

FIGURE 12.15

10. Design a finite state machine to recognize strings over the alphabet $\{a, b\}$ having at least three a's.

11. List the sequence of states when the machine of Exercise 10 is given each of the following inputs:

   a. *aa*   **b.** *ababa*

   c. *bbaba*   **d.** *bbabbabbaa*

12. Design a finite state machine to recognize strings over the alphabet $\{0, 1\}$ having an odd number of 1's and an even number of 0's.

13. List the sequence of states when the machine of Exercise 12 is given each of the following inputs:

    a. 101010   b. 11100   c. 00110   d. 110100

14. Design a finite state machine to recognize strings over the alphabet $\{a, b, c\}$ that either

    (i)   Are of the form $a^n b$ for some positive integer n, or
    (ii) Belong to the set $\{ac, acc, accc\}$

15. List the sequence of states when the machine of Exercise 14 is given each of the following inputs:

    a. *acc*   b. *aaaab*   c. *abbc*   d. *acb*

12.4 • FINITE STATE MACHINES AND REGULAR GRAMMARS

Our purpose here is to prove the following theorem.

Theorem 12.2   A language can be defined by a regular grammar if and only if it can be recognized by a finite state machine.

To prove the theorem, we will do two things:

1. Given a finite state machine, write a regular grammar that defines the language recognized by the machine.
2. Given a regular grammar, design a finite state machine to recognize the language defined by the grammar.   ●

For part (1), we will illustrate the techniques by a series of examples. Part (2) of the proof will be given in the next section.

Recall that in Example 12.9, we found a correspondence between the state diagram and the productions of the grammar. The nonterminal symbols of the grammar corresponded to the states of the machine. The four transitions in the state diagram corresponded to four of the six productions of the grammar:

$$E ::= 0E \qquad E ::= 1F$$
$$F ::= 0F \qquad F ::= 1E$$

There are two more productions in the grammar:

$$F ::= 1 \qquad \text{and} \qquad E ::= 0$$

These correspond to the two transitions into the accepting state E. Using these ideas, we will be able to build a grammar for any finite state machine.

EXAMPLE 12.10   Write a grammar for the finite state machine given by the diagram in Figure 12.16.

The machine accepts a string if and only if the string ends in b. Since there are two states and two symbols in the alphabet, there are four transitions. The productions

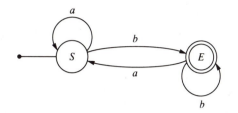

FIGURE 12.16   The machine of Example 12.10

for each of them are:

$$S ::= aS$$
$$S ::= bE$$
$$E ::= aS$$
$$E ::= bE$$

Two of the transitions go to an accepting state; we will write additional productions for each of them. The productions are:

$$S ::= b$$
$$E ::= b$$

The complete set of productions may be abbreviated

$$S ::= aS\,|\,bE\,|\,b$$
$$E ::= aS\,|\,bE\,|\,b$$

○

It is easy to see that the sentences of the grammar defined in Example 12.10 are exactly the strings that end in b. Moreover, there is an exact correspondence between the sequence of states that the machine passes through to accept an input and the sequence of productions used to generate the sentence. For the sentence $aaab$, the correspondence is as follows:

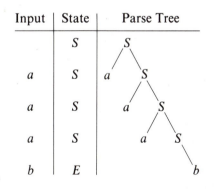

Of course, the grammar we constructed is not the simplest grammar for this language. The same language is also generated by a grammar with one nonterminal symbol S and the productions

$$S ::= aS \mid bS \mid b$$

EXAMPLE 12.11   Write a grammar for the machine in Figure 12.9.

According to our strategy, we will need five nonterminal symbols: S, A, B, E, and T. We will write ten productions (five states times two symbols), plus two productions for the two transitions to state E. The productions are:

$$S ::= aA \mid bB$$
$$A ::= aT \mid bE \mid b$$
$$B ::= bT \mid aE \mid a$$
$$E ::= aT \mid bT$$
$$T ::= aT \mid bT$$

○

The nonterminal symbol T plays the same role in the grammar as the state T in the machine: It is a trap. Once a production with a T is chosen, it is impossible to complete a sentence. In fact, we can simplify the grammar, without changing the language it generates, by discarding all productions that lead to T. We obtain

$$S ::= aA \mid bB$$
$$A ::= b$$
$$B ::= a$$

Note that the grammar we have constructed is not the same as the grammar given in Example 12.6.

EXAMPLE 12.12   Write a grammar for the machine of Figure 12.11.

We need four nonterminal symbols (corresponding to the states) and thirteen productions:

$$S ::= aA \mid bT \mid cE \mid c$$
$$A ::= aT \mid bS \mid cT$$
$$E ::= aT \mid bT \mid cT$$
$$T ::= aT \mid bT \mid cT$$

Again eliminating productions that lead only to the trap T, we get the simplified grammar

$$S ::= aA \mid c$$
$$A ::= bS$$

○

At this point, it is easy to see that the machine we start with, and the grammar we construct, accept the same sentences. For instance, if the input in Example 12.12 is *ababc*, the state sequence and corresponding parsing tree are:

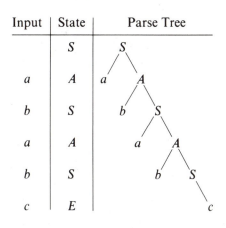

Input	State	Parse Tree
	S	
a	*A*	
b	*S*	
a	*A*	
b	*S*	
c	*E*	

The sequence of states that the machine passes through to accept a sentence again corresponds exactly to the sequence of productions to be applied to generate the sentence.

The reverse process—constructing a machine given a regular grammar—is more complex; it is considered in the next section.

EXERCISES 12.4

1. Write a grammar for the language recognized by the machine in Figure 12.17.

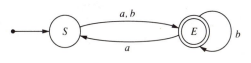

FIGURE 12.17

2. For the grammar of Exercise 1, draw parse trees for the following sentences:
   a. *abbab* **b.** *aaaabb*
3. For the grammar of Exercise 1, draw parse trees for the sentences:
   a. *bbab* **b.** *aaa*
4. Write a grammar for the language recognized by the machine in Figure 12.18.
5. For the grammar of Exercise 4, draw parse trees for the sentences:
   a. *bb* **b.** *ab*
6. For the grammar of Exercise 4, draw parse trees for the sentences:
   a. *baab* **b.** *aab*
7. Write a grammar for the language recognized by the machine in Figure 12.19.

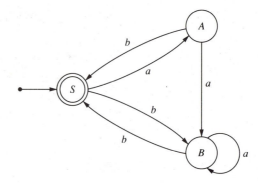

FIGURE 12.18

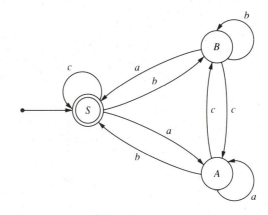

FIGURE 12.19

8.   Write a grammar for the language recognized by the machine in Figure 12.20.

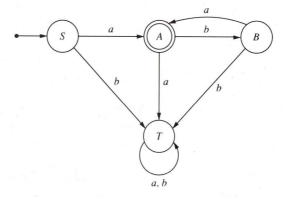

FIGURE 12.20

9.   Write a grammar for the language recognized by the machine in Figure 12.14 in Exercise 5 of Section 12.3.

10.   Write a grammar for the language recognized by the machine in Figure 12.15 in Exercise 7 of Section 12.3.

12.5 • DESIGNING A MACHINE TO RECOGNIZE A REGULAR LANGUAGE

Given a regular grammar, we would like to design a finite state machine to recognize the language defined by the grammar by reversing the process we used to construct a grammar from a machine. The plan is to use the nonterminal symbols of the grammar for the states of the machine. If the grammar contains a production of the form $A ::= bC$, we would like to draw a transition, labeled b, from state A to state C. And if the grammar contains a production of the form $A ::= b$, we would like to draw a transition, labeled b, from state A to some accepting state. These two building blocks are shown in Figure 12.21.

But what if the grammar contains productions of the form $X ::= a_0 a_1 \cdots a_n Y$? Then we can rewrite the grammar without using this form, but at the expense of introducing additional nonterminal symbols. We illustrate the technique with some examples.

(a) A transition for $A :: bC$ (b) A transition for $A :: b$

FIGURE 12.21   Building blocks for a machine to recognize a regular grammar

EXAMPLE 12.13   Rewrite the grammar whose productions are

$$S ::= abcS \mid de$$

using only productions of the form $X ::= aY$ and $X ::= a$. We need some additional nonterminal symbols, so choose K, L, M, and N. The production $S ::= abcS$ is equivalent to the productions

$$S ::= aK$$
$$K ::= bL$$
$$L ::= cS$$

Also, the production $S ::= de$ is equivalent to the productions

$$S ::= dM$$
$$M ::= e$$

○

The Normal Form of a Regular Grammar

Definition 12.3   A regular grammar is said to be in **normal form** if every production of the grammar is of the form $X ::= aY$ or $X ::= a$.

Theorem 12.3   Every regular grammar is equivalent to a regular grammar in normal form.

Proof   We can construct an equivalent grammar in normal form in two steps. First, we replace each production of the form $X ::= a_0 a_1 \ldots a_n Y$ with the productions

$$X ::= a_0 N_1$$
$$N_1 ::= a_1 N_2$$
$$\vdots$$
$$N_n ::= a_n Y$$

introducing a new set of additional nonterminal symbols for each such production. Second, eliminate productions of the form $X ::= Y$ by listing the right sides of productions for Y as productions for X.   ●

EXAMPLE 12.14   Write a normal form for the grammar whose productions are

$$S ::= abT \,|\, a \,|\, U$$
$$T ::= baS \,|\, bb$$
$$U ::= aS$$

Introducing new symbols as needed, we can write

$$S ::= aE \,|\, a \,|\, aS$$
$$E ::= bT$$
$$T ::= bF \,|\, bG$$
$$F ::= aS$$
$$G ::= b$$ ○

Note that we replaced $S ::= U$ with $S ::= aS$. The production for U could then be eliminated.

Thus, it is sufficient to find a way to design a finite state machine for regular grammars in normal form.

EXAMPLE 12.15   Design a machine for the grammar having productions

$$S ::= aS \,|\, bS \,|\, a$$

The sentences of the language generated by this grammar are those sequences of a's and b's that end in the symbol a. We will have a state S corresponding to the nonterminal symbol S. Using our strategy, we draw a transition, labeled a, from S to S, and another transition, labeled b, also from S to S. We must also draw a transition, labeled a, from S to an accepting state. What accepting state? It cannot be S, because then we would have a one-state machine, and a one-state machine can do nothing. We need to add an accepting state, call it E, and draw a transition from S to E. The result is shown in Figure 12.22.

Two things are wrong! First there are two transitions exiting state S and labeled a. Second, there are no transitions exiting state E. Our strategy has failed.

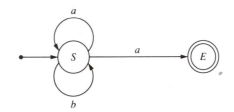

FIGURE 12.22   The machine of Example 12.15   ○

There is, of course, a way to salvage the situation and transform the "machine" of Figure 12.22 into a finite state machine. The solution will come at a price—greatly increasing the number of states. But first, to better understand the problem, let us consider another example.

EXAMPLE 12.16   Design a machine for a grammar having productions

$$S::= aA \mid bA$$
$$A::= bA \mid b$$

We need states S and A, and a terminal state E. For the productions for S, we draw two transitions to A. For the productions for A, we draw a transition to A and a transition to E. The result is shown in Figure 12.23.

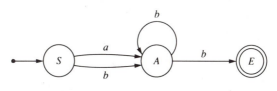

FIGURE 12.23   The machine of Example 12.16   ○

There are three problems with the diagram in Figure 12.23. First, there is no transition labeled a leaving state A. Second, there are two transitions leaving state A labeled b. Third, there are no transitions at all leaving state E.

Nevertheless, the diagrams of Figures 12.22 and 12.23 have a simple relationship to the languages defined by the corresponding grammars: A string of symbols belongs to the language if and only if there is a path through the diagram from the start state S to the end state E, whose transitions are labeled with the symbols of the string. For example, the string $abbb$ belongs to the language of Example 12.16. In Figure 12.23, you can get from S to E through states A, A, A by following the path in Figure 12.24.

The path can also be described by the table

	a	b	b	b
S	A	A	A	E

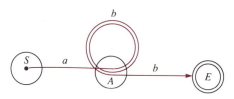

FIGURE 12.24   The state path for *abbb*

The language of Example 12.15 contains the string *aababa*. In Figure 12.22, you can get from S to E through states S, S, S, S, S by following the path in Figure 12.25. The path can also be described by the table

	a	a	b	a	b	a
S	S	S	S	S	S	E

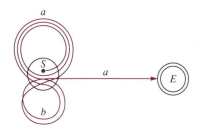

FIGURE 12.25   The state path for *aababa*

Nondeterministic Finite State Machines

You can interpret a diagram such as Figure 12.22 or 12.23 as assigning to each input symbol a and each state s, a set of states to which the machine can go. That is, the transition function F has as its range the power set $P(\mathbf{S})$ of the set of states $\mathbf{S}$, rather than $\mathbf{S}$ itself. $F(a, s)$ can be empty, or can contain one element, or can contain more than one element.

Definition 12.4

A **nondeterministic finite state machine** consists of

a. An **input alphabet** A, which is a finite set of symbols
b. A finite set $\mathbf{S}$, called the set of **states** of the machine
c. An element $s_0 \in \mathbf{S}$, called the **initial state**
d. A function $F: A \times \mathbf{S} \to P(\mathbf{S})$, the set of subsets of $\mathbf{S}$, called the **transition function**

Note that a nondeterministic finite state machine is *not* a finite state machine. In a finite state machine, $F(a, s)$ is a state, whereas in a nondeterministic finite state machine, $F(a, s)$ is a set of states. But if a nondeterministic finite state machine is such that every set $F(a, s)$ has exactly one element, that machine is equivalent to a finite state machine. A finite state machine is sometimes called a **deterministic** finite state machine.

The diagrams in Figures 12.22 and 12.23 define nondeterministic finite state machines. In Example 12.15, the transition function is given by the table

Symbols

		a	b
	S	$\{S, E\}$	$\{S\}$
States			
	E	$\varnothing$	$\varnothing$

In Example 12.16, the transition function is given by the table

Symbols

		a	b
	S	$\{A\}$	$\{A\}$
States	A	$\varnothing$	$\{A, E\}$
	E	$\varnothing$	$\varnothing$

As with a finite state machine, we can think of a nondeterministic finite state machine as accepting or rejecting input strings, provided that we distinguish a set of accepting states. We have seen that a nondeterministic finite state machine accepts a string if there is a path through its state diagram, from the start state to an accepting state, whose transitions are labeled with the symbols in the string. We can restate this in the following way. Let $a_1 a_2 a_3 \ldots a_n$ be a string. Let $S_1 = F(a_1, S)$. Then S_1 is the set of states that the machine can be in after processing the first symbol a_1. Now let S_2 be the set of all states that can be reached from a state in S_1 by a transition labeled a_2. Then S_2 is the set of all states that the machine can be in after processing the first two symbols a_1 and a_2. Let S_3 be the set of all states that can be reached from a state in S_2 by a transition labeled a_3. S_3 is the set of all states that the machine can be in after processing the first three symbols, a_1 through a_3. Continuing in this manner, we arrive eventually at S_n, which is the set of states that the machine can be in after processing the entire input string. The machine accepts the input if S_n contains any accepting state.

These considerations show us how to build a true finite state machine from a nondeterministic finite state machine. Given a nondeterministic finite state machine M, we construct a machine M^* as follows. The states of M^* are the subsets of the set of states of M. The start state of M^* is $\{S\}$, where S is the start state of M. The accepting states of M^* are those sets containing accepting states of M. The transition function F^* of M^* is defined as follows: If a is a symbol of the input alphabet and if X is a set of states of M, then $F^*(a, X)$ is the set of states that can be reached from a state in X by a transition of M labeled a.

EXAMPLE 12.17   Convert the nondeterministic machine of Example 12.15 to a deterministic machine. The states of the new machine are $\varnothing$, $\{S\}$, $\{E\}$, and $\{S, E\}$. The state diagram is shown in Figure 12.26.

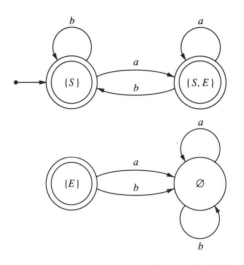

FIGURE 12.26   The machine of Example 12.17   o

Note that since state E in Figure 12.22 has no transitions leaving it, all transitions leaving $\{E\}$ go to the empty set. Also, all transitions from the empty set go to the empty set. The states $\{E\}$ and $\varnothing$ can never be reached by starting from $\{S\}$, so the machine may be simplified by discarding them. It is easy to see that the machine in the top half of Figure 12.26 does accept exactly those strings ending in a.

EXAMPLE 12.18   Convert the nondeterministic machine of Example 12.16 to a deterministic machine. The states of the new machine are $\varnothing$, $\{S\}$, $\{A\}$, $\{E\}$, $\{S, A\}$, $\{S, E\}$, $\{A, E\}$, and $\{S, A, E\}$. The state diagram is shown in Figure 12.27.   o

In Figure 12.27, the states $\{S, A\}$, $\{S, E\}$, and $\{S, A, E\}$ can never be reached by starting from $\{S\}$. The machine may be simplified by eliminating them. The result is the machine of Figure 12.28.

We have now completed the proof of Theorem 12.2, which states that a language is regular if and only if it can be recognized by a finite state machine. Given a regular grammar, you can build in three steps a finite state machine to recognize it. First, convert the grammar to normal form. Second, use the "building blocks" of Figure 12.21 to build a nondeterministic finite state machine. If the normal-form grammar has N symbols, this machine will have $N + 1$ states. Finally, convert the nondeterministic machine to a finite state machine using the construction shown in the two previous examples. The final machine will then have 2^{N+1} states.

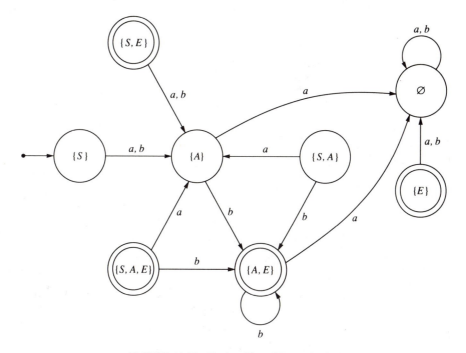

FIGURE 12.27   The machine of Example 12.18

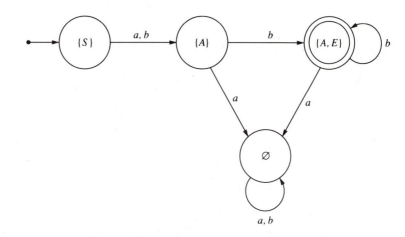

FIGURE 12.28   The machine of Example 12.18, simplified

EXERCISES 12.5

1. Convert each of the following grammars to normal form. In each case, capital letters are nonterminal symbols, small letters are terminal symbols, and S is the start symbol.

   a. $S ::= acbS \mid a \mid c$
                                       b. $S ::= cacS \mid acaS \mid b$

   c. $S ::= abB \mid baS \mid a$
      $B ::= aaaS \mid bbB$

2. Repeat Exercise 1 for the following grammars.

   a. $S ::= abbB \mid B$
      $B ::= cS \mid a$
                                 b. $S ::= baA \mid B$
      $A ::= B \mid aS$
      $B ::= S \mid c$

3. Refer to Figure 12.22. What is the set of states that the machine of this figure can be in after receiving each of the following inputs?

   a. *aba*   **b.** *bab*   **c.** *aab*   **d.** *bbab*

4. Refer to Figure 12.23. What is the set of states that the machine of this figure can be in after receiving each of the following inputs?

   a. *aabb*   **b.** *bbaa*   **c.** *abab*   **d.** *baba*

5. Draw state diagrams for nondeterministic finite state machines to recognize the languages generated by each of the following grammars. In each case, capital letters are nonterminal symbols, small letters are terminal symbols, and S is the start symbol

   a. $S ::= aB \mid c$
      $B ::= bC \mid a$
      $C ::= cS \mid b$
                                 b. $S ::= aB \mid bB \mid cB \mid a \mid c$
      $B ::= aS$

   c. $S ::= xA$
      $A ::= yB$
      $B ::= zC \mid w$
      $C ::= w$

12.6 • TURING MACHINES

We saw in the last section that the capabilities of a finite state machine are limited by its fixed "memory." The limitation is removed if we allow the machine to use its own input tape as a "scratch pad." If the input tape is to be useful as an auxiliary memory, the machine must be able to write on the input tape as well as read from it. The machine must also be able to move the tape backward as well as forward, so that it can read what it wrote. A machine with these capabilities is called a *Turing machine*. The name is in honor of British mathematician Alan Mathison Turing (1912–1954).

In the 1930s, Turing investigated the relationship between abstract computing machines and problems in mathematical logic.

Definition 12.5

A **Turing machine** consists of

a. An **alphabet** A, which is a finite set of symbols
b. A finite set S, called the set of **states** of the machine
c. Two distinguished states in S, called the **initial state** and the **halt state,** and
d. A function $F: A \times S \rightarrow A \times \{L, R\} \times S$, called the **transition function.**

The letters L and R refer to movements of the tape: left or right, respectively. The machine begins in the initial state, positioned at the beginning of the tape. For each cycle of machine operation, the function F specifies the symbol to be written on the tape, the direction the tape is to be moved (left or right), and the new state. The machine halts when it enters the halt state. The operation of a Turing machine can be illustrated by a simple example.

EXAMPLE 12.19   Let $A = \{a, b, c\}$ and $S = \{1, 2, H\}$. Choose 1 to be the initial state and H to be the halt state. Define F by the following table:

		Symbols		
		a	b	c
	1	$(b, R, 1)$	$(c, R, 1)$	$(c, L, 2)$
States	2	$(a, L, 2)$	(a, L, H)	$(a, L, 2)$
	H	—	—	—

(We have left the last row blank because once the machine enters the halt state, operation stops.)

The table is interpreted as follows. Suppose the tape contains the symbols $abcc$. The tape is positioned initially at the a. For the first move, look up $F(a, 1)$ in the table. It is $(b, R, 1)$. This means that the machine writes a b on the tape (replacing the a), moves the tape to the right, and enters state 1. The tape now contains the symbols $bbcc$, and the machine is positioned at the second b. Look up $F(b, 1)$ in the table. It is $(c, R, 1)$. The machine writes a c on the tape (replacing the b), moves right, and again remains in state 1. The operation of the machine through these and succeeding cycles may be listed as follows, where tape position is indicated by an underline:

State	Tape Contents
1	$\underline{a}$ b c c
1	b $\underline{b}$ c c
1	b c $\underline{c}$ c
2	b $\underline{c}$ c c
2	$\underline{b}$ a c c
H	$\underline{a}$ a c c

In Example 12.19, the machine halted after five moves, leaving the symbols *aacc* on the tape. Note that the machine was unable to carry out the last "move left," since it was already at the beginning of the tape. If the machine of Example 12.19 is given the input *bbcc*, then the operation is

State	Tape Contents
1	$\underline{b}$ b c c
1	c $\underline{b}$ c c
1	c c $\underline{c}$ c
2	c $\underline{c}$ c c
2	$\underline{c}$ a c c
2	$\underline{a}$ a c c
2	$\underline{a}$ a c c
2	$\underline{a}$ a c c

and so on. The machine is stuck in state 2 and will never halt.

The machine of Example 12.19 illustrates some of the ways in which Turing machines differ from finite state machines, but it does not perform any practical task. The machine in the next example does: It determines whether a string of symbols contains an equal number of *a*'s and *b*'s, a task that no finite state machine can do.

EXAMPLE 12.20   Let the alphabet *A* be

$$\{\text{BOT}, a, b, x, \text{EOI}, Y, N\}$$

The symbol BOT means "beginning of tape"; we will write this symbol at the beginning of the input tape and nowhere else. The symbol EOI means "end of information." We write the letter *Y* on the tape to answer "yes," and we write *N* on the tape to answer "no." The information on the input tape will consist of BOT, a sequence of *a*'s and *b*'s, then EOI. The strategy is to search the tape repeatedly, looking for an *a* and a matching *b*. When we find an *a*, we replace it with an *x* and look for a *b*, which we also replace with an *x*. Similarly, when we find a *b*, we replace it with an *x* and look for an *a*, which we replace with an *x*. The tape is rewound after each pass. We need the following states:

1. Searching for an a or a b
2. Found an a, looking for a b
3. Found a b, looking for an a
4. Rewinding the tape
H. Halt

The transition function is defined by the table

Symbols

		BOT	a	b	x	EOI
	1	(BOT, R, 1)	(x, R, 2)	(x, R, 3)	(x, R, 1)	(Y, —, H)
States	2	—	(a, R, 2)	(x, L, 4)	(x, R, 2)	(N, —, H)
	3	—	(x, L, 4)	(b, R, 3)	(x, R, 3)	(N, —, H)
	4	(BOT, R, 1)	(a, L, 4)	(b, L, 4)	(x, L, 4)	—

As in Example 12.19, we have omitted irrelevant information from the table. The transition function may also be described by the state diagram in Figure 12.29. Note that each transition in the diagram is marked with the output symbol and tape direction. To understand how the machine works, consider the input sequence

BOT $a\,a\,b$ EOI

where the underline indicates current tape position. The machine starts in state 1, finds the first a, replaces it with x, and enters state 2. The tape now contains

BOT $x\,\underline{a}\,b$ EOI

Now the machine passes over the second a, finds the b, replaces it with an x, and starts to rewind (state 4). The tape now contains

BOT $x\,\underline{a}\,x$ EOI

After rewinding the tape, the machine finds the second a but no matching b. When it encounters EOI while searching for a b, it writes an N and halts. The final state of the tape is

BOT $x\,x\,x\,\underline{N}$

The complete sequence of states and tape contents is as follows:

State	Tape Contents	Remarks
1	$\underline{\text{BOT}}\ a\,a\,b$ EOI	Search for a or b
1	BOT $\underline{a}\,a\,b$ EOI	Found a
2	BOT $x\,\underline{a}\,b$ EOI	Searching for b
2	BOT $x\,a\,\underline{b}$ EOI	Found b

State	Tape Contents	Remarks
4	BOT x $\underline{a}$ x EOI	Rewinding
4	BOT $\underline{x}$ a x EOI	Rewinding
4	$\underline{\text{BOT}}$ x a x EOI	Rewinding
1	BOT $\underline{x}$ a x EOI	Search for a or b
1	BOT x $\underline{a}$ x EOI	Found a
2	BOT x x $\underline{x}$ EOI	Searching for b
2	BOT x x x $\underline{\text{EOI}}$	Searching for b
H	BOT x x x $\underline{N}$	Not found

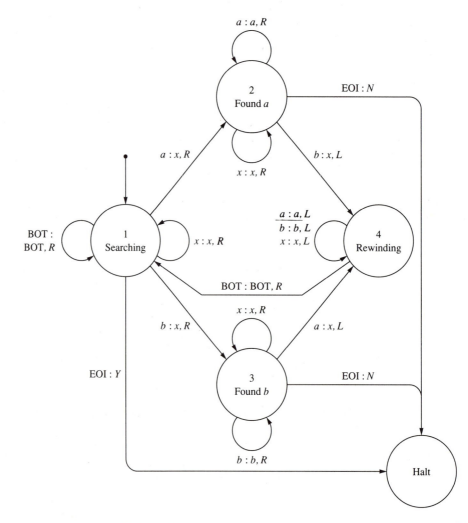

FIGURE 12.29   The state diagram for Example 12.20

Example 12.20 illustrates several more features of Turing machines. When a Turing machine halts, it produces an output: the information on the tape. In designing a Turing machine to perform a useful task, it is convenient to impose some restriction on the input. In Example 12.20, the problem concerned strings consisting only of a's and b's, but we introduced several additional characters (BOT, x, EOI, Y, N) for the machine's use. Then we assumed that the input began with BOT and ended with EOI. Note that the machine will never halt if either the BOT or EOI marks are left off the input. It is also customary to have a "blank" symbol in the alphabet and to assume that the input contains only a finite number of nonblank symbols. In Example 12.20, we could assume that all symbols beyond the EOI are blank.

EXAMPLE 12.21   Let A be the set of letters a through z. Design a machine that makes a copy of a string of letters. We choose the machine's alphabet to be

$$A \cup \{B, E, *, \#\}$$

We use B for beginning of tape and E for end of information. Also, we will assume that the input is of the form

$$Bx_1 x_2 \ldots x_n \# E$$

where $x_1, x_2, \ldots, x_n$ are letters. The task is to place a copy of the string $x_1 x_2 \ldots x_n$ between the $\#$ and the E. Rather than list a set of states and define a transition function by a table, we will describe a design strategy as follows:

Start at B and proceed to x_1. Remember the character x_1 (the machine must have enough states to do this) and replace it with $*$. Search forward for E. Write the character x_1 over the E, and write a new E to its right. Now search backward for the $*$, and replace it with x_1. Move right to go on to the x_2. Repeat the process until the $\#$ is encountered, then halt.

The strategy may also be expressed by the flowchart in Figure 12.30.

How many states does this machine have? There are 11 boxes in the flowchart, so the machine needs 11 states to keep track of what to do next. But there is also a "variable" X, that can hold any letter of the alphabet. So there are a total of $11 \times 26 = 286$ states!   ○

Example 12.21 shows that for all but the simplest of Turing machines, it is impractical to tabulate a transition function or to draw a state diagram. But it also illustrates that a Turing machine is a faithful mathematical model of a modern computer. You can think of a Turing machine as a computer with a fixed program (in read-only memory, or ROM) and a fixed amount of random-access memory (RAM). The memory required for the machine of Example 12.21 is 9 bits: 4 bits to keep track of the machine's current position in the flowchart (holds up to 16 states), and 5 bits for data storage (holds up to 32 characters).

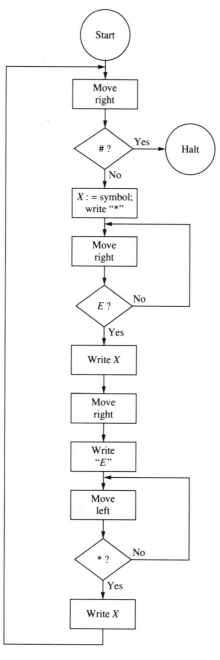

FIGURE 12.30   A flowchart for the machine of Example 12.21

EXERCISES 12.6

1. Trace the operation of the machine of Example 12.19 for each of the following inputs. List the state and tape contents for each cycle.

   a. *ababc*   **b.** *ccbbaa*

2. Repeat Exercise 1 for the following inputs:

   a. *cbaacccccb*   **b.** *bcabcabca*

3. Draw the state diagram for the machine of Example 12.19.

4. Trace the operation of the machine of Example 12.20 for each of the following inputs. List the states and tape contents for each cycle.

   a. EOI *a b a b* BOT   **b.** BOT *b b b* EOI

5. Design a Turing machine that will determine whether there are an equal number of *a*'s and *b*'s between the first BOT and the first EOI following the first BOT. Write down its transition function. (*Hint:* Modify the design of Example 12.20 to start out by searching for BOT.)

6. Design a Turing machine to make a copy of an input string in the reverse order. For instance, if the input is

   $$B\ a\ b\ c \# E$$

   then the output should be

   $$B\ a\ b\ c \# c\ b\ a\ E$$

7. Show that any task that can be accomplished by a machine similar to a Turing machine, but having two tapes, can be accomplished by a Turing machine.

8. Show that any task that can be accomplished by a Turing machine having an alphabet *A* of *N* symbols can also be accomplished by a Turing machine having the alphabet $\{0, 1\}$.

12.7 • THE HALTING PROBLEM

We have seen that a Turing machine, given a particular input on its tape, may or may not halt. In this respect also, a Turing machine is much like a real computer. Every programmer has had the experience of writing a program that inadvertently goes into an infinite loop for some input (or even for all inputs). We can ask, "Is there a way of determining in advance whether a Turing machine will halt on a given input?" Assuming as we do that Turing machines are faithful models of actual computers, we can rephrase the question as follows: "Can we write a computer program to determine whether another computer program, given a particular input, will halt?" An affirmative answer would be of considerable value for a university computing center. Presumably, much computer time could be saved if the center refused to run a student program that would never halt. The question raised here, in either formulation, is called the **halting problem.**

Programs as Data

In our second formulation of the halting problem, we considered a computer program, one of whose inputs was another computer program. That is, a program that is designed to process data can itself be data to another program. This is not an exotic situation. For example, a compiler is a program that reads a program written in one language, such as Pascal, and translates it into machine language.

Here is one way that the complete description of a Turing machine can be written as a set of symbols on a tape. For definiteness (and convenience), let us choose the alphabet A to be the set of printable ASCII characters. We define a machine using this alphabet by a string of symbols of the same alphabet, as follows. If there are N states in the state set S, number them from 1 to N with the start state as state 1. Write the numbers in base-10 notation. Then each state can be represented by a string of decimal digits. Now we can write the definition of the machine on a tape as a sequence of entries of the form

$$F(\text{``}<\text{input-symbol}>\text{''}, \text{state-number})$$
$$= (\text{``}<\text{output-symbol}>\text{''}, \text{move-direction}, \text{state-number})$$

separated by semicolons and ending with a period.

EXAMPLE 12.22   Encode the Turing machine of Example 12.19 on a tape with the input string $cbacc$. The states 1, 2, and H will be represented by 1, 2, and 3. We provide arbitrary definitions for $F(a, H)$, $F(b, H)$, and $F(c, H)$. The encoded version of the machine and the input data are:

$$F(\text{``}a\text{''}, 1) = (\text{``}b\text{''}, R, 1); F(\text{``}a\text{''}, 2) = (\text{``}a\text{''}, L, 2); F(\text{``}a\text{''}, 3) = (\text{``}a\text{''}, R, 1);$$
$$F(\text{``}b\text{''}, 1) = (\text{``}c\text{''}, R, 1); F(\text{``}b\text{''}, 2) = (\text{``}a\text{''}, L, 3); F(\text{``}b\text{''}, 3) = (\text{``}a\text{''}, R, 1);$$
$$F(\text{``}c\text{''}, 1) = (\text{``}c\text{''}, L, 2); F(\text{``}c\text{''}, 2) = (\text{``}a\text{''}, L, 2); F(\text{``}c\text{''}, 3) = (\text{``}a\text{''}, R, 1).$$
$$cbacc$$

In this way, every Turing machine can be represented as a string of characters on the tape. The end of the machine description and the beginning of the data can be recognized easily: It is the first period not enclosed in quotes. But not every string of symbols is a machine definition according to our scheme. In fact, the set of valid machine descriptions is a regular language. (The demonstration of this is left as an exercise.) If we wish every string of symbols to represent a machine, we can adopt the convention that if a string does not belong to the language, it represents the trivial machine that halts as soon as it is started, doing nothing.

Solution of the Halting Problem

We now have the tools to enable us to solve the halting problem.

Theorem 12.4   There does not exist any Turing machine H that examines a Turing machine definition T and an input X encoded on a tape, and determines (by writing an answer and halting) whether machine T will halt when given input X.

Proof   Define a propositional function HALT(x, y) by the rule: HALT(x, y) is true if machine x halts given input y. Now suppose a machine H with the given behavior exists. Then H computes the value of the propositional function HALT. That is,

a. If x halts when given input y, then H halts and prints "True" when given input (x, y).
b. If x does not halt when given input y, then H halts and prints "False" when given input (x, y).

Now design a new machine H^* from H, with the following behavior. When H^* is given input x, it first uses H to compute HALT(x, x). Then

a. If HALT(x, x) is true, H^* deliberately goes into an infinite loop and never halts.
b. If HALT(x, x) is false, H^* halts.

We can characterize the behavior of machine H^* succinctly as follows:

$$\forall x[\text{HALT}(H^*, x) \Leftrightarrow \neg\text{HALT}(x, x)]$$

Taking the special case $x = H^*$, we obtain

$$\text{HALT}(H^*, H^*) \Leftrightarrow \neg\text{HALT}(H^*, H^*)$$

which is a contradiction. We have constructed a machine that, when given its own description as input, halts if and only if it does not halt. This contradiction completes the proof.   ●

EXERCISES 12.7

1. Design a Turing machine that never halts, regardless of input.
2. Design a Turing machine that halts if the symbol "1" occurs in any of the first four positions on the input tape, and otherwise never halts.
3. Using the method described in this section, encode the Turing machine of Example 12.20 and the input string

    BOT *a a b* EOI

   on a single tape. (You will need to choose some single characters to represent BOT and EOI.)
4. Using the same method, encode the Turing machine that you designed in Exercise 1 on a single tape.
5. Using the same method, encode the Turing machine that you designed in Exercise 2, and the input "00101101," on a single tape.
6. Write a regular grammar for the language consisting of valid machine descriptions as defined in the text.

Computer Exercises for Chapter 12

1. Write a program to simulate the machine of Example 12.1.
2. Write a program to simulate the machine of Example 12.2.

3. Write a program to simulate the machine of Example 12.3. Include output statements signaling change made and items dispensed.

4. Write a program to simulate an arbitrary finite state machine (up to a certain size, which may depend on the memory size of your computer). The program should read the transition function of a machine and the list of accepting states (in a format you must define). Then the program should read an input string, simulate the operation of the machine, and determine whether the machine accepts or rejects the input.

5. Write a program to read a set of strings constituting a finite language, and write a normalized-form regular grammar defining the language.

6. Write a program to simulate the machine of Example 12.19.

7. Write a program to simulate the machine of Example 12.20.

• • • • CHAPTER REVIEW EXERCISES

1. Compare the characteristics of a CRT terminal to the tape of an abstract machine.

2. What is the result of running the machine of Figure 12.19 with each of the following inputs? Which are accepted by the machine?

   a. *abcccbaa*　　　**b.** *cbacbacc*　　　**c.** *bbbbbbbb*　　　**d.** *aaaaaaah*

3. What is the result of running the machine of Figure 12.20 with each of the following inputs? Which are accepted by the machine?

   a. *ababababa*　　　**b.** *bbbbbaaaaa*　　　**c.** *aaabbbaaa*　　　**d.** *bbbaaabba*

4. The state of many computer or terminal keyboards can be altered by striking the CAPS LOCK or NUM LOCK keys. Describe such a keyboard as a finite state machine with outputs.

5. Refer to Figure 12.31.
   a. Trace the sequence of states of the machine for each of the following inputs:
      1. *abcabcabca*　　**2.** *cabc*　　**3.** *abcc*　　**4.** *abcabcbaca*
   b. Write a grammar for the language consisting of strings accepted by the machine.
   c. Describe the language recognized by the machine.

6. Design a finite state machine to recognize each of the following languages:
   a. {*aaaab, aaaac, aaaad*}　　　**b.** {*abcc, abbb, abaa*}　　　**c.** {*a, b, ab, ba*}

7. Design a finite state machine to recognize strings over the alphabet {*a, b, c*} that contain exactly two *a*'s, or exactly two *b*'s, or exactly two *c*'s.

8. Refer to Figure 12.32.
   a. Trace the sequence of states of the machine for each of the following inputs:
      1. *abcdabcd*　　**2.** *eeeeabcd*　　**3.** *abced*　　**4.** *cedc*
   b. Write a grammar for the language consisting of strings accepted by the machine.

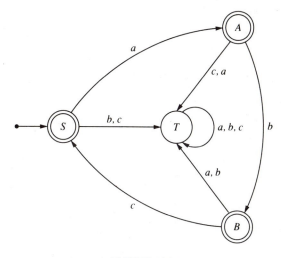

FIGURE 12.31

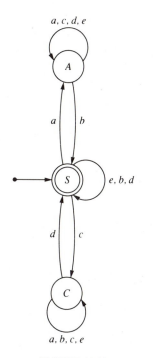

FIGURE 12.32

9. Rewrite the following regular grammars in normal form:

    a. $S ::= abS \mid bacS \mid bcS \mid a$ **b.** $S ::= abT \mid bT$

    $\qquad\qquad\qquad\qquad\qquad\qquad\qquad\qquad\qquad$ $T ::= cT \mid aS \mid S \mid a$

10. Design a nondeterministic finite state machine to recognize the languages generated by each of the following grammars:

    a. $S ::= aT \mid bU \mid cV$ **b.** $S ::= a \mid b \mid cZ$

    $\qquad$ $T ::= bS$ $\qquad\qquad\qquad\quad$ $Z ::= aS$

    $\qquad$ $U ::= cS$

    $\qquad$ $V ::= aT \mid b$

11. Design nondeterministic finite state machines to recognize the languages generated by the grammars of Exercise 9.

12. Design a (deterministic) finite state machine to recognize the language generated by the grammar of Exercise 10b.

13. Consider the Turing machine whose transition function is given by the table

		Symbols		
		a	b	(Blank)
	1	$(b, R, 1)$	$(b, R, 2)$	(a, L, H)
States	2	$(a, L, 2)$	(b, L, H)	$(a, R, 1)$
	H	(a, R, H)	(b, R, H)	(c, R, H)

Determine the sequence of operations of the machine when it is presented with input tapes with each of the following inputs:

a. *aaaaa*   **b.** *aaaabbb*   **c.** *ababa*   **d.** *bbbba*

Π (capital Greek pi) used to indicate a product of terms.

Σ (capital Greek sigma) used to indicate a sum of terms.

$\sigma(n)$ (lowercase Greek sigma) the sum of the positive divisors of n.

$\varnothing$ (lowercase Greek phi) the empty set.

$\varphi(n)$ (lowercase Greek phi) *see* **Euler φ-function.**

Absolute value (of a complex number $a + bi$) $\sqrt{a^2 + b^2}$.

Absolute value (of a real number a) a if a is positive or zero, and $-a$ if a is negative.

Addition rule the rule that if two mutually exclusive tasks can be performed in m and n ways, respectively, the number of ways to perform one or the other of the tasks is $m + n$.

Additive inverse (of a number a) the number, denoted by $-a$, such that $(-a) + a = 0$.

Adjacency list a way of representing a simple graph, in which for each vertex, the vertices adjacent to it are stored in a linked list.

Adjacency matrix a way of representing a graph, in which both rows and columns correspond to vertices, and an entry in the matrix is the number of edges connecting the corresponding two vertices.

Adjacent two vertices of a graph are adjacent if there is an edge connecting them.

Algorithm a procedure for carrying out a task that is specific enough to be reduced to a computer program.

Ambiguous context-free grammar a context-free grammar for which there is a sentence that can be parsed in two different ways.

Ancestor (of a vertex v in a rooted tree) a vertex w such that there is a directed path from w to v.

Antecedent (in a conditional proposition $p \Rightarrow q$) the proposition p.

Antisymmetric a relation R is antisymmetric if $x \, R \, y$ and $y \, R \, x$ imply $x = y$.

Argument (of a complex number z) the angle from the positive real axis to the line from the origin to z, measured in the counterclockwise direction.

Argument (of a function)   an element of the domain of the function.

Argument form   a structure for a sequence of statements constituting a logical argument.

Array   a rectangular arrangement of elements from a set, usually a set of numbers.

Array (in computer programming)   a data type that is a Cartesian product in which all the factors are equal.

Associative Law (for an operation *)   the statement $a * (b * c) = (a * b) * c$.

Asymptotically bounded   a function f is asymptotically bounded by a function g if there are positive numbers n_0 and M such that $|f(n)| \leq Mg(n)$ for all $n \geq n_0$.

Backus–Naur form   a method of writing a context-free grammar defining a programming language.

Base-b fraction   an expresssion of the form $a_1 \ldots a_k . b_1 b_2 \ldots$ where the a_i's and b_i's are digits in base b.

Big O   *see* $O(g)$.

Bijective (of a function)   both one-to-one and onto.

Binary operation (on a set A)   a function from $A \times A$ to A.

Binary relation   a relation having two domains, not necessarily distinct.

Binary search   a strategy for searching a sorted array, based on repeatedly dividing a selected portion of the array in two equal parts and selecting the part that contains the desired entry.

Binary search tree   a binary tree T with a one-to-one function f from the vertices of T to a totally ordered set, such that if v is any vertex, the values of f on the left subtree of v are all less than $f(v)$, and the values of f on the right subtree of v are all greater than $f(v)$.

Binary tree   a rooted tree in which each vertex has zero, one, or two children, and each child of a vertex is identified as either a left child or a right child.

Binomial coefficient   a coefficient of $x^r y^{n-r}$ in the expansion of $(x + y)^n$, equal to $C(n, r)$.

Binomial theorem   the theorem that $(x + y)^n = \Sigma\, C(n, r) x^r y^{n-r}$.

Bipartite graph   a graph whose set of vertices V can be written as the union of two nonempty disjoint sets V_1 and V_2, such that each edge of the graph connects a vertex of V_1 to a vertex of V_2.

Bit   the smallest unit of data; a bit can assume one of only two values, usually taken to be 0 and 1.

Bitwise operation   an operation performed on strings of bits of the same length, in which the value of the nth bit of the result depends only on the values of the nth bits of the operands.

BNF   *see* **Backus–Naur form.**

Boolean algebra   a set together with two operations, each of which is (a) commutative, (b) associative, (c) distributive over the other, and (d) has an identity element, and for which each element has a complement that is its inverse with respect to both operations.

Boolean data type   a data type (of a variable in a computer program) permitting only the values T (true) and F (false).

Boolean lattice   a lattice that is a Boolean algebra under the operations of meet and join.

Bound variable   an individual variable within the scope of a quantifier referring to it.

Breadth-first method   a strategy for constructing a spanning tree for a graph, in which at each step, as many vertices as possible that are adjacent to previously selected vertices are added.

Bridge   an edge in a connected graph which, if removed, leaves a nonconnected graph.

Byte   a sequence of eight bits.

C   the set of complex numbers.

Characteristic function (of a subset A of a set B)   the function on B defined by the rule $f(x) = 1$ if $x \in A$ and $f(x) = 0$ otherwise.

Characteristic polynomial (of a linear recurrence relation with constant coefficients)   the polynomial whose coefficients are the coefficients of the recurrence relation.

Child (of a vertex v in a rooted tree)   a vertex w such that there is an edge from v to w.

Chromatic number (of a graph)   the smallest number of subsets in a coloring of the graph.

Closed interval   an interval that contains both its endpoints.

Closed walk   a walk whose start and end vertices are the same.

Closure (of a subset Y of X, with respect to some property)   the smallest subset of X containing Y and having the specified property.

Codomain   see **Range**.

Coloring (of a graph)   a partitioning of the set of vertices of the graph into subsets so that no two adjacent vertices are in the same subset.

Combination (in counting problems)   The number of combinations of n things k at a time is the number of k-element subsets of a set of n elements.

Commutative Law (for an operation $*$)   the statement $a * b = b * a$.

Complement (of a set A relative to a larger set X)   the set of elements of X not in A. Denoted by A'.

Complete bipartite graph   a bipartite graph in which, for every $v_1 \in V_1$ and $v_2 \in V_2$, there is exactly one edge connecting v_1 to v_2.

Complete graph   a simple graph in which any two distinct vertices are adjacent.

Completeness (of predicate logic)   the theorem of K. Gödel that every valid formula of predicate logic may be derived from a specific short list of valid formulas, using validity-preserving operations.

Complex conjugate (of a complex number $a + bi$) $a - bi$.

Complex numbers   numbers of the form $a + bi$, where a and b are real and $i = \sqrt{-1}$.

Complex plane   a geometric representation of the complex numbers. The complex number $a + bi$ corresponds to the point in the plane with coordinates (a, b).

Composition (of functions f and g)   the function, denoted by $f \circ g$, defined by the rule $(f \circ g)(x) = f(g(x))$.

Compound data type   a data type that is a Cartesian product of other data types.

Conclusion   (a) the final statement of a logical argument; (b) *see* **Consequent.**

Conditional   *see* **Conditional proposition.**

Conditional probability   a probability measure on a subset of a sample space, derived from a probability measure on the whole sample space by dividing by the probability of the subset.

Conditional proposition   a proposition of the form "if p, then q."

Congruence modulo *n* a is congruent to b modulo n if n divides $a - b$.

Conjunction   the operation of combining two propositions using "and."

Connected component   a maximal connected subgraph of a graph.

Connected graph   a graph such that, for any two vertices, there is a path from one to the other.

Connective   a means of combining propositions to form new propositions.

Consequent (in a conditional proposition $p \Rightarrow q$)   the proposition q.

Constant function   a function f such that $f(a) = f(b)$ for all a and b in the domain of f.

Context-free grammar   *see* **Context-free phrase structure grammar.**

Context-free phrase structure grammar   a method of defining a language, consisting of a root symbol, nonterminal symbols, terminal symbols, and productions, in which the left side of each production is a single nonterminal symbol.

Continued fraction   an expression that is either a number or is of the form $a + \frac{1}{F}$, where a is a number and F is a continued fraction.

Contradiction   a formula of propositional logic that assumes the truth value F for every combination of truth values of its variables.

Contrapositive (of a conditional proposition $p \Rightarrow q$)   the conditional proposition $\neg q \Rightarrow \neg p$.

Convergent (of a continued fraction)   the value of a continued fraction obtained from the given fraction by discarding zero or more terms from the bottom.

Converse (of a conditional proposition $p \Rightarrow q$)   the conditional proposition $q \Rightarrow p$.

Countable set   a set for which there is a one-to-one correspondence with the set of natural numbers.

Countably infinite set   *see* **Countable set.**

Counterexample (of a formula of predicate logic)   an interpretation of the formula which assigns the formula the truth value F.

Counterexample (to a general statement)   an object to which the statement applies, but for which the statement is false.

Cycle   a closed path.

Cyclic graph   a graph consisting of a single cycle.

Data base   a body of information about elements of several sets, expressed as mathematical structures such as graphs, trees, or relations and stored in a computer.

Database management system   a computer program for creating, updating, and deriving reports from a data base.

Data type   the set of values that can be assumed by a variable in a computer program.

Decimal fraction   an expression of the form $a_1 \ldots a_k.b_1'b_2 \ldots$, where the a_i's and b_i's are decimal digits.

Degree (of a polynomial function in x)   the largest exponent n for which the coefficient of x^n is nonzero.

Degree (of a vertex in a graph)   the number of edges incident to the vertex, loops being counted twice.

DeMorgan's Laws (for propositional logic)   the laws $\neg(p \wedge q) \Leftrightarrow (\neg p \vee \neg q)$ and $\neg(p \vee q) \Leftrightarrow (\neg p \wedge \neg q)$; for sets, the laws $(A \cap B)' = A' \cup B'$ and $(A \cup B)' = A' \cap B'$.

Depth (of a rooted tree)   the maximum of the depths of its vertices.

Depth (of a vertex v in a rooted tree)   the length of the path from the root to v.

Depth-first method   a strategy for constructing a spanning tree for a graph, in which a path, once started, is continued as far as possible before returning to previously selected vertices.

Derangement   a permutation having no fixed point.

Descendent (of a vertex v in a rooted tree)   a vertex w such that there is a directed path from v to w.

Determinant (of a 2×2 matrix with entries a, b, c, and d)   the number $ad - bc$.

Difference (of sets A and B)   the set of objects that are in A but not in B; denoted by $A - B$.

Digraph   *see* **Directed graph.**

Diophantine equation   an equation for which integer solutions are desired.

Direct proof   mathematical proof in which the hypotheses are used in the form given.

Directed cycle   a directed path in which the start and end vertices are equal.

Directed Euler circuit (in a directed graph)   an Euler circuit in which every edge is traversed in its assigned direction.

Directed graph   a structure consisting of a set V of vertices or nodes, a set E of edges, and two functions (start-point and end-point) from E to V.

Directed path   a directed walk in which no vertex or edge is repeated (except for the first and last vertices in the case of a closed directed walk).

Directed walk   an alternating sequence of vertices and edges, starting and ending with vertices, in which each edge starts at the vertex immediately preceding it and ends at the vertex immediately following it.

Disjoint (of sets)   having empty intersection.

Disjunction   the operation of combining two propositions using "or."

Distributive Law (for an operation * over an operation #)   the rule $a * (b \# c) = (a * b) \# (a * c)$.

Domain (of a function or relation)   the set or sets over which the function or relation is defined.

D-type flip-flop   a flip-flop with two inputs, called the *clock input* and the *data input*, such that a pulse on the clock input causes the state of the flip-flop to become equal to the data input.

Dummy variable   *see* **Bound variable.**

Dynamical system   a physical system that evolves over time according to some law.

Edge table   a way of representing a graph by a table in which each row contains an edge and the two vertices adjacent to it.

Empty set   the set containing no elements; denoted by $\varnothing$.

End vertex   the second end of an edge in a directed graph.

Equality (of sets)   two sets are equal if they contain the same elements.

Equiprobable measure   a probability measure that assigns the same value to each outcome.

Equivalence   the operation of combining two propositions using "if and only if."

Equivalence class   if R is an equivalence relation on a set X, then an equivalence class is a subset of R consisting of an element $x \in X$ and all elements related to it.

Equivalence relation   a binary relation that is reflexive, symmetric, and transitive.

Equivalent propositional functions   propositional functions having the same truth-set.

Euclidean Algorithm   a method of finding the greatest common divisor of two integers by repeated division.

Euler circuit   a closed walk in which each edge occurs exactly once.

Euler φ-function (of a positive integer n)   the number of positive integers less than or equal to n and relatively prime to n.

Euler's Theorem   the theorem that when a graph is embedded in the plane, number of vertices − number of edges + number of faces = 2.

Event   a subset of a sample space.

Exclusive "or"   the interpretation of the word "or" that includes "but not both."

Existential quantifier   the quantifier that assigns to a propositional function P the proposition, "the truth set of P is not empty."

Expected value   the expected value of a function f on a sample space X with probability measure p is $\sum_{x \in X} f(x)p(x)$.

Face (of a graph embedded in the plane) a region of the plane bounded by edges of the graph.

Factorial the factorial of n is the product $1 \times 2 \times \cdots \times n$. Denoted by $n!$.

Favorable outcome an element of an event.

Feedback a feature of electronic circuits in which the output of a gate is connected, directly or through additional gates, to its own input.

Fiber if $f: X \rightarrow Y$ is a function and $y \in Y$, then the fiber of f over y is $\{x \in X \mid f(x) = y\}$. Denoted by $f^{-1}(y)$. Also known as the **level set** corresponding to y. *See also* **Inverse image.**

Fibonacci Sequence the sequence $0, 1, 1, 2, 3, 5, \ldots$, characterized by the initial conditions $F_0 = 0$ and $F_1 = 1$ and the recurrence relation $F_n = F_{n-1} + F_{n-2}$.

Field any of the domains of a relation; also, when the relation is presented in the form of a table, the corresponding column of the table.

Finite geometric series a geometric series containing a finite number of terms.

Finite set a set that is empty or for which there is a one-to-one correspondence with the set $\{1, 2, \ldots, n\}$ for some natural number n.

Finite state machine an abstract machine, consisting of an input alphabet, a finite set of states, and a transition function, having one input tape that can be read in the forward direction only.

First-order recurrence relation a recurrence relation in which the equation defining a term depends only on the previous term.

Fixed point (a) (of a function f) a value x such that $f(x) = x$; (b) (of a recurrence relation) a value x such that the constant sequence $x, x, x, \ldots$ is a solution of the relation.

Flip-flop an electronic circuit composed of gates, which has two stable states and so can remember one bit of information.

Flow diagram *see* **Syntax diagram.**

Formal power series an expression of the form $\Sigma a_n x^n$, where the sum may range over any finite or infinite sequence of consecutive integers.

Free variable an individual variable that is not a bound variable.

Full adder an electronic circuit that adds two 1-bit numbers and an incoming carry bit, producing a 1-bit sum and an outgoing carry bit.

Function a function from a set A to a set B consists of the sets A and B and a subset of $A \times B$ (the graph of the function), such that for each $a \in A$, there is exactly one point (a, b) in the graph.

Functional *see* **Function.**

Fundamental Theorem of Algebra the theorem that every polynomial $f(x) = x^n + a_1 x^{n-1} + \cdots + a_n$ with complex coefficients a_i can be factored into linear factors of the form $x - r$, where the complex numbers r are the roots of $f(x)$.

Gate an electronic circuit that implements a logical connective.

Gate delay the time elapsed between a change in the input to a gate and the resulting change to the output.

g.c.d.   see **Greatest common divisor.**

Generating function (of a sequence)   a function, the coefficients of whose power series expansion are the terms of the sequence.

Geometric series   a series (sum of terms) in which the ratio of two successive terms is constant.

Grammar   a method of defining a language, expressed in terms of operations on strings of symbols.

Graph   a structure consisting of a set V of vertices or nodes, a set E of edges, and a relation (the incidence relation) on $V \times E$, such that each edge is incident to exactly one or two vertices.

Graph (of a function)   a subset of $A \times B$, where A and B are the domain and range of the function; *see* **Function.**

Graph (of a relation)   a subset of $A_1 \times \cdots \times A_n$, where the sets A_i's are the domains of the relation; *see* **Relation.**

Greatest common divisor (of two integers a and b)   the largest integer dividing both a and b.

Greatest integer function   the function from the real numbers to the real numbers with the following rule: $f(x)$ is the greatest integer less than or equal to x. Denoted by $\lfloor x \rfloor$.

Half adder   an electronic circuit that adds two 1-bit numbers, producing a 1-bit sum and a carry bit.

Half-line   an interval for which one endpoint is a real number and the other endpoint is $-\infty$ or ∞.

Half-open interval   an interval that contains exactly one of its endpoints.

Halting problem   the problem of determining whether a given computer program, with a given input, will run forever or eventually halt.

Hamiltonian cycle   a cycle that includes each vertex of the graph.

Hypothesis   (a) *see* **Premise;** (b) *see* **Antecedent.**

Identity element   (for an operation *) e is an identity element if $e * x = x$ and $x * e = x$ for all x.

Identity function   the identity function on a set A is the function f with rule $f(a) = a$.

Image (of a function)   the set of all values actually assumed by the function; *see* **Range.**

Imaginary number   a complex number of the form bi where b is real and $i = \sqrt{-1}$.

Imaginary part   (of the complex number $a + bi$)   the number b.

Imaginary unit   the complex number $i = \sqrt{-1}$.

Implication   the operation of combining two propositions using "if . . . then."

Incidence matrix   a matrix of 1's and 0's, with the rows corresponding to the edges of a graph and the columns corresponding to the vertices, illustrating the incidence relation of a graph.

Incidence relation   a relation on the sets of vertices and edges of a graph, indicating which vertices are endpoints of which edges.

Incident   an edge of a graph is incident to a vertex if the vertex is an endpoint of the edge.

Inclusion–exclusion principle   *see* **Principle of inclusion–exclusion.**

Inclusive "or"   the interpretation of the word "or" that includes "or both."

Indegree (of a vertex in a directed graph)   the number of edges having the vertex as endpoint.

Independent events   a pair of events having the property that the probability of both is equal to the product of the probabilities of the events.

Index of summation (used with the summation symbol Σ)   a variable used to identify the terms to be summed. Its limits are written below and above the Σ symbol.

Indirect proof   a mathematical proof that is not direct. *See* **Proof by contradiction, Proof by contraposition.**

Individual variable   a symbol used to denote an element of a set.

Induced subgraph   the subgraph of a graph G induced by a subset V_1 of the set V of vertices of G is the largest subgraph of G whose set of vertices is V_1.

Induction   *see* **Principle of Mathematical Induction.**

Induction hypothesis (in a proof using mathematical induction)   the assumption, made in the second step of the proof, that all numbers less than n have the desired property.

Infinite geometric series   a geometric series containing an infinite number of terms.

Infinite set   a set that is not finite.

Infix notation   a method of writing algebraic expressions in which a binary operator is written between its operands.

Inherently ambiguous language   a language that cannot be defined by an unambiguous context-free grammar.

Initial condition   (a) a set of values for one or more terms of a sequence defined by a recurrence relation, from which all subsequent terms can be computed; (b) the state of a dynamical system at a specified time, from which the state of the system at later times can be computed.

Initial state   the state of an abstract machine in which it is started.

Injective   *see* **One-to-one.**

Input alphabet   the set of symbols allowed to be written on the input tape of a finite state machine.

Integers   the number system consisting of $\ldots, -2, -1, 0, 1, 2, \ldots$.

Internal vertex (in a rooted tree)   a vertex that is neither the root nor a terminal vertex.

Interpretation (of a formula of predicate logic)   a set of data, consisting of a set, values for the propositional and free individual variables, and values for the propositional functions, that determine a truth value for the formula.

Intersection (of sets)   the intersection of sets A and B is the set consisting of objects that are elements of both A and B; denoted by $A \cap B$.

Interval   a set of real numbers consisting of all numbers between two endpoints a and b. The endpoints may be numbers, $-\infty$, or ∞, and may or may not be included in the set.

Inverse (of a conditional proposition $p \Rightarrow q$)   the proposition $\neg p \Rightarrow \neg q$.

Inverse (of a function $f: A \to B$)   a function $g: B \to A$ such that $f \circ g$ and $g \circ f$ are the identity functions on A and B, respectively.

Inverse, additive   *see* **Additive inverse.**

Inverse image (of a subset X of the range of a function with domain A)   the set $\{a \in A \mid f(a) \in X\}$. Denoted by $f^{-1}(X)$. *See also* **Fiber.**

Inverse, multiplicative   *see* **Multiplicative inverse.**

Inverter   an electronic circuit that implements negation.

Irrational number   a real number that is not rational.

Isolated vertex   a vertex in a graph that is not incident to any edge.

Isomorphic (of graphs)   two graphs are isomorphic if there are one-to-one, onto functions between the sets of vertices and the sets of edges that preserve the incidence relation.

Kruskal's Algorithm   an algorithm for finding a minimal spanning tree of a graph, in which a subgraph (not necessarily a tree) is grown by addition at each step of an unused edge of minimal weight.

Language   a set of finite sequences of elements of a set of symbols, called a vocabulary.

Latch   *see* **D-type flip-flop.**

Lattice   a partially ordered set in which each pair of elements has a unique least upper bound (their "join") and a unique greatest lower bound (their "meet").

l.c.m.   *see* **Least common multiple.**

Leaf   a vertex of degree 1 in a tree.

Least common multiple (of two integers a and b)   the smallest positive integer that is a multiple of both a and b.

Left subtree (of a binary tree at a vertex v)   the subgraph containing the left child (if any) of v and all of its descendants.

Level set   *see* **Fiber.**

Linear function   a function of the form $f(x_1, \dots, x_n) = a_1 x_1 + a_2 x_2 + \cdots + a_n x_n$.

Linear graph   a tree with exactly two vertices of degree 1.

Linear order   *see* **Total order.**

Linear recurrence relation   a recurrence relation in which the equation defining a term is a linear function of preceding terms.

Logical data type   *see* **Boolean data type.**

Logical operation   a computer implementation of a logical connective.

Loop   an edge of a graph whose endpoints are the same vertex.

Lowest terms   a fraction $\frac{a}{b}$ (where a and b are integers) is in lowest terms if a and b have no common factors other than 1 and -1.

Map   *see* **Function.**

Mapping   *see* **Function.**

Matrix   a rectangular array of numbers.

Mersenne prime   a prime of the form $2^p - 1$, where p is prime.

Minimal spanning tree   a spanning tree of a weighted graph having minimal total weight.

Modus ponens   the argument form p; $p \Rightarrow q$; therefore q.

Multiple edges   two or more edges in a graph having the same endpoints.

Multiplication rule   the rule that if two independent tasks can be performed in m and n ways, respectively, the number of ways to perform both is mn.

Multiplicative inverse (of a number a)   the number, denoted by $\frac{1}{a}$, such that $\frac{1}{a} \times a = 1$.

Multiplicity (of a root a of a polynomial $P(x)$)   the largest integer n such that $(x - a)^n$ divides $P(x)$.

Mutually exclusive events   events with empty intersection.

N   the set of natural numbers.

NAND   the logical connective "not both p and q."

NAND gate   an electronic circuit that implements the NAND connective.

n-ary relation   a relation having n domains.

Natural numbers   the number system consisting of 1, 2, 3, ...

Necessary condition   *see* **Consequent.**

Negation   the operation of modifying a proposition by introducing "not."

Negative logic   a convention for representation of truth values in electrical circuits, in which the voltage corresponding to T is smaller than the voltage corresponding to F.

Neutral element   *see* **Identity element.**

Node   *see* **Vertex.**

Nondeterministic finite state machine   an abstract machine, similar to a finite state machine, in which the current state and the input define a set of possibilities for the next state.

Nonterminal symbol   a symbol, used in a grammar, that is not in the vocabulary of the language defined by the grammar.

NOR   the logical connective "neither . . . nor."

NOR gate   an electronic circuit that implements the NOR connective.

Normal form (of a regular grammar)   a form in which every production is of one of the forms $X ::= aY$ or $X ::= a$.

Number system   a set with two operations (addition and multiplication) in which the operations satisfy certain laws depending on the system.

O (g) the set of functions asymptotically bounded by g.

One's complement adder   an electronic circuit that adds numbers of a fixed size and adds the carry from the leftmost bit position (if any) back into the rightmost position.

One-to-one (of a function f)   having the property that $f(x) = f(y)$ implies $x = y$.

One-to-one correspondence   a function that is both one-to-one and onto.

Onto (of a function)   having the image equal to the range.

Open interval   an interval that does not contain either of its endpoints.

Open walk   a walk whose start and end vertices are different.

Order (of a complete graph)   the number of vertices.

Ordered rooted tree   a rooted tree in which, for each vertex, a total order relation is defined on its children.

Orientable graph   a graph in which it is possible to assign a direction to each edge in such a way that the resulting directed graph is strongly connected.

Outdegree (of a vertex in a directed graph)   the number of edges having the vertex as start point.

Overflow   the production of a wrong answer by an adder. *See* **One's complement adder, Two's complement adder.**

Parent (of a vertex v in a rooted tree)   the vertex w (if any) such that there is an edge from w to v.

Parsing tree   a tree diagram illustrating the derivation of a sentence of a language defined by a context-free grammar.

Partially ordered set   a set with a partial order relation.

Partial order relation   a binary relation that is reflexive, antisymmetric, and transitive.

Partition (of a set)   a decomposition of the set into nonempty, disjoint subsets.

Pascal's triangle   a triangular arrangement of the numbers $C(n, r)$ in which each number is the sum of the two numbers above it, illustrating the recurrence relation $C(n, r) = C(n - 1, r - 1) + C(n - 1, r)$.

Path   a walk in which no vertex or edge is repeated (except for the first and last vertices in the case of a closed walk).

Perfect number   a positive integer that is equal to the sum of its positive proper divisors.

Permutation   a one-to-one function from a finite set to itself.

Phrase-structure grammar   a method of defining a language in which the rules are definitions of types of phrases.

Pigeonhole Principle   the principle that if $n + 1$ (or more) objects are placed in n bins, at least one bin contains more than one object.

Planar graph   a graph that can be embedded in the plane without edges crossing.

Polish notation   *see* **Prefix notation.**

Polynomial function   a function of the form $f(x) = a_n x^n + a_{n-1} x^{n-1} + \cdots + a_1 x + a_0$.

Poset   *see* **Partially ordered set.**

Positive logic   a convention for representation of truth values in electrical circuits, in which the voltage corresponding to T is larger than the voltage corresponding to F.

Postfix notation   a method of writing algebraic expressions, in which an operator is written to the right of its operands. Also called reverse Polish notation.

Power set (of a set A)   the set of all subsets of A.

Predicate   *see* **Propositional function.**

Predicate variable   a symbol used to denote a propositional function.

Prefix notation   a method of writing algebraic expressions, in which an operator is written to the left of its operands. Also called Polish notation.

Premise   any of the starting points of a logical argument.

Prime   an integer having no factors other than itself, its negative, $+1$ and -1.

Primitive nth root of unity   the complex number of absolute value 1 and argument $\frac{360}{n}$ degrees.

Primitive set of connectives   a set of logical connectives in terms of which every other logical connective can be defined.

Prim's algorithm   an algorithm for finding a minimal spanning tree of a graph, in which a tree is grown by addition at each step of an unused edge of minimal weight.

Principle of inclusion–exclusion   a method of counting the elements in the union of two or more sets, taking into account the numbers of elements in their intersections.

Principle of Mathematical Induction (strong form)   the principle that, if A is a set of natural numbers such that (1) $1 \in A$ and (2) if all numbers less than n are in A, then $n \in A$, then A contains all natural numbers.

Principle of Mathematical Induction (weak form)   the principle that, if A is a set of natural numbers such that (1) $1 \in A$ and (2) if $n - 1 \in A$, then $n \in A$, then A contains all natural numbers.

Probability (of an event)   the sum of the values of a probability measure for all the outcomes in an event.

Probability measure   a nonnegative function on a sample space, such that the sum of all of the values of the function is 1.

Production   a rule of a grammar, consisting of two strings of symbols (left side and right side), stating that its right side may be substituted for its left side.

Proof by contradiction   a mathematical proof of a statement of the form $p \Rightarrow q$, in which one begins with $p \wedge \neg q$ and deduces a false statement.

Proof by contraposition   a mathematical proof of a statement of the form $p \Rightarrow q$, in which one begins with $\neg q$ and deduces $\neg p$.

Proper subset   the set A is a proper subset of B if $A \subset B$ and $A \neq B$.

Proposition   a statement that is either true or false.

Propositional function   a function whose range is a set of propositions.

Propositional variable   a symbol used to denote a proposition.

Pulse   a change in state of an electrical signal from one state to the other and back again.

Q   the set of rational numbers.

Qualified quantifier   a quantifier combined with a condition on the individual variable, such as $\forall x \in X$ or $\exists y > 0$.

Quantifier   a propositional function whose domain is a set of propositional functions.

R   the set of real numbers.

Range (of a function $f: A \to B$)   the set of B; *see* **Function; Image.**

Rational numbers   the number system consisting of all quotients $\frac{a}{b}$ where a and b are integers and $b \neq 0$.

Ray   *see* **Half-line.**

Real numbers   a number system containing the rational numbers, whose elements correspond to the points of a line.

Real part (of the complex number $a + bi$)   the number a.

Real-valued function   a function whose range is the set of real numbers or a subset of it.

Record (in computer programming)   a data type that is a Cartesian product in which the factors are not all equal.

Recurrence relation   a formula that defines a number in a sequence of numbers in terms of previous numbers in the sequence.

Recursive (of a definition or algorithm)   referencing itself.

Reflexive (of a binary relation R on a set A) R is reflexive if $a \, R \, a$ for every $a \in A$.

Reflexive closure (of a binary relation R on X)   the smallest binary relation on X that contains R and is reflexive.

Regular grammar   *see* **Type 3 grammar.**

Relation   a relation with domains $A_1, \ldots, A_n$ consists of the sets $A_1, \ldots, A_n$ and a subset (the graph of the relation) of the set $A_1 \times \cdots \times A_n$.

Relational model   a theory of database management which holds that data is most efficiently represented as a set of relations that are functions.

Relatively prime   having a g.c.d. equal to 1.

Reverse Polish notation   *see* **Postfix notation.**

Right subtree (of a binary tree at a vertex v)   the subgraph containing the right child (if any) of v and all of its descendants.

Root (of a rooted tree)   the vertex with indegree zero.

Rooted tree   a directed graph such that the underlying graph is a tree and the indegree of each vertex is either zero or one.

Root symbol (in a grammar)   the symbol with which one starts to build a sentence.

Rule (of a function or relation)   *see* **Graph.**

Sample space   the set of all possible outcomes of an experiment.

Satisfiable formula   a formula of predicate logic that has the truth value T for some interpretation of the formula.

Scope (of a quantifier)   the portion of a formula to which the quantifier applies.

Sentence   an element of a language.

Sequence   a function whose domain is the set of natural numbers or the set of nonnegative integers.

Set   a collection of objects.

Set–reset flip-flop   a flip-flop with two inputs, one of which is used to put the flip-flop into the T state and one of which is used to put it into the F state.

Sibling (of a vertex v in a rooted tree)   a vertex w such that v and w have the same parent.

Simple directed graph   a directed graph that has no loops and in which no two distinct edges have both the same start point and the same endpoint.

Simple graph   a graph with no loops and no multiple edges.

Spanning tree   a subgraph that is a tree and that contains every vertex of the graph.

S–R flip-flop   *see* **Set–reset flip-flop.**

Start vertex   the first end of an edge in a directed graph.

State diagram   a graphic method of representing the transition function of a machine.

Statement pattern   a method of defining a language, in which patterns or templates are provided for each of several types of sentences.

Strong induction   *see* **Principle of Mathematical Induction.**

Strongly connected (of a directed graph)   having the property that for every pair of vertices u and v, there is a directed path from u to v.

Subgraph   a subgraph of a graph G whose vertices and edges are among the vertices and edges of G, with incidence relation derived from that of G.

Subset   a set A is a subset of B if every element of A is also in B; written $A \subset B$.

Sufficient condition   *see* **Antecedent.**

Surjective (of a function)   *see* **Onto.**

Syllogism   a valid argument form.

Symmetric (of a binary relation R on a set A) R is symmetric if $a\,R\,b$ implies $b\,R\,a$ for every $a, b \in A$.

Symmetric closure (of a binary relation R on X)   the smallest binary relation on X that contains R and is symmetric.

Symmetric difference (of sets A and B)   the set $(A \cup B) - (A \cap B)$; denoted by $A \oplus B$.

Syntax diagram a method of defining a language, in which sentences of the language are constructed by traveling through a diagram.

Tautology a formula of propositional logic that assumes the truth value T for each combination of truth values of its variables.

Terminal symbol a symbol, used in a grammar, that is in the vocabulary of the language defined by the grammar.

Terminal vertex (in a rooted tree) a vertex with no children.

Ternary relation a relation having three domains.

Total order a partial order relation R such that for any x and y in its domain, either $x\,R\,y$ or $y\,R\,x$. Also called **linear order**.

Transistor–transistor logic a convention for representation of truth values in electrical circuits, in which F is represented by 0 to 0.8 volts and T is represented by 2.4 to 5 volts.

Transition diagram *see* **State diagram.**

Transition function the function that determines the next state of an abstract machine, depending on the current state and the inputs.

Transitive (of a binary relation R on a set A) R is transitive if $a\,R\,b$ and $b\,R\,c$ imply $a\,R\,c$ for every $a, b, c \in A$.

Transitive closure (of a binary relation R on X) the smallest binary relation on X that contains R and is transitive.

Tree a connected graph containing no cycles.

Truth set the set of points in the domain of a propositional function to which the function assigns the value T.

Truth table a table showing the truth values of a propositional formula corresponding to each combination of truth values of the variables.

Truth value either of the letters T or F, used to denote True and False, respectively.

TTL *see* **Transistor–transistor logic.**

Turing machine an abstract machine having one tape, used for both input and output, which can be moved in either direction.

Twin primes a pair of prime numbers differing by 2.

Two's complement adder an electronic circuit that adds numbers of a fixed size and discards the carry (if any) from the leftmost bit position.

Type 0 grammar a phrase-structure grammar in which the left side of a production is allowed to be any string of terminal and nonterminal symbols.

Type 1 grammar a type 0 grammar in which the length of the string on the right of each production is greater than or equal to the length of the string on the left side.

Type 2 grammar *see* **Context-free phrase structure grammar.**

Type 3 grammar a type 2 grammar in which the right side of each production has at most one nonterminal symbol, which must be at the extreme right.

Unambiguous context-free grammar   a context-free grammar in which each sentence has a unique parsing tree.

Uncountable set   a set that is neither finite nor countable.

Undecidability (of predicate logic)   the theorem of A. Church that there is no algorithm to determine whether a formula of predicate logic is valid.

Underlying graph (of a directed graph)   the graph obtained by "forgetting" the directions of the edges.

Union (of sets)   the union of sets A and B is the set consisting of objects that are elements of A or B or both; denoted by $A \cup B$.

Universal quantifier   the quantifier that assigns to a propositional function P the proposition "the truth set of P is all of its domain."

Universal set   a set X consisting of all objects relevant to the discussion of some topic. When a universal set is specified, the complement of any set $A \subset X$ is defined.

Universal Turing machine   a Turing machine that reads a tape with a description of a Turing machine followed by some data and carries out the actions that the machine of the description would carry out when given the data.

Valid formula   a formula of predicate logic which has the truth value T for every interpretation of the formula.

Venn diagram   a representation of sets by circles drawn in a plane.

Vertex   an endpoint of an edge.

Vocabulary   a set of symbols. (*See* **Language.**)

Walk   an alternating sequence of vertices and edges, starting and ending with vertices, in which each edge is incident to the vertices immediately preceding and following it.

Weak induction   *See* **Principle of Mathematical Induction.**

Weighted graph   a graph with a real-valued function defined on its set of vertices.

Weight function   a real-valued function defined on the set of vertices of a graph.

Well-formed formula   *see* **Sentence.**

Well-ordering principle   the statement that every nonempty set of natural numbers contains a least element.

wff   *see* **Well-formed formula.**

Word   a sequence of bits treated as a unit of information in a computer.

Z   the set of integers.

Z_n   the system of integers mod n, consisting of the equivalence classes (relative to congruence mod n) of the integers $0, 1, 2, \ldots, n - 1$.

SUGGESTED READINGS

Barr, Donald R., and Peter W. Zehna, *Probability*. Pacific Grove, Calif.: Brooks/Cole, 1971.

Biggs, N. L., E. K. Lloyd, and R. J. Wilson, *Graph Theory 1736–1936*. Oxford, U.K.: Clarendon Press, 1986.

Bogart, Kenneth P., and Peter G. Doyle, "Non-Sexist Solution of the Menage Problem." *American Mathematical Monthly* 93 (August–September 1986): 514–518.

Brown, J., L. C. Noll, B. K. Parady, J. F. Smith, S. E. Zarantonello, and G. W. Smith, letter, *American Mathematical Monthly* 97 (March 1990): 214.

Campbell, Stephen L., "Countability of Sets." *American Mathematical Monthly* 93 (June–July 1986): 480–481.

Chang, Hsi, and S. Sitharama Iyengar, "Efficient Algorithms to Globally Balance a Binary Search Tree." *Communications of the ACM* 27 (July 1984): 695–702.

Church, Alonzo, *Introduction to Mathematical Logic*. Princeton, N.J.: Princeton University Press, 1956.

Cohen, Daniel I. A., *Introduction to Computer Theory*. New York: Wiley, 1986.

Crypton, Dr., "The Mathematics of Connected Dots." *Science Digest* 93 (January 1985): 66–69.

———, "The Limits of Mathematical Knowledge." *Science Digest* 94 (March 1986): 72–75.

Cutland, N. J., *Computability: An Introduction to Recursive Function Theory*. Cambridge, U.K.: Cambridge University Press, 1986.

Date, C. J., *An Introduction to Database Systems*. Reading, Mass.: Addison-Wesley, 1975.

Davis, Martin, and Elaine J. Weyuker, *Computability, Complexity, and Languages*. Orlando, Fla.: Academic Press, 1983.

Dedekind, R., *Essays on the Theory of Numbers*. New York: Dover, 1963.

Dewdney, A. K., "An Ancient Rope-and-Pulley Computer Is Unearthed in the Jungle of Apraphul." *Scientific American* 258 (April 1988): 118–121.

Dixon, John, "Factorization and Primality Tests." *American Mathematical Monthly* 91 (June–July 1984): 333–352.

Eves, Howard, *Great Moments in Mathematics after 1650*. Washington, D.C.: Mathematical Association of America, 1983.

Eves, Howard, and Carroll V. Newsom, *An Introduction to the Foundations and Fundamental Concepts of Mathematics*. New York: Rinehart, 1958.

Fowler, Peter A., "The Konigsberg Bridges—250 Years Later." *American Mathematical Monthly* 258 (January 1988): 42–43.

Gallaire, Herve, Jack Minker, and Jean-Marie Nicolas, "Logic and Databases: A Deductive Approach." *ACM Computing Surveys* 16 (June 1984): 153–185.

Gardner, Martin, "Gauss's Congruence Theory Was Mod as Early as 1801." *Scientific American* 244 (February 1981): 17–20.

Gibbons, Alan, *Algorithmic Graph Theory*. Cambridge, U.K.: Cambridge University Press, 1985.

Gries, D., Compiler Construction for Digital Computers. New York: Wiley, 1971.

Harary, Frank, *Graph Theory*. Reading, Mass.: Addison-Wesley, 1969.

Hayes, Brian, "On the Finite-State Machine, a Minimal Model of Mousetraps, Ribosomes and the Human Soul." Scientific American 249 (December 1983): 19–28.

Hopcroft, John E., "Turing Machines." *Scientific American* 250 (May 1984): 86–98.

Humphreys, J. F., and M. Y. Prest, *Numbers, Groups and Codes*. Cambridge, U.K.: Cambridge University Press, 1989.

Jensen, K., and N. Wirth, *Pascal User Manual and Report*. New York: Springer-Verlag, 1974.

Kamke, E., *Theory of Sets*. New York: Dover, 1950.

Kemeny, J. G., J. L. Snell, and G. L. Thompson, *Introduction to Finite Mathematics*, 3rd ed. Englewood Cliffs, N.J.: Prentice-Hall, 1974.

Kent, William, "A Simple Guide to Five Normal Forms in Relational Database Theory." *Communications of the ACM* 26 (February 1983): 120–125.

Kleene, Stephen Cole, *Mathematical Logic*. New York: Wiley, 1968.

Lawson, Harold W., *Understanding Computer Systems*. Rockville, Md.: Computer Science Press, 1982.

Liu, C. L., *Elements of Discrete Mathematics*. New York: McGraw-Hill, 1985.

Long, Calvin T., *Elementary Introduction to Number Theory*. Englewood Cliffs, N.J.: Prentice-Hall, 1987.

Manber, Udi, "Using Induction to Design Algorithms." *Communications of the ACM* 31 (November 1988): 1300–1313.

Niven, Ivan, and Herbert S. Zuckerman, *An Introduction to the Theory of Numbers*, 4th ed. New York: Wiley, 1980.

Peterson, Ivars, "Exceptions to the Rule: At Least Two Human Languages Contain Grammatical Features That Put Them Outside Some Well-Known Grammars." *Science News* 128 (November 16, 1985): 314–315.

Pfleeger, Shari Lawrence, and David W. Straight, *Introduction to Discrete Structures*. New York: Wiley, 1985.

Pomerance, Carl, "The Search for Prime Numbers." *Scientific American* 247 (December 1982): 136–147.

———, "Recent Developments in Primality Testing." *Mathematical Intelligencer* 3 (January 1982): 97–105.

Ribenboim, Paulo, *The Book of Prime Number Records*. New York: Springer-Verlag, 1988.

Rosen, Kenneth H., *Elementary Number Theory and Its Applications*. Reading, Mass.: Addison-Wesley, 1984.

Singh, Jagjit, *Great Ideas of Modern Mathematics: Their Nature and Use*. New York: Dover, 1959.

Snell, J. Laurie, *Introduction to Probability*. New York: Random House, 1988.

Steen, Lynn Arthur (Ed.), *Mathematics Today: Twelve Informal Essays*. New York: Springer-Verlag, 1978.

Stout, Quentin F., and Patricia A. Woodworth, "Relational Databases." *American Mathematical Monthly* 90 (February 1983): 101–118.

Thomas, Gomer, *Finite Mathematics through Applications*. Seattle, Wash.: ASUW Lecture Notes, 1975.

Trakhtenbrot, B. A., *Algorithms and Automatic Computing Machines*. Boston: D. C. Heath, 1963.

Uspensky, J. V., and M. A. Heaslet, *Elementary Number Theory*. New York: McGraw-Hill, 1939.

Vanden Eynden, Charles, *Elementary Number Theory*. New York: Random House, 1987.

Velleman, D., letter, *American Mathematical Monthly* 96 (1989): 602–603.

Vinogradov, I. M., *Elements of Number Theory*. New York: Dover, 1954.

Waldrop, M. Mitchell, "Natural Language Understanding." *Science* 224 (April 1984): 372–374.

Weaver, Warren, *Lady Luck: The Theory of Probability*. New York: Doubleday, 1963.

Weil, Andre, *Number Theory for Beginners*. New York: Springer-Verlag, 1979.

ANSWERS TO ODD-NUMBERED EXERCISES

CHAPTER 1

Section 1.1

1. a. 10001 **b.** 10000000000 **c.** 11100100 **d.** 101010111100
3. a. 14711 **b.** 31 **c.** 38 **d.** 57005 **5. a.** 174 **b.** 7C **7. a.** 1221 **b.** 291
9. a. not prime **b.** prime **c.** not prime **d.** prime

Section 1.2

1. a. 110100 **b.** 110010 **3. a.** 1000111111 **b.** 1110111111 **5.** 100010
7. a. 4344 **b.** 46 **9. a.** 79BDF **b.** BC836
11. No. Suppose b is the base and $b > 2$. Then if $b - 1$ is a digit in base b, $(b - 1)(b - 1) > b - 1$; therefore, there will be a carry.
13. ᛉ ᛉ ᛉ   ᛉ ᛉ ᛉ ᛉ   ⟨ ⟨ ⟨ ⟨ ⟨

Section 1.3

1. a. $x_8^2 + x_9^2 + \cdots + x_{20}^2$ **b.** $(x_1 + y_1)^2 + (x_2 + y_2)^2 + \cdots + (x_{10} + y_{10})^2$
  c. $x_1^2 + y_1^2 + x_2^2 + y_2^2 + \cdots + x_{10}^2 + y_{10}^2$ **d.** 225 **e.** 39 **f.** 66

3. a. $\displaystyle\sum_{i=5}^{22} i$ **b.** $\displaystyle\sum_{i=1}^{20} x_i$

5. a. $\displaystyle\sum_{i=1}^{n} (x_i + 5) = \sum_{i=1}^{n} x_i + \sum_{i=1}^{n} 5$ **Rule 3**

  $\displaystyle\qquad\qquad\quad = \sum_{i=1}^{n} x_i + 5n$ **Rule 1**

7. a. $(x + 2)^2 (x + 4)^2 \cdots (x + 20)^2$ **b.** 0 **9. a.** $\displaystyle\sum_{i=1}^{10} \frac{1}{i^2}$ **b.** $\displaystyle\sum_{k=1}^{6} (k + 3)^2$
11. $(0 + 0)(0 + 1)(0 + 2)(0 + 3) + (1 + 0)(1 + 1)(1 + 2)(1 + 3) + (2 + 0)(2 + 1)(2 + 2)(2 + 3) = 144$

Section 1.4

1. 2 **3.** 30.8125 **5.** $10011.\overline{0011}$
7. Convert integer part as before. To convert fractional part, use successive multiplication by 8 until the fractional part is zero or the pattern repeats. Then write down the fractional part from top to bottom.
  a. $3.\overline{4}$ **b.** $14.2\overline{3146}$ **c.** 0.2 **d.** $7.\overline{005075341217270243656}$
9. a. 0.F0F8 **b.** 0.C00B **c.** 0.1FF8

Section 1.5

1. Assume $\sqrt{3}$ is rational and deduce a contradiction using the method shown in the text for the case of $\sqrt{2}$.

3.

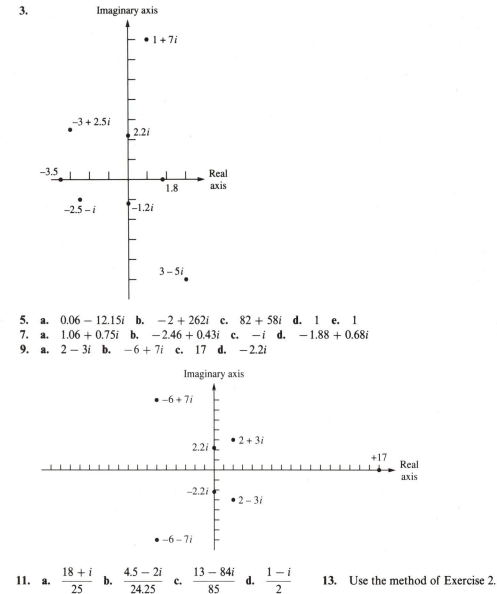

5. a. $0.06 - 12.15i$ **b.** $-2 + 262i$ **c.** $82 + 58i$ **d.** 1 **e.** 1

7. a. $1.06 + 0.75i$ **b.** $-2.46 + 0.43i$ **c.** $-i$ **d.** $-1.88 + 0.68i$

9. a. $2 - 3i$ **b.** $-6 + 7i$ **c.** 17 **d.** $-2.2i$

11. a. $\dfrac{18 + i}{25}$ **b.** $\dfrac{4.5 - 2i}{24.25}$ **c.** $\dfrac{13 - 84i}{85}$ **d.** $\dfrac{1 - i}{2}$ **13.** Use the method of Exercise 2.

15. By definition of cosine and sine, $\cos(\text{Arg}(z)) = \dfrac{a}{|z|}$, and $\sin(\text{Arg}(z)) = \dfrac{b}{|z|}$.

Section 1.6

1. a. $59, 59$ **b.** $177, -79$ **c.** $239, -17$ **3. a.** -2^{15} to $2^{15} - 1$ **b.** -2^{31} to $2^{31} - 1$

5. a. 7864 **b.** -8145 **7. a.** T **b.** T **c.** F

9. a. $[1], [3], [5], [7], [9], [11], [13], [15]$ **b.** $[1], [2], [3], [4], [5], [6]$

11. If $[a] = [a']$ and $[b] = [b']$, then $a - a' = kn$ and $b - b' = hn$. So

$$ab - a'b' = ab - a'b + a'b - a'b'$$
$$= (a - a')b + a'(b - b')$$
$$= (kn)b + a'(hn)$$
$$= n(kb + a'h)$$

Therefore, $[ab] = [a'b']$.

13. **a.** 01111001, carry and overflow **b.** 00011110, carry **c.** 00010101, carry **d.** 10111101, overflow

15. The two's complement negative is 1 plus the one's complement negative.

Chapter Review

1. **a.** 111001 **b.** 71 **c.** 39 **3.** **a.** 63 **b.** 18 **c.** 108 **5.** **a.** 3424 **b.** 714
7. 272 **9.** **a.** not prime **b.** not prime **c.** not prime **d.** prime
11. Use the method of Exercise 2, Section 1.5. **13.** **a.** $3 + 10i$ **b.** $-2 + 28i$
15. **a.** $\sqrt{13}, 34°$ **b.** $\sqrt{29}, -68°$ **c.** $\sqrt{17}, 166°$ **d.** $\sqrt{13}, -124°$ **e.** $5, 180°$ **f.** $2, 270°$
17. **a.** $-1 - 2i, \dfrac{1 - 2i}{5}$ **b.** $-2 + 2i, \dfrac{1 + i}{4}$ **c.** $3i, \dfrac{i}{3}$ **d.** $4 - i, \dfrac{-4 - i}{17}$ **e.** $8, \dfrac{-1}{8}$

f. $1 - \sqrt{2}\,i, \dfrac{(-1 - \sqrt{2}\,i)}{3}$

19. **a.** 1111100111011011 **b.** 1111100111011100 **21.** **a.** 61 **b.** -4
23. **a.** 01101001, carry and overflow **b.** 00001101, carry **c.** 10111011, overflow **d.** 10101010, carry
e. 01110110, carry

25. **a.** $1100.\overline{0110}$ **b.** $111.\overline{000101110000101000111}$ **27.** **a.** $[1], [3], [5], [7]$ **b.** $[1], [3], [7], [9]$
29. $(1^1 2^1 3^1)(1^2 2^2 3^2)(1^3 2^3 3^3) = 46{,}656$

CHAPTER 2

Section 2.1

1. **a.** $B \not\subset C$ **b.** $r \in C$ **c.** $m \notin B$ **d.** $C \subset A$ **e.** $\varnothing \subset B$ **f.** $B \subset B$
3. **a.** $\{1, 2, 3, 4, 5, 6, 7, 8, 9, 10\}$ **b.** $\{3, 6, 9, 12, \dots\}$ **c.** $\{-2, -1, 0, 1, 2\}$ **d.** $\{1, \frac{1}{2}, \frac{1}{3}, \frac{1}{4}, \dots\}$
e. $\{2, 4, 8, 16, \dots\}$ **f.** $\{-4, 4\}$ **g.** $\{-1, 1\}$
5. $C \subset D$ **7.** Since $A \subset B$, if $x \in A$, then $x \in B$ and $x \in C$, since $B \subset C$. Thus, $A \subset C$.
9. $A \subset B, B \not\subset A$

Section 2.2

1. **a.** B **b.** $\{e, r\}$ **c.** $\{c, o, r, e, t, m\}$ **d.** B **e.** $\{m, p, u, t\}$ **f.** $\{t, m\}$ **g.** $\varnothing$ **h.** $\{c, o\}$
i. $\{p, u\}$ **j.** $\{m, p, u, t, r, e\}$ **k.** $\{c, o, m, p, u, t, e\}$ **l.** $\{c, o\}$
3. **a.** $\{20, 22, 24, \dots, 36\}$ **b.** $\{19, 21, 23, 25, 27, 29, 31, 33, 35\}$ **c.** $\{2, 4, 6, \dots, 18\}$
d. $\{2, 4, 6, \dots, 18, 19, 20, 21, \dots, 36\}$
5. **a.** B **b.** B **c.** $\{a, b, \{a\}, \{a, b\}\}$ **d.** B **e.** $\{\{a\}, \{a, b\}\}$ **f.** $\{\{a, b\}\}$
7. No, since $A \cap B \cap C$ is a subset of $A \cap B$, $A \cap C$, and $B \cap C$.
9. $\{\varnothing, \{a\}, \{b\}, \{c\}, \{a, b\}, \{a, c\}, \{b, c\}, A\}$ **11.** $\{\varnothing, \{\varnothing\}, \{A\}, P(A)\}$
13. **a.** $\{0, 2, 5, 6, 7, 9\}$ **b.** $\{1, 2, 4, 5, 7, 9\}$ **c.** $\{0, 2, 5, 6, 7\}$ **d.** $\{1, 3, 4, 9\}$ **e.** A **f.** $\varnothing$

15.

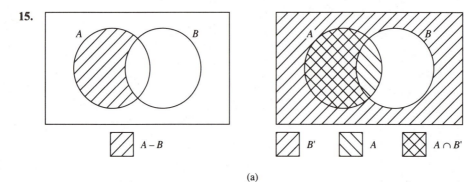

(a)

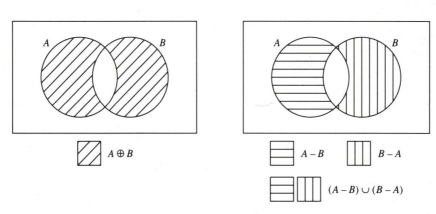

(b)

17. **a.** If $x \in (A - B)'$, then $x \notin A - B$, so either $x \notin A$ or $x \in B$ (that is, $x \in A' \cup B$). So
$(A - B)' \subset A' \cup B$. If $x \in A' \cup B$, then either $x \in A'$ or $x \in B$. If $x \in B$, then $x \notin A - B$, so
$x \in (A - B)'$. If $x \in A'$, then $x \notin A$, so $x \notin A - B$, and $x \in (A - B)'$.

b. If $x \in A \cap (A \cup B)$, then $x \in A$, so $A \cap (A \cup B) \subset A$. If $x \in A$, then $x \in A \cup B$, so $x \in A \cap (A \cup B)$,
and $A \subset A \cap (A \cup B)$.

19. **a.** If $x \in A \oplus B$, then exactly one of the following holds:

(1) $x \in A, x \notin B, x \in C$; then $x \in B \oplus C$

(2) $x \in A, x \notin B, x \notin C$; then $x \in A \oplus C$

(3) $x \notin A, x \in B, x \in C$; then $x \in A \oplus C$

(4) $x \notin A, x \in B, x \notin C$; then $x \in B \oplus C$

b. If $A \oplus B = \emptyset$, then $A \cup B = A \cap B$, so $A = B$.

c. If $x \in A \oplus B$, then either

(1) $x \in A$ and $x \notin B$; then $x \notin A'$ and $x \in B'$; or

(2) $x \in B$ and $x \notin A$; then $x \notin B'$ and $x \in A'$.

This shows that $A \oplus B \subset A' \oplus B'$. The converse is similar.

d. Similar to part (c); check each of eight cases.

21. **a.** False. $A = \{1, 2\}, B = \{1\}, C = \{2\}$.   **b.** False. $A = \{1\}, B = \{2\}, C = \{3\}$.

c. True. If $x \in A$, then $x \in A \cup B$, so $x \in A \cap B$. Therefore, $x \in B$. This shows that $A \subset B$. Similarly, $B \subset A$.

d. True. If $x \in A \cup C$, then either $x \in A$ or $x \in C$. If $x \in A$, then $x \in B$, since $A \subset B$, so
$x \in B \cup D$. If $x \in C$, then $x \in D$, since $C \subset D$, so $x \in B \cup D$. So $A \cup C \subset B \cup D$.

Section 2.3

1. $\{1, 2, 3\}$ 3. **a.** $[0, \infty)$ **b.** $(-3, -2) \cup (2, 3)$ 5. **a.** F **b.** F **c.** F **d.** F
7. **a.** $\{1\}$ **b.** $(-\infty, 1]$ **c.** $(-\infty, 1] \cup \{2\}$ **d.** $(0, 1) \cup (1, 3)$ **e.** $(1, 4]$
9. **a.** $(-\infty, 0] \cup [1, \infty)$ **b.** $(-\infty, -5]$

Section 2.4

1. **a.** $\{(1, 0), (1, 1), (2, 0), (2, 1), (3, 0), (3, 1), (4, 0), (4, 1)\}$
    b. $\{(0, 1), (0, 2), (0, 3), (0, 4), (1, 1), (1, 2), (1, 3), (1, 4)\}$
    c. $\{(0, 0), (0, 1), (1, 0), (1, 1)\}$
    d. $\{\varnothing, \{(0, 0)\}, \{(0, 1)\}, \{(1, 0)\}, \{(1, 1)\}, \{(0,0), (0, 1)\}, \{(0, 0), (1, 0)\}, \{(0, 0), (1, 1)\},$
    $\{(0, 1), (1, 0)\}, \{(0, 1), (1, 1)\}, \{(1, 0), (1, 1)\}, \{(0, 0), (0, 1), (1, 0)\}, \{(0, 0), (0, 1), (1, 1)\}, \{(0, 0), (1, 0), (1, 1)\},$
    $\{(0, 1), (1, 0), (1, 1)\}, B \times B\}$
3. $\{(1, 2), (1, 4), (2, 2), (3, 2)\}$
5. **a.** $\{\varnothing, \{(0, a)\}, \{(1, a)\}, A \times B\}$
    b. $\{(\varnothing, \varnothing), (\varnothing, B), (\{0\}, \varnothing), (\{0\}, B), (\{1\}, \varnothing), (\{1\}, B), (A, \varnothing), (A, B)\}$
7. The set of points inside the unit square   9. A and B are disjoint.
11. If $(x, y) \in A \times (B \cap C)$, then $x \in A$ and $y \in B \cap C$, so $y \in B$ and $y \in C$. Thus, $(x, y) \in (A \times B)$ and
    $(x, y) \in (A \times C)$, so $(x, y) \in (A \times B) \cap (A \times C)$. If $(x, y) \in (A \times B) \cap (A \times C)$, then $(x, y) \in A \times B$ and
    $(x, y) \in A \times C$. This means that $x \in A$ and $y \in B$ and $y \in C$. Thus, $y \in B \cap C$. So $(x, y) \in A \times (B \cap C)$.
13. Suppose $(x, y) \in A \times C$. Then $x \in A$, so $x \in B$, since $A \subset B$. And $y \in C$, so $y \in D$, since $C \subset D$. Therefore,
    $(x, y) \in B \times D$, so $A \times C \subset B \times D$.
15. **a.** $I \times I \times I \times I \times R \times R$ **b.** $R \times I \times R$

Section 2.5

1. $1 \leftrightarrow a, 2 \leftrightarrow b, 3 \leftrightarrow c, 4 \leftrightarrow d, 5 \leftrightarrow e$ 3. $f(n) = n^2$ 5. $f(b) = \frac{b}{3}$
7. For $x \in (0, 1)$, define $f(x) = (b - a)x + a$.
9. For $n \in \mathbf{Z}$, the function $f(n) = 2^n$ is a one-to-one correspondence between the set $\mathbf{Z}$ and the set of
    integral powers of 2.
11. Let $f(1)$ be the smallest element in the set, $f(2)$ the next smallest, etc. This defines a function, because
    for any a, $f(a)$ can be computed in a finite number of steps.

Section 2.6

1. **a.** *Step 1:* $n = 1$: $1 = 1^2$
    Step 2: Assume true for $n = k - 1$; that is,
    $$1 + 3 + 5 + \cdots + [2(k - 1) - 1] = (k - 1)^2$$
    Then for $n = k$,
    $$1 + 3 + \cdots + [2(k - 1) - 1] + (2k - 1) = (k - 1)^2 + (2k - 1)$$
    $$= k^2 - 2k + 1 + 2k - 1$$
    $$= k^2$$

    b. *Step 1:* $n = 1$: $1 = \dfrac{1(3 - 1)}{2}$
    Step 2: Assume true for $n = k - 1$:
    $$1 + 4 + 7 + \cdots + 3(k - 1) - 2 = \frac{(k - 1)[3(k - 1) - 1]}{2}$$

Then for $n = k$,

$$1 + 4 + \cdots + 3(k-1) - 2 + (3k-2) = \frac{(k-1)[3(k-1)-1]}{2} + (3k-2)$$

$$= \frac{(k-1)(3k-4)}{2} + (3k-2)$$

$$= \frac{(k-1)(3k-4) + 2(3k-2)}{2}$$

$$= \frac{3k^2 - 7k + 4 + 6k - 4}{2}$$

$$= \frac{3k^2 - k}{2}$$

$$= \frac{k(3k-1)}{2}$$

c. *Step 1:* $n = 1$: $1^2 = \dfrac{1(1+1)[2(1)+1]}{6}$

Step 2: Assume true for $n = k - 1$; that is,

$$1^2 + 2^2 + \cdots + (k-1)^2 = \frac{(k-1)[(k-1)+1][2(k-1)+1]}{6}$$

Then for $n = k$,

$$1^2 + 2^2 + \cdots + (k-1)^2 + k^2 = \frac{(k-1)[(k-1)+1][2(k-1)+1]}{6} + k^2$$

$$= \frac{(k-1)k(2k-1)}{6} + k^2$$

$$= \frac{(k-1)k(2k-1) + 6k^2}{6}$$

$$= \frac{(k^2 - k)(2k-1) + 6k^2}{6}$$

$$= \frac{2k^3 - 3k^2 - k^2 + k + 6k^2}{6}$$

$$= \frac{2k^3 + 3k^2 + k}{6}$$

$$= \frac{k(2k^2 + 3k + 1)}{6}$$

$$= \frac{k(k+1)(2k+1)}{6}$$

3. *Step 1:* $n = 1$: $2^1 = 2 > 1$

Step 2: Assume true for $n = k - 1$: $2^{k-1} > k - 1$; then for $n = k$:

$$2^k = 2(2^{k-1}) = 2^{k-1} + 2^{k-1}$$

$$> (k-1) + 2^{k-1} \qquad \text{\color{brown}By inductive hypotheses}$$

$$\geq (k-1) + 1 \qquad\quad \text{\color{brown}Since } 2^{k-1} \geq 1$$

$$= k$$

5. *Step 1:* $n = 1$: $a^1 < b^1$

Step 2: Assume true for $n = k - 1$: $a^{k-1} < b^{k-1}$ Then for $n = k$, $a^k = a(a^{k-1}) < b(b^{k-1}) = b^k$

7. *Step 1:* $n = 2$:

$$\frac{1}{\sqrt{1}} + \frac{1}{\sqrt{2}} = 1 + 0.707\ldots > 1.414\ldots = \sqrt{2}$$

Step 2: Assume true for $n = k - 1$:

$$\frac{1}{\sqrt{1}} + \frac{1}{\sqrt{2}} + \cdots + \frac{1}{\sqrt{k-1}} > \sqrt{k-1}$$

Then for $n = k$,

$$\frac{1}{\sqrt{1}} + \frac{1}{\sqrt{2}} + \cdots + \frac{1}{\sqrt{k-1}} + \frac{1}{\sqrt{k}} > \sqrt{k-1} + \frac{1}{\sqrt{k}}$$

Thus, it suffices to show that

$$\sqrt{(k-1)} + \frac{1}{\sqrt{k}} > \sqrt{k}$$

But

$$k(k-1) > (k-1)^2$$
$$\sqrt{k(k-1)} > k - 1$$
$$\sqrt{k(k-1)} + 1 > k$$

and the result follows upon dividing by $\sqrt{k}$.

9. a. *Step 1:* $n = 1$: $(A_1)' = A_1'$

Step 2: Assume true for $n = k - 1$:

$$(A_1 \cap A_2 \cap \cdots \cap A_{k-1})' = A_1' \cup A_2' \cup \cdots \cup A_{k-1}'$$

Then for $n = k$,

$$(A_1 \cap A_2 \cap \cdots \cap A_{k-1} \cap A_k)' = [(A_1 \cap A_2 \cap \cdots \cap A_{k-1}) \cap A_k)]'$$
$$= (A_1 \cap A_2 \cap \cdots \cap A_{k-1} \cap A_k)' \cup A_k'$$
$$= (A_1' \cup A_2' \cup \cdots \cup A_{k-1}') \cup A_k'$$
$$= A_1' \cup \cdots \cup A_k'$$

b. Interchange the $\cup$ and $\cap$ symbols in part (a).

11. The sum is $n + 1$.

Step 1: for $n = 0$, the equation reads $1 = 1$.

Step 2:

$$\sum_{i=0}^{k} 1^i = \sum_{i=0}^{k-1} 1^i + 1^k = [(k-1) + 1] + 1 = k + 1$$

13. If $n = 2$, $X_1 \cap X_2$ may be empty, and z does not exist.

15. Let $S' = S \cup \{1, 2, \ldots, n_0\}$. Then S' satisfies the condition of the weak form of mathematical induction, so $S' = \mathbf{N}$. Then S is $\mathbf{N}$ less the numbers less than n_0.

Section 2.7

1. A: 11000100; B: 00111110; C: 00101011; X: 11111111

3. a. 00000101 **b.** 4 and 5 **c.** 1, 3, and 6 **d.** 1, 2, 4, and 6 **e.** 3 and 4

Chapter Review

1. $\{(1, 0), (2, 0), (3, 0), (4, 0), (2, 1), (3, 1), (4, 1), (3, 2), (4, 2), (4, 3)\}$

3. $f(1) = a, f(2) = b$, etc., in order. There are many others.   **5.** $f(x) = x$

7. a. $(-1, 3)$ **b.** $(0, 1]$ **c.** $(-2, 5)$ **d.** $(-\infty, 0] \cup [1, 2)$ **e.** $[3, 5]$

9. If $x \in A - (B \cap C)$, then $x \in A$ and x is not in both B and C. If x is not in B, then $x \in A - B$, and if x is not in C, then $x \in A - C$. This shows that $A - (B \cup C) \subset (A - B) \cup (A - C)$. Now suppose that $x \in (A - B) \cup (A - C)$. Then either $x \in A - B$ or $x \in A - C$. If $x \in A - C$, then $x \in A - (B \cap C)$, and the other case is similar.

11. **a.** *Step 1:* If $n = 1$, $1 - \dfrac{1}{2} = \dfrac{1}{1+1}$

Step 2: If $k > 1$,

$$\left(1 - \frac{1}{2}\right)\left(1 - \frac{1}{3}\right) \cdots \left[1 - \frac{1}{k+1}\right] = \left(1 - \frac{1}{2}\right)\left(1 - \frac{1}{3}\right) \cdots \left(1 - \frac{1}{k}\right)\left(1 - \frac{1}{k+1}\right)$$

$$= \left(\frac{1}{k}\right)\left(\frac{k}{k+1}\right) = \frac{1}{k+1}$$

b. *Step 1:* If $n = 1$, $1 = \dfrac{3^1 - 1}{2}$

Step 2: $1 + 3 + \cdots + 3^{k-1} = 1 + 3 + \cdots + 3^{k-2} + 3^{k-1}$

$$= \frac{3^{k-1} - 1}{2} + 3^{k-1}$$

$$= \frac{3^{k-1} - 1 + 2(3^{k-1})}{2}$$

$$= \frac{3(3^{k-1}) - 1}{2} = \frac{3^k - 1}{2}$$

c. *Step 1:* If $n = 1$, $2(1) \le 2^1$

Step 2: If $k > 1$, $2k = 2(k-1) + 2 \le 2^{k-1} + 2 \le 2^{k-1} + 2^{k-1} = 2^k$

13. **a.** 000010 **b.** 100100 **c.** 100100 **d.** 010001 **e.** 111101 **f.** 100100 **g.** 110111

CHAPTER 3

Section 3.1

1. **a.** x **b.** x **c.** $z + x$ **d.** x **3.** $\{(1, u), (1, v)\}$

5. No—the range can be any set containing the image.   **7.** **a.** 3 **b.** 12 **c.** 7 **d.** 22

9. **a.** 1 **b.** 5   **11.** **a.** -8 **b.** 0 **c.** $-\frac{1}{8}$ **d.** $\frac{7}{2}$

13. Image is set of nonnegative real numbers.

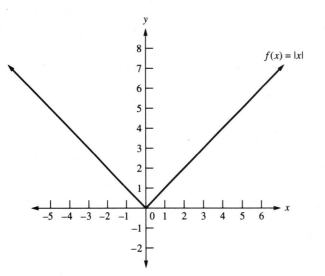

15. a. 5 **b.** 0 **c.** If $x = 5$, the left side is equal to 5 and the right side is equal to 6.

17. a. $(-1)^{n-1}$ **b.** $\frac{1}{n}$ **c.** $(\frac{1}{2})^n$ **d.** $(\frac{1}{10})^n$

19. a.
$$3(3(3(5) - 3) + 2) - 1 = 3(3(15 - 3) + 2) - 1$$
$$= 3(3(12) + 2) - 1$$
$$= 3(36 + 2) - 1$$
$$= 3(38) - 1 = 114 - 1 = 113$$

b. 2 compared to 6

21. 8 from A to B; 9 from B to A

23. Define $f: \{1, 2\} \to \{1\}$ by $f(1) = f(2) = 1$. Let $X = \{1\}$ and $Y = \{2\}$. Then $X \cap Y = \varnothing$ and $f[X \cap Y] = \varnothing$, but $f[X] \cap f[Y] = \{1\}$.

Section 3.2

1. a. onto **b.** both **c.** both **d.** one-to-one   **3. a.** neither **b.** one-to-one **c.** one-to-one

5. Not one-to-one, since $0! = 1! = 1$. Not onto, since the image is not equal to **N**.   **7.** no; no

9. no; no   **11.** No, since f is not one-to-one   **13.** $f^{-1}(x) = \dfrac{x+1}{3}$

15. $f \circ g = \{(a, e), (b, b), (c, d), (d, b), (e, b)\}$
$g \circ f = \{(1, 1), (2, 4), (3, 1), (4, 1)\}$

17. a. $(2n + 5)^2 + 1$ **b.** $(n^2 + 1)^2 + 1$ **c.** $2(n^2 + 1) + 5$ **d.** $2(2n + 5) + 5$

19. $(f + g)(x) = x^2 + 1 + 1 - x = x^2 - x + 2$
$(f - g)(x) = x^2 + 1 - (1 - x) = x^2 + x$
$(fg)(x) = (x^2 + 1)(1 - x) = -x^3 + x^2 - x + 1$

21. Suppose that $(g \circ f)(x) = (g \circ f)(y)$. That is, $g(f(x)) = g(f(y))$. Since g is one-to-one, $f(x) = f(y)$, and since f is one-to-one, $x = y$.

23. Suppose f is one-to-one. If f is not onto, then there is an $x \in B$ such that $x \neq f(a)$ for any $a \in A$. So there are at most $n - 1$ elements in the image of f. Since there are n elements in A, at least two elements in A must have the same image in B (by the Pigeonhole Principle). This is impossible, because f is one-to-one. Conversely, suppose f is onto and f is not one-to-one. Then there are distinct elements x and y in A such that $f(x) = f(y)$. So the image of f contains at most $n - 1$ elements. This is impossible, since f is onto.

25. a. $\{1\}$ **b.** $\{2, 3, 4\}$ **c.** $\varnothing$ **d.** X

27. Let $X = \{1, 2\}$, $Y = \mathbf{N}$, $f(1) = f(2) = 1$, $A = \{1\}$. Then $f[A'] = f[\{2\}] = \{1\}$, but $(f[A])' = \{1\}' = \mathbf{N} - \{1\}$.

Section 3.3

1. $2^9 = 512$ relations; 27 functions

3.

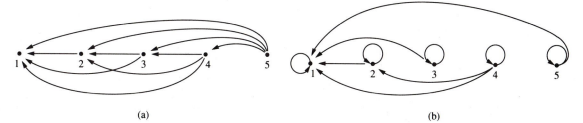

(a)   (b)

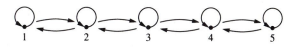

(c)

5. a. symmetric **b.** reflexive, symmetric, and transitive **c.** symmetric

7.

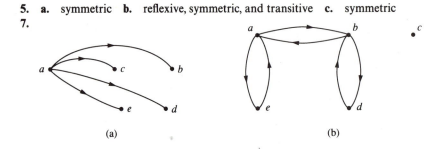

(a)   (b)

(c)

9. $(x, y, z) \in \mathbf{R} \times \mathbf{R} \times \mathbf{R}$ if $x + y + z < 1$ **11.** reflexive, symmetric, and transitive

13. Given a, there does not necessarily exist any b for which $a\,R\,b$. For instance, let $A = \{1, 2, 3\}$ and $R = \{(1, 1), (1, 2), (2, 1)\}$; nothing is related to 3.

15. relations a and b **17.** the relation of Exercise 6c   **19.** relation a

21. If R is asymmetric and if $(a, a) \in R$, then $(a, a) \notin R$, a contradiction. Therefore, $(a, a) \notin R$.

Section 3.4

1.

If	then
$a\,R\,a$ and $a\,R\,a$	$a\,R\,a$
$a\,R\,a$ and $a\,R\,b$	$a\,R\,b$
$a\,R\,b$ and $b\,R\,a$	$a\,R\,a$
$a\,R\,b$ and $b\,R\,b$	$a\,R\,b$
$b\,R\,a$ and $a\,R\,a$	$b\,R\,a$
$b\,R\,a$ and $a\,R\,b$	$b\,R\,b$
$b\,R\,b$ and $b\,R\,a$	$b\,R\,a$
$b\,R\,b$ and $b\,R\,b$	$b\,R\,b$
$c\,R\,c$ and $c\,R\,c$	$c\,R\,c$
$c\,R\,c$ and $c\,R\,d$	$c\,R\,d$
$c\,R\,d$ and $d\,R\,c$	$c\,R\,c$
$c\,R\,d$ and $d\,R\,d$	$c\,R\,d$
$d\,R\,c$ and $c\,R\,d$	$d\,R\,d$
$d\,R\,c$ and $c\,R\,c$	$d\,R\,c$
$d\,R\,d$ and $d\,R\,c$	$d\,R\,c$
$d\,R\,d$ and $d\,R\,d$	$d\,R\,d$
$e\,R\,e$ and $e\,R\,e$	$e\,R\,e$

3. Reflexive: $(x, y)\,S\,(x, y)$ is true, since $x + y = y + x$, by the Commutative Law.

Symmetric: If $(x, y)\,S\,(z, w)$, then $x + w = y + z$ and $y + z = x + w$, since equality is symmetric, and $z + y = w + x$, so $(z, w)\,S\,(x, y)$.

Transitive: Suppose $(x, y)\,S\,(z, w)$ and $(z, w)\,S\,(u, v)$. Then $x - y = z - w$ and $z - w = u - v$, so $x - y = u - v$, and $(x, y)\,S\,(u, v)$.

5.

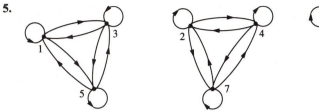

7. a. sets of form $\{a, -a\}$ where $a \in \mathbf{R}$ **b.** half-open intervals $[n, n+1)$ where $n \in \mathbf{N}$
  c. half-open intervals $(n, n+1]$ where $n \in \mathbf{N}$ **d.** lines of slope $\frac{2}{3}$
  e. If $x^2 + y^2 = b$, then the level set is the circle centered at $(0, 0)$ with radius $\sqrt{b}$.

9.

Reflexive closures	Symmetric closures	Transitive closures

(a)

(b)

(c)

11. "is a descendent of" **13.** Add the pairs $(b, a), (b, c), (c, a)$.
15. $a \equiv b \pmod 0$ means $a = b$, so each equivalence class is a set consisting of a single integer.

Section 3.5

1.

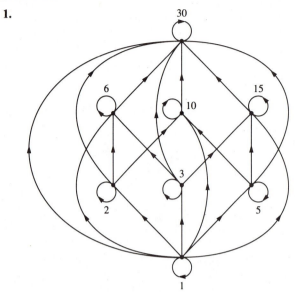

3. a, b, d **5.** Part c is a partial order.
7. Since $x\,R\,y$ is always false, there are no pairs (x, y) and (y, x) with $x \neq y$.
9. If R is a relation that is not antisymmetric, and if $R \subset S$, then S is not antisymmetric. (For if $a \neq b$, $a\,R\,b$, and $b\,R\,a$, then also $a\,S\,b$ and $b\,S\,a$.) Therefore, if R is not antisymmetric, antisymmetry cannot be restored by adding more ordered pairs to it.
11. $\geq$
13. Suppose that the relation R on the set X is not asymmetric. Then there exist $a \in X$ and $b \in X$ such that $a\,R\,b$ and $b\,R\,a$. By transitivity, $a\,R\,a$, contradicting irreflexivity.

Section 3.6

(In the following answers, * indicates a key field.)
1. R1: {*Employee SSN, Employee Name, Skill}
R2: {*Customer Name, Customer Address}
R3: {*Contract Number, Customer Name}
R4: {*Job Number, Contract Number}
R5: {*Employee SSN, *Week, *Job Number, Hours Worked}
3. R1: {*Item Number, Location Found, Date Found, Culture, Finder, Organization}
R2: {*Culture, Period}
R3: {*Finder, Finder's Current Address}
R4: {*Organization, Organization's Current Address}

Chapter Review

1. $\{(-1, -2), (0, -1), (\frac{1}{3}, -\frac{26}{27}), (2, 7)\}$ **3.** Domain is $\{1, 2, 3, 5\}$; image is $\{1, 4, 9, 25\}$
5. $\{(1, a), (2, b), (3, c)\}$; no   **7. a.** neither **b.** onto   **9. a.** no **b.** no
11. a. $\{(u, a), (w, b), (z, c), (x, d)\}$ **b.** $\{(1, -1), (0, 0), (-1, 1)\}$ **13.** No; f is not onto.
15. $\{(1, 1), (1, 2)\}$
17. Let $A = \{1, 2, 3\}$
    a. $\{(1, 1), (2, 2), (3, 3), (1, 2), (2, 1), (2, 3), (3, 2)\}$ **b.** $\{(1, 1), (2, 2), (3, 3), (1, 2)\}$
    c. $\{(1, 1), (1, 2), (2, 1)\}$ **d.** $\{(1, 1), (2, 2), (3, 3)\}$

19. Not reflexive, since 0 is not prime; symmetric; not transitive (12 *R* 7 and 7 *R* 4, but 12 *R* 4 is false).

21.

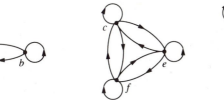

23. For $a, b \in B$, define R to be: $a \, R \, b$ if $a = b$ or a is the two's complement of b. The equivalence classes are $\{00000000\}$, $\{10000000\}$, and two-element sets consisting of the representations of n and $-n$ for $n = 1, \ldots, 127$.

25.

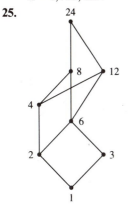

27. For X nonempty, R is symmetric, transitive, antisymmetric, asymmetric, and irreflexive. For X empty, R has all six properties.

29. Note that if a is of odd ancestry, then its first-order ancestor is of even ancestry. Let E_X, O_X, I_X consist of the elements of X having even, odd, and infinite ancestry, respectively; similarly, define E_Y, O_Y, I_Y. Define $h: X \rightarrow Y$ by $h(x) = f(x)$ if x is of even or infinite ancestry, and $h(x) = $ the first-order ancestor of x if x is of odd ancestry. Then h maps I_X to I_Y, E_X to O_Y, and O_X to E_Y.

CHAPTER 4

Section 4.1

1. a, c, e

3. a. 5 is prime and $\frac{1}{3}$ is rational.
   b. Nixon was elected president in 1968 and 1972 and Nixon did not complete two terms as president.

5. a. If Euclid was born in Sardinia, then Euclid was not six feet tall.
   b. Euclid was born in Sardinia if and only if Euclid was six feet tall.
   c. It is not true that Euclid was both not born in Sardinia and was six feet tall.
   d. Euclid was not born in Sardinia nor was Euclid six feet tall.

7. a. $q \Rightarrow p$ **b.** $\neg(q \wedge \neg p)$; or, $q \, \text{NAND} \, \neg p$ **c.** $\neg q \Rightarrow p$

9.

p	q	$p \veebar q$
T	T	F
T	F	T
F	T	T
F	F	F

11. a. $q \Rightarrow p$ **b.** $\neg(p \Rightarrow q)$ **c.** $\neg(q \Rightarrow p)$ **13.** 256

Section 4.2

1.

p	q	$\neg p \wedge q$	$p \vee \neg q$	$\neg p \Rightarrow q$	$\neg(p \Rightarrow q)$
T	T	F	T	T	F
T	F	F	T	T	T
F	T	T	F	T	F
F	F	F	T	F	F

3. a.

p	$p \Rightarrow p$
T	T
F	T

b.

p	q	$q \Rightarrow p$	$p \Rightarrow (q \Rightarrow p)$
T	T	T	T
T	F	T	T
F	T	F	T
F	F	T	T

c.

p	q	$p \Rightarrow q$	$\neg p$	$\neg p \vee q$	$(p \Rightarrow q) \Leftrightarrow (\neg p \vee q)$
T	T	T	F	T	T
T	F	F	F	F	T
F	T	T	T	T	T
F	F	T	T	T	T

d.

p	q	r	$p \wedge q$	$q \wedge r$	$(p \wedge q) \wedge r$	$p \wedge (q \wedge r)$	$(p \wedge q) \wedge r \Leftrightarrow p \wedge (q \wedge r)$
T	T	T	T	T	T	T	T
T	T	F	T	F	F	F	T
T	F	T	F	F	F	F	T
T	F	F	F	F	F	F	T
F	T	T	F	T	F	F	T
F	T	F	F	F	F	F	T
F	F	T	F	F	F	F	T
F	F	F	F	F	F	F	T

5. a. yes **b.** yes **c.** no

7. DeMorgan's Laws

p	q	$\neg(p \wedge q)$	$\neg p \vee \neg q$	$\neg(p \vee q)$	$\neg p \wedge \neg q$
T	T	F	F	F	F
T	F	T	T	F	F
F	T	T	T	F	F
F	F	T	T	T	T

9. F NAND (T NAND T) = T but (F NAND T) NAND T = F

11. If p is T and q is F, then $p \Rightarrow q$ is F. But $\neg p$ is F, so $\neg p \Rightarrow \neg q$ is T. Therefore, $p \Rightarrow q$ and $\neg p \Rightarrow \neg q$ are not equivalent.

13. yes; yes   **15.** yes; yes   **17.** yes—F   **19.** $(p$ NAND $p)$ NAND $(q$ NAND $q)$ _

21. $p \vee q \Leftrightarrow \neg p \Rightarrow q, p \wedge q \Leftrightarrow \neg(\neg p \vee \neg q)$, and all connectives can be written in terms of $\wedge$, $\vee$, and $\neg$.

23. $(p$ NOR $p)$ NOR $(q$ NOR $q)$ **25.** See solution to Exercise 9, Section 2.6.

Section 4.3

1. not valid   3. valid   5. contraposition
7. Let n be an even integer. Then $n = 2k$ for some integer k, and $n^2 = (2k)^2 = 4k^2 = 2(2k^2)$, which is even.
9. Let n and m be odd integers. Then $n = 2k + 1$ for some integer k, and $m = 2j + 1$ for some integer j. The product

$$nm = (2k + 1)(2j + 1)$$
$$= 4kj + 2k + 2j + 1$$
$$= 2(2kj + k + j) + 1$$

which is odd.
11. Suppose $f(x) = mx + b$ with $m \neq 0$. If $f(c) = f(d)$, then $mc + b = md + b$, so $mc = md$, and $c = d$.
13. Assume that R is asymmetric. We have to show that, if $x \, R \, y$ and $y \, R \, x$, then $x = y$. By asymmetry, the hypotheses "$x \, R \, y$ and $y \, R \, x$" of this conditional is false; therefore, the conditional "If $x \, R \, y$ and $y \, R \, x$, then $x = y$" is true.

Section 4.4

1. $\{e, o\}$ 3. the set of even integers
5. *Remainder of part 2 of theorem:* If x is in the intersections of the truth sets of P and Q, then both $P(x)$ and $Q(x)$ are true, so x is in the truth set of $P \wedge Q$.
   Part 3 of theorem: If x is an element of the truth set of $P \vee Q$, then $(P \vee Q)(x)$ is true. (That is, $P(x) \vee Q(x)$ is true.) This means that at least one of $P(x)$ and $Q(x)$ are true, so x is an element of the union of the truth sets of P and Q. Conversely, if x is in the union of the truth sets of P and Q, then either x is in the truth set of P or that of Q. In either case, $(P \vee Q)(x)$ is true.
7. **a.** $\{0\}$ **b.** $(-\infty, 1) \cup \{5\} \cup [7, 8]$
9.

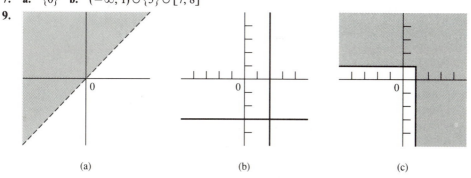

(a)   (b)   (c)

11. The truth set of P NAND Q is the complement of the intersection of the truth sets of P and Q.
13. **a.** equivalent   **b.** equivalent   **c.** not equivalent

Section 4.5

1. **a.** $\exists n P(n)$, where $P(x) =$ "x is an even perfect number."
   b. $\forall P(Q(P) \Rightarrow \exists x R(x, y))$, where $Q(x) =$ "x is a polynomial of odd degree" and $R(x, y)$ is "y is a root of the polynomial x."
   c. $\forall n(E(n)$ and $n > 2 \Rightarrow \exists p \exists q(P(p) \wedge P(q) \wedge p + q = n))$, where $E(x)$ is "x is even" and $P(x) =$ "x is prime."
   d. $\forall f(I(f) \Rightarrow N(f))$, where $I(f) =$ "f has an inverse" and $N(f) =$ "f is one-to-one."
3. **a.** Every integer is equal to itself; true.
   b. The equation $x^2 - x - 6 = 0$ has an integral solution; true.
   c. Every integer is either nonpositive or nonnegative; true.
   d. Every integer that is less than 0 is also less than 1; true.

5. If $\forall x[P(x) \Leftrightarrow Q(x)]$ is true, then the truth set of $P(x) \Leftrightarrow Q(x)$ is X. If x is in the truth set of P, then $P(x)$ is T, so $Q(x)$ is T, so x is in the truth set of Q; and conversely. Now suppose that P and Q have the same truth set A. If $x \in A$, then $P(x) \Leftrightarrow Q(x)$ is T $\Leftrightarrow$ T, which is T; and if $x \notin A$, then $P(x) \Leftrightarrow Q(x)$ is F $\Leftrightarrow$ F, which is T. Therefore, $\forall x[P(x) \Leftrightarrow Q(x)]$ is true.

7. a. $\varnothing$ **b.** R **c.** $(-1, 0)$ **9. a.** F **b.** T **c.** F **d.** T **e.** T

11.

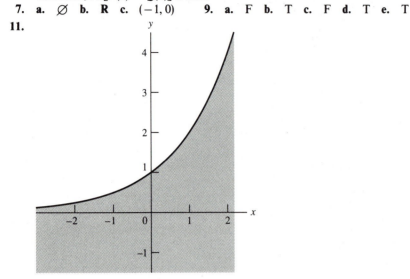

13. a. If the truth set of P contains a horizontal line, then it also contains a vertical line.
 b. The truth set of P contains the line $x = y$.
 c. The truth set of P is symmetric about the line $x = y$.

15. a. $P(x, y) =$ "$y = 0$" **b.** $P(x, y) = $ F **c.** $P(x, y) =$ "$y = 0$"

Section 4.6

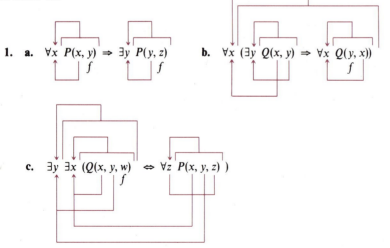

1. a. $\forall x \; P(x, y) \; \Rightarrow \; \exists y \; P(y, z)$ **b.** $\forall x \; (\exists y \; Q(x, y) \; \Rightarrow \; \forall x \; Q(y, x))$

c. $\exists y \; \exists x \; (Q(x, y, w) \; \Leftrightarrow \; \forall z \; P(x, y, z) \;)$

3. a. No; rule 3 violated; y is free in second formula. **b.** yes **c.** yes
5. a. $(\forall x \in A)(\exists y \in B)y^2 = x$ **b.** $(\exists x < 5)(\forall y > x)(2y > 1)$ **c.** $(\exists x < 0)(\exists y > 0)(xy = 5)$
7. a. $(x > y) \wedge (x^2 \leq z)$ **b.** $(x \geq y) \vee (x + y = 5)$ **c.** $(x < 7) \wedge (x \geq 3)$ **d.** $\exists x(x < 7)$ **e.** $\forall x(x \neq -1)$
9. a. $\exists x[\neg(x \; R \; x)]$ **b.** $\exists x \exists y[(x \; R \; y) \wedge \neg(y \; R \; x)]$ **c.** $\exists x \exists y \exists z[(x \; R \; y) \wedge (y \; R \; z) \wedge \neg(x \; R \; z)]$

Section 4.7

1. **a.** T **b.** F
3. **a.** Not valid; $p = $ F. **b.** Not valid; $p = $ F, $q = $ T.
   c. Not valid; $X = \mathbf{Z}$, $P(x) = $ "$x = 0$," $x = 1$. **d.** Not valid; $X = \mathbf{Z}$, $P(x) = $ "$x = 0$," $x = 0$, $y = 1$.
5. **a.** Not valid; $X = \mathbf{Z}$, $P(x) = $ "$x = 1$," $Q(x) = $ "$x = 2$." **b.** Not valid; $X = \mathbf{R}$, $P(x, y) = $ "$x = y$."
   c. Not valid; $X = \mathbf{Z}$, $P(x) = $ "$x = 1$." **d.** Not valid; $X = \mathbf{Z}$, $P(x) = $ "$x = 1$."
7. *Case 1:* The truth set of P is a subset of the truth set of Q. In this case, $\forall x[P(x) \Rightarrow Q(x)]$ is T, and the formula is then equivalent to $P(y) \Rightarrow Q(y)$, which is true in this case.
   Case 2: The truth set of P is not a subset of the truth set of Q. In this case, $\forall x[P(x) \Rightarrow Q(x)]$ is F, so the truth value of the formula is T.

Chapter Review

1. **a.** 6 is not a perfect number and 6 is prime. **b.** 6 is not a perfect number.
   c. 6 is not prime nor is 6 a perfect number. **d.** If 6 is a perfect number, then 6 is prime.

3.

p	q	$p \Rightarrow q$	$p \wedge (p \Rightarrow q) \vee q$
T	T	T	T
T	F	F	F
F	T	T	T
F	F	T	F

5. **a.** tautology **b.** neither **c.** neither **d.** tautology
7. **a.** $\neg[\neg(\neg p \vee q) \vee \neg r]$ **b.** $\neg(\neg p \vee \neg q) \vee r$ **c.** $\neg[\neg(p \vee \neg(\neg p \vee q)) \vee (\neg(\neg p \vee q) \vee \neg p)]$

9.

	Converse	Inverse	Contrapositive
a.	$p \Rightarrow q$	$\neg q \Rightarrow \neg p$	$\neg p \Rightarrow \neg q$
b.	$\neg q \Rightarrow \neg p$	$p \Rightarrow q$	$q \Rightarrow p$
c.	$\neg p \Rightarrow \neg q$	$q \Rightarrow p$	$p \Rightarrow q$
d.	$\neg q \Rightarrow p$	$\neg p \Rightarrow q$	$q \Rightarrow \neg p$

11. (T NOR T) NOR F = T, but T NOR (T NOR F) = F   13. direct proof
15. Let p and q be rational numbers. Then $p = \frac{a}{b}$ and $q = \frac{c}{d}$, where a, b, c, and d are integers. Then
$$p + q = \frac{a}{b} + \frac{c}{d} = \frac{ad + cb}{bd} = \frac{m}{n} \quad \text{for some integers } m \text{ and } n$$

17. $\{\ldots, -8, -7, -3, -2, 2, 3, 7, 8, \ldots\}$ 19. **a.** $(-\infty, -1)$ **b.** $[-1, 1]$
21. **a.** $\exists z(xz = y)$ **b.** $\forall a \forall b(a \, R \, b \Rightarrow b \, R \, a)$ **c.** $(\forall \varepsilon > 0)(\exists \delta > 0)\forall x[0 < |x - a| < \delta \Rightarrow |f(x) - b| < \varepsilon]$
23. **a.** $\exists x(x \, R \, x)$ **b.** $\exists x \exists y[(x \, R \, y) \wedge (y \, R \, z) \wedge (x \neq y)]$ **c.** $\exists x \exists y[(x \, R \, y) \wedge (y \, R \, x)]$
25. **a.** T **b.** T **c.** T

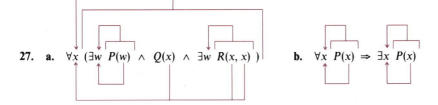

27. **a.** $\forall x \ (\exists w \ P(w) \ \wedge \ Q(x) \ \wedge \ \exists w \ R(x, x) \ )$ **b.** $\forall x \ P(x) \ \Rightarrow \ \exists x \ P(x)$

29. **a.** No; Rule 2 violated. **b.** yes
31. **a.** $\exists x\{x \in A \wedge \exists y[y \in B \wedge \forall z(z > 5 \Rightarrow x + y = z)]\}$ **b.** $\forall x\{x \in \mathbf{Z} \Rightarrow \exists y[y \in \mathbf{N} \wedge \exists z(z \in \mathbf{N} \wedge x = y - z)]\}$
33. **a.** $\exists y(y \geq 17)$ **b.** $\forall z(z^2 \neq -1)$ **c.** $\forall x \exists y(x + y \leq y)$ **d.** $\exists y \forall x(x \leq y^2 + 1)$
35. **a.** Not valid; $p = $ T. **b.** Valid **c.** Not valid; set $= \{0\}$, $P(0) = $ F, $x = 0$.

CHAPTER 5

Section 5.1

1. No; addition does not distribute over multiplication.   **3.** $2(8)(8) + 2(8)(8)(8) + 2(8)(8)(8) = 2176$

5. If 0_a and 0_b are identities for $\vee$, then $0_a = 0_b \vee 0_a = 0_a \vee 0_b = 0_b$

7. **a.** $(x \vee 0) \wedge (y \wedge 1)$ **b.** $[(x \vee y) \wedge z]'$

9. $x \vee x = (x \vee x) \wedge 1 = (x \vee x) \wedge (x \vee x')$
$$= x \vee (x \wedge x') = x \vee 0 = x$$

11. $(x \wedge y) \wedge (x' \vee y') = [(x \wedge y) \wedge x'] \vee [(x \wedge y) \wedge y']$
$$= [(x \wedge x') \wedge y] \vee [(y \wedge y') \wedge x]$$
$$= [0 \wedge y] \vee [0 \wedge x]$$
$$= 0 \vee 0 = 0$$

13. **a.** $x \wedge (x \vee y) = (x \vee 0) \wedge (x \vee y)$ **b.** $x \vee (x \wedge y) = (x \wedge 1) \vee (x \wedge y)$
$$= x \vee (0 \wedge y) = x \vee 0 = x \qquad\qquad = x \wedge (1 \vee y) = x \wedge 1 = x$$

15.   lub   glb   **17.**   lub   glb

  a. a j **a.** 10   1

  b. c k **b.**   6   3

  c. none f **c.** 35   1

  d. c none   **d.** 42   1

19. No; some glb's do not exist; lub's are not unique.   **21.** yes   **23.** 2, 3, 5   **25.** i, j, k

27. **a.** 1 and a **b.** 1; none

29. If $x \le y$, then $x = x \wedge y$ by definition, so
$$y \vee x = y \vee (x \wedge y) = (1 \wedge y) \vee (x \wedge y)$$
$$= (1 \vee x) \wedge y = 1 \wedge y = y$$
showing that $x \# y$. The converse is similar.

31. **a.** no **b.** yes **c.** yes

Section 5.2

1. **a.** no **b.** yes   **3.** $f: x\bar{y} \vee \bar{x}\bar{y}$ $g: x\bar{y} \vee \bar{x}y$ **5.** $x\bar{y}\bar{z} \vee \bar{x}yz \vee x\bar{y}z \vee \bar{x}\bar{y}z$

7.

	xy	$x\bar{y}$	$\bar{x}\bar{y}$	$\bar{x}y$
z	+		+	+
$\bar{z}$	+	+		+

$f(x, y, z) = y \vee \bar{x}z \vee x\bar{z}$

9.

1	5	13	9
2	6	14	10
4	8	16	12
3	7	15	11

z	+		+	+
$\bar{z}$		+		+

$g(x, y, z) = yz \vee \bar{x}z \vee \bar{x}y \vee x\bar{y}\bar{z}$

z			+	+
$\bar{z}$	+		+	+

$h(x, y, z) = \bar{x} \vee y\bar{z}$

11.

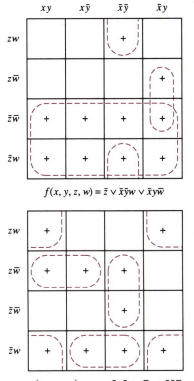

$$f(x, y, z, w) = \bar{z} \vee \bar{x}\bar{y}w \vee \bar{x}y\bar{w}$$

$$g(x, y, z, w) = yw \vee \bar{z}w\bar{y} + z\bar{w}x + \bar{x}\bar{y}\bar{w}$$

Section 5.3

1. **a.** Output $= \neg x \wedge \neg y$, where x and y are the inputs. It is equivalent to (b).
   b. Output $= \neg x \vee \neg y$, where x and y are the inputs. It is equivalent to (d).

3.

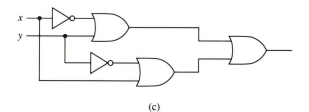

   (a)   (b)

   (c)

5. **a.** $\neg x \veebar \neg y$ **b.** $(x \text{ NAND } y) \vee (y \text{ NAND } z)$

7.

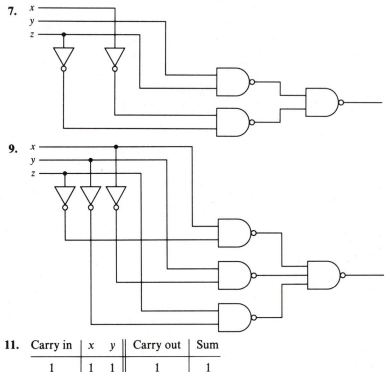

9.

11.

Carry in	x	y	Carry out	Sum
1	1	1	1	1
1	1	0	1	0
1	0	1	1	0
1	0	0	0	1
0	1	1	1	0
0	1	0	0	1
0	0	1	0	1
0	0	0	0	0

Section 5.4

1.

s	r	q	q'	$\neg(s \wedge q')$	$\neg(r \wedge q)$	$q \Leftrightarrow \neg(s \wedge q')$	$q' \Leftrightarrow \neg(r \wedge q)$
T	T	T	T	F	F	F	F
T	T	T	F	T	F	T	T
T	T	F	T	F	T	T	T
T	T	F	F	T	T	F	F
T	F	T	T	F	T	F	T
T	F	T	F	T	T	T	F
T	F	F	T	F	T	T	T
T	F	F	F	T	T	F	F
F	T	T	T	T	F	T	F
F	T	T	F	T	F	T	T
F	T	F	T	T	T	F	T
F	T	F	F	T	T	F	F
F	F	T	T	T	T	T	T
F	F	T	F	T	T	T	F
F	F	F	T	T	T	F	T
F	F	F	F	T	T	F	F

3. **a.** $q = \neg(q' \wedge \neg(d \wedge c))$; $q' = \neg(q \wedge \neg(\neg d \wedge c))$

b.

c	d	s	r	q	q'
T	T	F	T	T	F
T	F	T	F	F	T
F	T	T	T	T	F
F	T	T	T	F	T
F	F	T	T	T	F
F	F	T	T	F	T

Chapter Review

1. **a.** $(x \vee y') \wedge (z \vee 0)'$ **b.** $\neg(p \wedge T) \vee (F \wedge q)$ **c.** $(A \cap B)' \cup C$ **3.** always

5. **a.** yes

b. lub(z, w) is the complex number at the upper right corner of the rectangle (having sides parallel to the axes), with corners at z and w; glb(z, w) is the lower left corner of the same rectangle.

c. No; there are no identity elements for the operations of lub and glb.

7. To say that $x \vee y$ is the unique least upper bound of x and y means (a) $x \le x \vee y$, (b) $y \le x \vee y$, and (c) if $x \le w$ and $y \le w$, then $x \vee y \le w$.

a. $x \le x \vee y$ means that $x = x \wedge (x \vee y)$, which is true because it is one of the Absorption Laws.

b. similar to a

c. The statement to be proved means that if $x = x \wedge w$ and $y = y \wedge w$, then $x \vee y = (x \vee y) \wedge w$. If $x = x \wedge w$ and $y = y \wedge w$, then $x \vee y = (x \wedge w) \vee (y \wedge w) = (x \vee y) \wedge w$, by the Distributive Law.

The greatest lower bound part is even simpler:

a. $x \wedge y \le x$ means that $x \wedge y = (x \wedge y) \wedge x$, which is true by the Idempotent Law.

b. $x \wedge y \le y$ is similar.

c. If $w \le x$ and $w \le y$, then $w = w \wedge x$ and $w = w \wedge y$, so $w = w \wedge w = (w \wedge x) \wedge (w \wedge y) = w \wedge (x \wedge y)$, which means that $w \le x \wedge y$.

9. **a.** $xy\bar{z} \vee x\bar{y}\bar{z} \vee \bar{x}\bar{y}\bar{z} \vee \bar{x}\bar{y}z \vee \bar{x}y\bar{z}$ **b.** $pqr \vee \bar{p}qr \vee p\bar{q}\bar{r} \vee \bar{p}\bar{q}r \vee \bar{p}\bar{q}\bar{r}$

11. **a.** $[(p \veebar q) \veebar r] \veebar s$ **b.** $(\neg p \vee p) \wedge p$ **c.** $(p \text{ NOR } q) \text{ NOR } (\neg r \text{ NOR } q) \text{ NOR } (\neg p \text{ NOR } q)$

13. It is an S–R flip-flop.

15. $q = \neg(s \wedge q')$

$q' = \neg(r \wedge q)$

$y = \neg(x \wedge s)$

$s = \neg(y \wedge c)$

$r = \neg(s \wedge c \wedge x)$

$x = \neg(r \wedge d)$

The solutions are

c	d	x	y	s	r	q	q'
T	T	F	T	F	T	T	F
T	T	T	F	T	F	F	T
T	F	T	F	T	F	F	T
T	F	T	T	F	T	T	F
F	T	F	T	T	T	T	F
F	T	F	T	T	T	F	T
F	F	T	F	T	T	T	F
F	F	T	F	T	T	F	T

CHAPTER 6

Section 6.1

1. 35,536   3. 5^{20}
5. The total possible two- and three-letter initials is $(26)^2 + (26)^3 = 18,252 < 18,273$; apply the Pigeonhole Principle.
7. $(100)^5$ 9. $[(26)^3 - 137](10)^3 + (26)^2(10)^4 = 24,199,000$ 11. 2,598,960
13. For each of the n elements, there are two choices: It is in the subset, or it isn't. So there are $2 \times 2 \times \cdots \times 2 = 2^n$ subsets.
15. **a.** 4 **b.** 6   17. **a.** 99 **b.** 8 **c.** 9 **d.** 12   19. $2^{n^2 - n}$
21. *Step 1:* The formula holds for two sets.

   Step 2: Assume that the formula holds for n sets. Let $A = \bigcup_{i=1}^{n} A_i$, and let the $(n+1)$th set be B. Then by the Distributive Law,
   $$A \cap B = (\cup A_i) \cap B = \cup (A_i \cap B)$$
   By the induction hypotheses,
   $$n(A) = \sum n(A_i) - \sum n(A_i \cap A_j) + \sum (A_i \cap A_j \cap A_k) - \cdots$$
   and
   $$n(\cup (A_i \cap B)) = \sum n(A_i \cap B) - \sum n(A_i \cap A_j \cap B) + \cdots$$
   Therefore,
   $$n(A \cup B) = n(A) + n(B) - n(A \cap B)$$
   $$= \sum n(A_i) - \sum n(A_i \cap A_j) + \sum (A_i \cap A_j \cap A_k) - \cdots + n(B) - n(\cup (A_i \cap B))$$
   $$= \sum n(A_i) - \sum n(A_i \cap A_j) + \sum (A_i \cap A_j \cap A_k) - \cdots + n(B)$$
   $$- \sum n(A_i \cap B) + \sum n(A_i \cap A_j \cap B) - \cdots$$
   which is the inclusion–exclusion formula for the sets $A_1, \ldots, A_n, B$.

Section 6.2

1. **a.** 4 **b.** 20 **c.** 1 **d.** 362,880   3. 30,240; 27,216   5. 35,880,000; 32,292,000
7. $18!; 18! - 2(17!) = 16(17!)$ 9. 181,440; 840   11. 696,729,600   13. $7!/2! = 2520$
15. if n is even: $26^{n/2}$; if n is odd: $26^{(n+1)/2}$

Section 6.3

1. **a.** 3 **b.** 56 **c.** 1 **d.** 6
3. $C(n, n-1) = \dfrac{n!}{(n-1)![n-(n-1)]!}$
   $$= \dfrac{n!}{(n-1)!(1!)}$$
   $$= \dfrac{n!}{(n-1)!}$$
   $$= \dfrac{n(n-1)!}{(n-1)!}$$
   $$= n$$
5. 31,824   7. $C(52, 13)$ 9. 186^{80}
11. **a.** $C(12, 4)C(8, 4) = 34,650$ **b.** $3C(8, 4) = 210$ **c.** $(3)(2)C(10, 3)C(7, 3) = 25,200$
13. $C(n-1, r-1) = \dfrac{(n-1)!}{(r-1)![n-1-(r-1)]!}$
   $$= \dfrac{(n-1)!}{(r-1)!(n-r)!}$$

and $C(n-1, r) = \dfrac{(n-1)!}{r!(n-1-r)!}$

$\qquad\qquad\qquad = \dfrac{(n-1)!}{r!(n-r-1)!}$

So

$$C(n-1, r-1) + C(n-1, r) = \frac{(n-1)!r!(n-r-1)! + (n-1)!(r-1)!(n-r)!}{(r-1)!(n-r)!r!(n-r-1)!}$$

Since $r! = r(r-1)!$ and $(n-r)! = (n-r)(n-r-1)!$, the sum can be written as

$$\frac{(n-1)!(r-1)!(n-r-1)![r + (n-r)]}{(r-1)!(n-r)!r!(n-r-1)!}$$

Canceling like factors gives

$$\frac{n(n-1)!}{r!(n-r)!} = C(n, r)$$

15. a. $\dfrac{nC(n-1, r-1)}{r} = \dfrac{n(n-1)!}{r(r-1)!(n-1-r+1)!} = \dfrac{n!}{r!(n-r)!} = C(n, r)$

b. The intermediate result after the kth division is

$$\frac{(n-r+1)(n-r+2)\cdots(n-r+k)}{1 \times 2 \times \cdots \times k} = C(n-r+k, k)$$

Section 6.4

1. a. $x^6 + 6x^5 + 15x^4 + 20x^3 + 15x^2 + 6x + 1$ **b.** $625 - 1000a + 600a^2 - 160a^3 + 16a^4$ **3.** 1540

5. $-10500000x^3$

7. *Step 1:* If $n = 1$, $C(1, 0) + C(1, 1) = 1 + 1 = 2^1$.

Step 2:

$$\sum_{r=0}^{n+1} C(n+1, r) = C(n+1, 0) + \sum_{r=1}^{n} [C(n, r) + C(n, r-1)] + C(n+1, n+1)$$

$$= \sum_{r=0}^{n} C(n, r) + \sum_{r=1}^{n+1} C(n, r-1)$$

$$= 2^n + 2^n = 2^{n+1}$$

9. Choose $r = 1$.   **11.** This is the binomial expansion of $(1-1)^n$.

13. $C(n, r+1) = \dfrac{n!}{(r+1)![n-(r+1)]!} = \dfrac{n!}{(r+1)!(n-r-1)!}$

$\dfrac{n-r}{r+1} C(n, r) = \dfrac{(n-r)n!}{(r+1)r!(n-r)!} = \dfrac{(n-r)n!}{(r+1)r!(n-r)(n-r-1)!}$

$\qquad\qquad = \dfrac{n!}{(r+1)r!(n-r-1)!} = \dfrac{n!}{(r+1)!(n-r-1)!}$

$\qquad\qquad = C(n, r+1)$

15. $C(n, r)C(r, k) = \dfrac{n!}{r!(n-r)!} \times \dfrac{r!}{k!(r-k)!}$

$\qquad\qquad = \dfrac{n!r!}{r!k!(n-r)!(r-k)!} = \dfrac{n!}{k!(n-r)!(r-k)!}$

$C(n, k)C(n-k, r-k) = \dfrac{n!}{k!(n-k)!} \times \dfrac{(n-k)!}{(r-k)!(n-k-r+k)!}$

$\qquad\qquad = \dfrac{n!(n-k)!}{k!(n-k)!(r-k)!(n-r)!}$

$\qquad\qquad = \dfrac{n!}{k!(r-k)!(n-r)!} = C(n, r)C(r, k)$

Section 6.5

1. $\{rr, rg, ry, gy, yy\}$ **3.** p_1 and p_5 **5.** $\dfrac{1}{2}$ **7.** $\dfrac{13(48)}{C(52, 5)} = 0.00024$ **9.** $\dfrac{7}{32}$

11. $\dfrac{4}{9}$ **13.** 0.431   **15.** $\dfrac{30}{9^4} = 0.0046$ **17.** $\dfrac{7}{9}$ **19.** $\dfrac{1}{(5!)^4}$ **21.** $\dfrac{7}{64}$

23. **a.** 0.001441 **b.** 0.00198 **c.** 0.003940 **d.** 0.000015 **e.** 0.095078 **f.** 0.422569

Section 6.6

1.

Outcome	Probability
Both successful	0.0814
X successful only	0.2886
Y successful only	0.1386
Neither successful	0.4914

3.

Outcome	Probability
0	0.2215
1	0.4114
2	0.2743
3	0.0815
4	0.0107
5	0.0005

5. yes   **7.** 12.5   **9.** $20\left(\dfrac{1}{36}\right)\left(\dfrac{35}{36}\right)^{19} \approx 0.3253$ **11.** **a.** $\dfrac{13}{19}$ **b.** $\dfrac{13}{20}$

13. From the tree diagram, the probability that both are defective is the product $(\frac{2}{14})(\frac{1}{13}) = 0.011$. The probability that neither is defective is $(\frac{12}{14})(\frac{11}{13}) = 0.725$.

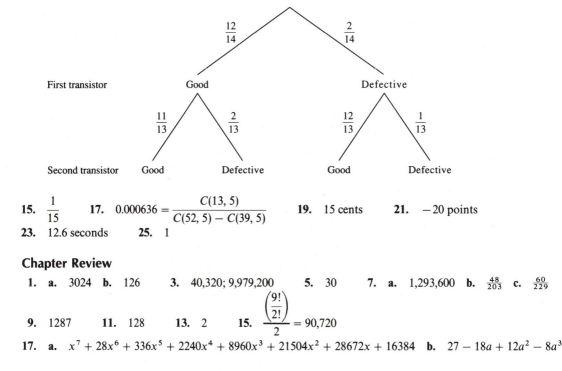

15. $\dfrac{1}{15}$ **17.** $0.000636 = \dfrac{C(13, 5)}{C(52, 5) - C(39, 5)}$ **19.** 15 cents   **21.** -20 points

23. 12.6 seconds   **25.** 1

Chapter Review

1. **a.** 3024 **b.** 126   **3.** 40,320; 9,979,200   **5.** 30   **7.** **a.** 1,293,600 **b.** $\frac{48}{203}$ **c.** $\frac{60}{229}$

9. 1287   **11.** 128   **13.** 2   **15.** $\dfrac{\left(\dfrac{9!}{2!}\right)}{2} = 90{,}720$

17. **a.** $x^7 + 28x^6 + 336x^5 + 2240x^4 + 8960x^3 + 21504x^2 + 28672x + 16384$ **b.** $27 - 18a + 12a^2 - 8a^3$

19. $9,496,093,750x^9$

21. $C(2n, 2) = \dfrac{(2n)!}{2!(2n-2)!} = \dfrac{(2n)(2n-1)(2n-2)!}{2!(2n-2)!}$

$\qquad\quad = \dfrac{(2n)(2n-1)}{2!} = 2n^2 - n$

$\quad 2C(n, 2) + n^2 = \dfrac{2n!}{2!(n-2)!} + n^2 = \dfrac{n(n-1)(n-2)!}{(n-2)!} + n^2$

$\qquad\qquad\qquad = n(n-1) + n^2 = 2n^2 - n$

$\qquad\qquad\qquad = C(2n, 2)$

23. $C(2n+2, n+1) = C(2n+1, n) + C(2n+1, n+1)$

$\qquad\qquad\qquad = [C(2n, n-1) + C(2n, n)] + [C(2n, n) + C(2n, n+1)]$

25. a. 7 **b.** 16,807 **c.** 0.000416 **27.** $\dfrac{48}{48 + C(4, 3)C(48, 2)} = \dfrac{1}{95}$ **29.** $\dfrac{91}{6}$

CHAPTER 7

Section 7.1

1.

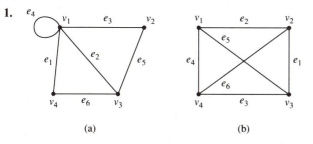

(a)   (b)

3. e_1 is incident to v_1 in graphs (a) and (c).

e_1 is incident to v_2 in graphs (b) and (c).

v_1 and v_2 are adjacent in graphs (b) and (c).

5. Degree 2: Grand Isle, Essex, and Windham

Degree 3: Rutland and Bennington

Degree 4: Franklin, Orleans, and Orange

Degree 5: all other counties

7. (a) and (b)

9.

11. m or n

13.

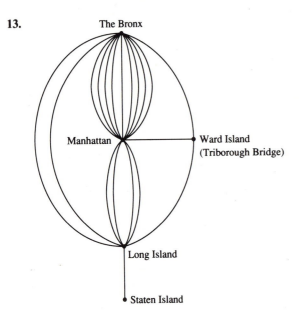

15.

17.

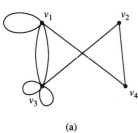

(a)

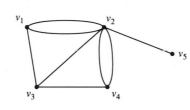

(b)

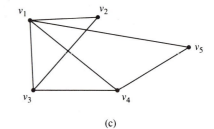

(c)

19.

Vertex Number	Item Pointer	Item Number	Vertex Number	Item Pointer
1	1	1	4	2
2	3	2	2	0
3	6	3	1	4
4	8	4	4	5
5	12	5	3	0
		6	2	7
		7	4	0
		8	1	9
		9	2	10
		10	3	11
		11	5	0
		12	4	0

21.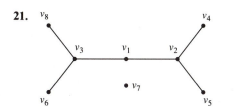

23. It has an $m \times n$ rectangle of 1's in the upper right corner; all other entries are zero.

Section 7.2

1.

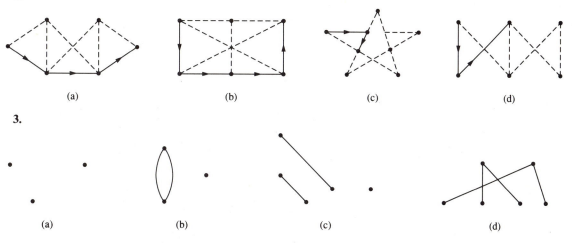

(a)	(b)	(c)	(d)

3.

(a)	(b)	(c)	(d)

5. R is reflexive, since for each vertex v, there is the trivial path (of length 0) from v to v. R is symmetric, since if there is a path from u to v, then there is a path from v to u. R is transitive, since a path from u to v can be combined with a path from v to w to form a path from u to w.

7. No. If G consists of a connected graph H with an Euler circuit and one or more isolated vertices, then the Euler circuit for H is also an Euler circuit for G.

9. m and n must both be even.

11. The proof is by induction on the number of edges in the walk.

Step 1: $n = 1$. The walk is a loop, which is a cycle.

Step 2: Assume that all closed walks of odd length less than n contain cycles, and let W be a closed walk of odd length n. If W is not itself a cycle, then two vertices of the walk are equal. The common vertex divides W into two closed walks, one of which must be of odd length and therefore contains a cycle.

Section 7.3

1.

	Edges	Vertices	Faces	Euler's Formula
(a)	3	2	3	$2 - 3 + 3 = 2$
(b)	5	5	2	$5 - 5 + 2 = 2$
(c)	9	10	1	$10 - 9 + 1 = 2$
(d)	17	12	7	$12 - 17 + 7 = 2$
(e)	9	5	6	$5 - 9 + 6 = 2$
(f)	3	2	3	$2 - 3 + 3 = 2$
(g)	3	1	4	$1 - 3 + 4 = 2$
(h)	0	1	1	$1 - 0 + 1 = 2$

3. Suppose H_1 has p_1 vertices, q_1 edges, and r_1 faces, and H_2 has p_2 vertices, q_2 edges, and r_2 faces, where $p_1 + p_2 = p$, $q_1 + q_2 + 1 = q$, and $r_1 + r_2 - 1 = r$. Since r_1 equals the number of inside faces $+ 1$ (the outside face), r_2 equals the number of inside faces $+ 1$; r equals the number of faces inside $H_1 +$ the number inside $H_2 + 1$, which is $r_1 + r_2 - 1$. So

$$
\begin{aligned}
p - q + r &= p_1 + p_2 - q_1 - q_2 - 1 + r_1 + r_2 - 1 \\
&= (p_1 - q_1 + r_1) + (p_2 - q_2 + r_2) - 2 \\
&= 2 + 2 - 2 \\
&= 2
\end{aligned}
$$

5. Each face must be bounded by at least four edges, and each edge has two sides. Therefore, $4r$ is at most the number of face boundaries, which is $2q$.

7.

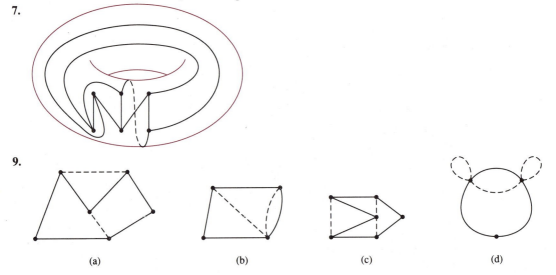

9.

(a)   (b)   (c)   (d)

11. Proof by induction on the number of circles.
Step 1: Trivial for one circle.
Step 2: When adding a new circle, reverse all colors inside the new circle.

13.

Section 7.4

1.

(a)   (b)

3. (a)

e	$v_s(e)$	$v_e(e)$
a	v	w
b	w	x
c	x	y
d	y	z
e	z	u

(b)

e	$v_s(e)$	$v_e(e)$
a	y	u
b	u	v
c	v	w
d	u	w
e	y	w
f	y	x
g	w	x
h	v	x

(c)

e	$v_s(e)$	$v_e(e)$
a	u	w
b	u	v
c	u	x
d	v	x
e	x	w
f	v	w

(d)

e	$v_s(e)$	$v_e(e)$
a	z	u
b	y	u
c	u	x
d	v	x

(e)

e	$v_s(e)$	$v_e(e)$
a	u	v
b	v	w
c	w	x
d	x	z
e	z	u
f	y	u
g	y	v
h	y	w
i	y	x
j	y	z

(f)

e	$v_s(e)$	$v_e(e)$
a	y	u
b	u	y
c	u	v
d	v	v
e	v	w
f	w	x
g	x	w
h	x	y

5.

(a)

Vertex	Indegree	Outdegree
u	1	0
v	0	1
w	1	1
x	1	1
y	1	1
z	1	1

(b)

Vertex	Indegree	Outdegree
u	1	2
v	1	2
w	3	1
x	2	1
y	1	2

(c)

Vertex	Indegree	Outdegree
u	0	3
v	1	2
w	3	0
x	2	1

(d)

Vertex	Indegree	Outdegree
u	2	1
v	0	1
w	0	0
x	2	0
y	0	1
z	0	1

(e)

Vertex	Indegree	Outdegree
u	2	1
v	2	1
w	2	1
x	2	1
y	0	5
z	2	1

(f)

Vertex	Indegree	Outdegree
u	1	2
v	3	1
w	1	2
x	1	2
y	2	1

7.

9.

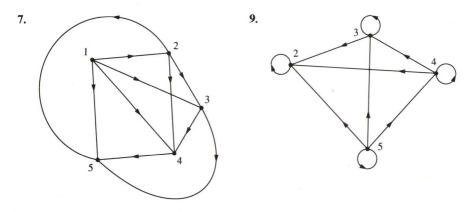

11. Each edge contributes 1 to the outdegree of its start vertex, so the sum of outdegrees is equal to the number of edges. Similar for indegrees.

13.

15. Let G be a connected graph without bridges, and let e be an edge of G. Remove e from G, leaving H. H is connected, so there is a path in H connecting the ends of e. This path, together with e, forms a cycle containing e. Again let G be a connected graph without bridges; we will construct a sequence $H_1, \ldots, H_n$ of successively larger subgraphs of G, each of which is strongly connected, and such that $H_n = G$. Let H_1 be any cycle and direct it either way. Now suppose that H_k has been constructed and directed such that H_k is strongly connected. Let e be an edge not in H_k but having one endpoint a in H_k. Let C be a cycle containing e. We may think of the cycle as beginning at a and leaving H along the edge e; the cycle must eventually return to H. Let b be the first vertex of the cycle that is in H. Direct the edges from a to b along the cycle in the direction that we just traveled from a to b, and add these edges to H_k. The result is a larger subgraph H_{k+1}, which is also strongly connected.

17. a and c

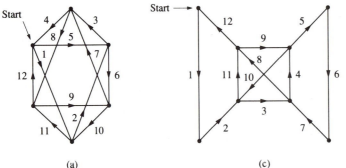

(a)   (c)

19. Counterexample: Let $X = \{a, b\}$ and let $R = \{(a, a), (b, b), (a, b)\}$. Then R is a total order, but there is no directed path from b to a.

21. (a)

Vertex Number	Out Pointer	In Pointer
1 (v)	1	0
2 (w)	3	2
3 (x)	5	4
4 (y)	7	6
5 (z)	9	8
6 (u)	0	10

Item Number	Vertex Number	Item Pointer
1	2	0
2	1	0
3	3	0
4	2	0
5	4	0
6	3	0
7	5	0
8	4	0
9	6	0
10	5	0

(b)

Vertex Number	Out Pointer	In Pointer
1 (u)	2	1
2 (v)	5	4
3 (w)	10	7
4 (x)	13	11
5 (y)	15	14

Item Number	Vertex Number	Item Pointer
1	5	0
2	2	3
3	3	0
4	1	0
5	3	6
6	4	0
7	1	8
8	2	9
9	5	0
10	4	0
11	2	12
12	3	0
13	5	0
14	4	0
15	1	16
16	3	0

23. **a.** outdegree(v) **b.** indegree(v)

Section 7.5

1.

Step	Sel.	a	b	c	d	e	f
1	c	∞	7	0	7	5	2
2	f	6	4	0	7	5	2
3	b	5	4	0	7	5	2
4	a	5	4	0	7	5	2
5	e	5	4	0	6	5	2
6	d	5	4	0	6	5	2

3.

Step	Sel.	a	b	c	d	e	f	g	h	z
1	a	0	1	∞	2	∞	∞	∞	∞	∞
2	b	0	1	7	2	6	∞	∞	∞	∞
3	d	0	1	7	2	3	∞	7	∞	∞
4	e	0	1	7	2	3	10	6	11	∞
5	g	0	1	7	2	3	10	6	10	∞
6	c	0	1	7	2	3	9	6	10	∞
7	f	0	1	7	2	3	9	6	10	11
8	h	0	1	7	2	3	9	6	10	11
9	z	0	1	7	2	3	9	6	10	11
	Pred.	—	a	b	a	d	c	e	g	f

5. *adeg*

7. The algorithm terminates after processing all vertices in the same connected component as the starting vertex.

9. No; consider a triangle with edge weights of 1, 2, and 3.

11. Yes; but first discard all loops, and where there are multiple edges, discard all but the shortest.

Chapter Review

1.

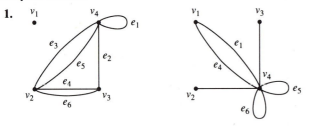

(a)   (b)

3.

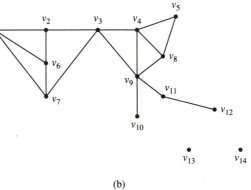

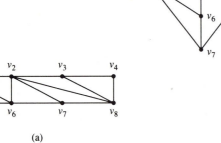

(a)   (b)

5.

	v_1	v_2	v_3	v_4	v_5	v_6	v_7	v_8
v_1	0	1	0	0	1	1	0	0
v_2	—	0	1	0	0	1	1	1
v_3	—	—	0	1	0	0	0	1
v_4	—	—	—	0	0	0	0	1
v_5	—	—	—	—	0	1	0	0
v_6	—	—	—	—	—	0	1	0
v_7	—	—	—	—	—	—	0	1
v_8	—	—	—	—	—	—	—	0

7. 2

9.

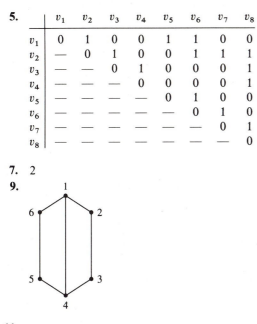

11. none

13. a.

	s	t	u	w	x
a	0	0	1	0	1
b	0	1	0	1	0
c	1	0	1	0	0
d	0	1	0	0	1

b.

	s	t	u	w	x	y	z
a	0	0	0	0	1	0	1
b	0	0	0	1	0	0	1
c	0	0	1	0	0	0	1
d	0	1	0	0	1	0	0
f	1	0	0	0	1	0	0
g	0	1	0	0	0	1	0

15. a.

	s	t	u	w	x
indegree:	1	1	1	0	1
outdegree:	0	1	1	1	1

b.

	s	t	u	w	x	y	z
indegree:	1	1	1	1	1	1	0
outdegree:	0	1	0	0	2	0	3

17. both   **19.** (c)   **21.** None; all have two components.

23.

	Vertices	Edges	Faces
(a)	7	8	3
(b)	9	12	5
(c)	8	6	2

(Euler's formula does not apply because graph (c) is not connected.)

25.

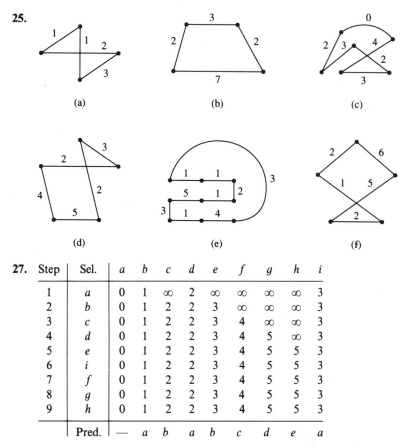

(a)   (b)   (c)

(d)   (e)   (f)

27.

Step	Sel.	a	b	c	d	e	f	g	h	i
1	a	0	1	∞	2	∞	∞	∞	∞	3
2	b	0	1	2	2	3	∞	∞	∞	3
3	c	0	1	2	2	3	4	∞	∞	3
4	d	0	1	2	2	3	4	5	∞	3
5	e	0	1	2	2	3	4	5	5	3
6	i	0	1	2	2	3	4	5	5	3
7	f	0	1	2	2	3	4	5	5	3
8	g	0	1	2	2	3	4	5	5	3
9	h	0	1	2	2	3	4	5	5	3
	Pred.	—	a	b	a	b	c	d	e	a

CHAPTER 8

Section 8.1

1. (a), (b)   3. 1 and 2

5. We will prove by induction on the number p of vertices that every tree can be embedded in the plane using straight lines for edges.

Step 1: If $p = 1$, the graph is a point.

Step 2: If G is a tree with n vertices, let v be a leaf. Let e be the edge to v and let w be the other end of e. Remove v and e. The remainder H is planar, by the induction hypothesis. Choose a point in the plane closer to w than any other vertex but not on any edge, call it v, and join it to w with a straight line. The number of faces is 1.

7. $C(n, 2)$

9.

   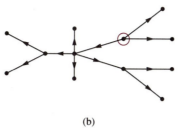

(a)   (b)

11.   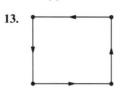   13.

15. (a)

Vertex	Parent	First Child	Next Sibling
a	0	b	0
b	a	e	c
c	a	i	d
d	a	j	0
e	b	0	f
f	b	0	g
g	b	0	h
h	b	0	0
i	c	m	0
j	d	0	k
k	d	0	l
l	d	0	0
m	i	0	n
n	i	0	0

(b)

Vertex	Parent	First Child	Next Sibling
a	0	b	0
b	a	0	c
c	a	0	d
d	a	0	e
e	a	0	f
f	a	j	g
g	a	0	h
h	a	0	i
i	a	0	0
j	f	0	k
k	f	0	l
l	f	0	0

(c)

Vertex	Parent	First Child	Next Sibling
a	0	b	0
b	a	0	c
c	a	d	0
d	c	0	e
e	c	f	0
f	e	0	g
g	e	h	0
h	g	0	i
i	g	0	0

(d)

Vertex	Parent	First Child	Next Sibling
a	0	b	0
b	a	c	0
c	b	d	0
d	c	e	0
e	d	0	f
f	d	h	g
g	d	0	0
h	f	i	0
i	h	0	0

17.

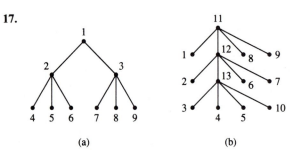

(a)   (b)

Section 8.2

1.

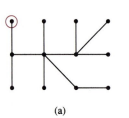   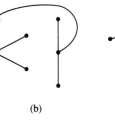   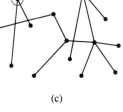   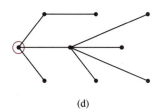

(a)   (b)   (c)   (d)

a.

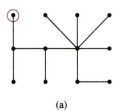   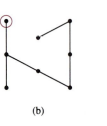   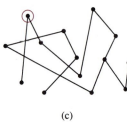   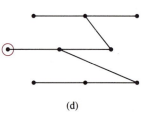

(a)   (b)   (c)   (d)

b.

3. a linear graph   **5.** a linear graph with a star attached to one end

7.

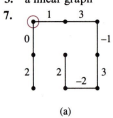   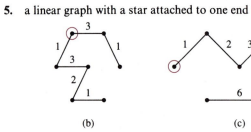

(a)   (b)   (c)

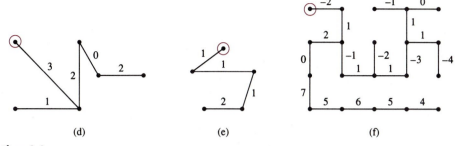

(d)   (e)   (f)

Section 8.3

1. greater than 19; greater than 37; less than 43; equal to 41   3. $2^{N+1} - 1$ vertices; $2^{N+1} - 2$ edges

5. **a.** less than Tell; greater than Boccanegra; less than Miller; greater than Budd; less than Lecouvreur; greater than Chenier; greater than Giovanni; greater than Herring—not found
   b. less than Tell; greater than Boccanegra; greater than Miller; greater than Schicchi—not found
   c. less than Tell; greater than Boccanegra; less than Miller; less than Budd—not found
   d. greater than Tell; greater than Troyens—not found

7. Insert Ivnor as right child of Herring; insert Stuarta as right child of Schicchi; insert Bolena as left child of Budd; insert Wozzeck as right child of Troyens.

9. a linear graph

Section 8.4

1. (a)   preorder: *abdefhigc*
          inorder: *dbhfiegac*
          postorder: *dhifgebca*
   (b)   preorder: *abdhlmiecfgjknp*
          inorder: *lhmdibeafcjgnkp*
          postorder: *lmhidebfjnpkgca*

3.

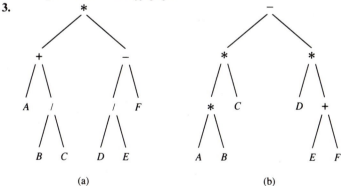

(a)   (b)

5. **a.** (1) $* + A/BC - /DEF$ (2) $ABC/ + DE/F - *$
   b. (1) $- *A * BC * D + EF$ (2) $AB * C * DEF + * -$

7. **a.** (1)

A	B	C	/	+	D	E	/	F	−	*
2	3	−1	−3	−1	5	−5	−1	11	−12	12
	2	3	2		−1	5	−1	−1	−1	
		2			−1			−1		

a. (2)

A	B	C		/	+	D	E		/	F		−		*
7	−12	5		−2.4	4.6	5	2		2.5	−21	23.5	108.1		
	7	−12			7		4.6	5	4.6	2.5	4.6			
		7						4.6		4.6				

b. (1)

A	B	*	C	*	D	E	F	+	*	−
2	3	6	−1	−6	5	−5	11	6	30	−36
	2		6		−6	5	−5	5	−6	
						−6	5	−6		
							−6			

b. (2)

A	B	•	*	C	*	D	E	F	+	*	−
7	−12	−84		5	−420	5	2	−21	−19	−95	−325
	7			−84		−420	5	2	5	−420	
							−420	5	−420		
								−420			

9. $AB − *$ and $AB − C*$. In the first, "−" must be processed as a unary operation by negating B; in the second, "−" must be processed as a binary operation. To resolve the ambiguity, it is necessary to use different symbols for unary − and binary −. Most calculators have a key labeled +/− for unary −; the "−" key is binary −.

11. left children for Boccanegra, Budd, Chenier, Giovanni; right children for Schicchi, Herring. (null), Boccanegra, (null), Budd, (null), Chenier, (null), Giovanni, Grimes, Herring, (null), Lecouvreur, Lescaut, Miller, Onegin, Schicchi, (null), Tell, Troyens

Chapter Review

1.

3.

Vertex	Parent	Children	Siblings
a	none	b, c	none
b	a	d, e, f	c
c	a	g, h	b
d	b	none	e, f
e	b	none	d, f
f	b	i	d, e
g	c	none	h
h	c	none	g
i	f	none	none

5. **c.** the entire graph   **d.** z to y **f.** 5 to 1 or 5 to 2

7.

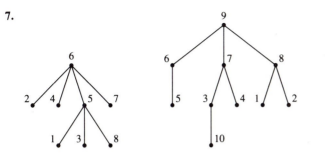

(a)   (b)

a.

Vertex	Parent	First Child	Next Sibling
1	5	0	3
2	6	0	4
3	5	0	8
4	6	0	5
5	6	1	7
6	0	2	0
7	6	0	0
8	5	0	0

b.

Vertex	Parent	First Child	Next Sibling
1	8	0	2
2	8	0	0
3	7	10	4
4	7	0	0
5	6	0	0
6	9	5	6
7	9	3	8
8	9	1	0
9	0	6	0
10	3	0	0

9.   The result depends on the weights you choose.
11.   **a.**   Chenier, Miller, Tell, Boccanegra, Troyens, Budd, Lecouvreur, Herring, Schicchi, Giovanni, Onegin, Lescaut, Grimes
   b.   Boccanegra, Tell, Budd, Lecouvreur, Troyens, Miller, Giovanni, Schicchi, Lescaut, Grimes, Onegin, Herring, Chenier
13.   **a.**   none; 6   **b.**   right child for 6; 3, 6, (null)   **c.**   left child for 3, right child for 6; (null), 3, 5, 6, (null)
15.   No; consider a triangle with edges of weights 2, 3, and 4, and start at the vertex opposite the edge of weight 2.

CHAPTER 9

Section 9.1

1.   **a.**   7, 19, 43, 91, 187, 379   **b.**   7, 19, 55, 163, 487, 1459   **c.**   2, 2, 2, 2, 2, 2   **d.**   6, 15, 34, 73, 152, 311
3.   1, 2, 3, 5; $s_n = s_{n-1} + s_{n-2}$ **5.** $s_n = s_{n-1} + s_{n-2} + s_{n-3}$ **7.** $s_n = 1.01s_{n-1} - 220$; $19,876.96
9. $s_n = 1.005s_{n-1} - 1000$ **11.** $D_n = (n-1)(D_{n-1} + D_{n-2})$
13.   *Step 1:* $b_0 = 1 = 2^0$
   Step 2: Assume $b_{n-1} = 2^{n-1}$. Then $b_n = 2(b_{n-1}) = 2(2^{n-1}) = 2^n$.
15.   **a.** $a_n = (1 - \alpha - \beta)a_{n-1} + 2\beta a_{n-2}$ **b.** $b_n = (1 - \alpha - \beta)b_{n-1} + 2\beta b_{n-2}$
   c. $s_n = (1 - \alpha - \beta)s_{n-1} + 2\beta s_{n-2}$
17.   The probability is $\dfrac{D_n}{n!}$; see Exercise 11.

Section 9.2

1.

	Order	Linear	Homogeneous	Const. Coeff.
a.	2	No		
b.	3	Yes	Yes	Yes
c.	2	No		

3. a. $s_n = 3^n(-3) + 1 = 1 - 3^{n+1}$ **b.** $s_n = 5 + 2n$ **c.** $s_n = -3(0.5)^n + 4$ **d.** $s_n = 10(-0.1)^n$

5. $s_n = -2000(1.01)^n + 22,000$

7. only when $b = 0$; then all points are fixed points, since $x = x + b$ is true for all x.

9. yes: $\dfrac{3 \pm \sqrt{5}}{2}$

11. The even and odd terms, considered separately, are defined by linear nonhomogeneous recurrence relations with constant coefficients. Therefore, the techniques of this section may be applied to the even and odd terms. The result is

$$\text{even terms:} \quad s_{2n} = 2^n\left(\frac{5}{4}\right) + \frac{3}{4}$$

$$\text{odd terms:} \quad s_{2n+1} = 2^n\left(\frac{-1}{4}\right) + \frac{3}{4}$$

Section 9.3

1. a. $x + 15$ **b.** $x^2 + 6x - 5$ **c.** $x^3 - x^2 - 7x + 13$

3. $5s_{n-1} - 6s_{n-2} = 5A(2)^{n-1} + 5B(3)^{n-1} - 6A(2)^{n-2} - 6B(3)^{n-2}$
$$= A(2)^{n-2}[5(2) - 6] + B(3)^{n-2}[5(3) - 6]$$
$$= A(2)^{n-2}(4) + B(3)^{n-2}(9)$$
$$= A(2)^{n-2}(2^2) + B(3)^{n-2}(3^2)$$
$$= A(2)^n + B(3)^n$$
$$= s_n$$

5. $s_n = 4(\frac{1}{2})^n - 2(\frac{-1}{2})^n$ **7.** $a_n = 1108.32(0.962)^n - 108.32(-0.312)^n$

9. $s_n = (1 + \frac{i}{2})(1 + i)^n + (1 - \frac{i}{2})(1 - i)^n$ **11.** $s_n = (\frac{2}{3})2^n + (\frac{1}{6})n2^n - (\frac{2}{3})(-1)^n$

13. With any set of initial conditions, the solution is a periodic sequence with period 3. This is because the roots of the characteristic equation are the cube roots of 1.

Section 9.4

1. $s_n = 2(\frac{1}{2})^n + n - 1$ **3. a.** $s_n = 5 + \dfrac{n^2 + n}{2}$ **b.** $s_n = -3 + \dfrac{n^2 + n}{2}$

5. $s_n = (-1 - \frac{i}{2})(1 + i)^n + (-1 + \frac{i}{2})(1 - i)^n - 2n + 4$ **7.** $a = 1$ **9.** $a = 2, b = -1$, and $c \neq 0$

Section 9.6

1. a. 2 **b.** $\frac{6}{5}$ **3. a.** $s_n = -1 + 2^n$ **b.** $s_n = \frac{3}{2}(\frac{1}{2})^n - \frac{1}{2}(\frac{-1}{2})^n$ **c.** $s_n = (\frac{3}{5})2^n + \frac{2}{5}(\frac{-1}{2})^n$

5. Choose $M = 14$. Then $13n + 7 \leq 14(n - 15) = 14n - 210$ whenever $n \geq 217$. Thus, you can choose $n_0 = 217$.

7. Choose $M = 4$. Then $3n^2 + 2 \leq 4(n^2 - 10) = 4n^2 - 40$ is equivalent to $-n^2 \leq -42$ (that is, $n^2 \geq 42$). So choose $n_0 = 7$.

9. By Exercise 2 of Section 2.7, $n \leq 2^n$ for $n \geq 1$. So, taking $n_0 = 0$ and $M = 1$,

$$\log_2(n) \leq 2^{\log_2(n)} = n$$

11. Given any n_0 and M, choose n such that $n \geq n_0$ and $n > M$.

13. We may always add more numbers so that the number of items to be sorted is a power of 2, so we will consider only the case $n = 2^m$. Let t_m be the number of steps required to sort 2^m items. Then $t_m = 2t_{m-1} + 2^m$. We want to show that $t_m \in \mathbf{O}(n \log_2(n)) = \mathbf{O}(m2^m)$. That is, we want to show that $t_m \le Mm2^m$ for all m and some M. This will be proved by induction.
Step 1: For $m = 1$, just take $M = t_1$.
Step 2: Assume that $t_{m-1} \le M2^{m-1}(m - 1)$. Then
$t_m = 2t_{m-1} + 2^m \le 2M2^{m-1}(m - 1) + 2^m = 2^m[M(m - 1) + 1] \le 2^m Mm$.

Chapter Review

1. **a.** 15.5, 19.25, 24.875, 33.3125 **b.** 1.5, 0.75, 1.125, 0.9375 **c.** $\frac{3}{2}, \frac{5}{3}, \frac{8}{5}, \frac{13}{8}$ **d.** 0, 3, $-2, 9$
3. 1, 2, 4, 7; $s_n = s_{n-1} + s_{n-2} + s_{n-3}$ **5.** $s_1 = 1, s_2 = 4, s_3 = 7, s_n = s_{n-1} + s_{n-2} + 1$
7. $a_n = 1.01a_{n-1} - 1100$, $a_0 = \$80,000$; $a_8 = \$77,514.30$
9. **a.** order 2, nonlinear
   b. order 5, linear, homogeneous, constant coefficients
   c. order 2, nonlinear
   d. order 2, linear, nonhomogeneous, constant coefficients
11. $2 \pm \sqrt{2}$
13. $F_n = \dfrac{1}{\sqrt{5}} \left(\dfrac{1 + \sqrt{5}}{2} \right)^n - \dfrac{1}{\sqrt{5}} \left(\dfrac{1 - \sqrt{5}}{2} \right)^n$; since $\dfrac{1 - \sqrt{5}}{2} = -0.62$, the second term is always less than $\frac{1}{2}$.
15. $s_n = \left(\dfrac{31 + 4\sqrt{31}}{62} \right) \left(\dfrac{1 + \sqrt{31}}{5} \right)^n - \left(\dfrac{31 - 4\sqrt{31}}{62} \right) \left(\dfrac{1 - \sqrt{31}}{5} \right)^n$
17. $a_0 = 1500$, and $a_1 = 0.6a_0 + 200 = 1100$
    $a_n = 1214.29 + 285.71(-0.4)^n$
19. $s_n = -\dfrac{2}{3} n \left(\dfrac{3}{2} \right)^n$ **21.** $s_n = 4 \left(\dfrac{1}{2} \right)^n - 2$ **23.** $s_n = 2 \left(\dfrac{1}{2} \right)^n + 2n - 2$
25. **a.** $s_n = -\dfrac{n^2 + n}{2}$ **b.** $s_n = 10 - \dfrac{n^2 + n}{2}$ **27.** $s_n = (\frac{i}{2} - 1)(1 + i)^n + (-\frac{i}{2} - 1)(1 - i)^n + n + 2$
29. **a.** $\frac{8}{3}$ **b.** $\frac{1}{30}$ **c.** $-\frac{7}{3}$ **d.** $\frac{16}{27}$
31. Choose $M = 2$. Then $3n^2 + 2n + 1 \le 2(2n^2 - 10)$ whenever $0 \le n^2 - 2n - 20$, and this is true provided $n \ge 6$. So choose $n_0 = 6$.
33. $|f(n)| \le 1$ for all n

CHAPTER 10

Section 10.1

1. b, c, e   **3. a.** 1 **b.** 1 **c.** 12   **5.** No; $4 | (2 \times 2)$, but 4 does not divide 2.
7. If d divides a, b, and c, then there exist integers u, v, and w such that $a = ud$, $b = vd$, and $c = wd$. Then $ax + by + cz = d(ux + vy + wz)$.
9. If $x | (a + b)$ and $x | (a - b)$, then by Theorem 10.1(b), $x | 2a$ and $x | 2b$. Since $(a, b) = 1$, $(2a, 2b) = 2$; so by Theorem 10.3, $x | 2$.
11. Clearly, any common divisor of a and b divides m and n. Solving the system for x and y yields $a = xm + un$, $b = ym + vn$, so any common divisor of m and n divides a and b.
13. The proof is by induction.
    Step 1: $(1, 1) = 1$
    Step 2: Assume that a and b are consecutive terms and that $(a, b) = 1$. The next term is $c = a + b$. If $x | b$ and $x | c$, then $x | (c - b)$ (that is, $x | a$). Therefore, $x = 1$.
15. $(a, b)[a, b] = |ab|$

17. Write out the algorithm for a and b:

$$a = bq_1 + r_1 \qquad \frac{a}{b} = q_1 + \frac{r_1}{b}$$

$$b = r_1 q_2 + r_2 \qquad \frac{b}{r_1} = q_2 + \frac{r_2}{r_1}$$

$$r_1 = r_2 q_3 + r_3 \qquad \frac{r_1}{r_2} = q_3 + \frac{r_3}{r_2}$$

$$\cdots \qquad\qquad \cdots$$

$$r_{n-2} = r_{n-1} q_n + r_n, \qquad \text{where } r_n = 1$$

$$r_{n-1} = r_n q_{n+1} + 0$$

Then

$$g\left(\frac{a}{b}\right) = 2^{q_1} g\left(\frac{r_1}{b}\right) = 2^{q_1}\left(1 + g\left(\frac{b}{r_1}\right)\right)$$

$$g\left(\frac{b}{r_1}\right) = 2^{q_2} g\left(\frac{r_2}{r_1}\right) = 2^{q_2}\left(1 + g\left(\frac{r_1}{r_2}\right)\right)$$

$$\cdots$$

$$g\left(\frac{r_{n-1}}{r_n}\right) = g(r_{n-1}) = 2^{r_{n-1}}$$

From this, $g(\frac{a}{b})$ can be computed by repeated substitution from bottom to top. For instance, to compute $g(\frac{13}{5})$,

$$13 = 5(2) + 1 \qquad g\left(\frac{13}{5}\right) = 2^2\left(1 + g\left(\frac{5}{3}\right)\right)$$

$$5 = 3(1) + 2 \qquad g\left(\frac{5}{3}\right) = 2^1\left(1 + g\left(\frac{3}{2}\right)\right)$$

$$3 = 2(1) + 1 \qquad g\left(\frac{3}{2}\right) = 2^1\left(1 + g\left(\frac{2}{1}\right)\right)$$

$$g(2) = 2^{2-1} = 2$$

and substitution yields $g(\frac{13}{5}) = 60$.

Section 10.2

1. **a.** $2 \cdot 3 \cdot 5 \cdot 7 \cdot 11$ **b.** $2^4 \cdot 3^3 \cdot 11$ **c.** $2 \cdot 3^2 \cdot 5 \cdot 7^2 \cdot 11$ **d.** $3^2 \cdot 5^3 \cdot 7 \cdot 11$
e. $2^2 \cdot 3^2 \cdot 5^3 \cdot 7^3$ **f.** $2^5 \cdot 3^4 \cdot 5^3 \cdot 7^2 \cdot 11$

3. *Step 1:* $n = 2$ is Theorem 10.9.
Step 2: Assume that if p divides a product of $n - 1$ factors, it divides one of them. If $x = a_1 \ldots a_{n-1} a_n$; then by Theorem 10.9, either $p \mid a_n$ or $p \mid a_1 \ldots a_{n-1}$; in the latter case, apply the induction hypothesis.

5. **a.** 6912 **b.** 14,880 **c.** 160,056 **d.** 194,688 **e.** 5,678,400 **f.** 813,404,592   **7.** 0

9. Let $n = 2^{p-1}(2^p - 1)$ and let $q = 2^p - 1$. Then $\sigma(n) = q(q+1) = q(2^p) = 2(2^{p-1})q = 2n$.

11. **a.**

$n =$	1	2	3	4	5	6	7	8	9	10	11	12
$\varphi(n) =$	1	1	2	2	4	2	6	4	6	4	10	4

b. $p - 1$
c. The numbers not relatively prime to p^k are the multiples of p, of which there are

$$\frac{p^k}{p} = p^{k-1}$$

in the range 1 to p^k.

13. n and 1 are divisors of n, so if $\sigma(n) = n + 1$, there can be no other positive divisors.

Section 10.3

1. **a.** no **b.** N **c.** no **d.** $x = -6, y = -37$ **e.** $x = 0, y = -3$

3. Since p is prime, every x with $1 \le x < p$ is relatively prime to p.

5. [1], [2], [4], [7], [8], [11], [13], [14] **7.** [4] and [5], respectively.

Section 10.4

1. **a.** $x = 1 + 17t, \quad y = 1 + 20t$

   b. $x = -6 + 13t, \quad y = -3 + 7t$

   c. $x = 14 + 17t, \quad y = -10 - 12t$

3. If x divides $a, b,$ and c, then $x \mid (a, b)$ and $x \mid c$, so $x \mid ((a, b), c)$. Conversely, if $x \mid ((a, b), c)$, then $x \mid (a, b)$, so x divides $a, b,$ and c.

5. 7 pads, 5 pens **7.** no **9.** 350 days after feeding the boa constrictor the first time

Section 10.5

1. **a.** $x = -8, \quad y = 25$ **b.** $x = -4, \quad y = 10$ **c.** $x = -30, \quad y = 69$

Section 10.6

1. **a.** $[7, 3, 5, 7]$ **b.** $[-3, 13, 2, 12, 4]$ **3.** **a.** $7, \frac{22}{3}, \frac{117}{16}, \frac{841}{115}$ **b.** $-3, -\frac{38}{13}, -\frac{79}{27}, -\frac{986}{337}, -\frac{4023}{1395}$

5. $[2, 4, 4, 4, 4, \ldots]$. The 4 repeats. The value x of this continued fraction satisfies $x = 2 + \frac{1}{x+2}$, which simplifies to $x^2 = 5$. Convergents and differences are

$\frac{2}{1}$	-0.236
$\frac{9}{4}$	0.014
$\frac{38}{17}$	-0.00077
$\frac{161}{72}$	0.000043
$\frac{682}{305}$	-0.0000024
$\frac{2889}{1292}$	0.00000013

7. *Step 1:* $h_{-1}k_0 - k_{-1}h_0 = 1 \cdot 1 - 0 \cdot 0 = 1 = (-1)^0$

   Step 2: The formula shows that each of these terms is the negative of the previous one.

9. $x = \frac{3 + \sqrt{13}}{2}$. Convergents are $3, \frac{10}{3}, \frac{33}{10}, \frac{109}{33}, \frac{360}{109}, \frac{1189}{360}$. **11.** $2, 3, \frac{8}{3}, \frac{11}{4}, \frac{19}{7}, \frac{87}{32}$

Chapter Review

1. b, c

3. **a.** $2 \cdot 3 \cdot 5 \cdot 7 \cdot 11 \cdot 123$ **b.** $2^2 \cdot 3 \cdot 5^2 \cdot 7 \cdot 11^2 \cdot 13$ **c.** $2 \cdot 7^2 \cdot 11^2$

   d. $2^7 \cdot 3^3$ **e.** $2 \cdot 5^5 \cdot 3^2$ **f.** $5 \cdot 997$

5. **a.** 64 **b.** 216 **c.** 18 **d.** 32 **e.** 36 **f.** 4

7. **a.** $x = 2 - 12t, \quad y = t$ **b.** $x = 21 + 48t, \quad y = 24 + 55t$ **c.** $x = 39 + 45t, \quad y = 852 + 983t$

9. [1], [3], [5], [9], [11], [13] **11.** [9]

13. **a.** $x = 2, \quad y = 0$

   b. as given above, with $t \ge 0$

   c. as given above, with $t \ge 0$

   d. as given above, with $t \ge 0$

15. 9 NAND gates and 5 inverters **17.** every 379.195 seconds

19. **a.** $[-1, 2, 8, 2]$ **b.** $[5, 7, 9, 11, 13]$ **21.** **a.** $-1, -\frac{1}{2}, -\frac{9}{17}, -\frac{19}{36}$ **b.** $5, \frac{36}{7}, \frac{329}{64}, \frac{3655}{711}, \frac{47,844}{9307}$

23. $[2, 2, 4, 2, 4, 2, \ldots]$. The pattern 2, 4 repeats. The value x of this continued fraction satisfies

$$x = 2 + \cfrac{1}{2 + \cfrac{1}{x + 2}}$$

which simplifies to $x^2 = 6$. Convergents and differences are

$$2 \qquad -0.45$$
$$\frac{5}{2} \qquad 0.05$$
$$\frac{22}{9} \qquad -0.005$$
$$\frac{49}{20} \qquad 0.0005$$
$$\frac{218}{89} \qquad -0.00005$$
$$\frac{485}{198} \qquad 0.000005$$

25. $x = 1 + \dfrac{\sqrt{2}}{2}$

$1, 2, \dfrac{5}{3}, \dfrac{12}{7}, \dfrac{29}{17}, \dfrac{70}{41}$

CHAPTER 11

Section 11.1

1. For instance, aa, bb, aab, bba
3. **a.** Sentence = noun phrase verb object; noun phrase = "the" noun, or "the" noun prepositional phrase; prepositional phrase = preposition noun phrase.
   b. The manager made the decision.
   The manager of the supervisor made the decision.
   The manager of the supervisor of the leader of the project made the decision.
5. **a.** noun phrase, verb in past tense, object
   b. object, "was," verb in past tense, "by," subject
   c. See b.
7. **a.** BASIC: each statement is a separate line; Pascal: separated by semicolon; C: each statement ends in semicolon.
   b. BASIC: REM statement; Pascal: enclosed in {...}; C: enclosed in /*···*/.
   c. BASIC: none (except in some dialects); Pascal: enclosed in begin ... end; C: enclosed in (...).

Section 11.2

1. any number of x's followed by one y
3. any possibly empty string of a's and b's, followed by the same string in reverse order
5.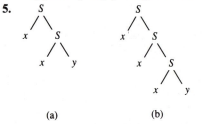

      (a)   (b)

7.

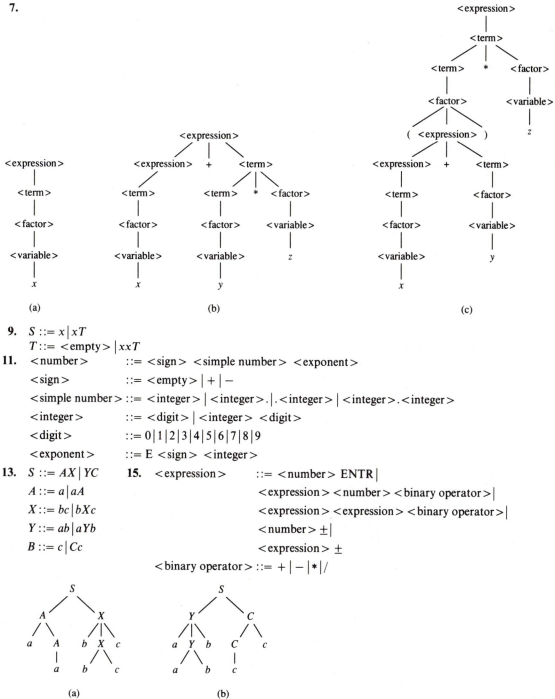

    (a)   (b)   (c)

9. $S ::= x \,|\, xT$
$T ::= <\text{empty}> \,|\, xxT$

11.

\<number\>	::= \<sign\> \<simple number\> \<exponent\>
\<sign\>	::= \<empty\> $\,|\,$ + $\,|\,$ −
\<simple number\>	::= \<integer\> $\,|\,$ \<integer\>. $\,|\,$.\<integer\> $\,|\,$ \<integer\>.\<integer\>
\<integer\>	::= \<digit\> $\,|\,$ \<integer\> \<digit\>
\<digit\>	::= 0$\,|\,$1$\,|\,$2$\,|\,$3$\,|\,$4$\,|\,$5$\,|\,$6$\,|\,$7$\,|\,$8$\,|\,$9
\<exponent\>	::= E \<sign\> \<integer\>

13. $S ::= AX \,|\, YC$
$A ::= a \,|\, aA$
$X ::= bc \,|\, bXc$
$Y ::= ab \,|\, aYb$
$B ::= c \,|\, Cc$

15.

\<expression\>	::= \<number\> ENTR $\,|\,$
	\<expression\> \<number\> \<binary operator\> $\,|\,$
	\<expression\> \<expression\> \<binary operator\> $\,|\,$
	\<number\> $\pm$ $\,|\,$
	\<expression\> $\pm$
\<binary operator\>	::= + $\,|\,$ − $\,|\,$ * $\,|\,$ /

    (a)   (b)

Section 11.3

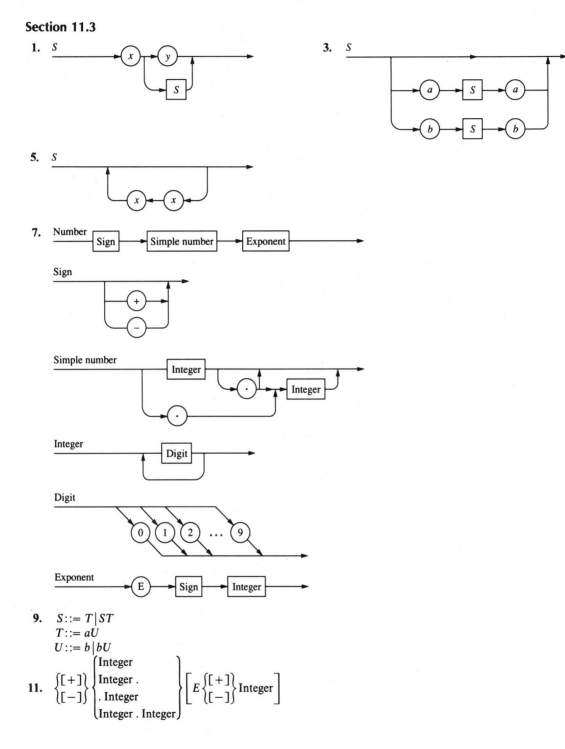

1. *S*

3. *S*

5. *S*

7. Number — Sign — Simple number — Exponent

Sign

Simple number

Integer

Digit

Exponent — E — Sign — Integer

9. $S ::= T \mid ST$
$T ::= aU$
$U ::= b \mid bU$

11. $\begin{Bmatrix} [+] \\ [-] \end{Bmatrix} \begin{Bmatrix} \text{Integer} \\ \text{Integer .} \\ \text{. Integer} \\ \text{Integer . Integer} \end{Bmatrix} \begin{bmatrix} E \begin{Bmatrix} [+] \\ [-] \end{Bmatrix} \text{Integer} \end{bmatrix}$

where Integer is defined by

$$\begin{Bmatrix} 0 \\ \cdots \\ 9 \end{Bmatrix} \begin{bmatrix} \begin{Bmatrix} 0 \\ \cdots \\ 9 \end{Bmatrix} \cdots \begin{Bmatrix} 0 \\ \cdots \\ 9 \end{Bmatrix} \end{bmatrix}$$

13.

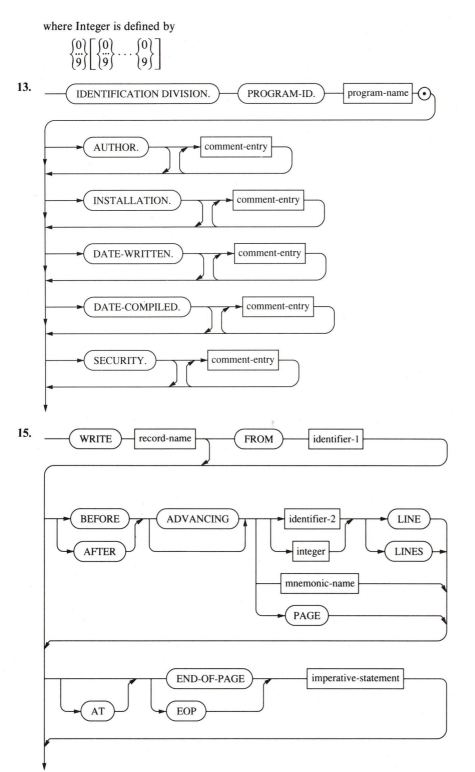

15.

Section 11.4

1. **a.** predicate logic, math   **b.** predicate logic, math   **c.** none
3. **a.** $z \notin x \vee z \in y$ **b.** $w \in x \wedge x \subset y \Rightarrow w \in y$

Chapter Review

1. Noun phrase, verb, object; where a noun phrase is either a noun (optionally preceded by an article or an adjective), an infinitive, or a gerund; an infinitive is the word "to" followed by a verb; object is an adjective or a noun phrase; and a gerund is a verb with "ing" attached followed by a noun phrase.
3. a string of x's with one b at either end
5. $S ::= xxS \,|\, xyE \,|\, yxE \,|\, yyS$

   $E ::= xxE \,|\, xyS \,|\, yxS \,|\, yyE \,|\, <empty>$
7.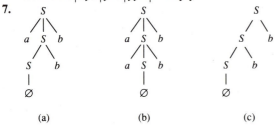

   (a)   (b)   (c)
9.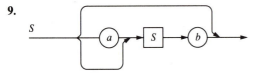
11. $S ::= T \,|\, ST$

    $T ::= a \,|\, abT$
13. $\begin{Bmatrix} b \, x \, [x] \dots \\ x \, [x] \dots b \end{Bmatrix}$

CHAPTER 12

Section 12.1

1. $2^{32} = 4{,}294{,}967{,}296$ **3.** printable characters, carriage return, line feed

Section 12.2

1. **a.** U, V, V, V **b.** U, T, T, V **c.** V, V, V
3. Input alphabet $= \{$nickel, dime, quarter$\}$
   States $= \{I, S5, S10, S15\}$
   Initial state $= I$
5. **a.** $S10, I, S10, I, S5$ **b.** $S5, S15, I, S10, S15, I, S5$

7.

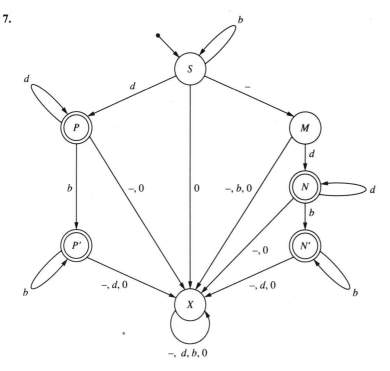

9. a. Input alphabet $= \{1, 2, 3, 4, 5, 6, 7, 8, 9, \text{Burglary}\}$

States $= \{\text{Armed}(A), A5, A54, A542, \text{Disarmed } (D), D7, \text{Alarm } (L), L5, L54, L542\}$

Transition function:

	1	2	3	4	5	6	7	8	9	B
A	A	A	A	A	$A5$	A	A	A	A	L
$A5$	A	A	A	$A54$	A	A	A	A	A	L
$A54$	A	$A542$	A	A	A	A	A	A	A	L
$A542$	A	A	A	A	A	A	D	A	A	L
D	D	D	D	D	D	D	$D7$	D	D	D
$D7$	D	D	D	D	D	D	D	D	A	D
L	L	L	L	L	$L5$	L	L	L	L	L
$L5$	L	L	L	$L54$	L	L	L	L	L	L
$L54$	L	$L542$	L	L	L	L	L	L	L	L
$L54$	L	L	L	L	L	L	D	L	L	L

b. Input alphabet = {SToP, PLaY, RECord, Fast Forward, REWind, PAUse, Beginning of Tape, End of Tape}

States = {SToP, PLaY, RECord, Fast Forward, REWind, PAUse-Play, PAUse-Record}

	STP	PLY	REC	FF	REW	PAU	BOT	EOT
STP	STP	PLY	REC	FF	REW	STP	STP	STP
PLY	STP	PLY	PLY	FF	REW	PAUP	PLY	STP
REC	STP	REC	REC	FF	REW	PAUR	REC	STP
FF	STP	PLY	REC	FF	REW	STP	FF	STP
REW	STP	PLY	REC	FF	REW	STP	STP	REW
PAUP	STP	PAUP	PAUP	FF	REW	PLY	PAUP	STP
PAUR	STP	PAUR	PAUR	FF	REW	REC	PAUR	STP

11.

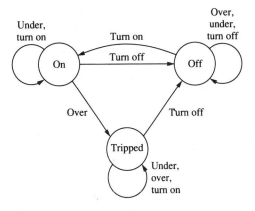

Section 12.3

1. a. *SA* **b.** *SAT* **c.** *SBT* **d.** *SAA'T*

3. a. *SA* **b.** *SAS* **c.** *SASAT* **d.** *SASASAS* **e.** *SASASASE* **f.** *STTTTTTT*

5. a. *SABSDSD*; accepted **b.** *SABSABSDSD*; accepted **c.** *SABSDS*; not accepted
   d. *SDSABTT*; not accepted

7. a. *SAE*; accepted **b.** *SBE*; accepted **c.** *SAED*; not accepted **d.** *SBEC*; not accepted
   e. *SBECDCDC*; not accepted **f.** *SADCDCD*; not accepted

9.

```
●──→( S )──a, b, c, d──(( E ))──a, b, c, d──( X )
```

(a)

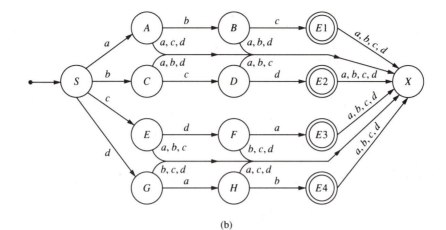

(b)

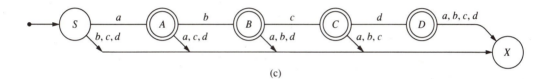

(c)

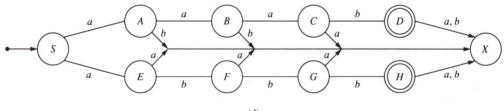

(d)

11. **a.** *SAB* **b.** *SAABBC* **c.** *SSSAAB* **c.** *SSSAAABBBCC*

13. **a.** *EE, EO, OO, OE, EE, EO, OO* **b.** *EE, EO, EE, EO, OO, EO* **c.** *EE, OE, EE, EO, EE, OE*
d. *EE, EO, EE, OE, OO, EO, OO*

15. **a.** *S, SA, AC, ACC* **b.** *S, SA, MA, MA, MA, MAB* **c.** *S, SA, MAB, X, X* **d.** *S, SA, AC, X*

Section 12.4

1. $S ::= aE \mid bE \mid a \mid b$
$E ::= aS \mid bE \mid b$

3.

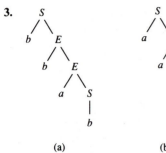

(a)   (b)

5.

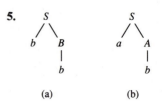

(a)   (b)

7. $S ::= aA \mid bB \mid cS \mid c$
$A ::= aA \mid bS \mid b \mid cB$
$B ::= aS \mid a \mid bB \mid cA$

9. $S ::= aA \mid dD \mid d$
$A ::= bB$
$B ::= cS$
$D ::= eS$

Section 12.5

1. **a.** $S ::= aA \mid a \mid c$
$A ::= cB$
$B ::= bS$

b. $S ::= cA \mid aC \mid b$
$A ::= aB$
$B ::= cS$
$C ::= cD$
$D ::= aS$

c. $S ::= aC \mid bD \mid a$
$B ::= aE \mid bC$
$C ::= bB$
$D ::= aS$
$E ::= aG$
$G ::= aS$

3. **a.** $\{S, E\}$ **b.** $\{S\}$ **c.** $\{S\}$ **d.** $\{S\}$

5.

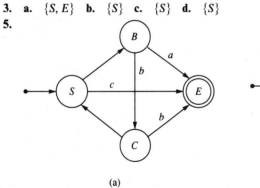

(a)

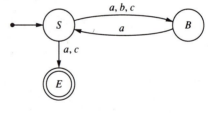

(b)

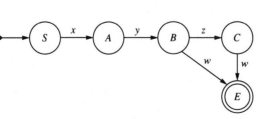

(c)

Section 12.6

1. a.

State	Tape
1	a̱babc
1	bḇabc
1	bca̱bc
1	bcbḇc
1	bcbcc̱
2	bcbc̱c
2	bcḇac
H	bca̱ac

b.

State	Tape
1	c̱cbbaa
2	cc̱bbaa
2	a̱cbbaa
2	a̱cbbaa
etc.	

3.

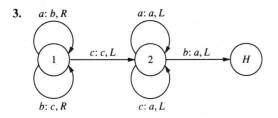

a: b, R   a: a, L

1   c: c, L   2   b: a, L   H

b: c, R   c: a, L

5.

Symbols

		BOT	a	b	x	EOI
	0	(BOT, R, 1)	(a, R, 0)	(b, R, 0)	(x, R, 0)	(EOI, R, 0)
	1	(BOT, R, 1)	(x, R, 2)	(x, R, 3)	(x, R, 1)	(Y, —, H)
States	2	—	(a, R, 2)	(x, L, 4)	(x, R, 2)	(N, —, H)
	3	—	(x, L, 4)	(b, R, 3)	(x, R, 3)	(N, —, H)
	4	(BOT, R, 1)	(a, L, 4)	(b, L, 4)	(x, L, 4)	

7. Represent two tapes (A and B) on one tape by putting tape A contents in even-numbered cells and tape B contents in odd-numbered cells. For each tape, introduce a "current location" marker into the alphabet. The machine must store in a memory the "true" contents of the positions occupied by the current location markers, and whether the current location is to the left or right of each marker. (Adding k bits of memory to a Turing machine is equivalent to multiplying the number of states by 2^k.) To do an operation on either tape, the machine must search for the current location marker, replace it with the true contents, modify the tape, note the contents of the new position, and rewrite the current location marker.

Section 12.7

1.

	x
1	x, R, 1

3. $F(\text{"BOT"}, 1) = (\text{"BOT"}, R, 1)$; $F(\text{"BOT"}, 2) = (\text{"BOT"}, R, 1)$;
$F(\text{"BOT"}, 3) = (\text{"BOT"}, R, 1)$; $F(\text{"BOT"}, 4) = (\text{"BOT"}, R, 1)$;
$F(\text{"a"}, 1) = (\text{"x"}, R, 2)$; $F(\text{"b"}, 1) = (\text{"x"}, R, 3)$; $F(\text{"x"}, 1) = (\text{"x"}, R, 1)$;
$F(\text{"a"}, 2) = (\text{"a"}, R, 2)$; $F(\text{"b"}, 2) = (\text{"x"}, L, 4)$; $F(\text{"x"}, 2) = (\text{"x"}, R, 2)$;
$F(\text{"a"}, 3) = (\text{"x"}, L, 4)$; $F(\text{"b"}, 3) = (\text{"b"}, R, 3)$; $F(\text{"x"}, 3) = (\text{"x"}, R, 3)$;

$F(\text{“}a\text{”}, 4) = (\text{“}a\text{”}, L, 4); F(\text{“}b\text{”}, 4) = (\text{“}b\text{”}, L, 4); F(\text{“}x\text{”}, 4) = (\text{“}x\text{”}, L, 4);$

$F(\text{“EOI”}, 1) = (\text{“}Y\text{”}, R, H); F(\text{“EOI”}, 2) = (\text{“}N\text{”}, R, H);$

$F(\text{“EOI”}, 3) = (\text{“}N\text{”}, R, H); F(\text{“EOI”}, 4) = (\text{“}N\text{”}, R, H).BaabE$

5. $F(\text{“1”}, 1) = (\text{“1”}, R, H); F(\text{“0”}, 1) = (\text{“0”}, R, 2);$

$F(\text{“1”}, 2) = (\text{“1”}, R, H); F(\text{“0”}, 2) = (\text{“0”}, R, 3);$

$F(\text{“1”}, 3) = (\text{“1”}, R, H); F(\text{“0”}, 3) = (\text{“0”}, R, 4);$

$F(\text{“1”}, 4) = (\text{“1”}, R, H); F(\text{“0”}, 4) = (\text{“0”}, R, 5);$

$F(\text{“1”}, 5) = (\text{“1”}, R, 5); F(\text{“0”}, 5) = (\text{“0”}, R, 5).00101101$

Chapter Review

1. The operator of a terminal can change the input sequence in response to the output, which is impossible with an abstract tape.

3. **a.** *SABABABABA*; accepted **b.** *STTTTTTTTTT*; not accepted **c.** *SATTTTTTTT*; not accepted **d.** *STTTTTTTTT*; not accepted

5. **a.** 1. *SABSABSABSA* 2. *STTTT* 3. *SABST* 4. *SABSABSTTTT*
    b. $S ::= \text{<empty>} \mid a \mid ab \mid abcS$ **c.** Any initial part of the string *abcabc* ......

7. There are 64 states, corresponding to 0, 1, 2, or >2 *a*'s, *b*'s, and *c*'s. The machine may be viewed as consisting of three counters. Encountering a letter increments the corresponding counter. The accepting states are those in which the value of at least one counter is 2.

9. **a.** $S ::= aA \mid bB \mid bC \mid a$ **b.** $S ::= aA \mid bT \mid bS$
    $A ::= bS$ $A ::= bT \mid bS$
    $B ::= aD$ $T ::= cT \mid aS \mid a$
    $D ::= cS$
    $C ::= cS$

11.

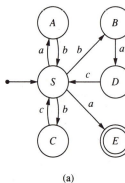

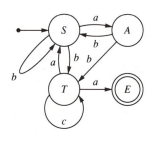

(a)   (b)

13. **a.**

States	Tape
1	aaaaa
1	baaaa
1	bbaaa
1	bbbaa
1	bbbba
1	bbbbb
H	bbbbba

b.

States	Tape
1	aaaabbb
1	baaabbb
1	bbaabbb
1	bbbabbb
1	bbbbbbb
2	bbbbbbb
H	bbbbbbb

c.

States	Tape
1	ababa
1	bbaba
2	bbaba
2	bbaba
H	bbaba

d.

States	Tape
1	bbbba
2	bbbba
H	bbbba

Index